Power System Dynamics and Stability

Power System Dynamics and Stability

Jan Machowski
Warsaw University of Technology, Poland

Janusz W. Bialek
James R. Bumby
University of Durham

John Wiley & Sons
Chichester • New York • Weinheim • Brisbane • Singapore • Toronto

National 01243 779777
International (+44) 1243 779777

e-mail (for orders and customer service enquiries): cs-books@wiley.co.uk

Visit our Home Page on http://www.wiley.co.uk
or
http://www.wiley.com

Other Wiley Editorial Offices

John Wiley & Sons, Inc., 605 Third Avenue,
New York, NY 10158-0012, USA

VCH Verlagsgesellschaft mbH, Pappelallee 3,
D-69469 Weinheim, Germany

Jacaranda Wiley Ltd, 33 Park Road, Milton,
Queensland 4064, Australia

John Wiley & Sons (Asia) Pte Ltd, 2 Clementi Loop #02-01,
Jin Xing Distripark, Singapore 129809

John Wiley & Sons (Canada) Ltd, 22 Worcester Road,
Rexdale, Ontario M9W 1L1, Canada

Library of Congress Cataloguing-in-Publication Data

Machowski, J.
Power system dynamics and stability/J. Machowski, J. Bialek, J.R. Bumby.
p. cm.
Includes bibliographical references and index.
ISBN 0 471 97174 X (PPC) 0 471 95643 0 (PR)
1. Electric power system stability. 2. Electric power systems.
I. Bialek, J. II. Bumby, J.R. (James Richard) III. Title.
TK1010.M33 1997
621.319'1 — dc20 96-39033
CIP

British Library Cataloguing in Publication Data

A catalogue record for this book is available from the British Library

ISBN 0 471 97174 X (PPC) 0 471 95643 0 (PR)

Typeset in 10/12pt Times by Laser Words, Madras, India
Printed and bound in Great Britain by Bookcraft (Bath) Ltd
This book is printed on acid-free paper responsibly manufactured from sustainable forestation, for which at least two trees are planted for each one used for paper production.

Contents

Foreword

Professor William Fairney, F. Eng, F.I.E.E

Director of Plant Development & Construction, National Power plc Visiting Professor in Engineering at Durham University

It gives me great pleasure to write this foreword as I have been associated with Durham University for over ten years and become well acquainted with the authors.

This work is a state-of-the-art exposition on the dynamics and stability of power systems, the complexity of which is not exceeded by any other dynamical system. Massive computer power can now be brought to bear on solving power system equations but this was not available when these problems first started to be addressed in the 1920's and 30's. It is a tribute to the pioneers of those days, who first formulated the complex equations, that they were able to solve them with paper, pen and mechanical calculators, and gain an understanding of the fundamental issues relating to power system dynamics. To enable calculations to be made at all, the models had to be simple, and conservative stability margins had to be used.

When I joined the electricity supply industry in 1965 and first became interested in power system dynamics, the stability of the CEGB network could only be solved on the huge mainframe computer based in Park Street, London. My interest was in the use of the newly invented thyristor for generator excitation, and the ability to explore a wide range of control settings was desirable. The frustration and high cost of sequential programming with punched card input data led to the search for alternative solutions. Necessity being the mother of invention, I was able to profit from the on-line satellite links to large mainframes in the USA which were just being introduced. Since that time processing power has mushroomed and computer size shrunk, to the point where today, a power system engineer can sit at his desk and optimise the performance of large power networks in real time.

Much of this development work in the UK has been carried out at Durham University, stimulated by research contracts with power system utilities. The privatisation of the electricity supply industry has resulted in a large number of generators and distribution utilities where power requirements are co-ordinated commercially through the electricity pool. The National Grid Company has the task of co-ordinating the technical performance of the network and of ensuring that technical needs are reflected in the commercial arrangements for the transmission of electricity.

This book will become a major reference work in future years for all involved in power system dynamics. Its comprehensive coverage of all aspects of the subject, plus its progressive approach from simplicity to complexity makes excellent reading on this fascinating subject.

Preface

The days when power systems were run by vertically integrated utilities and operated with large stability margins are all but over. In the present climate of deregulation and privatisation, the utilities are often separated into generation, transmission and distribution so as to help promote economic efficiency and encourage competition. Coupled with the difficulty of obtaining new rights of way for expanding the transmission system, this has resulted in power systems being operated much closer to their stability limits than ever before. These limits are to a large extent determined by the dynamic performance of the system and this is the main subject of this book.

This book has been written to try and answer some of our concerns about the education of power system engineers. With the widespread access to powerful computers, running ever more sophisticated simulation packages, there is a tendency to treat simulation as a substitute for understanding. This tendency is especially dangerous for students and young researchers who tend to think that simulation is a panacea for everything and always provides a true answer. What they do not realise is that, without a physical understanding of the underlying principles, they cannot be confident in understanding, or validating, the simulation results. It is by no means bad practice to treat the initial results of any computer software with a healthy pinch of scepticism.

Power system dynamics are not easy to understand. There are a number of good textbooks which deal with this topic and some of these are reviewed in Chapter 1. As the synchronous machine plays a decisive role in determining the dynamic response of the system, many of these books start with a detailed mathematical treatment of the synchronous generator in order to introduce Park's equations and produce a mathematical model of the generator. However, it is our experience that to begin a topic with such a detailed mathematical treatment can put many students off further study as they often find it difficult to see any practical relevance for the mathematics. This can be a major obstacle for those readers who are more practically inclined and who want to understand what is happening in the system without having to continuously refer to a complicated mathematical model of the generator.

Our approach is different. We first try to give a qualitative explanation of the underlying physical phenomena of power system dynamics using a simple model of the generator, coupled with the basic physical laws of electrical engineering. Having provided the student with a physical understanding of power system dynamics, we then introduce the full mathematical model of the generator, followed by more advanced topics such as system reduction, dynamic simulation and eigen-value analysis. In this way we hope that the material is made more accessible to the reader who wishes to understand the system operation without first tackling Park's equations.

In our work we have drawn information from many old, well known texts, particularly those by Kimbark and the Russian authors Zhdanov and Venikov. These latter texts are particularly well known in Eastern Europe but may be less familiar to enginners in the West. At the time these books were written there were no computers and researchers had to use other techniques to achieve understanding. It is ironical that nowadays one often hears at conferences of "discoveries", made using simulation packages, which were well known 30–40 years ago and had been made using only pen and paper. It is the welding of this traditional approach with modern computer techniques which appeals to us. Although our approach may be described as "back to basics" it is not our intention to live in the past. What we are trying to achieve is a basic understanding of the underlying physical phenomena of power system dynamics before tackling many of the more modern aspects such as simulation, eigen-value analysis, or the use of FACTS devices for power system stability enhancement.

Although this book naturally covers a lot of material known from other texts, we feel we have also made many important original contributions. This is particularly true in Chapters 5 and 6 when explaining the effect of rotor flux linkage variation on steady-state stability; in Chapter 7 when explaining voltage stability in a non-traditional way; in Chapter 8 when explaining the four stages of dynamics following a real power imbalance; and in Chapter 9 when introducing state-variable control of shunt and series elements based on local measurements. Also material from other texts has been re-examined and often adjusted to take into account recent findings.

The book may be conveniently divided into three major parts. The first part (Chapters 1–3) reviews the background for studying power system dynamics which, to large extent, may already be known to many students who have studied previous courses on power engineering. The second part (Chapters 4–9) attempts to explain the basic phenomena underlying power system dynamics using the classical model of the generator-infinite busbar system. The third part (Chapters 10–13) tackles some of the more advanced topics suitable for the modelling and dynamic simulation of large-scale power systems.

Examining the chapters in more detail, Chapter 1 classifies power system dynamics and provides a brief historical overview. Chapter 2 contains a brief description of the major power system components, including modern FACTS devices. Chapter 3 introduces steady-state models and their use in analysing the performance of the power system. Chapter 4 analyses the dynamics following a disturbance on the generator and introduces models suitable for analysing the dynamic performance of the synchronous generator. Chapter 5 explains the power system dynamics following a small disturbance (steady-state stability), while Chapter 6 examines the system dynamics following a large disturbance (transient stability). Chapter 7 provides an explanation of voltage stability together with some of the methods used for assessing such stability. Chapter 8 analyses the short and long term dynamics that follows a real power imbalance in the system. Chapter 9 provides an overview of the main methods of stability enhancement. Chapter 10 introduces advanced models of the different power system elements which are then used in Chapter 13 for power system dynamic simulation. Chapter 11 explains how to reduce the size of the simulation problem by using equivalents, while Chapter 12 examines the steady-state stability of multi-machine power systems using eigen-value analysis. The Appendix covers the per-unit system.

This book can serve as a textbook for final year undergraduate, and postgraduate, courses in power engineering, and also as a reference book for electric utility engineers.

Large parts of the material making up this book has been used for a number of years in final year lecture courses in power engineering at the Faculty of Electrical Engineering, Warsaw University of Technology. In addition some of the material has also been used for the final year power option in the School of Engineering, University of Durham, England.

Much of this book is based on an earlier Polish text by Machowski and Bernas (1989). In particular chapters 4, 6, 8, 12 and 13 are revised and extended versions of corresponding chapters from that book. We are grateful to the Polish publisher, Wydawnictwa Naukowo-Techniczne, for allowing us to make use of a number of diagrams from that book.

The authors are indebted to a number of organisations and individuals who helped make the writing of this book possible. Firstly we would like to acknowledge the financial support of the Commission of European Communities, TEMPUS/PHARE grant JEP-03026, without which the co-operation between the Polish and British co-authors would have been all but impossible. We are also grateful to University College, Durham, for a Fellowship which allowed Professor Machowski to join the other co-authors during the crucial final stage of writing, and to the School of Engineering, University of Durham, for providing the facilities that enabled the writing to take place. And last, but not least, we are grateful for the patience shown by our wives and families during the writing of this book.

J. Machowski, J. Bialek, and J. R. Bumby
Warsaw and Durham, August 1996.

List of symbols

Notation:

a bar on top of a symbol denotes a phasor or a complex number (e.g. $\overline{V}$, $\overline{S}$);
an arrow on top of a symbol denotes a spatial vector (e.g. $\vec{F}$);
lower case symbols normally denote instantaneous values (e.g. v, i);
upper case symbols normally denote rms or peak values (e.g. V, I);
bold face denotes a matrix or a vector (e.g. $\mathbf{Y}$);
subscripts, A, B, C refer to three-phase axes of a generator;
subscripts "d" and "q" refer to the direct- and quadrature-axis components.

Symbols:

The most important symbols are defined below:

B_{μ} — magnetising susceptance of a transformer;
B_{sh} — susceptance of a shunt element;
D — damping coefficient;
E_k — kinetic energy of the rotor relative to the synchronous speed;
E_p — potential energy of the rotor with respect to the equilibrium point
e_f — field voltage v_f referred to the fictitious q-axis armature coil;
e_q — steady-state emf induced in the fictitious q-axis armature coil proportional to the field winding self-flux linkages;
e'_d — transient emf induced in the fictitious d-axis armature coil proportional to the flux linkages of the q-axis coil representing the solid steel rotor body (round-rotor generators only);
e'_q — transient emf induced in the fictitious q-axis armature coil proportional to the field winding flux linkages;
e''_d — subtransient emf induced in the fictitious d-axis armature coil proportional to the total q-axis rotor flux linkages (q-axis damper winding and q-axis solid steel rotor body);
e''_q — subtransient emf induced in the fictitious q-axis armature coil proportional to the total d-axis rotor flux linkages (d-axis damper winding and field winding);
$\overline{E}$ — steady-state internal emf;
$\overline{E}_f$ — excitation emf proportional to the excitation voltage V_f;
E_{fm} — amplitude of the excitation emf;
E_d — d-axis component of the steady-state internal emf proportional to the rotor self-linkages due to currents induced in the q-axis solid steel rotor body (round-rotor generators only);

E_q — q-axis component of the steady-state internal emf proportional to the field winding self-flux linkages (i.e. proportional to the field current itself);
$\overline{E}'$ — transient internal emf proportional to the flux linkages of the field winding and solid steel rotor body (includes armature reaction);
E'_d — d-axis component of the transient internal emf proportional to flux linkages in the q-axis solid steel rotor body (round-rotor generators only);
E'_q — q-axis component of the transient internal emf proportional to the field winding flux linkages;
$\overline{E}''$ — subtransient internal emf proportional to the total rotor flux linkages (includes armature reaction);
E''_d — d-axis component of the subtransient internal emf proportional to the total flux linkages in the q-axis damper winding and q-axis solid steel rotor body;
E''_q — q-axis component of the subtransient internal emf proportional to the total flux linkages in the d-axis damper winding and the field winding;
$\overline{E}_r$ — resultant air-gap emf;
E_{rm} — amplitude of the resultant air-gap emf;
$\mathbf{E}_G$ — vector of the generator emf's
f — mains frequency;
f_n — rated frequency;
$\vec{F}_f$ — magnetomotive force (mmf) due to the field winding;
$\vec{F}_a$ — armature reaction mmf;
$F_{a\,ac}$ — ac armature reaction mmf (rotating);
$F_{a\,dc}$ — dc armature reaction mmf (stationary);
$\vec{F}_{ad}$, $\vec{F}_{aq}$ — d- and q-axis components of the armature reaction mmf;
$\vec{F}_r$ — resultant mmf;
G_{Fe} — core loss conductance of a transformer;
G_{sh} — conductance of a shunt element;
H — inertia constant;
i_A, i_B, i_C — instantaneous currents in phases A, B, and C;
$i_{A\,dc}$, $i_{B\,dc}$, $i_{C\,dc}$ — dc component of the current in phases A, B, and C;
$i_{A\,ac}$, $i_{B\,ac}$, $i_{C\,ac}$ — ac component of the current in phases A, B, and C;
i_d, i_q -currents flowing in the fictitious d- and q-axis armature coils;
i_D, i_Q — instantaneous d- and q-axis damper winding current;
i_f — instantaneous field current of a generator;
$\mathbf{i}_{ABC}$ — vector of instanteneous phase currents;
$\mathbf{i}_{fDQ}$ — vector of instantaneous currents in the field winding and the d- and q-axis damper windings;
$\mathbf{i}_{odq}$ — vector of armature currents in the rotor reference frame;
$\overline{I}$ — armature current;
$\overline{I}_d$, $\overline{I}_q$ — d- and q-axis component of the armature current;
$\overline{I}_S$, $\overline{I}_R$ — currents at the sending and receiving end of a transmission line;
$\overline{\mathbf{I}}_R$, $\overline{\mathbf{I}}_E$ — vector of complex current injections to the retained and eliminated nodes;
$\overline{\mathbf{I}}_G$, $\overline{\mathbf{I}}_L$ — vector of generator and load currents;
$\Delta\mathbf{I}_L$ — vector of load corrective currents;
J — moment of inertia;
k_{PV}, k_{QV} — voltage sensitivities of the load (the slopes of the real and reactive power demand characteristics as a function of voltage);
k_{Pf}, k_{Qf} — frequency sensitivities of the load (the slopes of the real and reactive power demand characteristics as a function of frequency);

K_{E_q} — steady-state synchronising power coefficient (the slope of the steady-state power angle curve $P_{E_q}(\delta)$);
$K_{E'_q}$ — transient synchronising power coefficient (the slope of the transient power angle curve $P_{E'_q}(\delta')$);
$K_{E'}$ — transient synchronising power coefficient (the slope of the transient power angle curve $P_{E'}(\delta')$);
K_i — reciprocal of droop for the i-th generating unit;
K_L — frequency sensitivity coefficient of the system real power demand;
K_T — reciprocal of droop for the total system generation characteristic;
l — length of a transmission line;
L_{AA}, L_{BB}, L_{CC}, L_{ff}, L_{DD}, L_{QQ} - self-inductances of the windings of the phase windings A, B, C, the field winding, and the d- and the q-axis damper winding;
L_d, L_q — inductances of the fictitious d- and q-axis armature windings;
L'_d, L'_q, L''_d, L''_q — d- and q-axis transient and subtransient inductances;
L_{xy}, where $x, y \in \{A, B, C, D, Q, f\}$ and $x \neq y$, are the mutual inductances between the windings denoted by the indices as described above;
L_s — minimum value of the self-inductance of a phase winding;
ΔL_s — amplitude of the variable part of the self-inductance of a phase winding;
$\mathbf{L}_R$ — submatrix of the rotor self and mutual inductances;
$\mathbf{L}_S$ — submatrix of the stator self and mutual inductances;
$\mathbf{L}_{SR}$, $\mathbf{L}_{RS}$ — submatrices of the stator-to-rotor and rotor-to-stator mutual inductances;
M — inertia coefficient;
M_f, M_D, M_Q — amplitude of the mutual inductance between a phase winding and, respectively, the field winding, and the d- and the q-axis damper winding;
N — generally, number of turns of a winding;
p — number of poles;
P_{acc} — accelerating power;
P_D — damping power;
P_e — electromagnetic air-gap power;
$P_{E_q cr}$ — critical (pull-out) air-gap power developed by a generator;
$P_{E_q}(\delta)$, $P_{E'}(\delta')$, $P_{E'_q}(\delta')$ — air-gap power curves assuming $E_q =$ constant, $E' =$ constant and $E'_q =$ constant;
P_L — real power absorbed by a load (Chapter 7) or total system load (Chapter 8);
P_m — mechanical power supplied by a prime mover to a generator;
P_n — real power demand at rated voltage;
P_R — real power at the receiving end of a transmission line;
P_{rI}, P_{rII}, P_{rIII}, P_{rIV} — contribution of the generating units remaining in operation towards covering the real power imbalance during the first, second, third and fourth stage of load frequency control;
P_{sI}, P_{sII}, P_{sIII}, P_{sIV} — contribution of the system towards covering the real power imbalance during the first, second, third and fourth stage of load frequency control;
P_S — real power at the sending end of a transmission line (Chapter 3) or real power supplied by a source to a load (Chapter 7);
P_{SIL} — surge impedance (natural) load;
P_s — real power supplied to the infinite busbar;
$P_{sE_q}(\delta)$ — curve of real power supplied to the infinite busbar assuming $E_q =$ constant;
P_T — total power generated in a system;
P_{tie} — net tie-line interchange power;

$P_{V_g}(\delta)$ — air-gap power curve assuming V_g = constant;
$P_{V_{gcr}}$ — critical value of $P_{V_g}(\delta)$;
Q_G — reactive power generated by a source (the sum of Q_L and the reactive power loss in the network);
Q_L — reactive power absorbed by a load;
Q_n — reactive power demand at rated voltage;
Q_R — reactive power at the receiving end of a transmission line;
Q_S — reactive power at the sending end of a transmission line (Chapter 3) or reactive power supplied by a source to a load (Chapter 7);
R — resistance of the armature winding of a generator;
r — total resistance between (and including) the generator and the infinite busbar;
R_A, R_B, R_C, R_D, R_Q, R_f — resistances of the phase windings A, B, C, the d- and q-axis damper winding, and the field winding;
$\mathbf{R}_{ABC}$ — diagonal matrix of phase winding resistances;
$\mathbf{R}_{fDQ}$ — diagonal matrix of resistances of the field winding and the d- and q-axis damper windings;
s — Laplace operator;
s — slip of induction motor;
s_{cr} — critical slip of induction motor;
S_n — rated apparent power of a generator;
S_{SHC} — short-circuit power;
t — time;
T'_d, T''_d — short-circuit d-axis transient and subtransient time constants;
T'_{do}, T''_{do} — open-circuit d-axis transient and subtransient time constants;
T'_q, T''_q — short-circuit q-axis transient and subtransient time constants;
T'_{qo}, T''_{qo} — open-circuit q-axis transient and subtransient time constants;
T_a — armature winding time constant;
$\mathbf{T}$ — transformation matrix between network (a,b) and generator (d,q) co-ordinates;
v_A, v_B, v_C, v_f — instantaneous voltages across phases, A, B, C and the field winding;
v_d, v_q — voltages across the fictitious d- and q-axis armature coils;
$\mathbf{v}_{ABC}$ — vector of instantaneous voltages across phases, A, B, and C;
$\mathbf{v}_{fDQ}$ — vector of instantaneous voltages across the field winding and the d- and q-axis damper windings;
V — Lyapunov function;
V_{cr} — critical value of the voltage;
$\overline{V}_d$, $\overline{V}_q$ — direct and quadrature-axis component of the generator terminal voltage;
V_f — voltage aplied to the field winding;
$\overline{V}_g$ — voltage at the generator terminals;
$\overline{V}_s$ — infinite busbar voltage;
$\overline{V}_{sd}$, $\overline{V}_{sq}$ — direct- and quadrature-axis component of the infinite busbar voltage;
$\overline{V}_S$, $\overline{V}_R$ — voltage at the sending and receiving end of a transmission line;
V_{sh} — local voltage at the point of installation of a shunt element;
$\overline{V}_i = V_i \angle \delta_i$ — complex voltage at node i;
$\overline{\mathbf{V}}_R$, $\overline{\mathbf{V}}_E$ — vector of complex voltages at the retained and eliminated nodes;
W — work;
$\mathbf{W}$ — Park's modified transformation matrix;
x_d, x'_d, x''_d — total d-axis synchronous, transient and subtransient reactance between (and including) the generator and the infinite busbar;

$x_{d\,PRE}$, $x'_{d\,F}$, $x'_{d\,POST}$ — pre-fault, fault and post-fault value of x'_d;
x_q, x'_q, x''_q — total q-axis synchronous, transient and subtransient reactance between (and including) the generator and the infinite busbar;
X_a — armature reaction reactance (round-rotor generator);
X_C — reactance of a series compensator;
X_l — armature leakage reactance of a generator;
X_D — reactance corresponding to the flux path around the damper winding;
X_f — reactance corresponding to the flux path around the field winding;
X_d, X'_d X''_d — d-axis synchronous, transient and subtransient reactance;
X_q, X'_q, X''_q — q-axis synchronous, transient and subtransient reactance;
X_{SHC} — short-circuit reactance of a system as seen from a node;
Y_T — admittance of a transformer;
$\overline{\mathbf{Y}}$ — admittance matrix;
$\mathbf{Y}_{GG}$, $\mathbf{Y}_{GL}$, $\mathbf{Y}_{LG}$ — admittance submatrices where subscript G corresponds to fictitious generator nodes and subscript L corresponds to all the other nodes (including generator terminal nodes);
$\overline{Y}_{ij} = G_{ij} + \mathrm{j}B_{ij}$ — element of the admittance matrix;
$\overline{\mathbf{Y}}_{RR}$, $\overline{\mathbf{Y}}_{EE}$, $\overline{\mathbf{Y}}_{RE}$, $\overline{\mathbf{Y}}_{ER}$ — complex admittance submatrices where subscript E refers to eliminated and subscript R to retained nodes;
$Z = \sqrt{r^2 + x_d x_q}$;
Z_c — characteristic impedance of a transmission line;
$\underline{Z_s} = R_s + \mathrm{j}X_s$ — internal impedance of the infinite busbar;
$\underline{Z_T} = R_T + \mathrm{j}X_T$ — series impedance of the transformer;
$\Delta\omega$ — rotor speed deviation equal to $(\omega - \omega_s)$;
γ — instantaneous position of the generator d-axis relative to phase A;
γ_0 — position of the generator d-axis at the instant of fault;
Φ_a — armature reaction flux;
Φ_{ad}, Φ_{aq} — d- and q-axis component of the armature reaction flux;
$\Phi_{a\,ac}$ — ac armature reaction flux (rotating);
$\Phi_{a\,dc}$ — dc armature reaction flux (stationary);
Φ_f — excitation (field) flux;
Ψ_A, Ψ_B, Ψ_C — total flux linkage of phases A, B, and C;
Ψ_{AA}, Ψ_{BB}, Ψ_{CC} — self-flux linkage of phases A, B, and C;
$\Psi_{a\,ac\,r}$ — rotor flux linkages produced by $\Phi_{a\,ac}$;
$\Psi_{a\,dc\,r}$ — rotor flux linkages produced by $\Phi_{a\,dc}$;
Ψ_{ar} — rotor flux linkages produced by the total armature reaction flux;
Ψ_D, Ψ_Q — total flux linkage of damper windings in axes d and q;
Ψ_d, Ψ_q — total d-and q-axis flux linkages;
Ψ_f — total flux linkage of the field winding;
Ψ_{fa} — amplitude of the excitation flux linkage with armature winding;
Ψ_{fA}, Ψ_{fB}, Ψ_{fC} — excitation flux linkage with phases A, B, and C;
$\mathbf{\Psi}_{ABC}$ — vector of phase flux linkages;
$\mathbf{\Psi}_{fDQ}$ — vector of flux linkages of the field winding and the d- and q-axis damper windings;
$\mathbf{\Psi}_{odq}$ — vector of armature flux linkages in the rotor reference frame;
τ_e — electromagnetic torque;
τ_ω — fundamental-frequency subtransient electromagnetic troque;
$\tau_{2\omega}$ — double-frequency subtransient electromagnetic torque;
τ_d, τ_q — direct- and quadrature-axis component of the electromagnetic torque;

τ_R, τ_r — subtransient electromagnetic torque due to stator and rotor resistances;
ε — rotor acceleration;
φ_g — power factor angle at the generator terminals;
δ — power (or rotor) angle with respect to the infinite busbar;
δ_g — power (or rotor) angle with respect to the voltage at the generator terminals;
$\hat{\delta}_s$ — stable equilibrium value of the rotor angle;
δ' — transient power (or rotor) angle between E' and V_s;
δ_{fr} — angle between the resultant and field mmf's;
λ_R — frequency bias factor;
ω — angular velocity of the generator (in electrical radians);
ω_s — synchronous angular velocity in electrical radians (equal to $2\pi f$);
ω_d — angular velocity of rotor swings in electrical radians;
$\mathbf{\Omega}$ — rotation matrix;
ρ — static droop of the turbine-governor characteristic;
ρ_T — droop of the total system generation characteristic;
ϑ — transformation ratio;
γ — propagation constant of a transmission line;
β — phase constant of a transmission line (Section 3.1);
$\Re$ — reluctance;
$\Re_d$, $\Re_q$ — reluctance along the direct- and quadrature-axis;

Abbreviations:

ac — alternating current;
ACE — area control error;
AGC — Automatic Generation Control;
AVR — Automatic Voltage Regulator;
d — direct axis of a generator;
dc — direct current;
emf — electro-motive force;
FACTS — Flexible AC Transmission Systems
HV — high voltage
LFC — load frequency control
mmf — magneto-motive force;
PSS — power system stabiliser
q — quadrature axis of a generator;
rms — root-mean-square;
rpm — revolutions per minute;
rhs — right-hand-side;
SMES — superconducting magnetic energy storage
STATCOM — static compensator;
SVC — Static VAR Compensators;
UPFC — unified power flow controller.

1

Introduction

1.1 General Classification Of Power System Dynamics

An electrical power system consists of many individual elements connected together to form a large, complex system capable of generating, transmitting and distributing electrical energy over a large geographical area. Because of this interconnection of elements, a large variety of dynamic interactions are possible, some of which will only affect some of the elements, others will affect fragments of the system, whilst others may affect the system as a whole. As each dynamic effect displays certain unique features power system dynamics can be conveniently divided into groups characterised either by their cause, consequence, time frame, physical character or the place in the system that they occur.

Of prime concern is the way the power system will respond to both a changing power demand and to various types of disturbance; the two main causes of power system dynamics. A changing power demand introduces a wide spectrum of dynamic changes into the system, each of which occurs on a different time scale. In this context the fastest dynamics are due to sudden changes in demand and are associated with the transfer of energy between the rotating masses in the generators and the loads. Slightly slower are the voltage and frequency control actions needed to maintain system operating conditions until finally the very slow dynamics corresponding to the way in which the generation is adjusted to meet the slow daily demand variations take effect. Similarly the way in which the system responds to disturbances also covers a wide spectrum of dynamics and associated time frames. In this case the fastest dynamics are those associated with the very fast wave phenomena that occur in high voltage transmission lines. These are followed by fast electromagnetic changes in the electrical machines themselves before the relatively slow electromechanical rotor oscillations occur. Finally the very slow prime mover and automatic generation control actions take effect.

Based on their physical character the different power system dynamics may be divided into four groups defined as: *wave*, *electromagnetic*, *electromechanical* and *thermodynamic*. This classification also corresponds to the time frame involved and is shown in Figure 1.1. Although this broad classification is convenient it is by no means absolute with some of the dynamics belonging to two or more groups whilst others lie on the boundary between groups. Figure 1.1 shows the fastest dynamics to be the *wave* effects, or surges, in high voltage transmission lines and corresponds to the propagation of electromagnetic waves caused by lightning strikes or switching operations. The time frame of these dynamics is

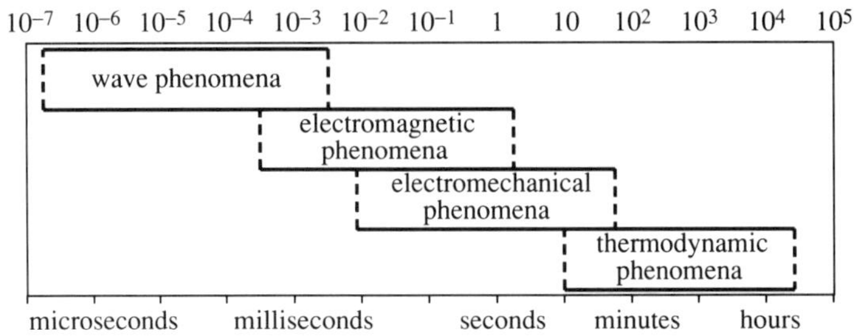

Figure 1.1 Time frame of the basic power system dynamic phenomena

from microseconds to milliseconds. Much slower are the *electromagnetic dynamics* that take place in the machine windings following a disturbance, operation of the protection system or the interaction between the electrical machines and the network. Their time frame is from milliseconds to a second. Slower still are the electromechanical dynamics due to the oscillation of the rotating masses of the generators and motors that occurs following a disturbance, operation of the protection system or voltage and prime mover control. The time frame of these dynamics is from around a second to several seconds. The slowest dynamics are the *thermodynamic changes* which result from boiler control action in steam power plants as the demands of the automatic generation control is implemented.

Careful inspection of Figure 1.1 shows the classification of power system dynamics with respect to time frame to be closely related to where the dynamics occur within the system. For example, moving from left to right along the time scale in Figure 1.1 corresponds to moving through the power system from the electrical *RLC* circuits of the transmission network, through the generator armature windings to the field and damper winding, then along the generator rotor to the turbine until finally the boiler is reached.

The fast wave phenomena, due to lightning and switching overvoltages, occur almost exclusively in the network and basically do not propagate beyond the transformer windings. The electromagnetic phenomena involve mainly the generator armature and damper windings and partly the network. The electromechanical phenomena, that is the rotor oscillations and accompanying network power swings, involve mainly the rotor field and damper windings and the rotor inertia. As the power system network connects the generators together this enables interactions between swinging generator rotors to take place. An important role is played here by the automatic voltage control and the prime mover control. Slightly slower than the electromechanical phenomena are the frequency oscillations, in which the rotor dynamics still play an important part, but which are influenced to a much greater extent by the action of the turbine governing systems and the automatic generation control. Automatic generation control also influences the thermodynamic changes due to boiler control action in steam power plants.

The fact that the time frame of the different dynamics is closely related to where they occur within the power system has important consequences for the modelling of the system elements. In particular moving from left to right along Figure 1.1 corresponds to a reduction in the accuracy required in the models used to represent the network elements but an increase in the accuracy in the models used first to represent the electrical components of the generating unit and then, further to the right, the mechanical and thermal parts of the unit. This important fact is taken into account in the general structure of this book when later chapters describe the different power system dynamic phenomena.

1.2 Brief Historical Overview

The first articles on power system dynamics began to appear in conference proceedings and technical journals at about the same time as the first interconnected power systems were constructed. As power systems developed the interest in their behaviour grew until power system dynamics became a scientific discipline in its own right.

Perhaps the greatest contribution in developing the theoretical foundations of power system dynamics has been from research workers in those countries whose power systems cover large geographic areas, most notably the USA, Canada and the former Soviet Union. However much excellent work has also been contributed from research workers in many other countries. With the mountain of research papers and books now available it is difficult to attempt to give a short historical overview of all the literature on power system dynamics and, out of necessity, the following overview is restricted to what the authors regard as some of the most important books dealing with power system dynamics.

Some of the first monographs on power system dynamics published in English were books by Dahl (1938), a two-volume textbook by Crary (1945 and 1947) and a large, three-volume, monograph by Kimbark (1948, 1950, 1956, reprinted in 1995). In all these books the main emphasis was on electromechanical phenomena. At the same time a Russian text was published by Zhdanov (Жданов, 1948) also dealing mainly with electromechanical phenomena. Zhdanov's work was later continued by V. A. Venikov (Веников), who published about a dozen books in Russian between the years 1958 and 1985; one of these books, again dealing mainly with electromechanical phenomena, was published in English by Pergamon Press (Venikov, 1964). An extended and modified version of this book was published in Russian in 1978 and later in the same year translated into English (Venikov, 1978). The main feature of Venikov's books is the emphasis placed on the physical interpretation of the dynamic behaviour.

One of the first books devoted to the general description of power system dynamics was written in Germany by Rüdenberg (1923). This book was later translated into many languages with an English edition appearing in 1950. Other important books that have dealt generally with power system dynamics have been written by Yao-nan Yu (1983), Rácz & Bókay (1988) and Kundur (1994). The comprehensive text by Kundur contains an excellent overview of modelling and analysis of power systems and constitutes the basic monograph on power system dynamics. The fast electromagnetic phenomena, like wave and switching transients, are described by Greenwood (1971).

From the 1940s until the 1960s power system dynamics were generally studied using physical (analogue) models of the system. However rapid developments in computer technology brought about an ever increasing interest in mathematical modelling of power systems with the main monographs on this topic being written by Anderson & Fouad (1977), Arillaga & Arnold (1983 and 1990) and Kundur (1994).

Another category of books use Lyapunov's direct method to analyse the electromechanical stability of power systems. The main texts here are those written by Pai (1981 and 1989), Fouad & Vittal (1992) and Pavella & Murthy (1994). It is worth stressing that a large number of excellent books on Lyapunov's direct method have been published in Russia (Lyapunov's homeland) but have not been translated into English.

A brief overview of the large number of papers published over the last twenty to thirty years shows the main emphasis of power system research to have been on the effective use of computers in power system analysis. Given the rapid developments in computer technology, and its fundamental importance in power system analysis, this is perhaps to be expected and understood. However there is a danger that young engineers and researchers become more concerned with the computer technology than in understanding the underlying physical principles of the power system dynamics themselves. In time this may endanger progress in the field. To try to combat this problem this book first describes the underlying physical process of the particular power system dynamic phenomena and only after a thorough understanding has been reached is a more rigorous mathematical treatment attempted. Once the mathematical treatment has been completed computers can be used to obtain the necessary quantitative results. For these reasons this book concentrates on developing a basic analysis of the different problem areas and often refers to more specialised publications.

2
Power System Components

Modern day society requires a large amount of energy for use in industry, commerce, agriculture, transportation, communications, domestic households, etc. The total energy required during one year is called the *annual energy demand* and is satisfied using naturally occurring primary energy resources, principally fossil fuels such as coal, oil, natural gas and uranium. In the current world energy scene these fossil fuels are also the main fuels used in the generation of electrical energy, with the renewable energy resources such as hydro, biogas, solar, wind, geothermal, wave and tidal energy being used to a lesser extent. In the future it is likely that the share of the energy market taken by the renewables will increase as environmental issues play a more dominant role on the political agenda.

Perhaps the most important, and unique, feature of an electrical power system is that *electrical energy cannot easily and conveniently be stored in large quantities.* This means that at any instant in time the energy demand has to be met by corresponding generation. Fortunately the combined load pattern of a power system normally changes in a relatively predictable manner even though individual consumer loads may vary quite rapidly and unpredictably. Such a predictable system demand pattern goes some way towards allowing the daily generation schedule to be planned and controlled in a predetermined manner.

If a power utility is to provide an acceptable supply of electrical energy to its consumers it must address the following issues:

Reliability of supply

High supply reliability is of fundamental importance, as any major interruption of supply will cause, at the very least, major inconvenience to the consumer, can lead to life threatening situations and, for the industrial consumer, may pose severe technical and production problems. Invariably in such situations the electrical supply utility also incurs a large loss in financial revenue. High reliability of supply can be ensured by:

- high quality of installed elements
- the provision of reserve generation

- employing large interconnected power systems capable of supplying each consumer via alternative routes
- a high level of system security

Supplying electrical energy of good quality

Electrical energy of good quality is provided by:

- regulated and defined voltage levels with low fluctuations
- a regulated and defined value of frequency with low fluctuations
- low harmonic content

Two basic methods can be used to ensure a high quality of electrical supply: firstly the proper use of automatic voltage and frequency control methods, and secondly by employing large, interconnected power systems which, by their very nature, are less susceptible to load variations and other disturbances.

Economic generation and transmission

The majority of electricity is generated by first converting the thermal energy stored in the fossil fuel into mechanical energy, and then converting this mechanical energy into electrical energy for transmission through the power system to the consumer. Unfortunately the efficiency of this overall process is relatively low, particularly the first stage conversion of thermal energy into mechanical energy. It is therefore vital that the operation of the overall system is optimised by minimising the generation and the transmission costs. Once again some saving can be achieved by connecting, and operating, a number of smaller systems as one larger, interconnected system.

Environmental issues

Modern society demands careful planning of generation and transmission to ensure as little effect as possible on the natural environment whilst meeting expectations for a secure electrical supply. Consequently air and water pollution produced by power generation plants are limited to prescribed quantities whilst the pathways for transmission lines are planned so as to cause minimal disturbance to the environment. In addition new plans for power stations and transmission lines are subject to close public scrutiny.

Environmental issues are now playing an increasingly important role on the political agenda and, therefore, in the planning of power system expansion. As a consequence of this, power utilities must continually seek ways of making better use of their existing system as obtaining planning permission for new transmission lines and generation sites becomes more difficult and stringent.

It is within this political and operational framework that an electrical power utility generates, transmits and distributes electrical energy to its consumers. Consequently the purpose of this chapter is to describe how the different elements of a power system function and the effect they have on both power system operation and control.

2.1 Structure of the Electrical Power System

The basic structure of a contemporary electrical power system is illustrated schematically in Figure 2.1. Conventionally the power system is divided into three parts, one concerned with generation, one with transmission and one with distribution. Historically power systems have tended to be vertically integrated, with each utility responsible for generation, transmission and distribution in its own service area. Recently, some countries (e.g. England and Wales or Chile) have split their system horizontally with separate companies being responsible for each part.

Different parts of the power system operate at different voltages. Generally voltages can be considered to be *low* if they are below the 1 kV mark whilst *medium voltages*, used in distribution systems, are typically between 1 kV and 100 kV. The *high voltages* used in sub-transmission networks are between 100 kV and 300 kV and the *extra high voltages* used in transmission networks are above 300 kV. This classification is loose and by no means strict.

Generation

Electricity is produced by converting the mechanical energy appearing on the output shaft of an engine, or more usually a turbine, into electrical energy. In most contemporary

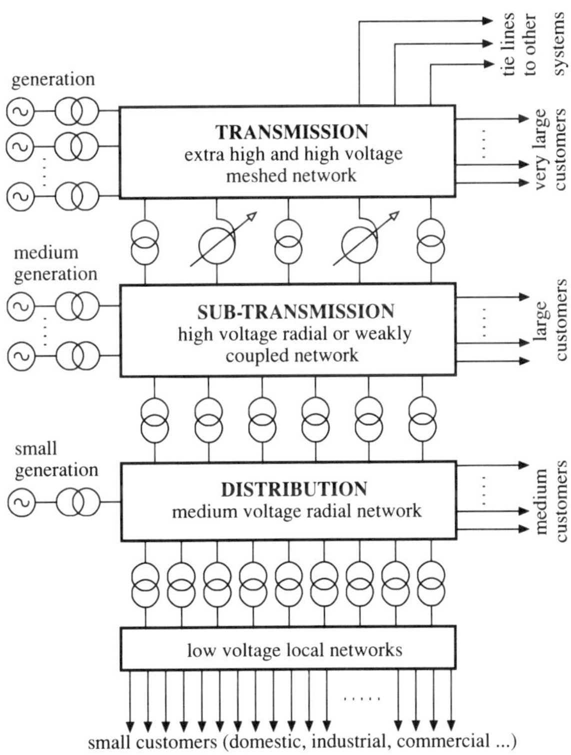

Figure 2.1 Structure of an electrical power system

power systems the mechanical energy is either obtained from thermal energy or extracted directly from flowing water (hydro-generation). The main thermal energy resources used commercially are coal, natural gas, nuclear fuel and oil. There is also an increasing use of non-fossil fuels such as wind, tidal, geothermal and biogas in the generation of electricity. The main non-thermal source of mechanical energy is water or hydro-power.

The conversion of mechanical to electrical energy is almost universally achieved by the use of a synchronous generator, although some wind generation systems may use an induction generator. The synchronous generator feeds its electrical power into the transmission system via a step-up transformer (Figure 2.1) in order to increase the voltage from the generation level (10–20 kV) to the transmission level (hundreds of kV).

Transmission

One significant advantage of electrical energy is that it can be generated near to the primary energy resource or water reservoirs and then be transmitted over long distances to the load centres. Since the energy lost in a transmission line is proportional to the current squared, transmission lines operate at high or very high voltages. The electrical network connects all the power stations into one system, and transmits and distributes power to the load centres in an optimal way. Usually the transmission network is connected in a mesh structure in order to provide many possible routes for electrical power to flow from the generators to the electrical consumers, thereby improving the flexibility and reliability of the system.

As the electrical energy gets closer to the load centre, it is directed from the transmission network into a sub-transmission network. When a power system expands with the addition of new, high voltage, transmission lines some of the older, lower voltage, lines may become part of the sub-transmission network. There is no strict division of the network into transmission and sub-transmission networks, and smaller power generation plants may feed directly into the sub-transmission network whilst bulk power consumers may be fed directly from the transmission or sub-transmission network (Figure 2.1).

Distribution

Most of the electrical energy is transferred from the transmission, or sub-transmission, network to distribution high voltage and medium voltage networks in order to bring it directly to the consumer. The distribution network is generally connected in a radial structure as opposed to the mesh structure used in the transmission system. Large consumers may be supplied from a weakly coupled, meshed, distribution network or, alternatively, they may be supplied from two radial feeders with a possibility of automatic switching between feeders in case of a power cut. Some industrial consumers may have their own embedded or on-site generation as a reserve or as a by-product of the technological process (e.g. steam generation). Ultimately power is transformed to a low voltage and distributed directly to consumers.

Typically about 8% of the electrical energy appearing at the generator terminals will be lost on its way to the consumers in the transmission and distribution level.

Demand

The demand for electrical power is never constant and changes continuously throughout the day and night. The changes in demand of individual consumers may be fast and frequent but as one moves up the power system structure (Figure 2.1) from individual consumers, through the distribution network, to the transmission level, the changes in demand become smaller and smoother as individual demands are aggregated. Consequently the total power demand at the transmission level changes in a more or less predictable way that depends on the season, weather conditions, the way of life of a particular society, etc. Fast global power demand changes on the generation level are usually small and are referred to as the *load fluctuations*.

2.2 Generating Units

The block diagram of a generating unit is shown in Figure 2.2. Electrical energy is produced by a synchronous generator driven by a prime mover, usually a turbine or a diesel engine. The turbine is equipped with *a turbine governor* which either controls the speed or the output power according to a pre-set power-frequency characteristic. The

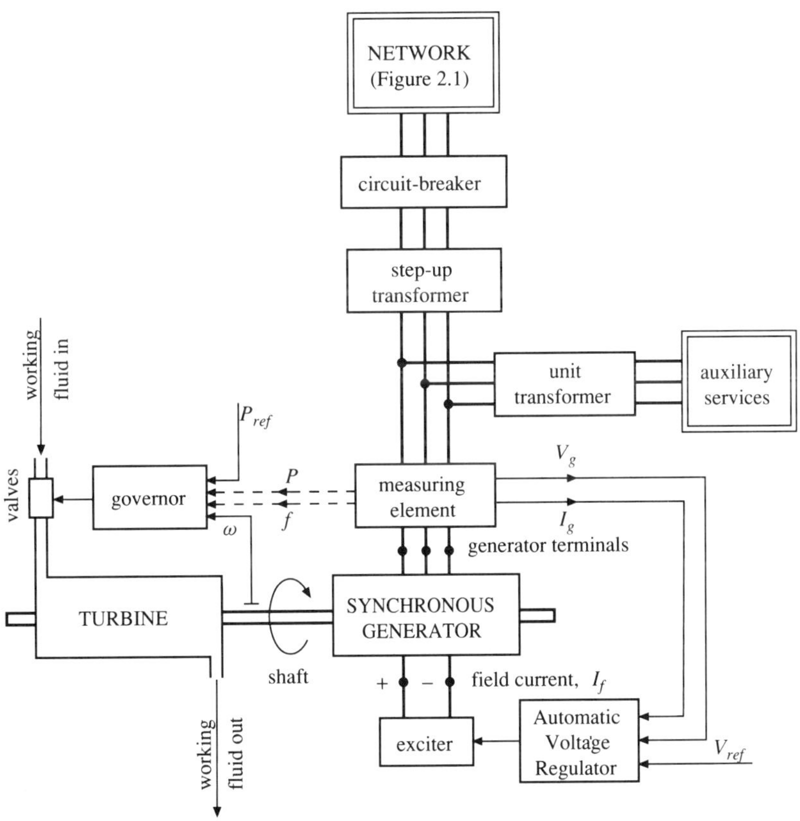

Figure 2.2 Block diagram of a power generating unit

generated power is fed into the transmission network via a step-up transformer. The dc excitation (or field) current, required to produce the magnetic field inside the generator, is provided by the *exciter*. The excitation current, and consequently the generator's terminal voltage, is controlled by an *automatic voltage regulator (AVR)*. In new generating units an additional *unit transformer* may be connected to the busbar between the generator and the step-up transformer in order to supply the power stations auxiliary services comprising motors, pumps, the exciter, etc.

2.2.1 Synchronous generators

Synchronous generators can be loosely classified as either high-speed generators, driven by steam or gas turbines (and often called *turbogenerators*), or low-speed generators, driven by water turbines. To reduce centrifugal forces high-speed turbogenerators have relatively low diameter but large axial length and are mounted horizontally. Typically they will have two or four electrical poles so that in a 50 Hz system a generator would be driven at 3000 rpm or 1500 rpm respectively. In contrast low-speed generators operate at typically 500 rpm and below, and have a large number of electrical poles, large diameter and shorter axial length. The actual number of magnetic poles depends on the required speed and nominal frequency of the power system.

All generators have two main magnetic parts termed the *stator* and the *rotor*, both of which are manufactured from magnetic steel. The *armature winding*, which carries the load current and supplies power to the system, is placed in equidistant slots on the inner surface of the stator and consists of three identical phase windings. The rotor of a high-speed generator also contains slots for the dc excitation winding, whilst the excitation winding for low-speed generators is wound on the salient poles of the rotor. The rotor also has additional short-circuited *damper*, or *amortisseur, windings*, to help damp mechanical oscillations of the rotor. In high-speed, non-salient pole, generators the damper windings are usually in the form of conductive wedges mounted in the same slots as the excitation winding. In low-speed generators the damper windings are mounted in axial slots in the pole face.

The rotor excitation winding is supplied with a dc current to produce a rotating magnetic flux the strength of which is proportional to the excitation current. This rotating magnetic flux then induces an electromotive force (emf) in each phase of the three-phase stator armature winding which forces ac currents to flow out to the power system. The combined effect of these ac armature currents is to produce their own *armature reaction* magnetic flux which is of constant magnitude but rotates at the same speed as the rotor. The excitation flux and the armature reaction flux then produce a resultant flux that is stationary with respect to the rotor but rotates at synchronous speed with respect to the stator. As the resultant flux rotates relative to the stator it is necessary to laminate the stator iron core axially in the shaft direction to limit the iron losses due to eddy currents. However as the magnetic flux is stationary with respect to the rotor the rotor is normally constructed from a solid steel forging.

If, for some reason, the rotor speed deviates from synchronous the flux will not be stationary with respect to the rotor and currents will be induced in the damper windings. According to Lenz's law these currents will oppose the flux change that has produced them and so help restore synchronous speed and damp the rotor oscillations.

Historically there has been a universal tendency to increase the rated power of new power stations and individual generators as capital cost and operating cost (per unit MW) decrease with increased MW rating. This economy of scale results in lower generator mass per unit MW, smaller buildings and power station area, and lower auxiliary equipment and staffing costs. However in countries such as the UK who have access to cheap supplies of natural gas this trend in building generators of ever increasing rating has been reversed in recent years. Here combined cycle gas turbine plant utilising air-cooled generators up to typically 250 MW have become the norm. Consequently modern synchronous generators have ratings ranging from about 100 MW to more than 1300 MW and operate at voltages of between 10 kV and 32 kV.

Generally a synchronous generator is connected to the transmission network via a step-up transformer. In the case of a small unit the generator and transformer are connected by cables whilst a large, high-power, generator may be connected to its transformer by a number of single-phase, screened busbars. The generator transformer is usually located outdoors and is of the tank type. Power from the transformer is fed to the substation busbars via high-voltage cables or a short overhead line.

2.2.2 Exciters and Automatic Voltage Regulators

The generator excitation system consists of an exciter and an automatic voltage regulator (AVR) and is necessary to supply the generator with dc field current as shown in Figure 2.2. The power rating of the exciter is usually in the range 0.2–0.8% of the generator's MW rating. In the case of a large generator this power is quite high, in the range of several MW. The voltage rating of the exciter will not normally exceed 1000 V as any higher voltage would require additional insulation of the field winding.

Excitation systems

Generally exciters can be classified as either rotating or static. Figure 2.3 shows some typical systems. In the rotating exciters of Figure 2.3a–c, the excitation current is supplied either by a dc generator or by an ac generator with rectifiers. As dc generators usually have relatively low power ratings, they are cascaded to obtain the necessary output, Figure 2.3a. Because of commutation problems with dc generators this type of exciter cannot be used for large generators which require large excitation currents.

As the number of cascaded dc generators increases, the dynamic properties of the exciter deteriorate resulting in an increase in the equivalent time constant. Nowadays dc generators have been almost entirely replaced by alternators which are simpler and more reliable. This change to alternators has been possible because of advances in power electronics which allow cheap, high-power rectifiers to be used in conjunction with the ac exciter.

The exciter shown in Figure 2.3b is a reluctance machine (inductor generator) operating at about 500–600 Hz so that the rectified current requires little smoothing. With this exciter both windings (ac and dc) are on the stator side. One disadvantage of this system is that slip rings are required to feed the rectified excitation current to the rotating field winding of the main generator. A further disadvantage is that the exciter itself tends to be quite large. This is a direct result of the way in which the sinusoidal flux changes,

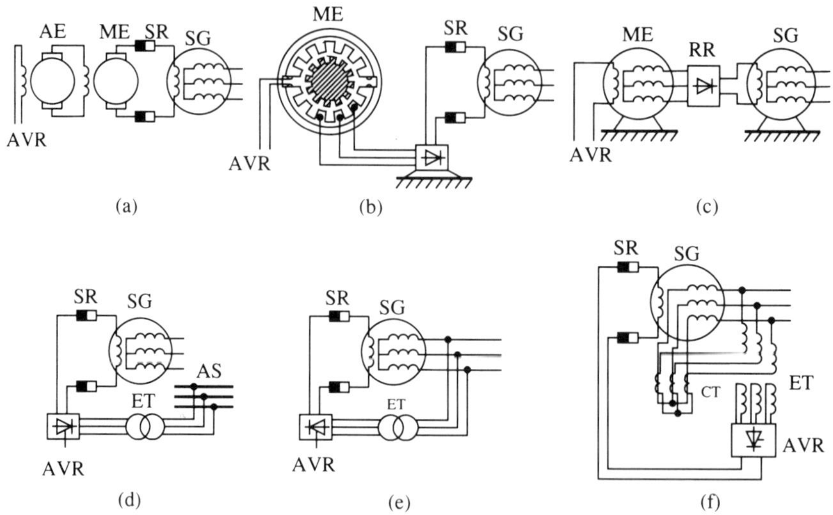

SG - synchronous generator; SR - slip rings; ME - main exciter; AE - auxiliary exciter;
RR - rotating rectifier; ET - excitation transformer; AS - auxiliary service busbars;
CT - current transformer; AVR - automatic voltage regulator

Figure 2.3 Typical exciter systems: (a) cascaded dc generators; (b) reluctance machine with rectifier; (c) inside-out synchronous generator with rotating rectifier; (d) controlled rectifier fed from the auxiliary supply; (e) controlled rectifier fed from the generator terminals; (f) controlled rectifier fed by the generator's voltage and current

necessary to induce the alternating emf in the armature, are produced solely by the changes in reluctance due to the rotation of the salient rotor teeth.

The exciter shown in Figure 2.3c has neither commutator nor slip rings. The principal excitation source is an inside-out synchronous machine with the field winding on the stator and armature winding on the rotor. The induced current is rectified by diodes, which are also mounted on the rotor, and fed directly to the excitation winding of the main generator. One limitation of this type of exciter is that the current supplied to the main generator can only be controlled indirectly via field control of the exciter. This tends to introduce a time constant of about 0.5 to 1 second into the exciter control system. One solution to this problem is to use rotating thyristors, rather than diodes, and control the exciter output via the firing angle of the thyristors. Unfortunately controlling the firing angle of a rotating thyristor is not easy and the reliability of such systems tends to be compromised by stray fields causing unscheduled thyristor firing.

Some alternative exciter systems using static thyristor converters are shown in Figure 2.3d–f. In these exciters the thyristor rectifiers are controlled directly by a voltage regulator. The main differences between the systems is in the type of supply used. Figure 2.3d shows an exciter supplied by an additional auxiliary service transformer. Figure 2.3e shows an alternative, and simpler, solution in which the exciter is fed from the generator output via a transformer. However, should a short circuit occur, particularly one close to the generator terminals, the decrease in the generator terminal voltage will result in a possible loss of excitation. With careful design the exciter can operate when the short circuit is further away from the generator terminals, e.g. at the high voltage terminals of the step-up transformer. More flexibility can be obtained by modifying the supply to the rectifier as shown in the exciter design of Figure 2.3f. In this system the generator does not lose excitation because its supply voltage is augmented, or compounded, by a component derived from the generator load current.

The main disadvantage of all static exciters is the necessity of using slip rings to feed current to the rotor of the main generator. This is offset to a large extent by the rapid speed with which they can react to control signals. As the cost of high-power rectifiers decreases, and reliability increases, static exciters are becoming the main source of excitation for high-power generators.

Automatic Voltage Regulators

The automatic voltage regulator (AVR) regulates the generator terminal voltage by controlling the amount of current supplied to the generator field winding by the exciter. The general block diagram of the AVR subsystem is shown in Figure 2.4. The measuring element senses the current, power, terminal voltage and frequency of the generator. The measured generator terminal voltage V_g is compensated for the load current I_g and compared with the desired reference voltage V_{ref} to produce the voltage error ΔV. This error is then amplified and used to alter the exciter output, and consequently the generator field current, so that the voltage error is eliminated. This represents a typical closed-loop control system. The regulation process is stabilised using a negative feedback-loop taken directly from either the amplifier or the exciter.

The load compensation element, together with the comparator, is shown in Figure 2.5. The voltage drop across the compensating impedance Z_C due to the generator current I_g is added to the generator voltage V_g to produce the compensated voltage V_C according to the function (2.1)

$$V_C = \left|\overline{V}_g + (R_C + \mathrm{j}X_C)\overline{I}_g\right| \tag{2.1}$$

where $\overline{Z}_C = R_C + \mathrm{j}X_C$ is the compensation impedance. If load compensation is not employed $\overline{Z}_C = \overline{0}$, $V_C = V_g$, and the AVR subsystem maintains constant generator terminal voltage. The use of load compensation ($\overline{Z}_C \neq \overline{0}$) effectively means that the

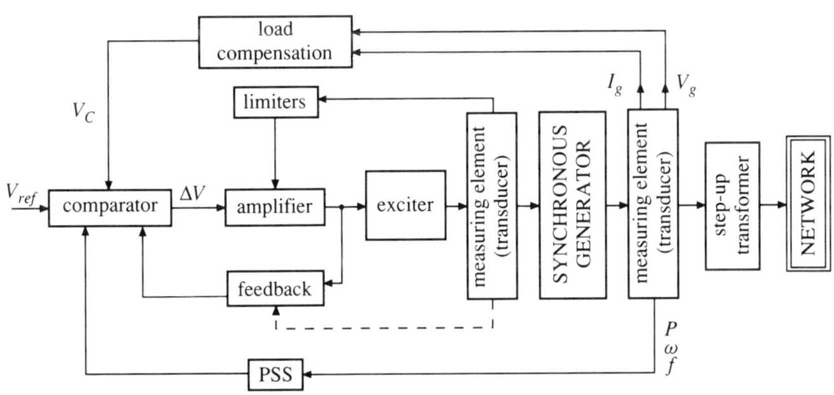

Figure 2.4 Block diagram of the excitation and AVR system

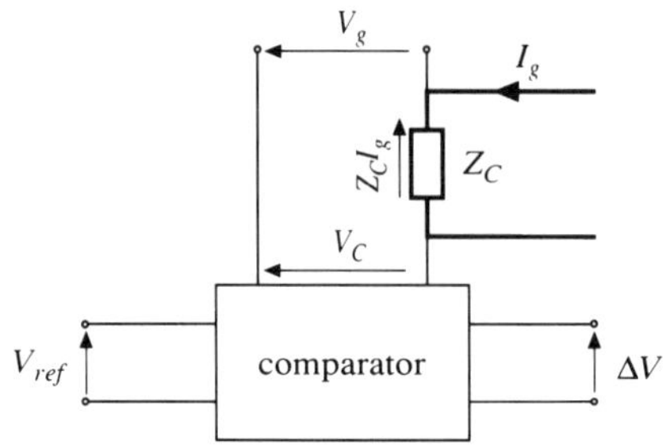

Figure 2.5 Load compensation element together with the comparator

point at which constant voltage is maintained is "pushed back" into the network by a distance that is electrically equal to the compensation impedance. In the case of parallel generators supplying a common busbar the compensation impedance must be smaller than the impedance of the step-up transformer in order to maintain stable reactive power dispatch between the parallel generators. Usually $X_C \approx 0.85X_T$, where X_T is the reactance of the step-up transformer.

The AVR subsystem also includes a number of limiters whose function is to protect the AVR, exciter and generator from excessive voltages and currents. They do this by maintaining the AVR signals between preset limits. Thus the amplifier is protected against excessively high input signals; the exciter and the generator against too high a field current; and the generator against too high an armature current and too high a power angle. The last three limiters have in-built time delays to reflect the thermal time constant associated with the temperature rise in the winding.

A power system stabiliser (PSS) is sometimes added to the AVR subsystem to help damp power swings in the system. PSS is typically a differentiating element with phase shifting corrective elements. Its input signals may be proportional to rotor speed, generator output frequency or the electrical real power output of the generator.

2.2.3 Turbines and their governing systems

In a power system, the synchronous generators are normally driven by either steam-, gas- or hydro-turbines as shown in Figure 2.2. Each turbine is equipped with a governing system to provide a means by which the turbine can be started, run-up to the operating speed, and operate on load with the required power output.

Steam turbines

In coal-burn, oil-burn and nuclear power plants the energy contained in the fuel is used to produce high pressure, high temperature steam in the boiler. The energy in the steam is then converted to mechanical energy in axial flow steam turbines. Each turbine consists of a number of stationary and rotating blades concentrated into groups, or stages. As the high pressure steam enters the fixed set of stationary blades it is accelerated and acquires increased kinetic energy as it expands to a lower pressure. The stream of fluid is then guided onto the rotating blades where it experiences a change in momentum and direction, thereby exerting a tangential force on the turbine blade and output torque on the turbine shaft. As the steam passes axially along the turbine shaft its pressure reduces so its volume increases and the length of the blades must increase from the steam entrance to exhaust to accommodate this change. Typically a complete steam turbine will be divided into three or more stages, with each turbine stage being connected in *tandem* on a common shaft. Dividing the turbine into stages in this way allows the steam to be reheated between stages to increase its enthalpy and consequently increase the overall efficiency of the steam cycle. Modern coal-fired steam turbines have thermal efficiency reaching 45%.

Steam turbines can be classified as either *non-reheat, single reheat*, or *double reheat* systems. Non-reheat turbines have one turbine stage and are usually built for use in units of below 100 MW. The most common turbine configuration used for large steam turbines is the single tandem reheat arrangement shown diagramatically in Figure 2.6. In this arrangement the turbine has three sections: the *high-pressure* (HP), *intermediate pressure* (IP), and *low-pressure* (LP) stages. Steam leaving the boiler enters the steam chest and flows through the *main emergency stop valves* (MSV) and the *governor control valves* (GV) to the high pressure turbine. After partial expansion the steam is directed back to the boiler to be reheated in the *heat-exchanger* to increase its enthalpy. The steam then flows through the *reheat emergency stop valves* (RSV) and the *intercept control valves* (IV) to the intermediate-pressure turbine where it is again expanded and made to do work. On leaving the IP stage the steam flows through the crossover piping for final expansion in the low pressure turbine. Finally the steam flows to the *condenser* to complete the cycle. Typically the individual turbine stages contribute to the total turbine torque in the ratio: 30% (HP), 40% (IP), 30% (LP). Note that the governor valves are also referred to as the main control valves or the HP control valves whilst the intercept valves are also referred to as the IP intercept valves or simply the IP control valves.

The steam flow in the turbine is controlled by the *governing system* (GOV). When the generator is synchronised the emergency stop valves are kept fully open and the turbine speed and power regulated by controlling the position of the *governor control valve(s)* (GV) and the *intercept control valve(s)* (IV). The speed signal to the governor is provided by the *speed measuring device* (SD). The main amplifier of the governing system and the valve mover is an *oil servomotor* controlled by the *pilot valve*. When the generator is

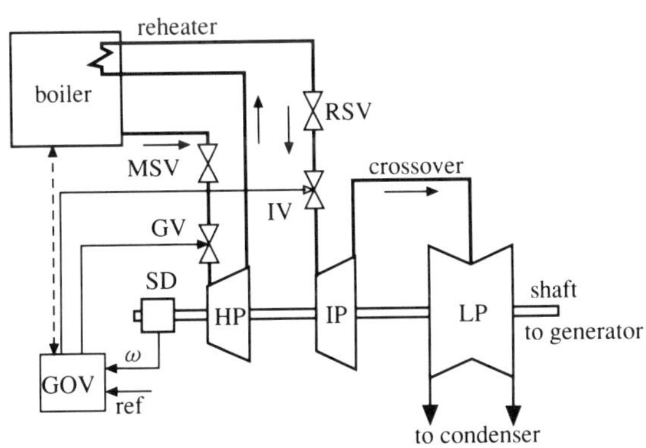

Figure 2.6 Steam configuration of a tandem compound single-reheat turbine

synchronised the *emergency stop valves* are only used to stop the generator under emergency conditions although they are often used to control the initial start-up of the turbine.

Besides the tandem compound single reheat turbine shown in Figure 2.6 other turbine arrangements are also used. Double reheat turbines have their first, high pressure section divided into the very-high-pressure (VHP) turbine and the high-pressure (HP) turbine with reheat between them. In this arrangement the individual turbines typically contribute to the total torque in the ratio: 20% (VHP), 20% (HP), 30% (IP), 30% (LP). Control valves are mounted after each of the reheaters and before the VHP section. In contrast to the single shaft arrangements just described *cross-compound, two-shaft turbines* are sometimes used where one of the shafts rotates at half the speed of the other. These turbines may have single or double reheat steam cycles.

Gas turbines

Unlike steam turbines, gas turbines do not require an intermediate working fluid; instead the fuel thermal energy is converted into mechanical energy using the hot turbine exhaust gases. Air is normally used as the working fluid with the fuel being natural gas or heavy/medium fuel oil. The most popular system for gas turbines is the open regenerative cycle shown in Figure 2.7 which consists of a *compressor* (C), *combustion chamber* (CH), and *turbine* (T). The fuel is supplied through the governor valve to the *combustion chamber* (CH) to be burnt in the presence of air supplied by the *compressor* (C). The hot, compressed air, mixed with the combustion products, is then directed into the *turbine* (T) where it expands and transfers its energy to the moving blades in much the same way as in the steam turbine. The exhaust gases are then used to heat the air delivered by the compressor. There are also other, more complicated cycles that use either compressor intercooling and reheating, or intercooling with regeneration and reheating. The typical efficiency of a gas turbine plant is about 35%.

Combined cycle gas turbines (CCGT)

A significant technological step forward in the use of gas turbines came with the introduction of the combined cycle illustrated in Figure 2.8. In this system the exhaust heat from

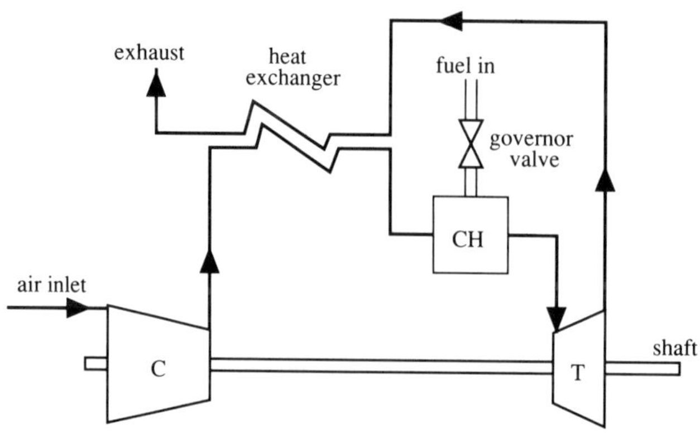

Figure 2.7 Open regenerative cycle of the gas turbine

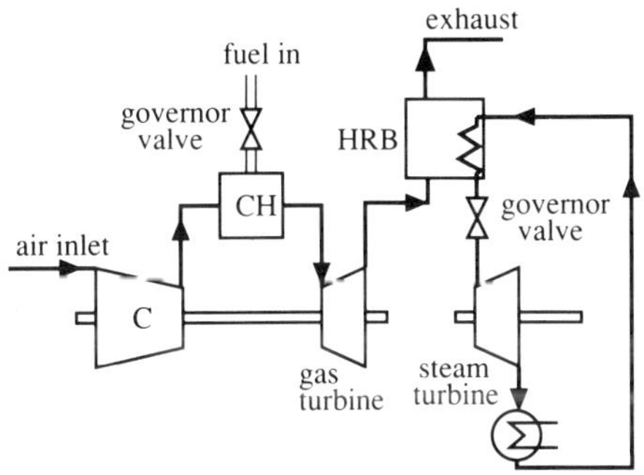

Figure 2.8 Example of a combined cycle gas turbine

the gas turbine is directed into a heat-recovery boiler (HRB) to raise steam, which is then used to generate more electricity in a steam-driven generating unit. Generally the temperature of the gas turbine exhaust gases is quite high, typically around 535°C, so by adding a steam turbine cycle at the bottom end of the gas cycle the otherwise wasted heat can be utilised and the overall cycle efficiency significantly increased. Modern CCGT plant can have an efficiency approaching, or even exceeding, 55%. Usually CCGT power stations utilise the exhaust gases from two or three gas turbines to raise steam for one steam turbine, with both types of turbine driving separate generators. More recently *single-shaft modes* have become popular where both the gas and the steam turbines are mounted on the same shaft and drive the same generator. In some CCGT designs the heat-recovery boiler may be equipped with supplementary firing to increase the temperature of the high pressure steam. In addition some combined cycle plants are designed to produce steam for district heating or for use in the process industry. CCGT plants are at the forefront of technology with significant developments currently taking place so that it is probable that by the time this book is published new generating plants of higher efficiencies will have been constructed.

CCGT plants, apart from higher thermal efficiency, also have other important advantages over more traditional coal-fired plants. They have a short construction time and low capital construction cost, both about half that of the equivalent coal-fired plant; they are relatively clean with almost no SO_2 emission; they require little staffing; and the materials-handling problem of gas versus coal and ash is much simpler.

Hydro turbines

The oldest form of power generation is by the use of water power. Hydraulic turbines derive power from the force exerted by water as it falls from an upper to a lower reservoir. The vertical distance between the upper reservoir and the level of the turbine is called the *head*. The size of the head is used to classify hydroelectric power plants as *high-head, medium-head*, and *low-head (run-of-river)* although there is no strict demarcation line.

Low and medium head hydro-electric plant is built using *reaction turbines* such as the *Francis turbine* shown in Figure 2.9a. Because of the relatively low head, reaction turbines typically use a large volume of water, require large water passages and operate at low speed. The low rotational speed requires that the generators have a large diameter. In operation water enters the turbine from the intake passage or *penstock* through a spiral

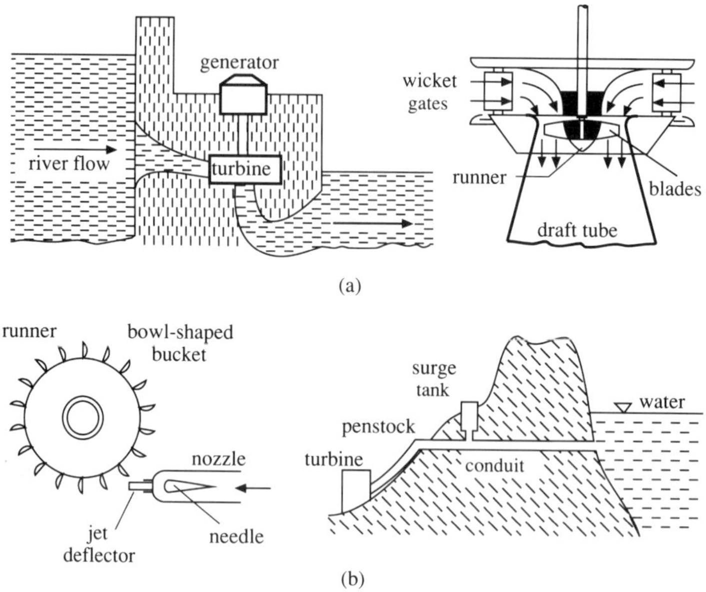

Figure 2.9 Hydro-turbines: (a) low- and medium-head reaction turbine; (b) high-head Pelton wheel

case, passes through the stay ring and the movable *wicket gates* onto the *runner*. On leaving the runner, the water flows through the *draft tube* into the *tail-water reservoir*. The movable wicket gates, with their axis parallel to the main shaft, control the power output of the turbine. Francis turbine runners have the upper ends of their blades attached to a crown and the lower ends attached to a band. For low-head operation the runner has no crown or band so that the blades are unshrouded. The blades themselves may be either fixed or adjustable. For adjustable-blade runners the governor can change both the blade angle and the wicket gate opening *(Kaplan type turbine)*. The blades are adjusted by means of an oil-operated piston located within the main shaft.

In high-head hydroelectric power plants *Pelton wheel* impulse turbines, shown in Figure 2.9b, are used. In these turbines the high pressure water is converted into high velocity jets by a set of fixed nozzles. The high velocity water jets impact on a set of bowl-shaped buckets attached around the periphery of the runner which turn back the water so impacting the full effect of the water jet to the runner. The size of the jet, and thus the power output of the turbine, is controlled by a needle in the centre of the nozzle. The movement of the needle is controlled by the governor. A jet deflector is located just outside the nozzle tip to deflect the jet from the buckets in the event of sudden load reduction.

Turbine governing systems

For many years the turbine governing systems were of a *mechanical-hydraulic* type and used the *Watt centrifugal mechanism* as the speed governor. The original Watt mechanism used two *flyballs* as the speed responsive device but on new machines the Watt governor

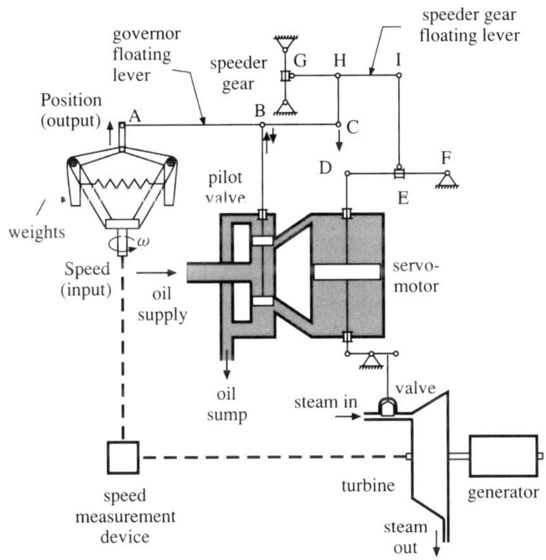

Figure 2.10 Mechanical-hydraulic governing system of the steam turbine

has been replaced by an *electro-hydraulic* one. However it is useful to understand the operation of the traditional mechanical-hydraulic system, shown in Figure 2.10, as it is still in use in various forms on older machines and it is a good way to illustrate the general principle of turbine control.

The pair of spring loaded weights in the centrifugal governor are driven by a motor that receives its power from the turbine shaft such that the height of the weights depends on the speed. When the turbine mechanical torque is equal to the counter-acting generator electromagnetic torque, the rotational speed of the turbine-generator is constant and the position of the weights does not change. If the electrical torque increases, due to a change in load, so that it is greater than the mechanical driving torque, the rotational speed ω decreases and the weights move radially inwards under centrifugal action. This causes point A on the governor floating lever to rise and the floating lever *A-B-C* rotates around point *C*. This rotation results in point *B* and the pilot valve moving upwards, allowing high pressure oil to flow into the upper chamber of the main servomotor. The differential pressure across the piston forces it to move downwards, partially opening the turbine valve and increasing the turbine power. The displacement of the main servomotor piston downwards causes points *D*, *E*, *I* and *H* to lower and the speeder gear floating lever to rotate downwards about point *G*. This lowers point *C*, around which the lever *A-B-C* rotates, and partially closes the pilot valve to reduce the oil flow into the upper chamber.

This governing system has two negative feedback loops: the main speed feedback loop through the turbine speed measuring device and the centrifugal governor, and the second valve position feedback loop through the steam valve, piston and points *D*, *E*, *I*, *H*, and *C*. This latter feedback loop ensures that the static speed-power characteristic of the turbine has a negative slope. As will be explained later in this section, such a characteristic is fundamental to the turbine control as it ensures that any speed increase will be met by a corresponding reduction in turbine torque and vice versa. The slope, or gain, of the characteristic may be changed by moving point *E* horizontally on the lever *D-E-F*.

The purpose of the *speeder gear* is twofold. Firstly it controls the speed of the unsynchronised generator and secondly it controls the power output of the synchronised generator. To see how the speeder gear works assume that the generator is synchronised and that it is required to increase the power output. As the generator is synchronised its speed will be constant and equal to synchronous speed. If the speeder gear is used to raise point G then points C and B and the pilot valve will also rise. High pressure oil will then enter the upper chamber of the main servomotor, the piston will lower and the steam valve will be opened thereby increasing the steam flow through the turbine and the power output. As the servomotor piston lowers so do points D, E, I and H. This movement lowers point C and returns the pilot valve to its equilibrium position. The schematic diagram of the mechanical-hydraulic governor is shown in Figure 2.11a with the position of the speeder gear setting the *load reference set point*.

The main disadvantages of the Watt centrifugal governor are the presence of deadbands and a relatively low accuracy. The size of the deadbands also tends to increase with time due to wear in the moving mechanical elements. Newer solutions replace the Watt centrifugal mechanism with an electronic regulator. In these systems the turbine rotor speed is measured electronically, with high accuracy, using a toothed wheel and a probe. The resulting electrical signal is amplified and acts on the pilot valve via an electro-hydraulic converter. The schematic diagram of the electro-hydraulic system in

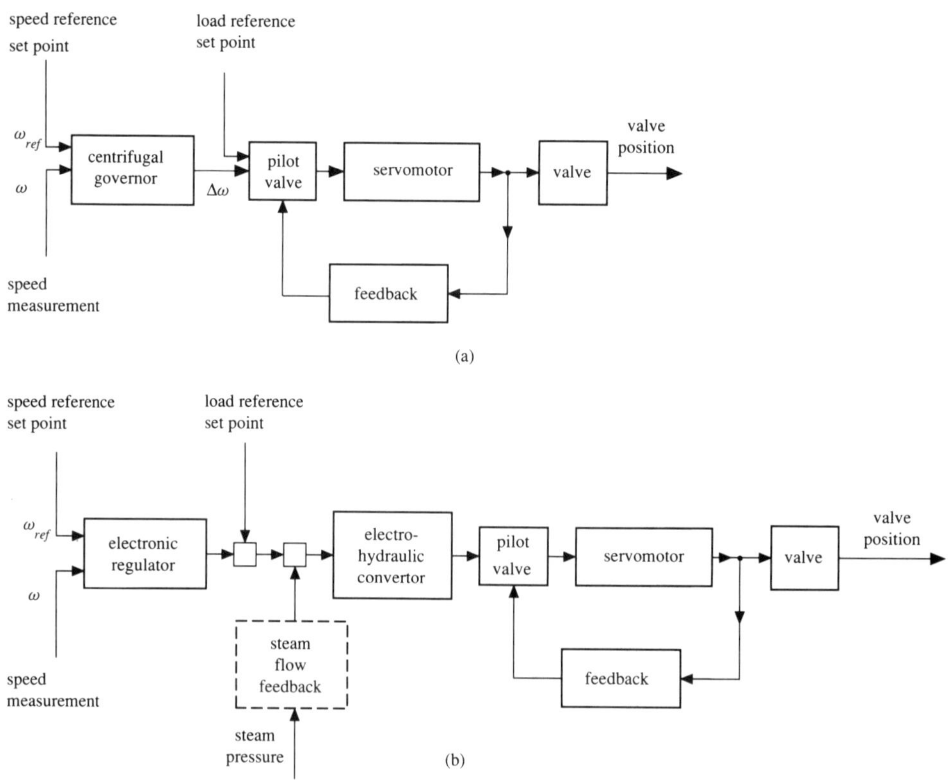

Figure 2.11 Schematic diagram of the governing system: (a) mechanical-hydraulic; (b) electro-hydraulic

Figure 2.11b shows that its operation does not differ much from that of the mechanical-hydraulic system shown in Figure 2.11a, but the flexibility of electronic regulators enables additional control loops to be introduced linking the boiler and the turbine control systems. The dashed line in Figure 2.11b symbolises the steam flow feedback whose function is to prevent the valves being opened by the speed regulator when the steam inlet pressure is too low. The reference speed is set electronically in the *speed reference set point*. It is also possible to change the turbine power using an additional signal that is added to the control circuit at the *load reference set point*.

Higher forces are required to move the control gates in hydro-turbines than the valves in steam turbines and, as a result, hydro-turbine governing systems usually employ two cascaded servomotors. The first, low power, *pilot servomotor* operates the *distributor* or *relay valve* of the second, high power, *main gate servomotor*, just as in the steam turbine the pilot servomotor has a pilot valve which is controlled either by a mechanical Watt-type governor or by an electronic regulator via an electro-hydraulic converter. The turbine governing system is similar to that used in steam turbines (Figure 2.11) but the number of servomotors is higher and the feedback loop has an additional dashpot for reasons described in Chapter 10.

Turbine characteristics

Of importance here is the way the power output of the turbine varies with speed. For stable operation the turbine must have a power-speed characteristic such that as the speed increases the mechanical input power reduces. This will restore the balance between the electrical output power and mechanical input power. Similarly a decrease in speed should result in an increase in the mechanical power.

To examine how such a characteristic can be achieved Figure 2.12 shows the idealised power-speed characteristics for an unregulated (dashed line) and a regulated (solid line) turbine. Point A is the rated point which corresponds to the optimal steam flow through

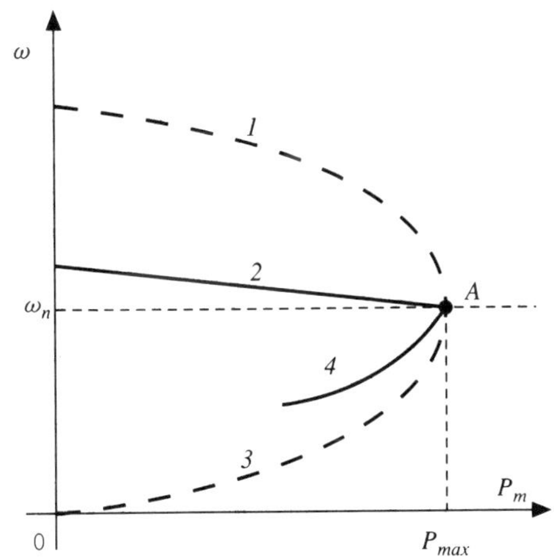

Figure 2.12 Turbine power-speed characteristic for the unregulated (lines 1 and 3) and the regulated turbine (lines 2 and 4)

the turbine, as determined by the turbine designers. Consider first the unregulated characteristic and assume that the turbine is initially operating at point A with the turbine control valve fully open. The generator is assumed to be synchronised with the system and its speed can only change if the system frequency changes. If, due to some reason, the system frequency rises, so does the speed of the rotor. As the main valve is fully open the speed increase causes additional losses in the turbine and the efficiency of the steam flow drops (with respect to the optimal point A) with a corresponding reduction in power as shown by the dashed line *1*. Similarly, a decrease in the system frequency causes the rotor speed to drop with a corresponding drop in power as shown by the dashed line *3*. In practice as the speed drops the actual change in the turbine power will be slightly greater than that indicated by line *3* because the steam flow through the turbine also depends on the performance of the boiler and the boiler feed pumps. The performance of these pumps is strongly frequency-dependent, and hence will be reduced by a reduction in system frequency (and rotor speed). This causes a decrease in the amount of steam flowing through the turbine and a further drop in the turbine torque as shown by the solid line *4*. The way the turbine speed-power characteristic can be improved by the use of a governor will now be discussed by reference to the governor functional diagrams in Figure 2.11.

If the steam flow feedback in the electro-hydraulic governing system is neglected and the governor response assumed to be dominated by the time constant of the servomotor, both the mechanical-hydraulic and the electro-hydraulic governors shown in Figure 2.11 may be represented by the simplified block diagram shown in Figure 2.13. The coefficient K_A in Figure 2.13a corresponds to the amplification gain of the servomotor, while coefficient R corresponds to the gain of the feedback loop. Transformation of the block diagram allows R to be eliminated from the feedback loop by moving it into the main loop to obtain the block diagram shown in Figure 2.13b where $T_G = 1/(K_A R)$ and is the effective governor time constant.

The block diagram of Figure 2.13b allows an approximate analysis of the static and dynamic properties of the turbine-governor system. In the steady-state $t \to \infty$, $s \to 0$, and

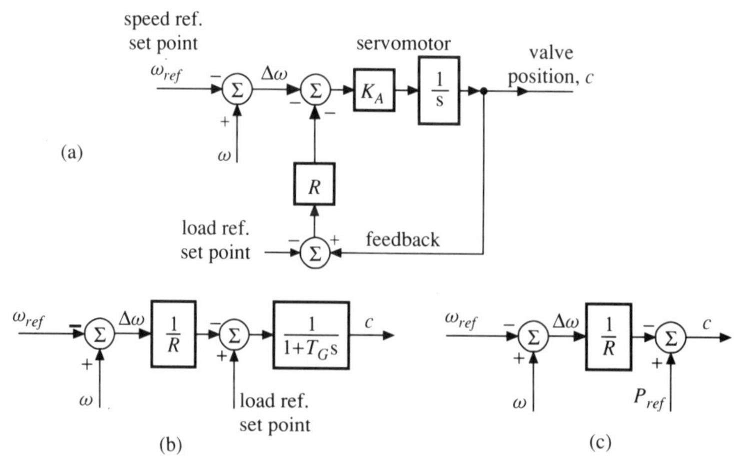

Figure 2.13 Simplified model of the steam turbine governing system: (a) block diagram with negative feedback; (b) equivalent block diagram; (c) equivalent block diagram for the steady-state

the turbine block diagram can be simplified to that shown in Figure 2.13c where P_{ref} is the load reference set point expressed as a fraction of the nominal or rated power, P_n. If the valve position c is assumed to vary between 0 (fully closed) and 1 (fully open) then a small change in turbine speed $\Delta\omega = \omega - \omega_{ref}$ will produce a corresponding change in valve position $\Delta c = -\Delta\omega/R$. Normally $\Delta\omega$ is expressed as a fraction of rated speed ω_n so that

$$\Delta c = -\frac{1}{\rho}\frac{\Delta\omega}{\omega_n} \quad \text{or} \quad \frac{\Delta\omega}{\omega_n} = -\rho\Delta c \tag{2.2}$$

where $\rho = R/\omega_n$ is termed the *speed-droop coefficient* or simply the *droop*. The reciprocal of droop $K = 1/\rho$ is the *effective gain* of the governing system. Physically droop can be interpreted as the percentage change in speed required to move the valves from fully open to fully closed. If a linear relationship is assumed between the valve position and mechanical power then the turbine power output ΔP_m expressed as a fraction of the nominal or rated power output P_n is given by $\Delta P_m/P_n = \Delta c$ and

$$\frac{\Delta\omega}{\omega_n} = -\rho\frac{\Delta P_m}{P_n} \quad \text{or} \quad \frac{\Delta P_m}{P_n} = -K\frac{\Delta\omega}{\omega_n} \tag{2.3}$$

Equation 2.3 describes an idealised turbine power-speed characteristic. In (P_m, ω) co-ordinates this gives a straight line of gradient ρ shown in Figure 2.12 by line *2*. However it is important to realise that once the steam valves are fully open no more control can be exerted over the turbine so that should the speed drop the turbine would follow the characteristic *4* in the same way as the unregulated turbine.

A good control system should ensure that any load fluctuation ΔP_m would only produce a small speed change $\Delta\omega$. This is achieved by making the droop ρ small. However it should be emphasised that the *droop cannot be zero or negative*. Figure 2.14 illustrates this point. The system demand is dominated by electrical loads for which an increment in the active power ΔP_{load} is weakly dependent on the change in the system frequency and therefore on the change in the rotational speed of the synchronous generators. As a result the static load characteristic $\omega(P_{load})$ is almost a vertical line in the (P, ω)-plane but with a slight positive slope that reflects the frequency dependence. The point of intersection of the load characteristic $\omega(P_{load})$ and the turbine characteristic $\omega(P_m)$ is the equilibrium point where the opposing electromagnetic and mechanical torque acting on the shaft are equal in magnitude and the rotational speed is constant.

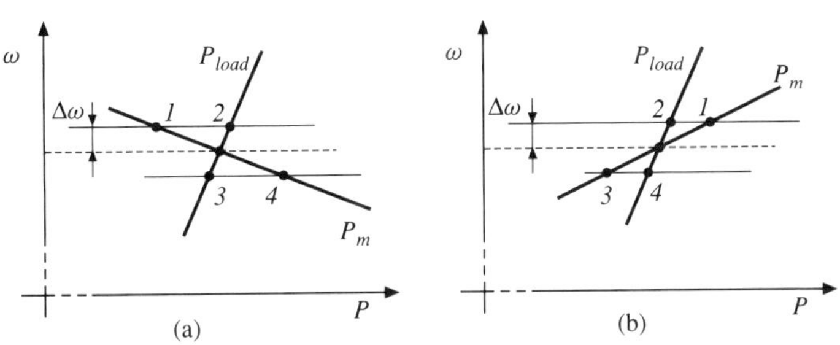

Figure 2.14 The equilibrium point between the turbine power and the load power: (a) stable point; (b) unstable point

Figure 2.14a shows the case where the droop is positive ($\rho > 0$) so that, according to equation (2.3), the turbine power increases when its rotational speed decreases. In this case a small disturbance in frequency (turbine speed) causes the system to return automatically to the equilibrium point. For example, if there is a momentary increase in frequency, then the disturbed load power (point *2*) is greater than the disturbed turbine power (point *1*). This means that the opposing electromagnetic torque is greater than the driving torque and the system is forced to return to the equilibrium point. Similarly if frequency decreases then the load power (point *3*) is less than the turbine power (point *4*) and this excess mechanical power causes the turbine speed to increase and the system to return to the equilibrium point.

On the other hand Figure 2.14b shows a case when the droop is assumed to be negative ($\rho < 0$). The situation is then reversed as any increase in frequency will cause the turbine power (point *1*) to be greater than the load power (point *2*). This would result in an increase in turbine speed and a further increase in the power imbalance. Similarly, a decrease in frequency would result in the turbine power (point *3*) being less than the load power (point *4*) with further movement away from the equilibrium point. Such a droop characteristic is not resistant to disturbances and the system is unstable.

The case when $\rho = 0$ is the marginal stability case and corresponds to the absence of the negative feedback loop on valve position. For the Watt regulator of Figure 2.10 this corresponds to the absence of the lever H-E which realises the position feedback between the servomotor piston and the pilot valve. Without this negative feedback the governing system would be of the constant speed type. Such a governor cannot be used if two or more generators are electrically connected to the same system since each generator would be required to have precisely the same speed setting (which is technically impossible) or they would "fight" each other, each trying to pull the system frequency to its own setting. If generators connected to the system have speed-droop characteristics with $\rho > 0$, then there will always be a unique frequency at which they will share load. This will be described in Chapter 8.

Typical values of the speed-droop coefficient (in per unit) are between 0.04 and 0.09 with the lower value corresponding to turbogenerators and the higher to hydrogenerators. This is connected with the relative ease with which hydro-plant can accept a change in load compared to thermal plant.

Governor control functions

Having established the basic workings of the governor it is now prudent to look at the overall control functions required of a practical turbine governor. These control functions can be sub-divided into *run-up control, load/frequency control, overspeed control* and *emergency overspeed trip. Run-up control* of the unsynchronised generator is not considered further other than to comment that this is one area where control may be carried out using the main stop valves with both sets of control valves and the interceptor stop valves fully open. Primary *load/speed control* and secondary *frequency/tie-line power control* is achieved via the governor control valves (GV) with the intercept valves (IV) fully open. This control action is fundamental to turbine operation and is discussed in more detail in Chapter 8. If a severe disturbance occurs, the turbine speed may increase quickly and the aim of the *overspeed control* is to limit the maximum overspeed to about 110%. If overspeed control were only possible via the governor control valves then, depending on the droop setting, the generator speed would increase to, say, 105% (5% droop) before the main valves were shut. Although this would quickly reduce the HP torque the entrapped steam in the reheater would reduce only slowly, typically with a time constant of 5 seconds or more, resulting in a slow decay of both the IP and the LP torque. As the IP and LP typically contribute 70% of the torque, the turbine speed would continue to increase until the steam flow had time to reduce. Consequently the purpose of the overspeed control is to shut the intercept valves and, as these are at the inlet to the IP turbine, they have an immediate effect on reducing the IP and LP torque so limiting the overspeed. Typically the IV valves will be kept fully open until the generator speed has reached an overspeed of, say 104%, when the IV valves will be closed.

In addition to IV valve closure electro-hydraulic governors may also be equipped with additional *fast-valving* control logic which uses auxiliary control signals such as acceleration, electrical power, generator current etc. to fast close the control valves when a large disturbance close to the generator is sensed. These fast valving control functions are discussed in more detail in Section 9.2. The final stage of protection, the *emergency overspeed trip*, is independent of the overspeed control. If this trip function is activated both sets of control valves and the emergency stop valves are shut and the boiler tripped to ensure that the turbine is quickly stopped.

2.3 Substations

A substation can be regarded as a point of electrical connection where the transmission lines, transformers, generating units, system monitoring and control equipment etc. are connected together. Consequently it is at substations that the flow of electrical power is controlled, voltages are transformed from one level to another and system security is provided by automatic protective devices.

All substations consist of a number of incoming and outgoing circuits located in *bays*. These incoming and outgoing circuits are connected to a common *busbar* system and are equipped with apparatus to switch electrical currents, conduct measurements and protect against lightning. Each electrical circuit can be divided into a *primary* and a *secondary* circuit. The primary circuit includes the transmission line, power transformer, busbars etc. and the high-voltage side of voltage and current transformers. The secondary circuit consists of the measurement circuits on the low-voltage side of the voltage and current transformers and the control circuits for the circuit-breakers, protection circuits, and isolators.

The busbar constitutes a point of electrical contact for individual lines and transformers. In indoor substations the busbar consists of flat conductors made of aluminium or copper and supported by insulators, whilst in outdoor substations the busbars are stranded conductors made of steel and aluminium and suspended on insulators. A number of different busbar arrangements are possible, each of which differ in the flexibility of possible electrical connections and the ease at which maintenance can be carried out without disturbing either the operation of the substation or the system security. The type of busbar system used will depend on the role and importance of the substation in the power system, the voltage level, the installed capacity and on the expected reliability of network operation. Bigger substations tend to use more elaborate busbar systems requiring higher capital investment and operation cost. A description of different types of substation layout can be found in Giles (1970).

2.4 Transmission and Distribution Network

The transmission and distribution network connects all the power stations into one supplying system and transmits and distributes power to individual consumers. The basic elements of the network are the overhead power lines, underground cables, transformers and substations. Auxiliary elements are the series capacitors, shunt reactors and compensators, switching elements, metering elements and protection equipment.

2.4.1 Overhead lines and underground cables

Overhead lines are universally used to transmit electrical energy in high voltage transmission systems whilst underground cables are normally only used in low and medium voltage urban distribution networks. Because of their high cost, and the technical problems associated with the capacitive charging current, the use of high voltage underground cables can only be justified under special circumstances such as in densely populated urban areas, wide river crossings or in areas of major environmental concern etc. For example short distance cables are sometimes used to connect a power station to a substation.

Whenever current flows through any network element real power is lost. As this power loss is proportional to the square of the current, transmission lines operate at high voltage and low current. Generally the more power that is sent over a transmission line the higher will be its voltage. For practical reasons there is a standardisation of voltage levels within different regions of the world. Unfortunately these standard voltages tend to vary slightly between regions but are not too dissimilar. The typical transmission voltage levels are 110, 220, 400, 750 kV for Continental Europe; 132, 275, 400 kV for the UK; and 115, 230, 345, 500, 765 kV for USA.

The maximum theoretical voltage value at which an overhead transmission line can be built is limited by the electrical strength of air and is estimated to be about 2 400 kV. Currently the maximum voltage at which commercial lines have been built is 765 kV (Canada) and 750 kV (former Soviet Union). Experimental lines have been built to operate at 1 100 kV in Japan and 1 200 kV in the former Soviet Union (CIGRE, 1994).

Because of the high right-of-way costs associated with overhead lines multi-circuit lines are usually built where more than one three-phase circuit is supported on the same tower. If a large increase in transmitted power is predicted for the future, space may be left on the transmission towers for extra circuits to be added later.

Distribution networks generally operate at lower voltages than the transmission network. Here the voltage standards used, both by different countries and by different areas in one country, can be quite varied partly because of the way the system has developed. Historically different parts of a network may have belonged to different private companies each of which would have followed its own standardisation procedures. For example there are 12 different standard distribution voltages in the USA, in the range between 2.4 and 69 kV. In the UK the distribution voltages are 6.6, 11, 33 and 66 kV.

2.4.2 Transformers

Transformers are necessary to link parts of the power systems that operate at different voltage levels. In addition to changing voltage levels, transformers are also used to control voltage and are almost invariably equipped with taps on one or more windings to allow the turns ratio to be changed. Power system transformers can be classified by their function into three general categories:

- *generator step-up transformers* (which connect the generator to the transmission network) and *unit transformers* (which supply the auxiliary service - Figure 2.2)
- *transmission transformers* which are used to connect different parts of the transmission network, usually at different voltage levels, or connect the transmission and distribution networks
- *distribution transformers* which reduce the voltage at load centres to a low voltage level required by the consumer

Generator and transmission transformers have ratings from several tens of MVA to more than 1 000 MVA and are usually oil-cooled. The transformer core is placed inside a tank filled with oil which acts as both a coolant and insulator for the transformer windings. Heat, due to core loss and ohmic loss in the transformer windings themselves, is removed from the oil through external radiators. The circulation of oil inside the transformer is either natural or forced. The air circulation outside the transformer is usually forced using fans. Because of transportation problems large, high-power, transformers are usually constructed as three, separate, single-phase transformers. Smaller power transformers are usually of an integrated three-phase design.

Generator transformers step the voltage up from the generator level of typically 10–20 kV to the transmission or sub-transmission voltage. In a power station employing large generators of typically 200–500 MW and above each generator may have its own transformer consisting of three interconnected *two-winding transformers*. In contrast to this the generators in a smaller power station may operate with two generators connected to one *three-winding* three-phase transformer.

Generator step-up transformers are usually Δ-Y connected with the neutral grounded. The delta low voltage winding closes the path for the circulating current resulting from asymmetrical loading and the undesirable third harmonic magnetising current, caused by the non-linear B-H characteristic of the transformer core, so that they remain trapped inside it. In a large power station with many generating units some of the transformer neutrals may not be grounded to limit the single-phase short-circuit currents in the transmission network.

Transmission transformers connect different parts of the transmission and sub-transmission networks operating at different voltage levels, supply distribution networks and connect large industrial consumers directly to the transmission network as shown in Figure 2.1. The windings of the transformers tying transmission and sub-transmission networks are normally Y-Y connected with the neutral grounded. These transformers often also have a low-power, medium voltage, Δ-connected tertiary winding to provide the path for the circulating current when the high voltage winding is asymmetrically loaded. This additional winding can also be used to supply local loads inside a substation or to connect a reactive power compensator.

If the required transformation ratio is not too high the two-winding transformer shown in Figure 2.15a can be replaced by the one-winding *autotransformer* shown in Figure 2.15b. In the autotransformer, part of the primary winding, w_1, and the secondary winding, w_2, are common giving an obvious economy. Autotransformers are normally

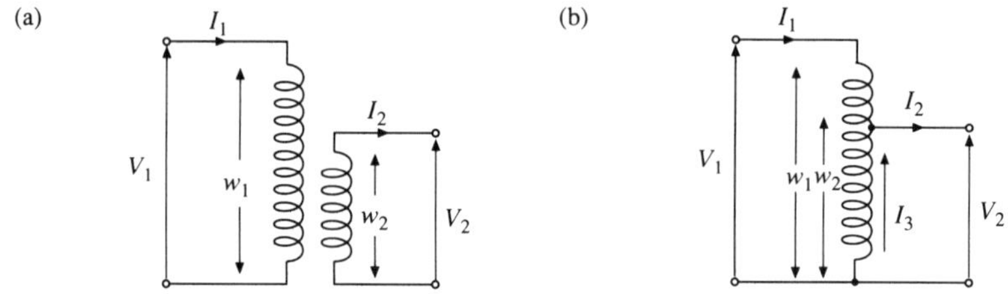

Figure 2.15 Transmission transformers: (a) two-winding transformer (b) autotransformer

used to connect networks at consecutive voltage levels, e.g. 132/275 kV, 275/400 kV in UK, 138/230 kV, 230/345 kV, 345/500 kV in USA, and 110/220 kV, 220/400 kV in continental Europe.

Distribution networks are normally supplied from transmission and sub-transmission networks by transformers with the high-voltage side connected in star and the medium-voltage side connected in delta to help minimise any possible load asymmetry. Autotransformers linking parts of distribution networks operating at different, but close, voltage level are usually star connected with the neutral grounded.

Tap-changing and regulating transformers

The main task of any transformer is to link parts of the network at different voltage levels. However changes in the system load often cause undesirable voltage fluctuations and to help alleviate this problem transformers are often equipped with *tap changers* to provide a means of voltage control, typically in the range of $\pm$ 10%.

Transformer taps may be situated in the same tank as the main winding or in a separate tank. When situated in the same tank, the taps are usually on the high voltage side of the transformer (because of the lower current) and near the neutral end of the winding (where the voltage with respect to ground is least). In autotransformers taps are also on the high voltage side but near the common part of the winding.

When a separate tank is used that part of the winding with the taps can be made as a tertiary winding of the main transformer. The voltage from the tertiary winding can then be injected, using an additional *series transformer*, into the main circuit as shown schematically in Figure 2.16a. Another solution, shown in Figure 2.16b, is to use a separate *excitation transformer* (equipped with the tap changer) which supplies the series transformer.

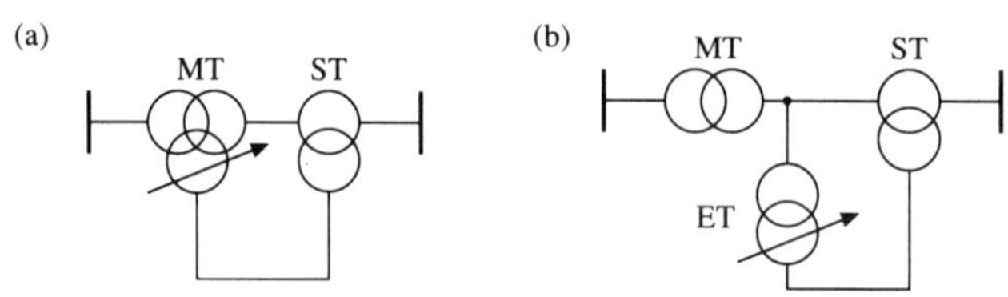

MT - main transformer, ST - series transformer, ET - excitation transformer

Figure 2.16 Two ways of supplying the series transformer: (a) from the tertiary winding of the main transformer (b) from a separate excitation transformer

Tap changing facilities can be made to operate either off-load or on-load. The off-load tap changer requires the transformer to be de-energised whilst tap changing takes place. The *under-load tap-changing transformer* (ULTC, also called *on-load tap changing*, OLTC, or *load tap changing*, LTC) allows the taps to be changed whilst the transformer is energised. The tap changing is automatic and operated by motors which respond to relays set to hold voltage at the prescribed level. The number, or frequency, of tap changes per day is usually minimised in order to prolong the life of the tap-changer.

With the transformer connections shown in Figure 2.16 the voltage can be regulated in both magnitude and phase with the degree of regulation depending on the connection made between the tertiary winding of the main transformer (or the excitation transformer) and the series transformer. Some possible schemes are illustrated in Figure 2.17.

Figure 2.17a shows the series transformer windings where ΔV_A, ΔV_B and ΔV_C are the voltages supplied from the excitation transformer (not shown) to be injected into each phase of the main circuit. Figure 2.17b shows the phase relationship of the phase, and phase-to-phase, voltages used to supply the excitation transformer. If the excitation transformer is constructed so that the secondary-side voltages ΔV_A, ΔV_B and ΔV_C are proportional to, and in-phase with, the primary-side phase voltages V_A, V_B and V_C then the series transformer will produce a change in the voltage magnitude as shown in Figure 2.17c. The secondary voltages of the series transformer are $V_{A'}$, $V_{B'}$ and $V_{C'}$. Alternatively, as shown in Figure 2.17d, the excitation transformer can be constructed so that the voltages ΔV_A, ΔV_B and ΔV_C are proportional to the phase-to-phase values V_{BC}, V_{CA} and V_{AB} of the primary-side voltages. As the line voltage between two phases

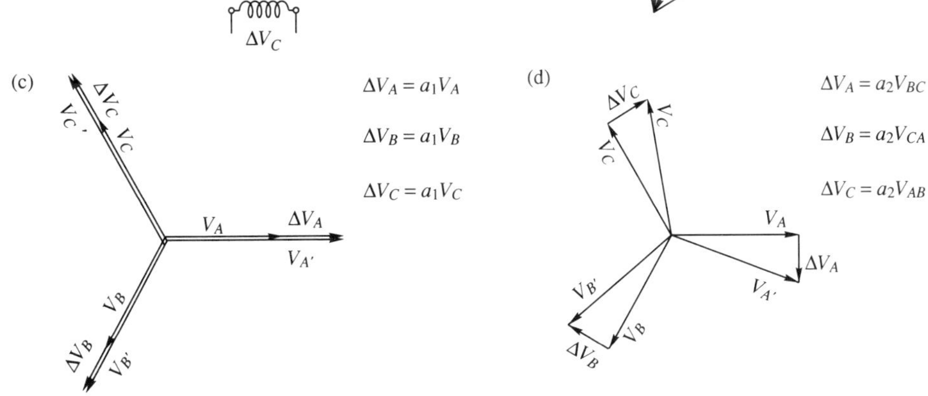

Figure 2.17 Complex transformation ratio: (a) windings of the series transformer; (b) the triangle of phase and line voltages; (c) in-phase booster voltages; (d) quadrature booster voltages

in a three-phase system is always in quadrature with the voltage of the third phase, the series transformer will now introduce a small change in voltage magnitude but a large change in the voltage angle. Generally the excitation transformer can be constructed so that it supplies the series transformer with a net voltage made up of an in-phase component and a quadrature component. The voltage change in all three phases can be then expressed as

$$\Delta\overline{V}_A = a_1\overline{V}_A + a_2\overline{V}_{BC}, \quad \Delta\overline{V}_B = a_1\overline{V}_B + a_2\overline{V}_{CA}, \quad \Delta\overline{V}_C = a_1\overline{V}_C + a_2\overline{V}_{AB} \qquad (2.4)$$

where a_1 and a_2 are the voltage transformation ratios associated with the in-phase component and quadrature component respectively. Both these voltage ratios can be adjusted to allow control of both the voltage magnitude and angle. In this case the transformation ratio is a complex number, $\overline{\vartheta} = \overline{V}_{A'}/\overline{V}_A = \overline{V}_{B'}/\overline{V}_B = \overline{V}_{C'}/\overline{V}_C = \vartheta e^{j\theta}$, where θ is a phase shift angle.

Transformers with complex transformation ratio are called *phase shifting transformers*. Sometimes it is necessary to re-direct the flow of power in a given line without changing the voltage level. This can be done by inserting a *quadrature booster*, consisting of the supply transformer and the series transformer, whose sole task is to change the voltage angle.

2.4.3 Shunt and series elements

Shunt and series elements, such as series capacitors and shunt compensators (static and rotating), are used in transmission networks for a number of purposes. From the point of view of this book their use will be considered for the purposes of reactive power compensation and stability improvement.

Shunt elements

Section 3.1 will show how the ever-changing real and reactive power demand may cause large variations in the network voltage profile. Generally reactive power cannot be transmitted over long distances and should be compensated for close to the point of consumption. The simplest, and cheapest, way of achieving this is by providing *shunt compensation*, that is by installing capacitors and/or inductors connected either directly to a busbar or to the tertiary winding of a transformer. Shunt elements may also be located along the transmission route to minimise losses and voltage drops. Traditionally static shunt elements are breaker-switched either manually or automatically by a voltage relay. Modern solutions, involving the use of thyristors, will be described in the next sub-section.

When the system power demand is low reactive power produced by the transmission line capacitance may prevail over the reactive power consumed by the inductance, and the transmission line may be a net source of reactive power. The effect of this may be to boost the network voltages to unacceptably high values. In such circumstances *shunt reactors* can be used to consume the surplus reactive power and depress the voltages. Usually shunt reactors are required for transmission lines longer than about 200 km. During heavy loading conditions some of the reactors may have to be disconnected and *shunt capacitors* used to supply reactive power and boost local voltages.

Another traditional means of providing shunt compensation is by the use of a *synchronous compensator*. This is a salient-pole synchronous motor running at no-load whose field is controlled so as to either generate or absorb reactive power. When overexcited the synchronous compensator is a source of reactive power, while when underexcited it absorbs reactive power. Although relatively expensive synchronous compensators play an important role in the control of voltage and reactive power at the transmission, and especially sub-transmission, voltage levels. They are used in order to improve stability and maintain voltages within desired limits under varying load conditions and contingency situations. When used in new substations synchronous compensators are often supplemented by switched shunt capacitor banks and reactors so as to reduce installation and operating costs. The majority of synchronous compensator installations are designed for outdoor operation and operate unattended with automatic control of start-up and shutdown. Small synchronous compensators of several MVA are usually connected to the tertiary winding of the transmission transformer whilst larger units of up to few hundred MVA are connected by individual step-up transformers to the busbars of a HV substation. The small units are generally air-cooled whilst the bigger units are either water-cooled or hydrogen-cooled.

Series elements

Series capacitors are connected in series with transmission line conductors in order to offset the inductive reactance of the line. This tends to improve electromechanical and voltage stability, limit voltage dips at network nodes and minimises the real and reactive power loss. Typically the inductive reactance of a transmission line is compensated to between 25 and 70%. A full 100% compensation is never considered as it would make the line flows extremely sensitive to changes in angle between the voltages at the line terminals, and the circuit would be series resonant at fundamental frequency. Moreover, high compensation increases the complexity of protection equipment and increases the probability of subsynchronous resonance as discussed in sub-Section 6.6.3.

Normally series capacitors are located either at the line terminals or at the middle of the line. Although fault currents are lower, and line protection easier, when the capacitors are located at the mid-point the access necessary for maintenance, control and monitoring is significantly eased if the capacitor banks are positioned at the line terminals. For this reason the compensating capacitors and associated shunt reactors are usually split into two equal banks positioned at each end of the line Typically each bank will be capable of compensating the line to a maximum of 30%. A detailed discussion of the benefits and problems of locating series capacitors at different points along the line can be found in Ashok Kumar *et al*(1970) and Iliceto *et al.* (1977).

Sometimes during power swings or heavy power transfers the line reactive current is high and the voltage may rise excessively on one side of the series capacitor. In this case the system must be designed to limit the voltage to acceptable levels or capacitors of appropriately high voltage rating must be used. Normally the voltage drop across a series capacitor is only a few percent of the rated line voltage. A short-circuit on one side of the capacitor may, however, produce a temporary voltage across the element approximately equal to the line rated voltage. As such a fault is rare it is uneconomical to design the element to withstand such a high voltage so that normally provision is made for the capacitor to be bypassed during such a fault and re-inserted after fault clearing.

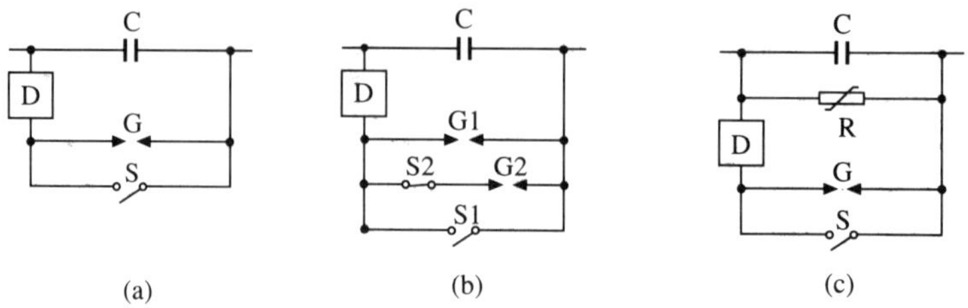

Figure 2.18 Series capacitors protective schemes; (a) single-gap scheme; (b) dual-gap scheme; (c) zinc-oxide scheme

The traditional way of bypassing the capacitor is to include a spark gap across either the capacitor bank itself or across each module of the bank. A better solution is to use non-linear, zinc oxide resistors which provide almost instantaneous re-insertion. Figure 2.18 shows some alternative bypass schemes (ABB, 1991). The single-gap protective scheme in Figure 2.18a uses a single spark gap G which bypasses the capacitor if the voltage exceeds a pre-set value, normally equal to about 3–4 times the rated voltage of the capacitor. The short circuit-current flowing through the capacitor is damped in the damper D. When the gap current is detected the bypass breaker S is closed diverting the current from the gap. When the line current returns to normal the by-pass breaker opens within 200–400 ms and the capacitor is re-inserted.

A faster re-insertion time, in the range of 80 ms, is provided by the dual-gap scheme, Figure 2.18b. When the fault occurs the spark gap $G2$, which is set lower than $G1$, sparks over first bypassing the capacitor. Breaker $S2$, which is normally closed, opens immediately upon sensing the normal line current and re-inserts the capacitor. In this way capacitor re-insertion is not delayed by de-ionisation time. The other gap $G1$ and the bypass breaker $S1$ serve as back-up protection.

Due to its non-linear properties the zinc-oxide resistor shown in Figure 2.18c limits the voltage across the capacitor bank during a fault and re-inserts the bank immediately when the current returns to normal. The spark gap G does not normally operate and is provided only as back-up overvoltage protection for the resistor.

2.4.4 Flexible AC Transmission Systems (FACTS)

Traditionally the main control actions in a power system, such as changing transformer taps or switching current, have been achieved using mechanical devices and were therefore rather slow. However the continuing progress in the development of power electronics has enabled a number of devices to be developed providing the same functions but with much faster operation and with less technical problems. At the heart of these devices is the *thyristor valve* which is an assembly of high power, line-commutated, thyristors connected back-to-back. The conventional line-commutated thyristor has been around since the early 1970s but has one main limitation: whilst its turn-on can be controlled its turn-off cannot. Recently semiconductor manufacturers have succeeded in producing high power *Gate Turn-Off Thyristors (GTOs)* which overcome this drawback.

In recent years environmental concerns and high costs have delayed, or sometimes even rendered impossible, construction of new overhead lines in many countries. The philosophy of *Flexible AC Transmission Systems (FACTS)*, developed in the late 1980s, is to use thyristor-based devices to control power flows in a transmission network. This allows parallel circuits in a network to be loaded up to their full thermal capability and/or enforce a pattern of flows that are required for economic operation. Due to their fast speed of operation FACTS devices can also be used to control system dynamics and their use for transient stability enhancement will be described in Chapter 9.

FACTS technology is relatively new and still quite expensive. However as all FACTS controllers utilise similar thyristor-based technology their production should eventually benefit from the economies of scale with a corresponding reduction in cost.

Static VAR compensators (SVCs)

SVCs have been used in power systems since the 1970s, long before the concept of FACTS was formulated. The role of SVCs is to adjust the amount of reactive power compensation to the actual system needs. A flexible and continuous reactive power compensation scheme that operates in both the capacitive and inductive regions can be constructed using thyristor-switched and/or thyristor controlled shunt elements.

Figure 2.19a shows a thyristor valve comprising two thyristors connected in anti-parallel that will allow smooth regulation of the current flowing through a shunt reactor. It is worth noting that the current will contain high harmonics and therefore the device must be equipped with additional harmonic filters to help smooth the current waveform. Such filters are quite expensive and constitute a substantial part of the overall compensator cost.

In the case of the capacitor it is not possible to obtain smooth control of the current due to the long time constant associated with the capacitor charge/discharge cycle so that the thyristor valve can only switch the capacitor on or off as shown in Figure 2.19b. When the flow of the current is blocked, the capacitor discharges via the discharge resistor.

It is possible to design a variety of SVC systems using the basic principles described above. Some typical configurations are shown in Figure 2.20. In Figure 2.20a one reactor is thyristor-controlled and the other is thyristor-switched. When the inductive VAR demand is low only the thyristor-controlled reactor operates. When demand increases both reactors are switched on with the thyristor-controlled reactor being used to control the actual amount of reactive power drawn from the system.

Figure 2.20b shows a thyristor-switched bank of capacitors. The reactive power control (in the capacitive region only) can be accomplished in steps by switching consecutive capacitors in or out. The SVC shown in Figure 2.20c consists of a bank of shunt capacitors connected in parallel with a thyristor-controlled shunt reactor. The thyristor valve enables

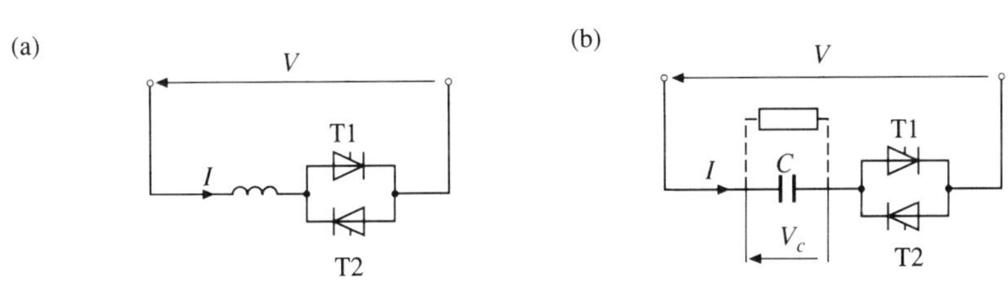

Figure 2.19 Principles of SVC compensation: (a) thyristor-controlled reactor; (b) thyristor-switched capacitor

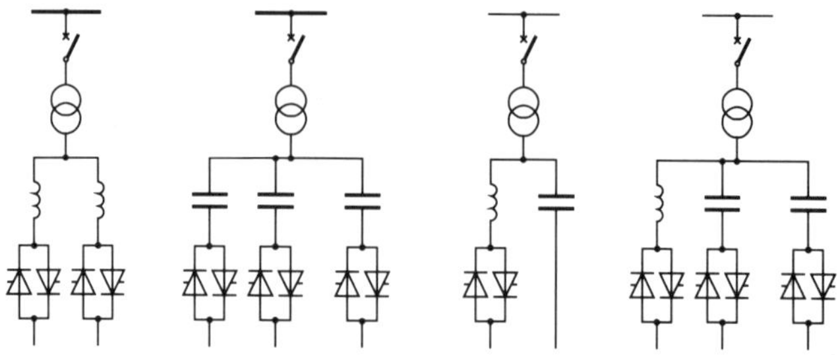

TCR - thyristor-controlled reactor; TSR - thyristor-switched reactor;
TSC - thyristor-switched capacitor; FC - fixed capacitor

Figure 2.20 Types of Static VAR Compensators (SVC)

smooth control of the lagging VARs drawn by the reactor. With the reactors switched fully on the parallel reactor-capacitor bank appears to be net inductive, while with the reactors fully off, the bank is net capacitive. By controlling the reactor current it is possible to achieve a full control range between these two extremes. A similar principle is used in the system shown in Figure 2.20d which additionally contains a bank of thyristor-switched capacitors.

The thyristor firing circuits used in SVCs are usually controlled by a voltage regulator which attempts to keep the busbar voltage constant by controlling the amount, and polarity, of the reactive power injected into the busbar. The regulator can also be used to damp power swings in the system following a network disturbance.

The cost of SVCs is typically several times that of an uncontrolled bank of shunt reactors or fixed capacitors. Consequently their use tends to be limited to those parts of a system where heavy fluctuations of real power occur and compensation of both inductive and capacitive VARs is necessary.

The static compensator (STATCOM)

The *static compensator (STATCOM)*, also called *GTO-SVC* or *Advanced SVC*, is still in the experimental stage (at the time of writing this book) but is widely predicted to reach commercial service within the next few years (Hingorani, 1994, Ralls, 1995). STATCON provides shunt compensation in a similar way to the SVCs but utilises a voltage source converter rather than shunt capacitors and reactors. Consequently it incorporates a very high content of power electronics but its conventional components are reduced to only a transformer and a capacitor. This leads to a compact, modular construction that can be easily shipped to site, fully assembled, and, if necessary, easily re-located. Its operating principle is shown in Figure 2.21. The GTO converter produces a fundamental frequency voltage V_2 which is in phase with the power system voltage V_1. As V_2 and V_1 are in phase, the difference between them results in a reactive current I flowing through the transformer reactance X:

$$\bar{I} = (V_1 - V_2)/\mathrm{j}X \tag{2.5}$$

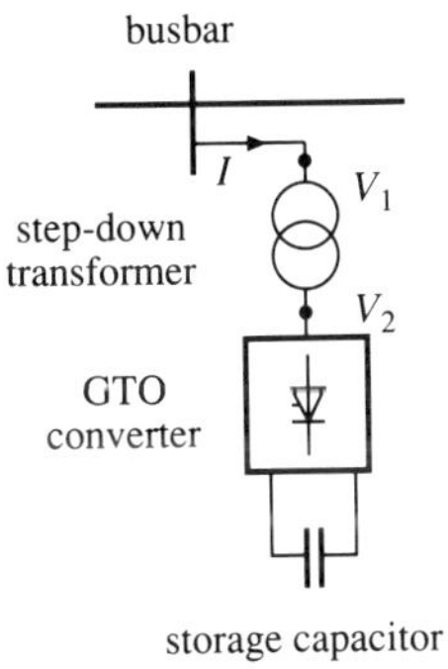

Figure 2.21 STATCOM (Advanced SVC)

If $V_2 > V_1$ then $\overline{I}$ leads $\overline{V}_1$, reactive power is delivered to the busbar and the converter behaves like a large capacitor. Conversely if $V_2 < V_1$ then $\overline{I}$ lags $\overline{V}_1$, reactive power is drawn from the busbar and the converter acts like a reactor. For a transformer reactance of 0.1 p.u. a $\pm$ 10% change in V_2 produces ± 1 p.u. change in the inserted reactive power. Besides being used for voltage control and VAR compensation purposes STATCOM can also be used to help damp electromechanical oscillations.

Thyristor-controlled braking resistor

The *thyristor-controlled braking resistor*, shown in Figure 2.22, is used exclusively for transient stability enhancement. It acts as an additional resistive load capable of absorbing some of the generation in the case of a severe fault occurring near a generator, thus preventing loss of synchronism.

Superconducting magnetic energy storage (SMES)

Fast progress in power electronics and superconductivity research has renewed interest in the potential application of superconducting magnetic energy storage (SMES) units as a means of energy storage and/or damping power system oscillations. SMES can be seen as a controllable current source whose magnitude and phase can be changed

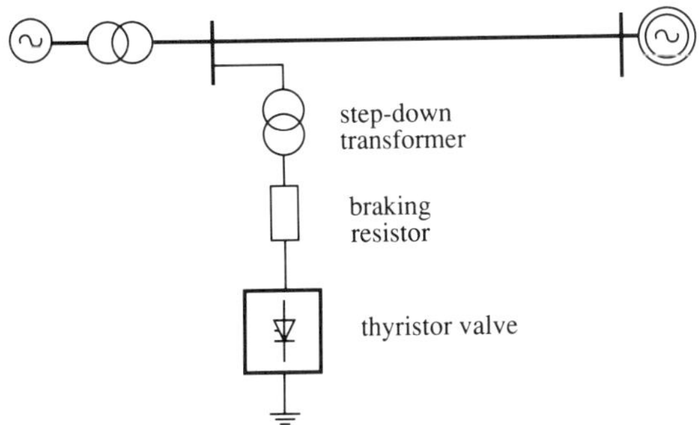

Figure 2.22 Thyristor-controlled braking resistor

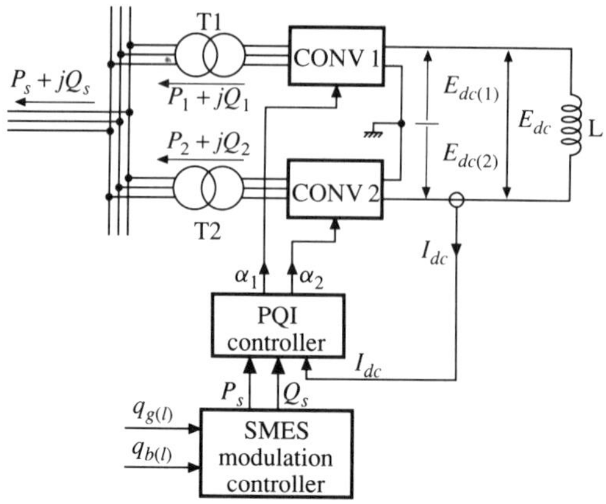

Figure 2.23 SMES unit with double GTO thyristor bridge

within one cycle; the upper limit of this source is imposed by the dc current in the superconducting coil.

The typical configuration of a SMES unit is shown in Figure 2.23. The superconducting coil L is coupled to the transmission system via two converters $CONV1$ and $CONV2$ and transformers $T1$ and $T2$. The converter firing angles α_1 and α_2 are determined by the PQI controller so as to control the output real and reactive power and the dc current in the coil (hence the name PQI controller). The control strategy is determined by the *modulation controller* in such a way as to damp power swings in the network.

The active and reactive power available from SMES depends on the type of ac/dc converter used. Two, 6-pulse, line commutated converters allow two quadrant operation with both the real and reactive power (lagging) being controlled. Alternatively with the forced commutated, double GTO, thyristor bridge converter described by Ise *et al*(1986) the SMES power can be regulated in four quadrants of the complex power domain within a circular range limited by

$$[P_s(t)]^2 + [Q_s(t)]^2 \leq |S_{MAX}|^2 \tag{2.6}$$

where S_{MAX} is the maximum available apparent power.

Experimental and simulation results obtained for a two-machine system model have shown that SMES can constitute a powerful tool for transient stability enhancement (Rogers *et al*, 1983, Boening & Hauer 1985) and can support primary frequency regulation (Handschin & Stephenblome, 1992).

Series compensator and SSR damper

Figure 2.24 shows the application of FACTS technology to control the impedance of a transmission line. Figure 2.24a shows the *thyristor-controlled series capacitors (TCSC)* where the sections of the capacitor bank are switched on and off as required. The thyristors are protected against overvoltages by non-linear zinc-oxide resistors as shown in Figure 2.18.

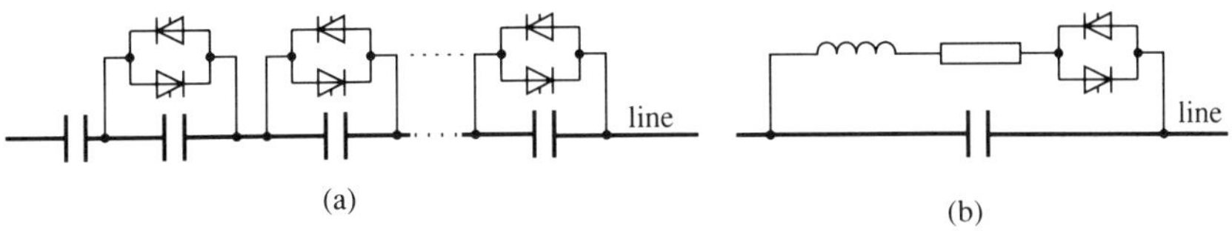

Figure 2.24 FACTS line impedance control devices: (a) thyristor-controlled series capacitor; (b) subsynchronous resonance damper

The *subsynchronous resonance damper (NGH-SSR Damper)*, Figure 2.24b, allows additional series resistance to be inserted in the line to inhibit slow oscillations on a long heavily loaded line. Such oscillations may arise due to series compensation of the line and will be discussed in sub-Section 6.6.3.

Phase-angle regulator and unified power flow controller

The mechanical tap changers for the in-phase and quadrature voltage regulation previously shown in Figure 2.17 can be replaced by FACTS technology. The *thyristor-controlled phase angle regulator (TCPAR)* is a thyristor-based version of the quadrature booster discussed in sub-Section 2.4.2. Here the thyristors are used as electronic switches to replace the traditional mechanical tap changers as shown in Figure 2.25a. The supply (excitation) transformer ET has three sets of windings on the secondary with the turns ratio 1:3:9. The thyristor-switched system THS allows any combination of these three sets of windings to be switched in series in either a positive or negative direction. Consequently each of these sets of windings can therefore be in one of three states giving $3^3 = 27$ possible values of the voltage supplying the series transformer. This corresponds to 27 taps in a traditional quadrature booster. As TCPAR uses thyristors as simple switches to switch between different windings their full capabilities for the continuous control of current are not utilised.

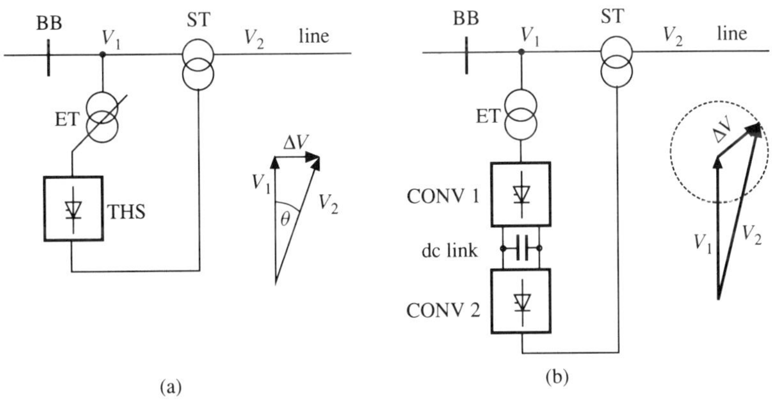

BB - busbar, ET - excitation (supply) transformer, ST-series transformer, THS - thyristor-switched system, CONV1 - rectifier, CONV2 - inverter

Figure 2.25 FACTS control devices (a) phase angle regulator; (b) the unified power flow controller

In contrast GTO thyristors are fully utilised in the *unified power flow controller (UPFC)* where the voltage is continuously regulated in both magnitude and phase, Figure 2.25b. The excitation transformer ET feeds the thyristor-controlled rectifier CONV 1. This rectifier is connected directly with the thyristor-controlled inverter CONV 2 via a short dc link and smoothing capacitor. The inverter regulates the phase angle and magnitude of the voltage ΔV supplied to the series transformer (Hingorani, 1994) so that the transmission line voltage V_2 can take any value within the circle around V_1 shown by the dashed line. This allows the flow of real and reactive power in the network to be controlled and electromechanical oscillations to be damped. The rectifier can also operate as a controllable current source when the dc link must also contain an inductor to ensure a constant dc current (Jäger, 1995).

The UPFC is currently attracting a substantial amount of research interest but, at the time of writing, is still at the experimental phase.

2.5 Protection

No system element is completely reliable and can be damaged due to some internal or external fault. If the damaged element is not immediately switched out of circuit it may suffer further damage and be completely destroyed. A damaged element may also disturb operation of the neighbouring elements so as to threaten the operation of the whole power system and the continuity of energy supply to the consumer. Protection equipment is therefore needed to detect a fault and switch the faulty element out of the circuit. Typically power system protection equipment consists of current and/or voltage transformers, relays, secondary circuits supplying the relays and controlling the circuit-breakers, and auxiliary power supplies for the relays.

Operation of the protection must be fast, reliable and selective. A fast *speed of response* and *high reliability* are vital to limit the damage that could be caused by a fault. In addition the protection must be *selective* so that only the faulty element is switched off. Reliability is achieved by having high quality equipment and by using two different protection schemes for each element: the *main protection* and the *back-up protection*. The main protection should operate according to different physical principles than its back-up. If back-up protection is placed in the same substation bay as the main protection then it is termed *local back-up*. If the main protection of a neighbouring element is used as the back-up protection of the given element then it is called *remote back-up*.

There are a number of books which deal in detail with power system protection (Phadke & Thorap, 1988, Wright & Christopoulos, 1993, Ungrad, Winkler & Wiszniewski, 1995). This section will contain only a brief overview of some of the main protection schemes and will introduce terms necessary for the remainder of the book.

Protection of transmission lines

The main faults on transmission lines are short-circuits. Overhead transmission lines are shielded from lightning strokes by ground wires, hung above the phase conductors, and surge diverters connected to the conductors themselves. Nevertheless lightning is still the most common cause of faults on overhead transmission lines, with single-phase faults contributing 75–90% of all faults. In contrast multi-phase-to-ground faults constitute

5–15% of faults whilst multi-phase faults with no ground connection are the rarest at 5–10%. Other rare causes of faults are insulator breakages, swinging of wires caused by strong winds and temporary contact with foreign objects.

The majority (80–90%) of the faults on overhead lines are of a temporary nature and are caused by flashovers between phase conductors or between one or more of the phase conductors and earthed metal or the ground caused by, for example, a lightning stroke. The remaining 10–20% of faults are either semi-temporary or permanent. Temporary faults can be dealt with by switching off the line until the arc or arcs are extinguished, and then switching it on again after a certain period, termed the *dead time*. The whole procedure is referred to as *auto-reclosing* and significantly improves the continuity of energy supply. Obviously in the case of a permanent fault the re-energised line will be tripped again by its protection. There may be two or three such attempts, but usually only a single-shot reclosure is used in high-voltage transmission networks.

The oldest type of protection is the *differential-current protection* based on Kirchhoff's Current Law which states that the sum of currents flowing into and out of a circuit is equal to zero. Figure 2.26 shows the basic differential-current protection scheme where the current transformers installed at both terminals of a line are connected via resistors and pilot wires so as to oppose each other. Under healthy conditions or external faults (like fault F_2), no current flows in the pilot wires. A fault occuring within the protected zone (fault F_1) creates a potential difference between the resistors and a small circulating current flows through the pilot wires energising the overvoltage relays and operating the circuit breakers.

Because of the need to continuously transmit signals with fairly high accuracy between the ends of the protected line, the differential scheme with pilot wires is only used to protect short transmission lines with a maximum length of about 20–30 km. For longer lines the *interlock scheme* is used which employs directional relays sited at each end of the protected zone. These relays initiate the opening of the circuit breakers at each end of a line if both sets of relays indicate that currents are flowing into the line. The information which must be transmitted between the relays is a logical yes or no, as opposed to the analogue signals used in the differential scheme. To avoid the expense of the pilot cables, the logic signal is usually transmitted over the conductors of the protected line using high-frequency signals and is called the *power line carrier*.

Another, more popular, protection scheme utilising the power line carrier is the phase-comparison scheme in which the phase of the current at the two ends of the protected circuit is compared. Under normal operating conditions, or in a case of an external fault,

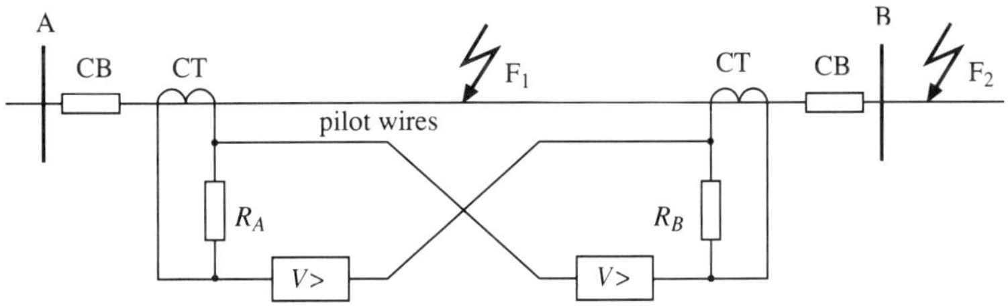

A, B - substation busbars; CB - circuit-breaker; CT - current transformer;
V> - overvoltage relay; F_1 , F_2 - internal and external faults

Figure 2.26 Differential-current protection

the currents at the opposite ends of the line are almost in phase with each other whereas they are displaced by large angles when internal faults are present.

The frequency used in the power line carriers is typically in the range 20–200 kHz. Lower frequencies increase the cost of the carrier, while higher frequencies cause too high an attenuation of the signal. The lines must be equipped with *line traps*, which are tuned circuits to block the high-frequency signals so that they do not enter other lines.

The *directional-comparison* protection scheme utilises the travelling waves which accompany the short-circuit. Any short-circuit produces voltage and current travelling waves which travel in both directions from the point of fault. The fault is detected by comparing the direction of the travelling waves at both ends of the line. An internal fault will cause the waves to travel in opposite directions, whilst an external fault will result in waves travelling in the same direction through the protected zone.

The development of optical-fibre technology has opened new possibilities for transmitting information over long distances. Optical-fibre links placed inside ground wires of transmission lines are currently used in directional-comparison and phase-comparison schemes, replacing pilot wires or power line carriers.

The most popular scheme for protecting transmission lines is *distance protection*. Its main advantage is that it does not require pilot wires or power line carriers. The principle of operation recognises that the impedance of a high-voltage transmission line is approximately proportional to its length. This means that the apparent impedance measured during a fault by the relay is proportional to the distance between the point of fault and the relay. If this impedance is less than the series impedance of the line then it may be concluded that a fault has occurred inside the protected zone and the line should be tripped. Unfortunately, the measurement of the apparent impedance has low accuracy due to the errors introduced by the current and voltage transformers, the relay itself and other factors such as the impedance of the fault. Consequently the relay must have a time characteristic with a few zones corresponding to different impedance settings and tripping times, as shown in Figure 2.27. Distance protection of line *A-B* at busbar *A* (marked by the small solid rectangle) has three zones, Z_{A1}, Z_{A2} and Z_{A3}, with tripping times t_{A1}, t_{A2} and t_{A3}, respectively.

To make sure that a distance relay will not over-reach the protected zone, that is unnecessarily trip for faults outside the zone, the first protection zone is usually set at between 85% and 90% of the line length. As this first zone, Z_{A1}, cannot protect the entire line the distance relay is equipped with a second zone, Z_{A2}, which deliberately reaches

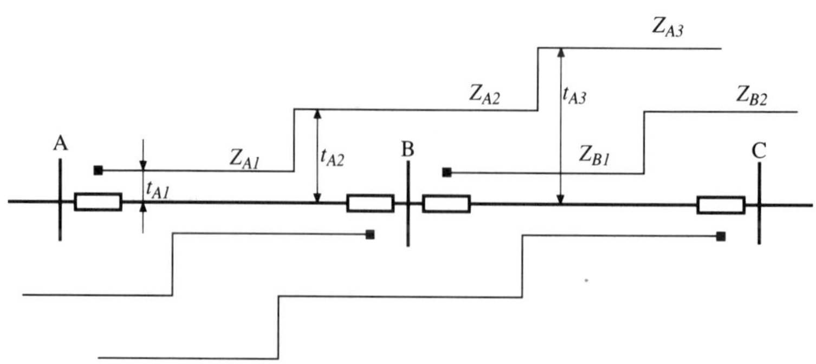

Figure 2.27 Distance protection zones of two neighbouring lines

beyond the remote terminal of the transmission line. The second zone is slowed down in order that, for faults in the next line section (B-C), the protection of the next line (located at B) will operate before the second zone of the distance relay at A. The second zone of A also partially backs up the distance relay of the neighbouring line. In order to provide a back up of the entire neighbouring line, it is customary to provide yet another zone, Z_{A3}, for the relay at A, extending to about 150% of the next line section. This third protection zone is obviously the slowest.

Protection of transformers

Power transformers are an important link in high-voltage transmission networks and must be protected against both external and internal faults. Internal faults may be due to earth faults on windings, inter-phase faults, inter-turn faults and inter-winding faults. This classification is important as it also corresponds to the different types of protection used.

The main form of transformer protection is the differential-current protection. Its principle of operation is similar to the current differential protection used to protect a transmission line, Figure 2.26. The transformer protection scheme has, however, some special features due to the fact that the magnitude and phase of the current are different on the primary and the secondary side. The transformer protection scheme must also cope with the possible presence of a large *magnetising inrush current*. Depending on the instant an unloaded transformer is energised a magnetising current several times the rated current of the transformer may flow and, because the losses in the transformer are small, will decay slowly with a time constant of several seconds to its normal small value. Because of the non-linearity of the magnetising characteristic this inrush current has a high second harmonic and this distortion can be used to distinguish it from a normal fault current.

Differential protection, despite its sensitivity, is not able to detect inter-turn faults. The transformer is protected against these faults by the *Buchholz protection* installed in the pipe between the main tank of the oil-filled transformer and the conservator. Localised heating or arcing associated with inter-turn or inter-winding faults decomposes the oil and produces gases such as hydrogen and carbon monoxide. The gases rise from the transformer and pass up the sloping pipe through the Buchholz relay, filled with oil, towards the conservator. On their way the gases rise inside the Buchholz relay and become trapped in the top of the casing so displacing oil. As a result a pivoted float or bucket falls and, depending on the amount of gas released, causes an alarm or trips the relay.

Large transformers are also equipped with two distance protection devices, one on each side of the transformer. The first and third zone of both relays are directed toward the impedance of the transformer and their tripping signals are passed to the trip-coil circuits of both circuit-breakers. This constitutes a local back-up of the transformer's differential protection. The second zone is directed in the opposite direction towards the network. This constitutes the main protection from external faults and the local back-up for busbar protection.

In addition transformers may also be equipped with earth fault protection (supplied with zero-sequence currents), combined differential and restricted earth fault protection, overload protection in the form of an overcurrent relay (supplied from the current transformers), and a thermal relay reacting to the temperature inside the transformer tank.

Protection of busbars

Faults on substation busbars are relatively rare compared to those on overhead transmission lines. The most common causes of busbar faults are flashover on the insulators, power apparatus failure and, quite often, human error due, for example, to opening or earthing the isolator when on-load. The consequences of a fault on a substation busbar may be far more severe than for a fault on a transmission line. The busbars of less important substations, in which all the outgoing circuits are protected with distance protection, are themselves protected by the second zone of the distance protection at neighbouring stations. This is illustrated in Figure 2.27 where the busbars at substation B are protected by the distance protection at substations A and C. An obvious drawback of this scheme is the long fault clearing time associated with the second zone of the distance protection. This cannot be tolerated for more important substations so they are normally equipped with current-differential protection for each of their circuits, or with modern phase-comparison protection schemes. In this case the distance protection of outgoing circuits constitutes a remote back-up for the busbar protection.

Protection of generating units

As shown in Figure 2.2 a generating unit is a complex system comprising the generator, exciter, prime mover, step-up transformer and often a unit transformer supplying the auxiliary services. As the generating unit may be subject to a variety of faults or disturbances, it is normally protected by a number of protection systems. The most important is the differential protection against faults inside the generator and transformer. This normally consists of three differential protection systems: one for the generator, a second for the unit transformer, and a third for the generator with its step-up transformer (or for the generator with both step-up and unit transformer). Large generating units are protected from external faults by the distance protection directed towards the network, in a similar manner as for transformers. The unit transformer is usually protected from external faults by an overcurrent relay. Similar overcurrent protection is used to protect the stator winding from both overloading and load asymmetry, and the rotor winding from overloading. Additional protection systems are used to protect the generator from loss of excitation, loss of synchronism, failure of the prime mover (motoring protection), and from earth faults in the rotor and stator windings.

The generator is also equipped with protection from non-electrical disturbances due to low vacuum, lubrication oil failure, loss of boiler fire, overspeeding, rotor distortion, excessive vibration and difference in expansion between rotating and stationary parts.

3

The Power System in the Steady-State

One of the characteristic features of power system operation is the need to continually adjust the system operating conditions in order to meet the ever-changing load demand. Although the demand from any particular load can vary quite significantly the total demand, consisting of a very large number of individual loads, changes rather more slowly and in a predictable manner. This characteristic is important as it means that within any small time period the transmission and sub-transmission systems (shown in Figure 2.1) can be regarded as being in the steady-state and, as time progresses, considered to move slowly from one steady-state condition to another.

To help describe and quantify the system behaviour under these steady-state conditions this chapter develops steady-state mathematical models of all the main system elements such as transmission lines, transmission cables, transformers, generators etc. The behaviour of the main types of system loads is also considered so that the way in which the demand may change with both voltage and frequency is understood. Having established such component models the way they can be linked together to form the full set of network equations is described.

3.1 Transmission Lines

Figure 3.1 shows the single-phase equivalent circuit of a transmission line suitable for analysing its symmetrical three-phase operation. In this circuit the voltages and the currents at the line terminals are the variables whilst the line parameters are assumed to be uniformly distributed over the line length.
The parameters describing this circuit are:

r	series resistance per unit length per phase(Ω/km)
$x = \omega L$	series reactance per unit length per phase(Ω/km) and L is the series inductance per phase (H/km)
g	shunt conductance per unit length per phase (S/km)

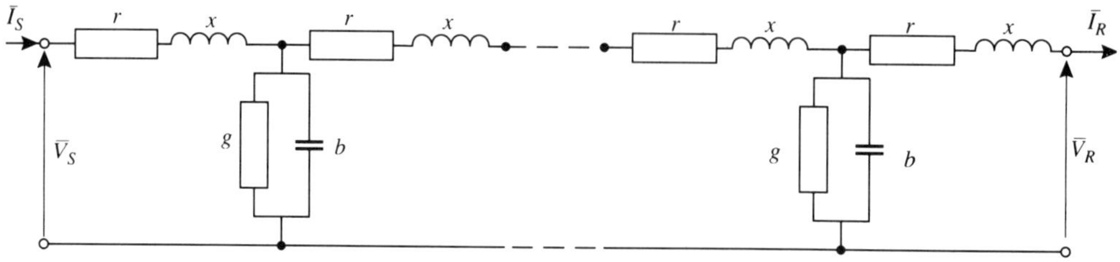

Figure 3.1 Single-phase equivalent circuit of a transmission line with distributed parameters

$b = \omega C$	shunt susceptance per unit length per phase (S/km) and C is the shunt capacitance per phase (F/km)
l	line length (km)

where $\omega = 2\pi f$ and f is the frequency. The series impedance and shunt admittance per phase per unit length are defined as $\overline{z} = r + jx$ and $\overline{y} = g + jb$. Each of the parameters in the equivalent circuit has a physical meaning and is linked with a particular aspect of the transmission line behaviour. Thus the resistance r represents the Joule loss that occurs due to current flow and depends on the type, construction and diameter of the conductor used. The series inductance L depends on the partial flux linkages within the conductor cross-section itself and the external flux linkages with the other conductors. The shunt conductance g represents the corona loss and the leakage current on the insulators. This value is not usually constant as the corona loss depends on air humidity while the leakage current depends on dirt and salt contamination on the surface of insulators. In power lines, g is small and is usually neglected. The shunt capacitance C is due to the potential difference between the conductors. As the voltages are ac this shunt capacitance is alternately charged and discharged resulting in the flow of the *line charging current*.

3.1.1 Line equations and the pi-equivalent circuit

For steady-state analysis the variables of interest are the voltages and currents at the line terminals V_R, I_R, V_S, and I_S where the subscripts R and S signify the receiving end and sending end of the line respectively. The voltages and currents are linked by the *long line equation*:

$$\begin{bmatrix} \overline{V}_S \\ \overline{I}_S \end{bmatrix} = \begin{bmatrix} \cosh \overline{\gamma} l & \overline{Z}_C \sinh \overline{\gamma} l \\ \sinh \overline{\gamma} l / \overline{Z}_C & \cosh \overline{\gamma} l \end{bmatrix} \begin{bmatrix} \overline{V}_R \\ \overline{I}_R \end{bmatrix} \tag{3.1}$$

where $\overline{Z}_C = \sqrt{\overline{z}/\overline{y}}$ is the *characteristic* (or *surge*) *impedance* of the line and $\overline{\gamma} = \sqrt{\overline{z}\overline{y}}$ is the *propagation constant*. As both the constants $\overline{\gamma}$ and $\overline{Z}_C$ are complex quantities the propagation constant can be expressed as $\overline{\gamma} = \alpha + j\beta$ where α is termed the *attenuation constant* and β the *phase constant*. The four elements of the matrix linking the sending-end and the receiving-end voltages and currents are often referred to as the *ABCD* constants where $\overline{A} = \overline{D} = \cosh \overline{\gamma} l$, $\overline{B} = \overline{Z}_C \sinh \overline{\gamma} l$ and $\overline{C} = \sinh \overline{\gamma} l / \overline{Z}_C$. Interested readers are referred to Grainger & Stevenson (1994) or Gross (1986) for the detailed derivation of equation 3.1.

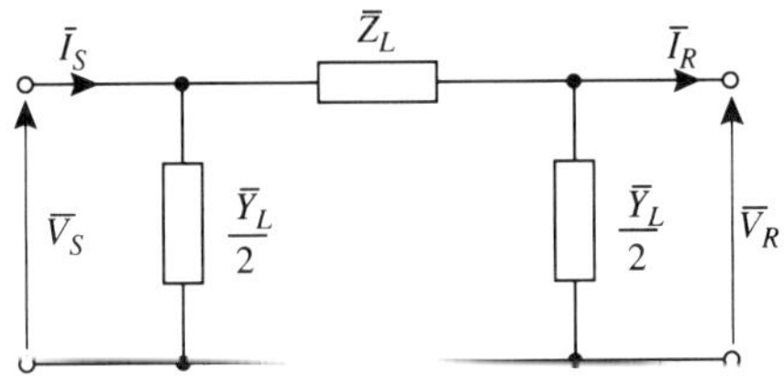

Figure 3.2 Equivalent-π circuit of a transmission line

As power system networks consist of many lines the use of equation 3.1 for analysis purposes is very inconvenient and a simpler way is to replace each line by its π – *equivalent* shown in Figure 3.2. A little circuit analysis shows that the π-equivalent parameters $\overline{Z}_L$ and $\overline{Y}_L$ are given by

$$\overline{Z}_L = \overline{Z}\frac{\sinh \overline{\gamma} l}{\overline{\gamma} l}, \qquad \overline{Y}_L = \overline{Y}\frac{\tanh(\overline{\gamma} l/2)}{\overline{\gamma} l/2} \tag{3.2}$$

where $\overline{Z} = \overline{z}l$ is the line total series impedance per phase and $\overline{Y} = \overline{y}l$ the total shunt admittance per phase.

It should be emphasised that the π-equivalent parameters $\overline{Z}_L$ and $\overline{Y}_L$ are generally not equal to the total impedance and admittance of the line, i.e. $\overline{Z}_L \neq \overline{Z} = \overline{z}l$ and $\overline{Y}_L \neq \overline{Y} = \overline{y}l$. However, for a typical power line $\overline{\gamma} l$ is small and the hyperbolic functions can be approximated as $\sinh \overline{\gamma} l \approx \overline{\gamma} l$ and $\tanh(\overline{\gamma} l/2) \approx \overline{\gamma} l/2$. Substituting these values into equation 3.2 gives the *medium-length line* (80 km $< l <$ about 200 km) parameters

$$\overline{Z}_L = \overline{Z}, \qquad \overline{Y}_L = \overline{Y} \tag{3.3}$$

For a *short line* ($l <$ 80 km) the charging current (and the capacitance C) may be neglected when the parameters are

$$\overline{Z}_L = \overline{Z}, \qquad \overline{Y}_L = 0 \tag{3.4}$$

Nowadays, with the wide use of computers, such approximations have limited practical value as the parameters can be easily calculated using equation 3.2. For manual calculations the most convenient, and practical, formulae to use are those of the medium-length line, equation 3.3, when the parameters in the nominal π-equivalent circuit represent the total series impedance Z and the total shunt admittance Y.

3.1.2 Performance of the transmission line

For a typical high-voltage transmission line g can be neglected whilst $r \ll x$. An additional insight into the line performance may be obtained by considering the *lossless line*, that is by neglecting r altogether. With r and g neglected, the characteristic impedance is purely resistive

$$\overline{Z}_C = \sqrt{\overline{z}/\overline{y}} = \sqrt{L/C} \tag{3.5}$$

and the propagation constant, $\overline{\gamma}$, is purely imaginary

$$\overline{\gamma} = \sqrt{\overline{z}\overline{y}} = j\omega\sqrt{LC} \tag{3.6}$$

so that $\alpha = 0$ and $\beta = \omega\sqrt{LC}$. With these values of Z_C and γ, the hyperbolic functions become $\sinh \overline{\gamma}l = \mathrm{j}\sin\beta l$ and $\cosh\overline{\gamma}l = \cos\beta l$ so that equations (3.1) simplify to

$$\begin{aligned}\overline{V}_S &= \overline{V}_R\cos\beta l + \mathrm{j}Z_C\overline{I}_R\sin\beta l\\ \overline{I}_S &= \overline{I}_R\cos\beta l + \mathrm{j}(\overline{V}_R/Z_C)\sin\beta l\end{aligned} \tag{3.7}$$

and the voltage and current are seen to vary harmonically along the line length. The *wavelength* of the full cycle can be calculated as $\lambda = 2\pi/\beta$. For 50 Hz lines the wavelength is nearly 6 000 km while for 60 Hz lines it is $60/50 = 1.2$ times shorter and is equal to about 5 000 km.

Power engineers often find it convenient to compare the actual loading on the line to the *natural load*, or *surge impedance load* (SIL), where SIL is defined as the power delivered at rated voltage to a load impedance equal to Z_C. That is

$$P_{SIL} = \frac{V_n^2}{Z_C} \tag{3.8}$$

Substituting $\overline{I}_R = \overline{V}_R/Z_C$ into the first, and $\overline{V}_R = Z_C\overline{I}_R$ into the second, of the equations 3.7 shows that when a lossless transmission line is loaded at its SIL

$$\overline{V}_S = \overline{V}_R e^{j\beta l} \text{ and } \overline{I}_S = \overline{I}_R e^{j\beta l} \tag{3.9}$$

indicating that:

- the voltage and current profiles are flat, $V_S = V_R$ and $I_S = I_R$
- the voltage and current at both ends (and at any point along the line) are in phase so that the reactive power loss in the line is zero (in other words the reactive power generated by C is compensated by the reactive power absorbed by L)

As the reactive power loss of the line is zero, the natural impedance loading is an optimum condition with respect to voltage and reactive power control. Table 3.1 shows typical values of SIL and other characteristic parameters of overhead lines.

Unfortunately the loading on the power line is rarely equal to the natural load. At night the loading may be a small fraction of SIL, whereas during peak load periods the loading

Table 3.1 Examples of overhead lines parameters (NGC, 1994, Kundur, 1994)

f_n Hz	V_n kV	r Ω/km	$x = \omega L$ Ω/km	$b = \omega C$ μS/km	β rad/km	Z_C Ω	P_{SIL} MW
50	275	0.067	0.304	4.14	0.00112	271	279
(British)	400	0.018	0.265	5.36	0.00119	222	720
	230	0.05	0.488	3.371	0.00128	380	140
60	345	0.037	0.367	4.518	0.00129	285	420
(USA)	500	0.028	0.325	5.2	0.0013	250	1,000
	765	0.012	0.329	4.978	0.00128	257	2280
	1 100	0.005	0.292	5.544	0.00127	230	5260

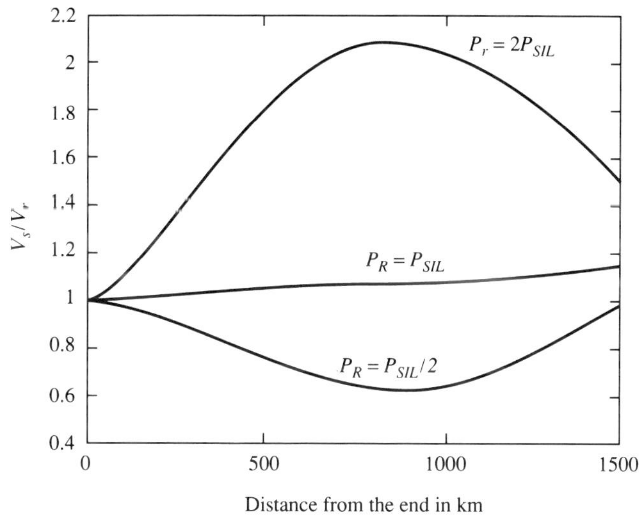

Figure 3.3 Sending end voltage required to supply a given real power load at rated voltage and at unity power factor

may be significantly greater than SIL. To examine the effect this has on the line voltage, Figure 3.3 shows three plots of the sending end voltage V_S required to deliver power P_R at rated voltage and unity power factor at the receiving-end, each plotted as a function of line length. These calculations have been done using the full line model described by equation 3.1 (with resistance included) assuming the parameters of the 400 kV line given in Table 3.1. When the delivered power P_R is less than SIL, V_S is smaller than V_R whilst $P_R > P_{SIL}$ requires V_S to be larger than V_R. Delivering $P_R = P_{SIL}$ does not actually result in the voltage along the line being constant but it is seen to increase slightly towards the sending-end. This is due to including the line resistance in the calculation. Note the harmonic variation of the voltage with length.

Real power transmission

Let $\overline{V}_R$ be the reference phasor and assume that $\overline{V}_S$ leads $\overline{V}_R$ by angle δ_{SR}, that is $\overline{V}_R = V_R$ and $\overline{V}_S = V_S e^{j\delta_{SR}}$. The angle δ_{SR} is referred to as the *load angle* or *transmission angle*. Assuming a lossless line, the receiving-end current can be calculated from equation 3.7 as

$$\overline{I}_R = \frac{\overline{V}_S - \overline{V}_R \cos\beta l}{\mathrm{j}Z_C \sin\beta l} = \frac{V_S}{Z_C \sin\beta l} e^{\mathrm{j}(\delta_{SR}-\pi/2)} - \frac{V_R \cos\beta l}{Z_C \sin\beta l} e^{-\mathrm{j}\pi/2} \tag{3.10}$$

allowing the complex apparent power at the receiving end to be calculated as

$$\overline{S}_R = \overline{V}_R \overline{I}_R^* = \frac{V_R V_S}{Z_C \sin\beta l} e^{\mathrm{j}(\pi/2-\delta_{SR})} - \frac{V_R^2 \cos\beta l}{Z_C \sin\beta l} e^{\mathrm{j}\pi/2} \tag{3.11}$$

The real power at the receiving end P_R can now be obtained as

$$P_R = \mathrm{Re}[\overline{S}_R] = \frac{V_S V_R}{Z_C \sin \beta l} \sin \delta_{SR} \tag{3.12}$$

showing that for sending-end and receiving-end voltages of specified magnitude *an increase in the receiving-end power P_R leads to an increase in the load angle δ_{SR}* whilst the maximum power transfer occurs when $\delta_{SR} = \pi/2$ and is equal to

$$P_{R,\max} = \frac{V_S V_R}{Z_C \sin \beta l} \approx \frac{P_{SIL}}{\sin \beta l} \tag{3.13}$$

Table 3.1 shows that βl is small for medium-length lines ($l <$ about 200 km). Thus, it can be assumed that the following simplifications hold

$$\sin \beta l \cong \beta l; \quad \cos \beta l \approx 1; \quad Z_C \sin \beta l \cong \sqrt{\frac{L}{C}} \omega \sqrt{LC} l = \omega L l = X \tag{3.14}$$

where X is the total inductive reactance of the line. In this case equation 3.12 simplifies to

$$P_R \cong \frac{V_S V_R}{X} \sin \delta_{SR} \tag{3.15}$$

showing that the power transfer limit is approximately inversely proportional to the line series inductance and the length. Any attempt to increase power transfer above $P_{R,\max}$ will result in instability and is connected with the *steady-state stability* discussed in detail in Chapter 5.

It is worth noting that a transmission line constitutes only part of the transmission link that connects a generator with the system or connects two parts of a system together. This means that it is the equivalent voltages of the generator and the system which should be considered to be constant and used in equation 3.12, not the terminal voltages of the line itself. Obviously the equivalent impedance of the generator and the system will then have to be added to the line impedance. As a result δ_{SR} is smaller than the angle between the equivalent voltages and, in practice, the power transfer limit occurs at $\delta_{SR} < \pi/2$. In addition when transient stability effects, discussed in Chapter 6, are taken into account the practical limit on δ_{SR} is significantly less than $\pi/2$.

Although equation 3.13 suggests that the maximum line loading is due to the power transfer limit set by the load angle δ_{SR} this is not always the case and other effects such as the thermal rating of the line and the maximum allowable voltage drop between the ends of the line must also be considered. For short lines, less than about 100 km, the maximum line loading is usually limited by the thermal rating rather than the maximum power transfer limit. On the other hand for lines longer than about 300 km, the maximum line loading is determined by the power transfer limit which may be substantially below the limit set by thermal considerations. Typically utilities will give the maximum power loading of a line as a fraction (or multiple) of P_{SIL}.

Reactive power considerations

If for this discussion the sending-end voltage is assumed constant the question then arises as to what happens to the receiving-end voltage if (a) the reactive power Q_R at the receiving-end changes and (b) if the real power P_R at the receiving-end changes.

Changes in Q_R will be considered first. The reactive power at the receiving-end may be found from the imaginary part of equation 3.11 as

$$Q_R = \text{Im}\,[\overline{V}_R \overline{I}_R^*] = \frac{V_S V_R}{Z_C \sin\beta l}\cos\delta_{SR} - \frac{V_R^2 \cos\beta l}{Z_C \sin\beta l}$$

$$- \frac{V_R}{Z_C \sin\beta l}(V_S \cos\delta_{SR} - V_R \cos\beta l) \qquad (3.16)$$

The simplifications of equation 3.14 allow equation 3.16 to be approximated as

$$Q_R \cong \frac{V_R}{X}(V_S \cos\delta_{SR} - V_R) \qquad (3.17)$$

Normally, for the stability reasons considered earlier, the angle δ_{SR} is kept small so that $\cos\delta_{SR} \approx 1$. Strictly speaking this assumption corresponds to considering the reactive power transfer only so that $P_R = 0$ and, according to equation 3.12, $\delta_{SR} = 0$. This simplification gives

$$Q_R \approx \frac{V_R(V_S - V_R)}{X} \qquad (3.18)$$

and shows very clearly that the reactive power, Q_R, is strongly dependent on the magnitude of the voltage at both the sending-end and at the receiving-end of the line and that it flows from the higher voltage to the lower voltage. Assuming that $V_S =$ constant and plotting the variation of Q_R against the receiving-end voltage V_R produces the parabolic relationship shown in Figure 3.4a with a maximum at $V_R = V_S/2$. As system voltages must be kept close to V_n, the operating point is always to the right of this peak with the condition $V_R > V_S/2$ always being satisfied. Thus, an *increase in Q_R leads to a decrease in V_R* whilst a *decrease in Q_R leads to an increase in V_R*. This observation has important repercussions with regards the introduction of reactive power compensation.

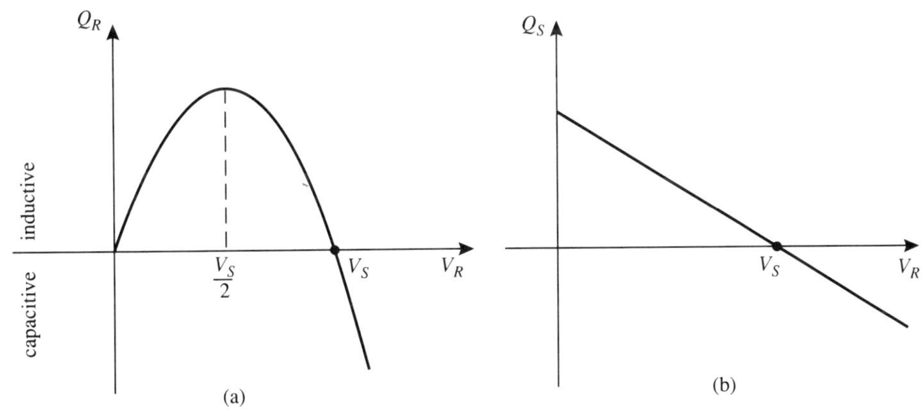

Figure 3.4 The approximate effect on receiving-end voltage of (a) changes in Q_R (b) changes in Q_S, when the influence of the real power flow is neglected

The sending-end reactive power Q_S can be calculated using a similar derivation to that for equation 3.16 to give

$$Q_S = \frac{V_S}{Z_C \sin\beta l}(V_S \cos\beta l - V_R \cos\delta_{SR}) \tag{3.19}$$

which, assuming that βl is small, gives

$$Q_S \cong \frac{V_S}{X}(V_S - V_R \cos\delta_{SR}) \tag{3.20}$$

Assuming δ_{SR} is also small gives

$$Q_S \approx \frac{V_S}{X}(V_S - V_R) \tag{3.21}$$

Again the strong dependency of reactive power with voltage magnitude is apparent and, as V_S is assumed constant, the approximate variation of Q_S with V_R is linear as shown in Figure 3.4b.

Now consider how including the real power transfer ($P_R \neq 0$ and hence $\delta_{SR} \neq 0$) modifies the above considerations. To investigate this consider the reactive power loss in the line itself due to flow of P_R. Generally the reactive line loss can be positive or negative so that the line can be a net source or sink of reactive power. The reactive power is generated in the line capacitance and depends on the voltage which is normally kept close to the rated value. This relatively constant reactive power generation is offset by the consumption in the line inductance. This consumption varies as the square function of the line current ($Q = I^2X$) and strongly depends on the actual loading of the line. To quantify this effect the reactive power loss $\Delta Q = Q_s - Q_R$ can be calculated from equations 3.16 and 3.19 as

$$\Delta Q = Q_S - Q_R = \frac{V_S^2 \cos\beta l - 2V_S V_R \cos\delta_{SR} + V_R^2 \cos\beta l}{Z_C \sin\beta l} \tag{3.22}$$

Assuming $V_S \approx V_R \approx V_n$ and eliminating $\cos\delta_{SR}$ using equation 3.12 gives

$$\Delta Q(P_R) \approx \frac{2P_{SIL}}{\sin\beta l}\left(\cos\beta l - \sqrt{1 - \left(\frac{P_R \sin\beta l}{P_{SIL}}\right)^2}\right) \tag{3.23}$$

The characteristics produced by equation 3.23 are shown in Figure 3.5 for transmission lines that operate at different voltage levels. The reactive power loss is zero only when $P_R = P_{SIL}$. If $P_R < P_{SIL}$ then the line is a net source of reactive power, whilst $P_R > P_{SIL}$ if the situation is reversed and the line is a net sink of reactive power.

The consequence of Figures 3.4 and 3.5 is profound. According to Figure 3.5 as the real power increases the reactive power loss in the line also increases. This loss must be supplied from the sending end giving an increase in Q_S. Figure 3.4b shows that for this to occur with V_S constant, V_R must reduce. If the resulting reduction in V_R is unacceptable then, according to Figure 3.4a, this can be compensated for by reducing Q_R in some way, perhaps by the introduction of some form of reactive power compensation, see sub-Section 2.4.3.

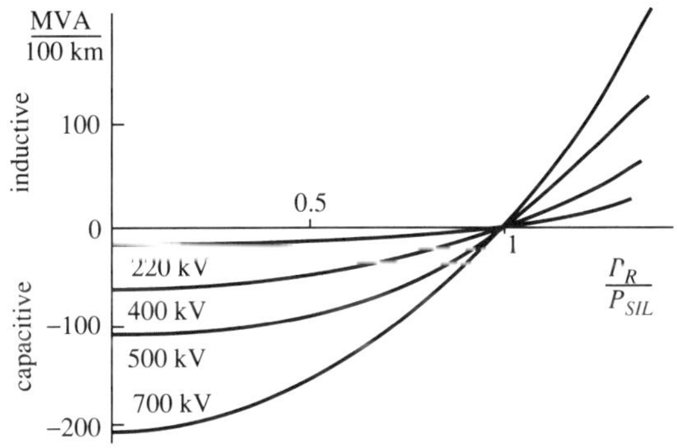

Figure 3.5 Examples of reactive power absorbed by a lossless line as a function of its real load for various voltage ratings

3.1.3 Underground cables

From a mathematical point of view an underground cable can be modelled exactly the same way as an overhead transmission line. The only difference is in the values of the characteristic parameters. Because there are many methods of cable construction the parameters of these different cables can also be quite different. In particular the shunt capacitance of the cable depends strongly on whether the three-phase conductors are screened or if they constitute separate single-phase cables. Typically the per-unit-length series reactance of a cable is about half that of a similarly rated overhead line. On the other hand the per-unit-length charging current is about 30 times more. This means that even for a short cable run of several tens of kilometres the charging current in the cable constitutes a substantial portion of its thermally admissible maximum current and severely limits its transmission capacity. In an extreme case of a critically long cable the charging current would be equal to the maximum current and there would be no capacity left for any power transmission. The charging current barrier is the main obstacle for practical application of ac cables in power transmission networks.

3.2 Transformers

The main types of transformers were discussed in Section 2.4.2. In this section the equivalent circuit of the transformer will be derived together with a method of dealing with off-nominal transformation ratio.

3.2.1 Equivalent circuit

The equivalent circuit of a transformer is shown in Figure 3.6a. The main element of the circuit is the *ideal transformer* with the transformation ratio $\vartheta = N_1/N_2$, where N_1 and N_2 are respectively the number of turns on the primary and secondary winding. The resistance R_1 and R_2 account for the I^2R loss in the primary and secondary winding whilst the reactances X_1 and X_2 account for the *leakage flux*, that is the flux component that does not link the two windings but links only the primary or the secondary winding. The

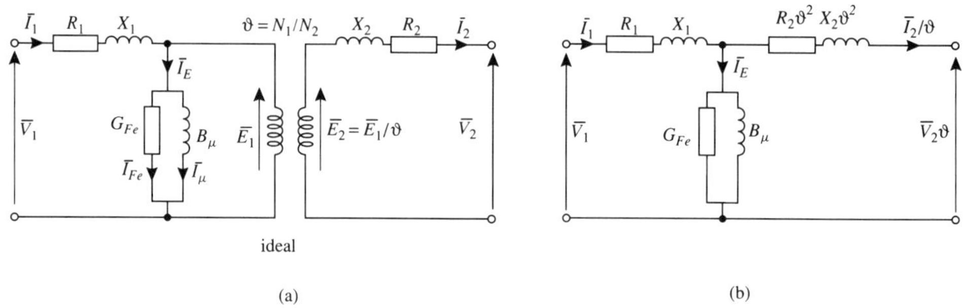

Figure 3.6 Practical single-phase two-winding transformer: (a) equivalent circuit; (b) equivalent circuit with secondary referred to the primary

sinusoidal changes in the supply voltage cause cyclic re-magnetisation of the transformer core as determined by the hysteresis curve of the core steel. This results in the flow of a *magnetisation current* I_μ even at no-load. I_μ is in phase with the flux and is therefore delayed with respect to the induced emf E_1 by $\pi/2$. This current is modelled by the shunt susceptance B_μ. The cyclic changes of the core flux also dissipate a certain amount of energy in the form of heat in the transformer core. This component of the core loss is referred to as the *hysteresis loss*. As the core itself is an electrical conductor the flux changes induce emfs which result in the flow of circulating *eddy currents*. These eddy currents produce a thermal energy loss called the *eddy-current loss*. The sum of the eddy-current loss and the hysteresis loss is the *core loss* P_{Fe}. This energy loss in the core is provided for by an additional current I_{Fe} flowing in the primary but this time in phase with the induced emf. This is modelled by inserting a shunt conductance G_{Fe} in the equivalent circuit of Figure 3.6a. At no-load the primary-side current is equal to the phasor sum of the magnetisation current and the core loss current and is referred to as the *excitation current* I_E. When the transformer is loaded the excitation current is superimposed on the load current.

The transformation ratio ϑ can be eliminated from the diagram if all the quantities in the equivalent circuit are referred to either the primary or secondary side of the transformer. For example Figure 3.6b shows the case when all voltages, currents, and impedances of the circuit of Figure 3.6a are referred to the primary. To produce this model of the transformer the secondary voltage is multiplied by ϑ whilst the secondary current is divided by ϑ. Consequently the secondary impedance must be multiplied by ϑ^2.

The equivalent circuit shown is of the T type but it is often more convenient in network analysis to deal with π-type circuits. If the shunt element in the transformer equivalent circuit cannot be neglected then the circuit of Figure 3.6b can be transformed into the π form using the standard star-delta transformation. However the circuit will not be symmetrical if the series impedances of the primary and the secondary branches in Figure 3.6b are not equal. To avoid this the following approximations are usually made:

- the secondary series impedance referred to the primary ($\overline{Z}_2 = R_2\vartheta^2 + jX_2\vartheta^2$) is equal to the primary series impedance ($\overline{Z}_1 = R_1 + jX_1$);
- the shunt impedance ($1/(G_{FE} + jB_\mu)$) is much greater than the total series impedance ($\overline{Z}_T = \overline{Z}_1 + \overline{Z}_2$).

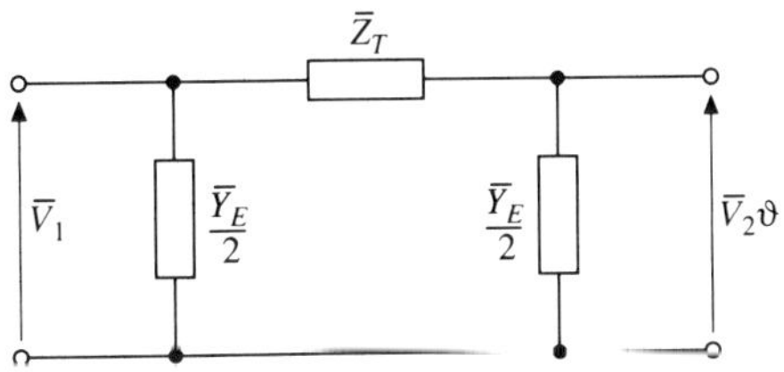

Figure 3.7 The approximate, symmetrical, equivalent π-circuit of a two-winding transformer

With these assumptions the parameters of the π-equivalent circuit, shown in Figure 3.7, are given by:

$$\overline{Z}_T = \overline{Z}_1 + \overline{Z}_2 = R + jX, \qquad \overline{Y}_E = G_{FE} + jB_{\mu} \tag{3.24}$$

The value of the parameters in the transformer equivalent circuit can be determined from the *no-load test* and the *short-circuit* test. In both of these tests the supply voltage, current and real power are measured. Table 3.2 shows some examples of test data.

On no-load the secondary winding of the transformer is open-circuited and the primary supplied with rated voltage $V_1 = V_n$. In this condition the series winding impedance is small compared to the shunt admittance and can be neglected so that the measured current and power correspond entirely to the shunt branch and relate to the excitation current I_E and iron loss P_{Fe}. Parameters of the shunt branch can be then computed, in per-unit to the unit rating, from:

$$G_{Fe} = P_{Fe(pu)}, \qquad Y_E = I_{E(pu)}, \qquad B_{\mu} = \sqrt{Y_E^2 - G_{Fe}^2} \tag{3.25}$$

where $P_{Fe(pu)}$ and $I_{E(pu)}$ are also expressed in pu.

During the short-circuit test the secondary winding is short-circuited. The primary winding is supplied with a voltage that is of sufficient magnitude to circulate a current in the short-circuited secondary that is equal to the rated load current ($I_2 = I_n$). This primary voltage is usually referred to as *the short-circuit voltage* V_{SHC} and is much smaller than the rated voltage: $V_{SHC} \ll V_n$. Small primary voltage means that the excitation current I_E is much smaller than $I_2 = I_n$ and it can be neglected. Hence, the parameters of the series branch, in per-unit to the unit rating, can be found from:

$$Z_T = V_{SHC(pu)}, \qquad R_T = P_{Cu(pu)}, \qquad X_T = \sqrt{Z_T^2 - R_T^2} \tag{3.26}$$

where $V_{SHC(pu)}$ and $P_{Cu(pu)}$ are given in pu.

Table 3.2 Typical values of transformer parameters

S_n MVA	V_{SHC} pu	P_{Cu} pu	I_E pu	P_{Fe} pu
150	0.11	0.0031	0.003	0.001
240	0.15	0.0030	0.0025	0.0006
426	0.145	0.0029	0.002	0.0006
630	0.143	0.0028	0.004	0.0007

The equivalent circuit for a multi-winding transformer is formed in a similar way. For example the equivalent circuit of a three-winding transformer consists of three winding impedances connected in star with the shunt admittances connected to the star point.

3.2.2 Off-nominal transformation ratio

The power system usually operates at a number of voltage levels. The equivalent circuit shown in Figure 3.7 is not very convenient as the secondary voltage must be referred to the primary, resulting in the equivalent circuit including the transformer turns ratio. If the parameters of the equivalent circuit are expressed in pu, as explained in Appendix A.6, the transformation ratio ϑ must be related to the nominal network voltages on both sides of the transformer. If the transformer rated voltages are equal to the nominal network voltages then the pu nominal transformation ratio is equal to unity, $\vartheta = 1$, and may be neglected in the transformer equivalent circuit shown in Figure 3.6a. However the pu transformation ratio may not be unity for two reasons: (i) the rated transformer voltages are slightly different from the nominal network voltages and (ii) the tap changer adjusts the turns ratio away from the nominal setting thereby changing the transformation ratio.

A convenient way to account for off-nominal turns ratio is to replace the actual turns ratio by some fictitious reactive shunt elements in such a way that these elements change the voltage up or down, as required. As an example consider the transformer from Figure 3.6a with the shunt branch neglected and the secondary series impedance referred to the primary, as shown in Figure 3.8. The series impedance Z_T has been replaced by its reciprocal Y_T in order to use the node voltage analysis. The transformer turns ratio will be taken as a complex number as this allows the general case of the phase-shifting transformer to be considered.

For the ideal part of the transformer equivalent circuit shown in Figure 3.8a the apparent power on both sides is the same and the current in the primary is equal to the secondary current divided by the conjugate of the turns ratio i.e. $\underline{I}_2/\underline{\vartheta}^*$. The two-port network corresponding to the branch with admittance Y_T can be described by the following nodal equation:

$$\begin{bmatrix} \underline{I}_1 \\ -\underline{I}_2/\underline{\vartheta}^* \end{bmatrix} = \begin{bmatrix} \underline{Y}_T & -\underline{Y}_T \\ -\underline{Y}_T & \underline{Y}_T \end{bmatrix} \begin{bmatrix} \underline{V}_1 \\ \underline{\vartheta}\underline{V}_2 \end{bmatrix} \tag{3.27}$$

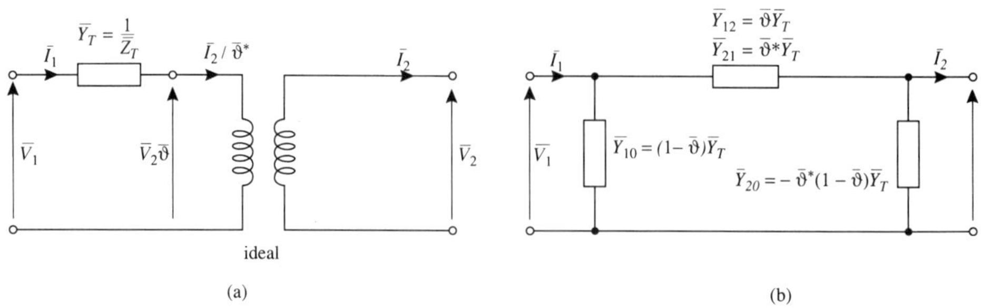

Figure 3.8 Equivalent circuit of a two-winding transformer: (a) using the ideal transformer; (b) using fictitious shunt elements

Eliminating the turns ratio $\overline{\vartheta}$ from the voltage and current vectors gives

$$\begin{bmatrix} \overline{I}_1 \\ -\overline{I}_2 \end{bmatrix} = \begin{bmatrix} \overline{Y}_T & -\overline{\vartheta}\overline{Y}_T \\ -\overline{\vartheta}^*\overline{Y}_T & \overline{\vartheta}^*\overline{\vartheta}\overline{Y}_T \end{bmatrix} \begin{bmatrix} \overline{V}_1 \\ \overline{V}_2 \end{bmatrix} \tag{3.28}$$

The above equation can be interpreted as a nodal voltage equation $\overline{I} = \overline{YV}$ describing a π-equivalent network. The admittance of the series branch of this network is equal to the off-diagonal element (with the sign inverted) in the nodal admittance matrix, i.e.

$$\overline{Y}_{12} = \overline{\vartheta}\overline{Y}_T, \qquad \overline{Y}_{21} = \overline{\vartheta}^*\overline{Y}_T \tag{3.29}$$

while the admittances of the shunt branches are equal to the sum of elements in corresponding rows of the nodal admittance matrix, i.e.

$$\overline{Y}_{10} = (1 - \overline{\vartheta})\overline{Y}_T, \quad \overline{Y}_{20} = \overline{\vartheta}^*(1 - \overline{\vartheta})\overline{Y}_T \tag{3.30}$$

The equivalent network corresponding to this nodal admittance matrix is shown in Figure 3.8b. It may not be symmetric as generally $\overline{\vartheta} \neq 1$. In the case of a phase-shifting transformer the transformer turns ratio ϑ is a complex number and the series branch of the equivalent network is *anisotropic*, that is as $\overline{Y}_{12} \neq \overline{Y}_{21}$ it has a different impedance value depending on from which side the transformer is viewed. The transformer ratio in the circuit diagram of Figure 3.8b is modelled by the shunt elements. Flow of current in these elements causes a change in the voltages in the series branch that corresponds to the voltage transformation in a real transformer.

If it is necessary to include the shunt branch corresponding to the exciting current then the equivalent circuit must be modified slightly by the addition of two shunt admittances to the circuit (similarly as in Figure 3.7): shunt admittance $\overline{Y}_E/2$ to the left-hand-side of the circuit and shunt admittance $\vartheta^2\overline{Y}_E/2$ to the right-hand-side.

3.3 Synchronous Generators

Chapter 2 described how a synchronous generator consists of a stator, on which the three-phase armature winding is normally wound, and a rotor, on which the dc field winding is wound. The rotor damper windings do not affect the steady-state operation of the generator and will not be considered in this chapter. The rotor is on the same shaft as the prime mover and may have salient or non-salient magnetic poles.

The steady-state performance of a synchronous machine with a non-salient round-rotor will be analysed first before generalising the analysis to include the effects of rotor saliency. Of particular importance in this analysis is the way the rotor magnetic field interacts with the stator magnetic field to produce electromagnetic torque. The understanding of this interaction is more important than the detailed equations themselves. Having established the mechanisms by which torque and emf are created, the role of the generator in the power system as a source of both active and reactive power will be analysed.

3.3.1 Round-rotor machines

The schematic diagram of a two-pole generator on no-load is shown in Figure 3.9. For simplicity, only the centre conductor of each of the distributed windings is shown. The

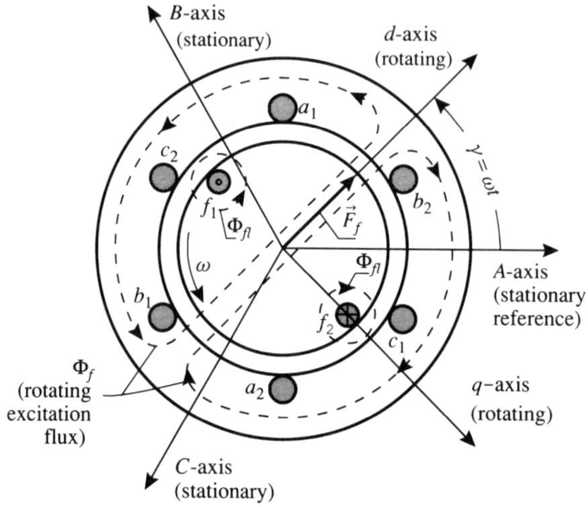

Figure 3.9 Symbolic representation of the generator and its fluxes at no-load

beginning and the end of the field winding is denoted by f_1 and f_2, whilst the beginning and the end of each of the phase windings is denoted by a_1 and a_2 (phase A), b_1 and b_2 (phase B), and c_1 and c_2 (phase C). The stator has three axes A, B, and C each corresponding to one of the phase windings. The rotor has two axes: the *direct axis* (d-axis), which is the main magnetic axis of the field winding, and the *quadrature axis* (q-axis), $\pi/2$ electrical radians behind the d-axis. The dashed lines show the path taken by the rotating *field* (or *excitation*) *flux* Φ_f produced by the field winding, and the field leakage flux Φ_{fl}. $\vec{F}_f$ shows the direction (or the peak value) of the *magnetomotive force* (mmf) wave produced by the field current. The angle $\gamma_m = \omega_m t$, where ω_m is the rotor angular velocity, defines the instantaneous position of the rotor d-axis with respect to the stationary reference assumed here to be along the A-axis.

For a two-pole machine one complete mechanical revolution corresponds to one electrical cycle and one electrical radian is equal to one mechanical radian. However if the generator has p-poles then one mechanical revolution corresponds to $p/2$ electrical cycles. In this general case one mechanical radian is equal to $p/2$ electrical radians and

$$\gamma_e = \frac{p}{2}\gamma_m \tag{3.31}$$

where γ_m is the angle γ expressed in mechanical radians (or degrees) and γ_e is the same angle expressed in electrical radians (or degrees). Similarly the rotor speed ω_e expressed in electrical radians per second can be related to the rotor speed in mechanical radians per second ω_m by

$$\omega_e = 2\pi f = \frac{p}{2}\omega_m \tag{3.32}$$

where f is the system nominal frequency (50 Hz in Europe, 60 Hz in USA). Often the rotor speed is expressed in revolutions per minute (rpm) when the relation between the rotor speed n and the system frequency is $n = 120f/p$.

For simplicity the generator equations will be developed by reference to the two-pole machine shown in Figure 3.9 when both the speed and the angle are the same in electrical and mechanical units and the subscript "e" and "m" can be dropped. The generator will be analysed using only the fundamental component of the stator and rotor spatial mmf waves. Although the equations will be developed for a two-pole generator, they are equally applicable to a p-pole machine when all angles and speeds are expressed in electrical units. All the equations are valid whether expressed in MKSA or per unit notation. If MKSA units are used it must be remembered that any power expression refers to the power per phase. In per unit they can be interpreted as generator power (See Appendix A).

The generator on no-load

To begin the analysis assume that the generator is on *no-load*, that is it is not generating any power and the current in the armature is zero.

The dc field current i_f produces an mmf wave which is approximately sinusoidally distributed around the circumference of the stator. The peak value of the mmf

$$F_f = N_f i_f \tag{3.33}$$

lies along the d-axis and is shown in Figure 3.9 by the vector $\vec{F}_f$. In this equation N_f is the effective number of field winding turns per pole and is smaller than the actual number of turns N_F in order to account for winding geometry and the actual trapezoidal mmf distribution of the field winding. It can be shown (McPherson & Laramore, 1990) that $N_f = (1/p)(4/\pi)N_F k_{wF}$ where the winding geometry is accounted for by the winding factor k_{wF} and the mmf distribution by the factor $4/\pi$.

The field winding mmf drives the *excitation* (or *field*) *flux* Φ_f around the magnetic circuit. The flux per pole is

$$\Phi_f = \frac{F_f}{\Re} = \frac{N_f i_f}{\Re} \tag{3.34}$$

where $\Re$ is the *reluctance* of the path per pole. As the reluctance of the path in iron is negligibly small compared with that in air $\Re$ is approximately directly proportional to the width of the air-gap.

The flux density produced by the field mmf is sinusoidally distributed around the circumference of the stator and, for a round-rotor machine, its peak coincides with the peak of the mmf wave. As the rotor rotates at synchronous speed, the excitation flux rotates with it and produces a time-varying flux linkage with each phase of the armature winding. Each of the phase flux linkages Ψ_{fA}, Ψ_{fB} and Ψ_{fC} reaches a maximum when the d-axis of the rotor aligns with the magnetic axis of the respective phase winding. Taking phase A as the reference gives

$$\begin{gathered}\Psi_{fA}(t) = \Psi_{fa}\cos\omega t = N_\phi \Phi_f \cos\omega t = N_\phi \frac{N_f i_f}{\Re}\cos\omega t = M_f i_f \cos\omega t \\ \Psi_{fB}(t) = M_f i_f \cos\left(\omega t - \frac{2\pi}{3}\right), \quad \Psi_{fC}(t) = M_f i_f \cos\left(\omega t - \frac{4\pi}{3}\right)\end{gathered} \tag{3.35}$$

In these equations $\Psi_{fa} = N_\phi \Phi_f$ is the amplitude of the excitation flux linkage of an armature phase winding, $M_f = N_\phi N_f/\Re$ is the *mutual inductance* between the field and the armature winding and $N_\phi = k_w N$ where N is the number of turns in series in each phase winding and k_w is the armature winding factor.

The time-varying flux linkages induce an *excitation emf* (also called the *internal voltage*) in each of the phase windings. Faraday's Law gives:

$$
\begin{aligned}
e_{fA} &= -\frac{d\Psi_{fA}(t)}{dt} = \omega M_f i_f \sin \omega t, \\
e_{fB} &= -\frac{d\Psi_{fB}(t)}{dt} = \omega M_f i_f \sin\left(\omega t - \frac{2\pi}{3}\right) \\
e_{fC} &= -\frac{d\Psi_{fC}(t)}{dt} = \omega M_i f_i \sin\left(\omega t - \frac{4\pi}{3}\right)
\end{aligned}
\tag{3.36}
$$

In the absence of any armature current these emfs appear at the generator terminals as the *no-load terminal voltage*. Figure 3.10a shows the time variation of the phase flux linkages and the reference emf e_{fA} as a function of $\gamma = \omega t$, whilst the phasor representation of the flux linkages $\overline{\Psi}_{fA}$, $\overline{\Psi}_{fB}$, $\overline{\Psi}_{fC}$ and the emfs $\overline{E}_{fA}$, $\overline{E}_{fB}$, $\overline{E}_{fC}$ induced by them are shown in Figure 3.10b. The root-mean-square (rms) value of each of these emfs (or the length of the phasors $\overline{E}_{fA}$, $\overline{E}_{fB}$, $\overline{E}_{fC}$) is

$$
E_f = \frac{1}{\sqrt{2}}\omega\Psi_{fa} = \frac{1}{\sqrt{2}}\omega N_\phi \Phi_f = \frac{1}{\sqrt{2}}\omega M_f i_f \cong 4.44 f M_f i_f \tag{3.37}
$$

This equation is the well-known transformer equation and illustrates how the primary (field) winding current induces an emf in the secondary (armature) winding. The emf is proportional to both the frequency and the field current i_f. The mutual inductance M_f is in practice not constant but depends on the saturation of the magnetic circuit.

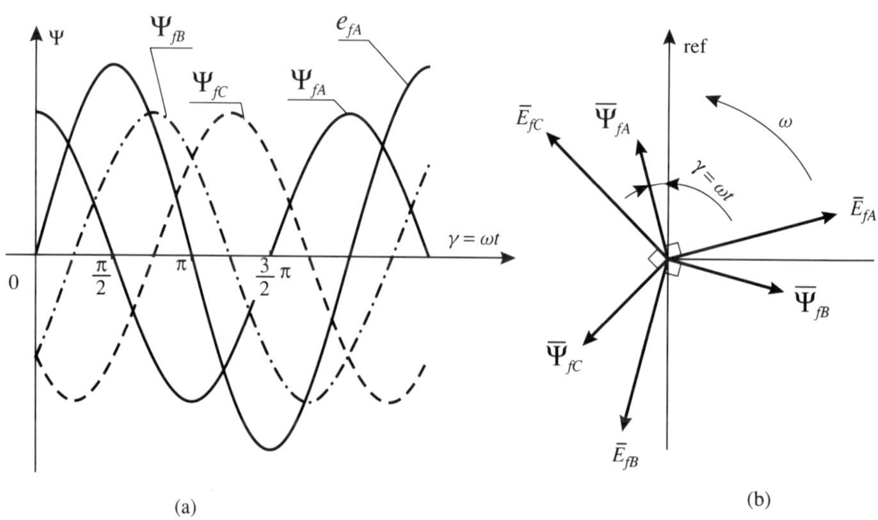

Figure 3.10 The phase excitation flux linkages and the emf's induced by them shown as: (a) time-varying waveforms; (b) rotating phasors

Armature reaction and the air-gap flux

Now consider the effect of loading the generator when an *armature current* flows in each phase of the stator winding. As the three phase windings are magnetically coupled Chapter 10 will show how the equivalent circuit can be developed by considering their self and mutual inductances. In this chapter another, simpler, approach will be followed when the combined magnetic effect of all three phase windings is considered. This is referred to as the *armature reaction*.

Generally the stator phase currents will be delayed by an angle λ with respect to the reference flux linkages $\Psi_{fA}(t)$ and may be written as:

$$i_A = I_m \cos(\omega t - \lambda),\ i_B = I_m \cos\left(\omega t - \lambda - \frac{2\pi}{3}\right),\ i_C = I_m \cos\left(\omega t - \lambda - \frac{4\pi}{3}\right) \quad (3.38)$$

where I_m is the maximum value of the armature current. Each of the phase currents produces a pulsating phase mmf per pole:

$$\begin{aligned} &F_A(t) = N_a I_m \cos(\omega t - \lambda), \quad F_B(t) = N_a I_m \cos\left(\omega t - \lambda - \frac{2\pi}{3}\right) \\ &F_C(t) = N_a I_m \cos\left(\omega t - \lambda - \frac{4\pi}{3}\right) \end{aligned} \quad (3.39)$$

where $N_a = (1/p)(4/\pi)N_\phi$ is the effective number of turns per phase, per pole and, as before, $N_\phi = k_w N$.

Because the magnetic axes of the phase windings are shifted in space from each other by $2\pi/3$ electrical radians the phase mmfs are shifted in both space and time. A convenient way to account for the shift in space is to represent the mmfs as *space vectors* directed along their respective phase axes with instantaneous values given by equations 3.39. A space vector will be denoted by an arrow on top of the symbol. The vector of the resultant *armature reaction mmf* $\vec{F}_a$ can then be obtained by adding the component phase vectors.

A neat way to analyse the space position of the phase mmfs is to introduce a complex plane which has its real axis directed along the A-axis and its imaginary axis $\pi/2$ ahead (counter-clockwise). The space operator $e^{j\theta}$ then introduces a phase shift in the complex plane which, when multiplied by the value of the phase mmf, will direct it in space along the magnetic axes of the winding. Axis B is at the angle $2\pi/3$ with respect to A, while axis C is at the angle $4\pi/3$. The value of F_B must be multiplied by $e^{j2\pi/3}$ to obtain a vector directed along the B-axis, while the value of F_C must be multiplied by $e^{j4\pi/3}$ to obtain a vector directed along the C-axis. The vector of the resulting armature mmf per pole $\vec{F}_a$ is then

$$\begin{aligned} \vec{F}_a &= \vec{F}_A + \vec{F}_B + \vec{F}_C = N_a i_A e^{j0} + N_a i_B e^{j2\pi/3} + N_a i_C e^{j4\pi/3} \\ &= N_a I_m \left[\cos(\omega t - \lambda) + \cos\left(\omega t - \lambda - \frac{2\pi}{3}\right) e^{j2\pi/3} \right. \\ &\quad \left. + \cos\left(\omega t - \lambda - \frac{4\pi}{3}\right) e^{j4\pi/3}\right] \end{aligned} \quad (3.40)$$

Using the identity

$$\cos(\alpha - \beta) = \cos\alpha\ \cos\beta + \sin\alpha\ \sin\beta \tag{3.41}$$

gives

$$\begin{aligned}\vec{F}_a &= N_a I_m \{\cos(\omega t - \lambda) + [-0.5\cos(\omega t - \lambda) + 0.866\sin(\omega t - \beta)](-0.5 + \mathrm{j}0.866) \\ &\quad + [-0.5\cos(\omega t - \lambda) - 0.866\sin(\omega t - \lambda)](-0.5 - \mathrm{j}0.866)\} = \\ &= N_a I_m[1.5\cos(\omega t - \lambda) + 1.5\mathrm{j}\sin(\omega t - \lambda)] = 1.5 N_a I_m e^{\mathrm{j}(\omega t - \lambda)}\end{aligned} \tag{3.42}$$

This equation shows that $\vec{F}_a$ is a vector of constant magnitude $F_a = 1.5N_a I_m$ which rotates in the complex plane with an angular velocity ω. As this is also the rotational speed of the generator the rotor and stator mmfs are stationary with respect to each other. To find the relative spatial position of the mmfs, recall that the angle λ was defined as a time delay with respect to the linkages $\Psi_{fA}(t)$. As $\vec{F}_f$ is in phase with $\Psi_{fA}(t)$ (that is Ψ_{fA} reaches maximum when $\vec{F}_f$ aligns with A-axis), equation 3.42 shows that $\vec{F}_a$ must lag $\vec{F}_f$ by the spatial angle λ.

As the two rotating mmfs $\vec{F}_a$ and $\vec{F}_f$ are stationary with respect to each other they can be combined to give a *resultant mmf* $\vec{F}_r = \vec{F}_f + \vec{F}_a$ which drives the *resultant air-gap flux* Φ_r. This is shown in Figure 3.11. It is important to realise that the magnetic circuit does not "see" $\vec{F}_a$ or $\vec{F}_f$ alone but $\vec{F}_r = \vec{F}_f + \vec{F}_a$. The peak flux density in the air-gap of a round-rotor generator coincides with the peak of $\vec{F}_r$ and it is in the same direction. Figure 3.11 shows the typical relative position of the mmfs with $\pi/2 < \lambda < \pi$. It can be seen that the armature reaction field demagnetises the generator and the resultant mmf is smaller than the excitation mmf alone.

As the resultant mmf is the sum of two rotating, sinusoidally distributed mmfs, both the resultant mmf per pole and the density of the air-gap flux driven by it are sinusoidally

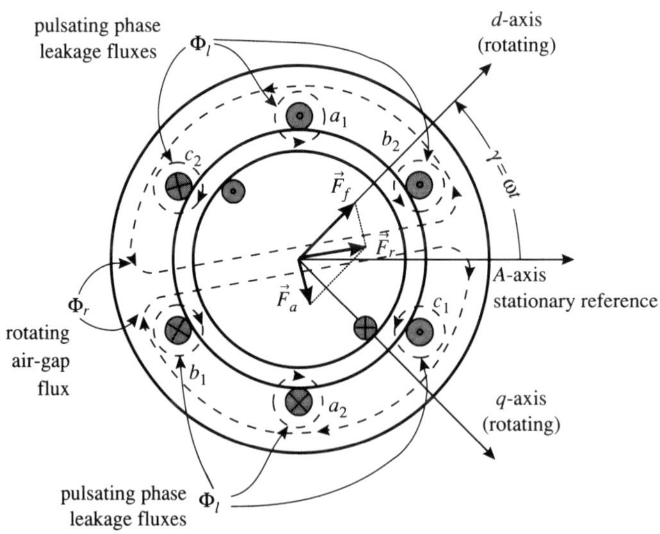

Figure 3.11 Resultant flux in the round-rotor generator operating at lagging power factor

distributed around the air-gap. In practice, magnetic saturation will cause a slight flattening of the flux density wave which results in third-harmonic voltages being induced in the armature winding. Connecting the generator windings in Δ or Y with the star point not earthed prevents the third-harmonic currents appearing at the generator terminals.

Equivalent circuit and the phasor/vector diagram

The rotating air-gap flux produces sinusoidally changing flux linkages with each of the stator phase windings. The linkages with phase A, F_{rA}, can be calculated by projecting the component rotor and stator mmf waves onto the A-axis in order to obtain the resultant mmf along this axis. Recall that F_f is assumed to align with the A-axis at $t = 0$ and that F_a is delayed with respect to F_f by λ. This gives

$$F_{rA}(t) = F_f \cos\omega t + F_a \cos(\omega t - \lambda) = N_f i_f \cos\omega t + 1.5 N_a I_m \cos(\omega t - \lambda) \quad (3.43)$$

This resultant mmf must be divided by the reluctance $\Re$ and multiplied by the number of armature turns N_ϕ to obtain the resultant flux linkages with phase A

$$\Psi_{rA}(t) = N_\phi \frac{F_{rA}(t)}{\Re} = M_f i_f \cos\omega t + L_a I_m \cos(\omega t - \lambda) \quad (3.44)$$

where $M_f = N_\phi N_f/\Re$ is the mutual inductance between the field and the armature winding and $L_a = 1.5 N_a N_\phi/\Re$ is the *armature reaction inductance* or *magnetising inductance*. In the round-rotor machines with a uniform air gap the reluctance $\Re$ does not depend on the flux position.

The linkages $\Psi_{rA}(t)$ induce *the air-gap emf* in phase A equal to

$$e_{rA} = -\frac{d\Psi_{rA}}{dt} = \omega M_f i_f \sin\omega t + \omega L_a I_m \sin(\omega t - \lambda) = e_{fA}(t) + e_{aA}(t) \quad (3.45)$$

where $e_{fA}(t) = \omega M_f i_f \sin\omega t$ and $e_{aA}(t) = \omega L_a I_m \sin(\omega t - \lambda)$. Equation 3.45 shows that the trick in modelling the synchronous generator is to represent the resultant air-gap emf by a sum of two fictitious emfs. The first is the *excitation* (or *internal*) *emf* $e_{fA}(t)$ due to the rotor field which is equal to the no-load terminal voltage, equation 3.36. The second is the *armature reaction emf* $e_{aA}(t)$, due to the armature reaction field, which lags behind the A-phase current by $\pi/2$ (compare equation 3.45 with equation 3.38). Using phasor notation this corresponds to multiplying the current phasor by -j so that the armature reaction emf phasor is $\overline{E}_a = -\mathrm{j}X_a\overline{I}$ where $X_a = \omega L_a$ is the *armature reaction reactance* or *magnetising reactance* and $\overline{I}$ is the current phasor of magnitude $I = I_m/\sqrt{2}$. The phasor of air-gap emf can therefore be expressed as

$$\overline{E}_r = \overline{E}_f + \overline{E}_a = \overline{E}_f - \mathrm{j}X_a\overline{I} \quad (3.46)$$

where $\overline{E}_f$ is the phasor of internal emf with magnitude given by equation 3.37. Thus $-\overline{E}_a = \mathrm{j}X_a\overline{I}$ has the properties of a reactance drop resulting from the effect of the armature current. When $-\overline{E}_a$ is replaced by the reactance voltage drop $\mathrm{j}X_a\overline{I}$, the circuit model becomes that to the left of the phasor $\overline{E}_r$ in Figure 3.12.

Figure 3.12a shows how the full equivalent circuit can be obtained by accounting for electrical and magnetic imperfections of the machine. Firstly, each armature winding has some resistance R which produces the voltage drop $R\overline{I}$. In the case of the synchronous generator this armature resistance is very small and can often be neglected. Secondly,

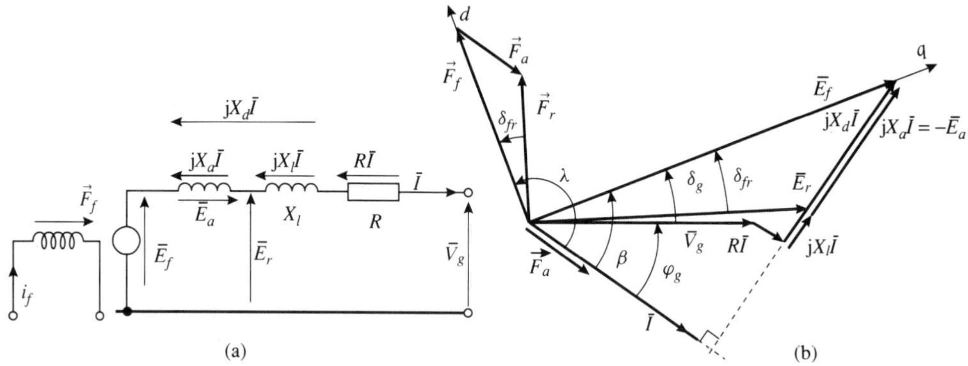

Figure 3.12 Round-rotor synchronous generator: (a) equivalent circuit diagram; (b) phasor/vector diagram for operation at lagging power factor

although most of the magnetic flux produced by the armature crosses the air-gap to link with the rotor winding, a part of it, called the *leakage flux*, does not. This pulsating leakage flux is shown as Φ_l on Figure 3.11 and only links with the armature winding. The leakage flux can be accounted for by inserting a *leakage reactance* X_l into the circuit. As the leakage flux closes its path largely through air $X_l \ll X_a$.

Figure 3.12a shows the total equivalent circuit of the round-rotor synchronous machine with all the imperfections included. The terminal voltage $\overline{V}_g$ is

$$\overline{V}_g = \overline{E}_f - jX_a\overline{I} - jX_l\overline{I} - R\overline{I} = \overline{E}_f - jX_d\overline{I} - R\overline{I} \tag{3.47}$$

where $X_d = X_a + X_l$ is the *synchronous reactance*, or more precisely the *direct-axis synchronous reactance*. As $L_l \ll L_a$, $X_l \ll X_a$ and to all practical purposes $X_d \approx X_a$. The internal emf E_f is sometimes referred to as the *voltage behind the synchronous reactance*. The phasor diagram resulting from equation 3.47 is shown in Figure 3.12b for a generator operating at lagging power factor.

The phasor diagram is normally constructed knowing only the generator voltage V_g and its per-phase active and reactive load powers P and Q. From these the current and the power factor angle are calculated

$$I = \frac{\sqrt{P^2 + Q^2}}{V_g}, \qquad \varphi_g = \arctan\frac{Q}{P} \tag{3.48}$$

Knowing the length and the direction of the current phasor relative to the reference voltage V_g the length and direction of the voltage drops $\overline{I}R$, $j\overline{I}X_l$ and $j\overline{I}X_a$ can be found and the phasor diagram constructed.

Apart from the usual voltage and current phasors, Figure 3.12b also shows the space vectors of the mmfs and the position of the (d, q)-axes. Combining both space vectors and phasors on the same diagram is very useful and is accomplished by drawing the voltage (and current) *phasors* and the mmf *vectors* on the same complex plane at $t = 0$, taking phase A as the reference.

To find the relative displacement between the phasors and the vectors on the diagram consider the relative position of the armature current I and the armature mmf $\vec{F}_a$. At

$t = 0$ the space angle of $\vec{F}_a$ is $(-\lambda)$ with respect to the A-axis, equation 3.42. At the same time the phase angle of the A-axis current is also $(-\lambda)$, equation 3.38, indicating that the current phasor $\overline{I}$ and the mmf vector $\vec{F}_a$ are in phase on the phasor/vector diagram. As $\overline{E}_a = -jX_a\overline{I}$, it follows that $\overline{E}_a$ lags $\vec{F}_a$ by $\pi/2$. A similar argument shows that all the emfs must lag their mmfs by $\pi/2$.

The phasor/vector diagram of Figure 3.12b provides additional insight into the meaning of all the angles on it. As all the emfs are perpendicular to their mmfs the mmf triangle (F_f, F_a, F_r) is similar to the voltage triangle (E_f, E_a, E_r) and all the angles shown in Figure 3.12 have a dual meaning: they are, at the same time, the spatial angles between the rotating mmfs and the phase shifts between the ac voltages. For example the angle δ_{fr}, which is the space angle between the field and the resultant air-gap mmf, is at the same time the phase shift between $\overline{E}_f$ and $\overline{E}_r$.

The torque creation mechanism

The rotor of the synchronous generator is driven by a prime mover which exerts a mechanical torque τ_m on it. For the speed of the rotor to remain constant, the machine must develop an equal, but opposing, electromagnetic torque τ_e. Resolution of the air-gap mmf into the stator and rotor component mmfs provides a means of understanding how the electromagnetic torque is developed. The component stator and rotor mmfs can be compared to two magnets rotating at the same speed and trying to co-align, with the north magnetic pole of one attracting the south pole of the other and vice versa. The torque produced by these attraction forces can be calculated from the three-phase air-gap power P_{ag}, which in its electrical form is equal to

$$P_{ag} = 3E_f I \cos\beta \tag{3.49}$$

where β is the angle between $\overline{E}_f$ and $\overline{I}$ and is referred to as the *internal power factor angle* (see Figure 3.12b). Neglecting mechanical losses, P_{ag} must be equal to the mechanical power $\tau_m\omega_m$ supplied by the prime mover so that for a p-pole machine

$$\tau = \frac{1}{\omega_m}P_{ag} = \frac{p}{2}\frac{1}{\omega}P_{ag} \tag{3.50}$$

Substituting for E_f from equation 3.37 and noting that $I = I_m/\sqrt{2}$ gives

$$\tau = \frac{3}{4}p\Phi_f N_\phi I_m \cos\beta \tag{3.51}$$

This equation can be written in terms of the angle λ and the armature mmf F_a by noting that $N_a = \dfrac{1}{p}\dfrac{4}{\pi}N_\phi$ when, from equation 3.42

$$F_a = \frac{3}{2}N_a I_m = \frac{3}{2}\left(\frac{1}{p}\frac{4}{\pi}N_\phi\right)I_m \tag{3.52}$$

Figure 3.12b shows that $\lambda = \pi/2 + \beta$ so that $\sin\lambda = \cos\beta$ and

$$\tau = \frac{\pi}{8} p^2 \Phi_f F_a \sin\lambda = \frac{\pi}{8} p^2 \frac{F_f F_a}{\Re} \sin\lambda \tag{3.53}$$

where λ is the angle between F_f and F_a. Inspection of Figure 3.12b shows that $F_a \sin\lambda = F_r \sin\delta_{fr}$ and equation 3.53 can be re-written as

$$\tau = \frac{\pi}{8} p^2 F_r \frac{F_f}{\Re} \sin\delta_{fr} = \frac{\pi}{8} p^2 F_r \Phi_f \sin\delta_{fr} \tag{3.54}$$

where the angle δ_{fr} is referred to as the *torque angle*. For a two-pole machine this equation simplifies to

$$\tau = \frac{\pi}{2} F_r \Phi_f \sin\delta_{fr} \tag{3.55}$$

If the rotor field leads the air-gap field, as in Figure 3.11, the electromagnetic torque acts in the opposite direction to the rotation and opposes the mechanical driving torque, and the machine acts as a generator. On the other hand, if the rotor field lags behind the air-gap field, then the electromagnetic torque acts in the direction of rotation and the machine acts as a motor.

3.3.2 Salient-pole machines

Because of the relative ease of balancing round-rotors, and their ability to stand high centrifugal forces, round-rotor generators are normally used for turbo-units driven by high-speed steam or gas turbines. Generators operating at a lower speed, such as those driven by water turbines, need many magnetic poles in order to operate at 50 or 60 Hz. As the centrifugal forces experienced by the rotors of these low-speed machines are lower than in the corresponding turbo-generators, salient poles can be used and the rotor diameter increased. Normally salient-pole rotors have more than two poles and the angles (and speed) expressed in mechanical units are related to those expressed in electrical units by equations 3.31 and 3.32. To simplify considerations a two-pole salient generator will be considered here so that the angles expressed in electrical and mechanical radians are the same. The simplified cross section of such a generator is shown in Figure 3.13.

The main problem with modelling the salient-pole machine is that the width of the air-gap varies circumferentially around the generator with the narrowest gap being along the d-axis and the widest along the q-axis. Consequently the reluctance of the air-gap is not uniform but varies between its minimum value $\Re_d$ and its maximum value $\Re_q$ as shown by the approximated function in Figure 3.14. This creates a problem as one of the fundamental assumptions behind modelling the uniform-gap, round-rotor, generator was that the peak density of the flux wave coincides with the peak density of the mmf wave or, in other words, the mmf and the flux vectors are in phase. It is important to realise that this assumption is no longer valid for the salient-rotor machines. As flux tends to take the path of least reluctance, the flux vector is shifted towards the d-axis (where reluctance is the least) when compared with the mmf vector. The only instance when the two vectors are in phase is when the mmf vector lies along either the d-axis (which is the position of minimum reluctance) or the q-axis (when the attraction forces that tend to shift the flux towards the d-axis balance each other). At any other position the flux

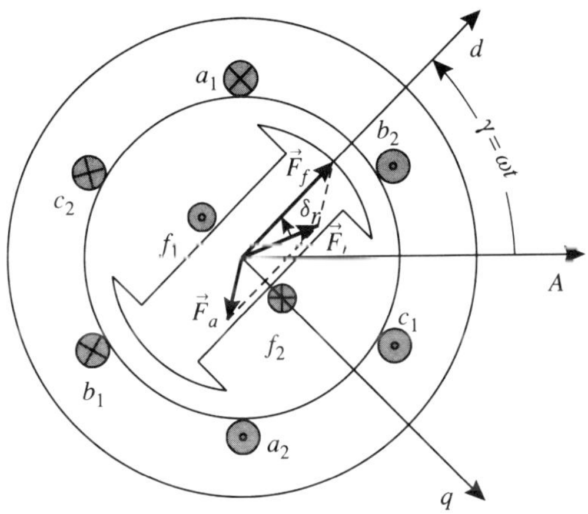

Figure 3.13 A simplified salient-pole generator

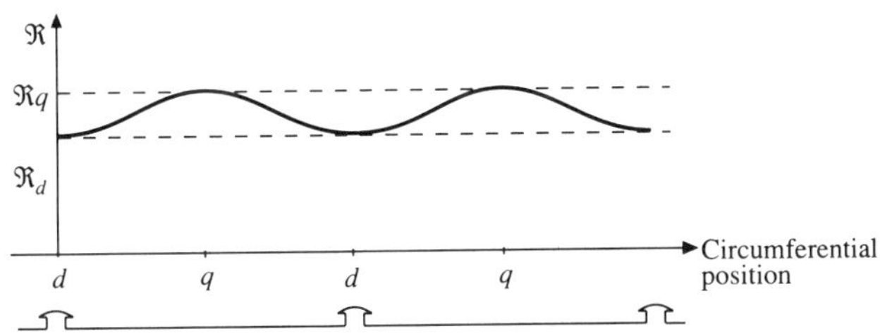

Figure 3.14 Approximate variation of reluctance with circumferential position

and mmf are out of phase and the simple analysis used for the round-rotor machine is no longer valid. To overcome this problem A. Blondel developed his *two-reaction theory* which resolves the mmfs acting in the machine along the (d, q)-axes. The emfs due to these component mmfs are then considered separately assigning different, but constant, values of reactances to the fluxes acting along these axes.

This concept is illustrated in Figure 3.15. The armature mmf $\vec{F}_a$ and the armature current $\overline{I}$ are resolved into two components: one acting along the d-axis ($\vec{F}_{ad}, \overline{I}_d$) and the other acting along the q-axis ($\vec{F}_{aq}, \overline{I}_q$). There is no need to resolve the excitation mmf $\vec{F}_f$ as this always acts only along the d-axis. The resultant mmf $\vec{F}_r$ may then be expressed as

$$\vec{F}_r = \vec{F}_d + \vec{F}_q \tag{3.56}$$

where $\vec{F}_d = \vec{F}_f + \vec{F}_{ad}$ and $\vec{F}_q = \vec{F}_{aq}$. Similarly the current $\overline{I}$ may be expressed as

$$\overline{I} = \overline{I}_d + \overline{I}_q \tag{3.57}$$

As for the round-rotor generator the resultant air-gap emf is equal to the sum of the component emfs, each of which is due to a corresponding mmf. In this case the resultant emf is assumed to be equal to the sum of three components due to $\vec{F}_f$, $\vec{F}_{ad}$, and $\vec{F}_{aq}$ respectively. Because the excitation mmf $\vec{F}_f$ always acts along the d-axis, the internal

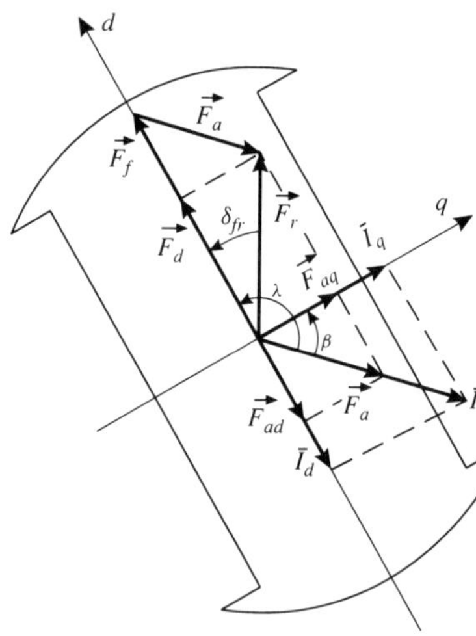

Figure 3.15 Resolution of the mmfs and the current into the (d, q)-components

emf E_f due to $\vec{F}_f$ depends only on the d-axis reluctance $\Re_d$ and is constant for a given value of field current. As the emfs lag their mmfs by $\pi/2$, $\overline{E}_f$ is directed along the q-axis.

The emf due to the d-axis armature mmf $\vec{F}_{ad}$ is proportional to $\overline{I}_d$ and delayed by $\pi/2$ with respect to it. This emf is therefore directed along the q-axis and may be expressed as

$$\overline{E}_{aq} = -\mathrm{j}X_{ad}\overline{I}_d \tag{3.58}$$

where X_{ad} is the *direct-axis armature reaction reactance*. As $\vec{F}_{ad}$ acts across the shortest gap, X_{ad} is inversely proportional to $\Re_d$.

The emf due to the q-axis armature mmf $\vec{F}_{aq}$ is proportional to $\overline{I}_q$ and delayed by $\pi/2$. This emf is therefore directed along the d-axis and may be expressed as

$$\overline{E}_{ad} = -\mathrm{j}X_{aq}\overline{I}_q \tag{3.59}$$

where X_{aq} is the *quadrature-axis armature reaction reactance*. As $\vec{F}_{aq}$ acts across the widest gap, X_{aq} is inversely proportional to $\Re_q$. The resultant air-gap emf is then

$$\overline{E}_r = \overline{E}_f + \overline{E}_{aq} + \overline{E}_{ad} = \overline{E}_f - \mathrm{j}X_{ad}\overline{I}_d - \mathrm{j}X_{aq}\overline{I}_q \tag{3.60}$$

The terminal voltage $\overline{V}_g$ is obtained by subtracting from this emf the voltage drops due to the armature leakage reactance and resistance to give

$$\begin{aligned} \overline{V}_g &= \overline{E}_r - \mathrm{j}X_l\overline{I} - R\overline{I} = \overline{E}_f - \mathrm{j}X_{ad}\overline{I}_d - \mathrm{j}X_{aq}\overline{I}_q - \mathrm{j}X_l(\overline{I}_d + \overline{I}_q) - R\overline{I} \\ &= \overline{E}_f - \mathrm{j}(X_{ad} + X_l)\overline{I}_d - \mathrm{j}(X_{aq} + X_l)\overline{I}_q - R\overline{I} \end{aligned} \tag{3.61}$$

or

$$\overline{E}_f = \overline{V}_g + \mathrm{j}X_d\overline{I}_d + \mathrm{j}X_q\overline{I}_q + R\overline{I} \tag{3.62}$$

where $X_d = X_{ad} + X_l$ is the *direct axis synchronous reactance* and $X_q = X_{aq} + X_l$ is the *quadrature axis synchronous reactance*. As the reluctance along the q-axis is the highest (because the gap is the widest) $X_d > X_q$.

Phasor diagram and the equivalent circuit

Figure 3.16 shows the phasor/vector diagram resulting from equation 3.62. Its construction is more complicated than that for the round-rotor generator shown in Figure 3.12. In order to determine $\overline{I}_d$ and $\overline{I}_q$, it is necessary first to determine the angle δ_g so as to locate the q-axis relative to $\overline{V}_g$. To solve this problem recall that the emf $\overline{E}_f$ lies along the q-axis. Re-arranging equation 3.62 gives

$$\overline{E}_f = \overline{V}_g + R\overline{I} + \mathrm{j}X_q\overline{I} + \mathrm{j}(X_d - X_q)\overline{I}_d = \overline{E}_Q + \mathrm{j}(X_d - X_q)\overline{I}_d \tag{3.63}$$

where

$$\overline{E}_Q = \overline{V}_g + (R + \mathrm{j}X_q)\overline{I} \tag{3.64}$$

Because multiplication of $\overline{I}_d$ by j shifts the resultant phasor $\pi/2$ ahead, the second term in equation 3.63, equal to $\mathrm{j}(X_d - X_q)\overline{I}_d$, is directed along the q-axis. As $\overline{E}_f$ itself lies on the q-axis, $\overline{E}_Q$ must also lie on the q-axis. Knowing $\overline{V}_g$ and $\overline{I}$, the phasor $(R + \mathrm{j}X_q)\overline{I}$ can be added to $\overline{V}_g$ to obtain $\overline{E}_Q$. This determines the direction of the q-axis and the angle δ_g. Once the location of the q-axis is known, $\overline{I}$ can be resolved into its components $\overline{I}_d$ and $\overline{I}_q$ and the phasor diagram completed. For the round-rotor generator $X_d = X_q$ and $\overline{E}_Q$ becomes equal to $\overline{E}_f$ (and $\overline{E}_q$).

Resolving equation 3.62 into its (d, q) components allows the generator equivalent circuit to be constructed. As multiplication of a phasor by j gives a phasor shifted $\pi/2$ counter-clockwise, the phasor $(\mathrm{j}X_q\overline{I}_q)$ is directed along the d-axis. Similarly the phasor $(\mathrm{j}X_d\overline{I}_d)$ is $\pi/2$ ahead of the d-axis, i.e. it is directed in the opposite direction to the q-axis so that its q-axis component is negative. Note that Figure 3.16 shows the phasor of $(\mathrm{j}X_d\overline{I}_d)$ to be along the q-axis; this is due to I_d being negative. Consequently, the (d, q) components of equation (3.62) are

$$\begin{aligned} d-\text{axis:} \quad & E_d = V_d + RI_d + X_q I_q = 0 \\ q-\text{axis:} \quad & E_q = V_q + RI_q - X_d I_d = E_f \end{aligned} \tag{3.65}$$

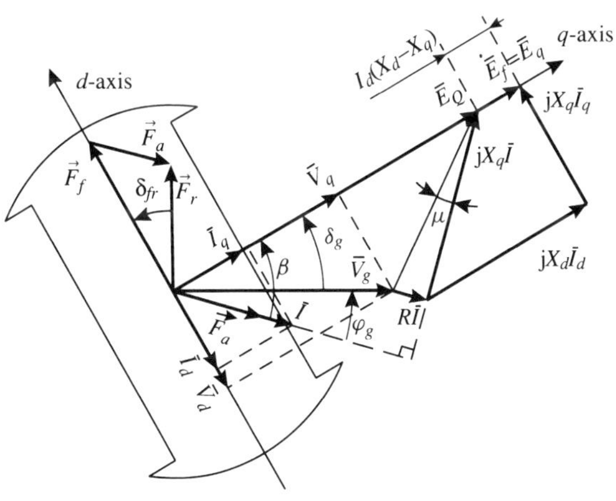

Figure 3.16 Phasor/vector diagram for a salient-pole generator

It is important to realise that all the variables in each of these equations are in phase with each other and are real (not complex) numbers. The (d,q) components of the terminal voltage and current are

$$\begin{aligned} V_d &= -V_g \sin\delta_g \quad & V_q &= V_g \cos\delta_g \\ I_d &= -I \sin\beta \quad & I_q &= I\cos\beta \end{aligned} \tag{3.66}$$

where $\beta = \delta_g + \varphi_g$. The minus sign in the d-components of the voltage and the current indicates the fact that they are directed in the opposite direction to the d-axis. The equivalent circuit for the salient-pole generator can now be drawn in two parts: one corresponding to the d-axis and the other to the q-axis, as shown in Figure 3.17. As all the variables in each of the equations 3.65 are in phase, the reactances X_d and X_q are shown using the symbol of resistance, rather than reactance, to give a voltage drop that is in phase with the current.

A slightly misleading feature of the equivalent circuit shown in Figure 3.17 is that the d-axis current flows into, rather than out of, the generator. This is a consequence of assuming that the d-axis leads the q-axis. As δ_g is assumed to be positive for generator action, this assumption gives negative values for the d-components of the terminal voltage and current, see equation 3.66. The selection of the d-axis leading the q-axis is purely arbitrary and is recommended by IEEE (1969) on the grounds that the d-axis current usually causes demagnetisation of the machine and should be negative in the generator reference frame. Many authors, however, assume the d-axis to lag the q-axis which removes the minus sign from equation 3.66 and changes the sign of the reactive terms in equations 3.65.

d and q axis armature coils

As will be seen in later chapters it is often helpful to consider the effect of the three-phase armature winding to be produced by two equivalent windings phase displaced by 90°. If one of these equivalent armature coils is located along the d-axis and the other along the q-axis then the equivalent circuit of Figure 3.17 can be seen to take on a more direct physical meaning. These two equivalent armature coils are known as the d and q axis armature coils and are assumed to have the same number of turns as an actual armature phase winding. This assumption is not strictly necessary but is the assumption made in

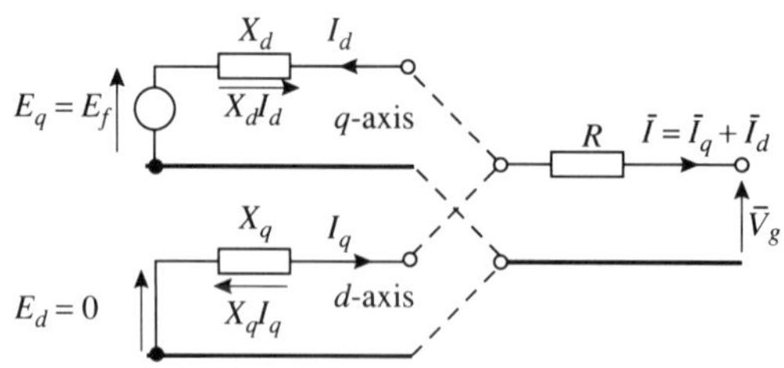

Figure 3.17 Equivalent (d,q) circuit diagrams of the salient-pole generator

this book. A fuller explanation of this is given in Chapter 10. To account for the dc nature of the equations in 3.65 these two equivalent armature coils are also assumed to rotate with the generator rotor with the voltages $E_q = E_f$ and E_d being injected into the respective armature coil. As E_f and E_d are proportional to the rotational speed ω these voltages are known as *rotational voltages*. This concept of (d,q) axis armature coils will be examined in detail in Chapter 10.

Torque in the salient pole machine

The mechanism by which electromagnetic torque is produced was explained earlier with regards to the round-rotor generator when it was shown that torque was proportional to the product of the stator mmf and rotor flux and the sine of the angle between them, equation 3.53. In the case of the salient-pole generator a similar torque creation mechanism applies but now d and q-axis components of mmf must be considered. Resolution of the armature mmf in Figure 3.15 gives

$$F_{ad} = F_a \cos\lambda \quad F_{aq} = F_a \sin\lambda \tag{3.67}$$

As the angle between Φ_f and F_{aq} is $\pi/2$, application of equation 3.53 leads to the electromagnetic torque in a two-pole generator equal to

$$\tau_q = \frac{\pi}{2}\Phi_f F_{aq} \tag{3.68}$$

This equation represents only part of the tangential force on the armature as there will be other forces due to the interaction of other components of the d-axis flux with the q-axis mmf. The effect of these other d-axis flux components is included by algebraically adding all the d-axis fluxes before multiplying by the q-axis mmf. In the steady state the only other flux along the d-axis is the flux produced by the d-axis mmf itself so that the torque expression becomes

$$\tau_q = \frac{\pi}{2}(\Phi_f + \Phi_{ad})F_{aq} \tag{3.69}$$

where $\Phi_{ad} = F_{ad}/\Re_d$. There will also be a similar interaction between any q-axis flux and the d-axis mmf that will produce an additional torque component

$$\tau_d = \frac{\pi}{2}\Phi_{aq}F_{ad} \tag{3.70}$$

where $\Phi_{aq} = F_{aq}/\Re_q$. As F_{ad} is produced by the d-axis current which flows into, rather than out of, the generator — see Figure 3.17 — the torque τ_d acts in the opposite direction to τ_q. The total torque is equal to the difference between the two components giving

$$\tau = \tau_q - \tau_d = \frac{\pi}{2}(\Phi_f + \Phi_{ad})F_{aq} - \frac{\pi}{2}\Phi_{aq}F_{ad} \tag{3.71}$$

Regrouping the terms and substituting $\Phi_{ad} = F_{ad}/\Re_d$, $\Phi_{aq} = F_{aq}/\Re_q$, $F_{aq} = F_a \sin\lambda = F_r \sin\delta_{fr}$ and $F_{ad} = F_a \cos\lambda = -F_f + F_r \cos\delta_{fr}$ finally gives

$$\tau = \frac{\pi}{2}\Phi_f \frac{\Re_d}{\Re_q} F_r \sin\delta_{fr} + \frac{\pi}{4} F_r^2 \frac{\Re_q - \Re_d}{\Re_q \Re_d} \sin 2\delta_{fr} \tag{3.72}$$

This equation shows that the torque developed in a salient pole generator consists of two components. The first component, proportional to $\sin\delta_{fr}$, is identical to the torque expressed by equation 3.55 for a round-rotor generator and is termed the *synchronous torque*. The second component, termed the *reluctance torque*, arises as the rotor tries to assume a position of minimum magnetic reluctance by moving towards the air-gap mmf. This additional torque arises due to the non-uniform air gap and is a direct consequence of the air-gap mmf and flux not being in phase. The reluctance torque is present even without any field excitation and is proportional to $\sin 2\delta_{fr}$. It vanishes for both $\delta_{fr} = 0$ and $\delta_{fr} = \pi/2$. In both these positions the air-gap mmf and flux are in phase, and the tangential forces acting on both poles balance each other. In the round-rotor generator $\Re_d = \Re_q$ and the torque expression reverts to that given in equation 3.55.

3.3.3 Including the equivalent network impedance

Synchronous generators are rarely used to supply individual loads but are connected to a power system that consists of a large number of other synchronous generators and loads linked by the transmission network (Figure 2.1). The power rating of an individual generator is usually many times smaller than the sum of the power ratings of all the remaining generators in the system. Therefore, for a simplified analysis, these remaining generators in the system can be treated as one equivalent, very large, generating unit with an infinite power rating. This equivalent generating unit is referred to as the *infinite busbar* and can be represented on the circuit diagram as an ideal voltage source behind an equivalent system impedance. The infinite busbar maintains a constant terminal voltage and is capable of absorbing all the active and reactive power output of the generator in question.

This concept is illustrated in Figure 3.18 where the generator is connected to the system via a step-up transformer, represented by the series impedance $\overline{Z}_T = R_T + jX_T$ (the shunt admittance is neglected). It is assumed that the ideal transformer has been eliminated from

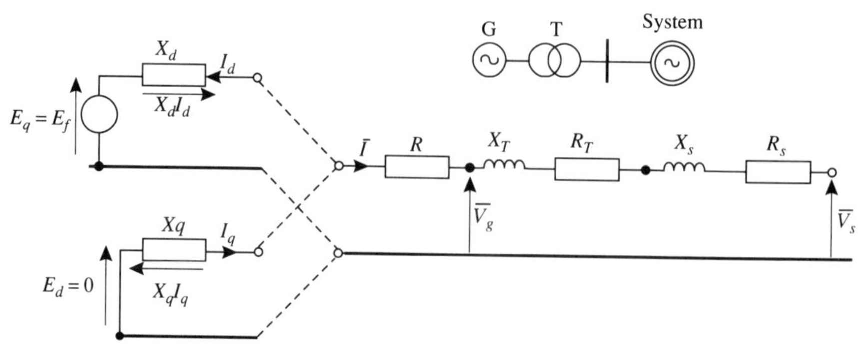

Figure 3.18 Equivalent circuit of the generator operating on infinite busbar

the circuit diagram by either using per-units or re-calculating all the quantities using a common voltage level. The rest of the power system is represented by the infinite busbar, i.e. by the ideal voltage source $\overline{V}_s$ behind the equivalent system impedance $\overline{Z}_s = R_s + jX_s$ (again the shunt admittance is neglected). The impedance $\overline{Z}_s$ combines the transmission network and remaining generators in the system.

The infinite busbar is assumed to have constant voltage and frequency, neither of which is influenced by the action of an individual generator. This means that the voltage $\overline{V}_s$ can be used as a reference and the phase angles of all the other voltages and currents in the circuit measured with respect to it. Of particular importance is the *power angle* δ, which defines the *phase shift* between $\overline{E}_f = \overline{E}_q$ and $\overline{V}_s$. As all the angles on the phasor/vector diagram have a dual time/space meaning, δ is also the *spatial* angle between the two synchronously rotating rotors, that of the generator under consideration and that of the fictitious generator replacing the power system. This spatial angle is referred to as the *rotor angle* and has the same numerical value (in electrical radians) as the power angle. As the "rotor" of the infinite busbar is not affected by an individual generator, this "rotor" also provides a synchronously rotating reference axis with respect to which the *space position* of all the rotors may be defined. This dual meaning of rotor/power angle is very important and will be extensively used throughout this book.

The elements of the equivalent circuit may be combined to give the total parameters

$$x_d = X_d + X_T + X_s, \quad x_q = X_q + X_T + X_s, \quad r = R + R_T + R_s \tag{3.73}$$

All the equations derived earlier in this chapter and the phasor diagram shown in Figure 3.16 can still be used by replacing X_d by x_d, X_q by x_q, R by r, and $\overline{V}_g$ by $\overline{V}_s$. Also the angle δ_g must be replaced by the phase shift δ between $\overline{E}_f$ and $\overline{V}_s$. With these modifications the round-rotor generator can be described by the following equation (corresponding to equation 3.47):

$$\overline{E}_f = \overline{V}_s + r\overline{I} + \mathrm{j}x_d\overline{I} \tag{3.74}$$

while the salient-pole generator can be described by an equation similar to equation 3.62:

$$\overline{E}_f = \overline{V}_s + r\overline{I} + \mathrm{j}x_d\overline{I}_d + \mathrm{j}x_q\overline{I}_q \tag{3.75}$$

This equation can be broken down into its (d,q) components to obtain equations similar to equation 3.65

$$\begin{aligned} E_d &= V_{sd} + rI_d + x_qI_q = 0 \\ E_q &= E_f = V_{sq} + rI_q - x_dI_d \end{aligned} \tag{3.76}$$

which can be expressed in matrix form as

$$\begin{bmatrix} E_d = 0 \\ E_q = E_f \end{bmatrix} = \begin{bmatrix} V_{sd} \\ V_{sq} \end{bmatrix} + \begin{bmatrix} r & x_q \\ -x_d & r \end{bmatrix} \begin{bmatrix} I_d \\ I_q \end{bmatrix} \quad \text{or} \quad \mathbf{E}_{dq} = \mathbf{V}_{sdq} + \mathbf{Z}_{dq}\mathbf{I}_{dq} \tag{3.77}$$

where

$$\begin{aligned} V_{sd} &= -V_s \sin\delta, \quad & V_{sq} &= V_s \cos\delta \\ I_d &= -I \sin\beta, \quad & I_q &= I \cos\beta \end{aligned} \tag{3.78}$$

and $\beta = \delta + \varphi$.

3.3.4 The synchronous generator as a power source

Equations 3.77 and 3.78 allow formulae for the active and reactive power supplied by the generator to the system to be derived.

Active power

Solving equation 3.77 with respect to the currents gives

$$\begin{bmatrix} I_d \\ I_q \end{bmatrix} = \frac{1}{Z^2} \begin{bmatrix} r & -x_q \\ x_d & r \end{bmatrix} \begin{bmatrix} 0 - V_{sd} \\ E_q - V_{sq} \end{bmatrix} \tag{3.79}$$

where $Z^2 = \det \mathbf{Z}_{dq} = r^2 + x_d x_q$

The real power supplied by each phase of the generator to the network (the infinite busbar in Figure 3.18) is given by

$$\begin{aligned} P_s &= V_s I \cos\varphi = V_s I \cos(\beta - \delta) = V_s I \sin\beta \sin\delta + V_s I \cos\beta \cos\delta \\ &= V_{sd} I_d + V_{sq} I_q = \frac{1}{Z^2} \left\{ V_{sd} \left[-r V_{sd} - x_q \left(E_q - V_{sq} \right) \right] \right. \\ &\quad \left. + V_{sq} \left[-x_d V_{sd} + r \left(E_q - V_{sq} \right) \right] \right\} \\ &= \frac{1}{Z^2} \left[-r(V_{sq}^2 + V_{sd}^2) - (x_d - x_q) V_{sd} V_{sq} + E_q (r V_{sq} - x_q V_{sd}) \right] \end{aligned} \tag{3.80}$$

which, after substituting for V_{sq} and V_{sd} from equation (3.78), gives

$$P_s = \frac{E_q V_s}{Z} \frac{x_q}{Z} \sin\delta + \frac{V_s^2}{2} \frac{x_d - x_q}{Z^2} \sin 2\delta + \frac{E_q V_s}{Z} \frac{r}{Z} \cos\delta - \frac{V_s^2}{Z} \frac{r}{Z} \tag{3.81}$$

The second component in equation 3.81 is referred to as the *reluctance power* and corresponds to the reluctance torque in the salient-pole generator. It depends on $\sin 2\delta$, vanishes for $\delta = 0°$ and $\delta = 90°$ and is proportional to $x_d - x_q = X_d - X_q$ and inversely proportional to Z^2. As $x_d - x_q = X_d - X_q$ is independent of X_s the Z^2 term tends to dominate this expression so that the reluctance power of a generator connected to the system through a long transmission link (large X_s and Z) can often be neglected.

For round-rotor generators $X_d = X_q$ and the reluctance term vanishes irrespective of the value of the power angle δ. For such generators $Z^2 = r^2 + x_d^2$ and equation 3.81 can be re-arranged as

$$\begin{aligned} P_s &= \frac{E_q V_s}{Z} \cos\mu \sin\delta + \frac{E_q V_s}{Z} \sin\mu \cos\delta - \frac{V_s^2}{Z} \sin\mu \\ &= \frac{E_q V_s}{Z} \sin(\delta + \mu) - \frac{V_s^2}{Z} \sin\mu \end{aligned} \tag{3.82}$$

where $\sin\mu = r/Z$ and $\cos\mu = x_d/Z$. The active power load on the generator is larger than the real power supplied to the system because of the $I^2 r$ loss in the resistance. If the resistance of the generator and the network is neglected ($r \ll Z$), P_s is equal to the real power load on the generator. Then for the salient-pole generator

$$P_s = \frac{E_q V_s}{x_d} \sin\delta + \frac{V_s^2}{2} \frac{x_d - x_q}{x_d x_q} \sin 2\delta \tag{3.83}$$

while for the round-rotor generator the real power is

$$P_s = \frac{E_q V_s}{x_d} \sin\delta \tag{3.84}$$

In per unit these equations can also be used to express the generator air-gap torque.

Reactive power

The reactive power supplied to the network is given by

$$\begin{aligned} Q_s &= V_s I \sin\varphi = V_s I \sin(\beta - \delta) = V_s I \sin\beta\cos\delta - V_s I \cos\beta \sin\delta \\ &= -V_{sq} I_d + V_{sd} I_q = \frac{1}{Z^2}\left\{V_{sq}\left[rV_{sd} + x_q\left(E_q - V_{sq}\right)\right]\right. \\ &\left. + V_{sd}\left[-x_d V_{sd} + r\left(E_q - V_{sq}\right)\right]\right\} \\ &= \frac{1}{Z^2}\left(E_q V_{sq} x_q - V_{sq}^2 x_q - V_{sd}^2 x_d + E_q V_{sd} r\right) \end{aligned} \tag{3.85}$$

which, after substituting for V_{sq} and V_{sd} from equation 3.78, gives

$$Q_s = \frac{E_q V_s}{Z} \frac{x_q}{Z} \cos\delta - \frac{V_s^2}{Z} \frac{x_d \sin^2\delta + x_q \cos^2\delta}{Z} - \frac{E_q V_s}{Z} \frac{r}{Z} \sin\delta \tag{3.86}$$

For the round-rotor generator $x_d = x_q$ and the second term is independent of δ. Equation 3.86 can be then written as

$$Q_s = \frac{E_q V_s}{Z} \cos(\delta - \mu) - \frac{V_s^2}{Z} \cos\mu \tag{3.87}$$

If the resistance r is neglected then $Z = x_d$, $\mu = 0$, $\cos\mu = 1$ and this equation simplifies to

$$Q_s = \frac{E_q V_s}{x_d} \cos\delta - \frac{V_s^2}{x_d} \tag{3.88}$$

A generator supplies reactive power to the system, and operates at a lagging power factor when the field current, and consequently E_q, is high enough to make the first component in equation 3.88 larger than the second. This state of operation is referred to as *over excitation*.

On the other hand a low value of the field current, and E_q, will make the second component of equation 3.88 larger than the first; the generator will then supply negative reactive power to the system and will operate at a leading power factor. This state of operation is referred to as *under excitation*.

As will be explained in Section 3.4, a typical power system load contains a high proportion of induction motors. Such a load operates at a lagging power factor and consumes positive reactive power. For this reason generators are usually overexcited, operate at lagging power factor, and supply positive reactive power to the system.

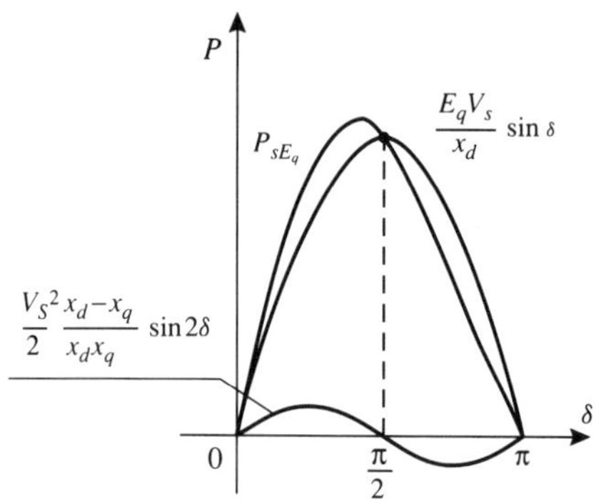

Figure 3.19 Power-angle characteristic for $P_{sEq}(\delta)$ for E_q = constant

Overexcited generators operate at high value of E_q which is also important for stability reasons as explained in Chapters 5 and 6.

Steady-state power-angle characteristic

Figure 3.16 shows the phasor diagram of a generator operating under a particular load as defined by the length and direction of the current phasor $\underline{I}$. Obviously the diagram will change if the load is changed. However the way in which the diagram will change depends on whether or not the Automatic Voltage Regulator (AVR) is active. When the AVR is active it will try to maintain a constant voltage at some point after the generator terminals by changing the excitation (see Section 2.2.2). This mode of operation is closely related to steady-state stability and will be considered in Chapter 5. In this chapter only the case when the AVR is inactive will be considered. The excitation is then constant, $E_f = E_q =$ constant, and any change in load will change the generator terminal voltage V_g, the angle δ_g and the power angle δ. If the resistance r is neglected then the real power supplied by the generator to the system is given by equation 3.83 which is repeated here for convenience

$$P_{sE_q}(\delta) = \frac{E_q V_s}{x_d} \sin\delta + \frac{V_s^2}{2}\frac{x_d - x_q}{x_d x_q} \sin 2\delta \qquad (3.89)$$

In this equation the subscript E_q in P_{sE_q} has been added to emphasise that this formula is valid for the case E_q = constant. Note that P_{sE_q} is a function of the power angle δ only. Function $P_{sE_q}(\delta)$ is referred to as the *power-angle characteristic* of the generator operating on the infinite busbar and is shown in Figure 3.19. The reluctance power term deforms the sinusoidal characteristic so that the maximum of $P_{sE_q}(\delta)$ occurs at $\delta < \pi/2$. For the round rotor generators the maximum of $P_{sE_q}(\delta)$ occurs at $\delta = \pi/2$.

3.4 Power System Loads

The word load can have several meanings in power system engineering, including:

- a device connected to the power system that consumes power
- the total active and/or reactive power consumed by all devices connected to the power system

- the power output of a particular generator or plant
- a portion of the system that is not explicitly represented in the system model, but treated as if it were a single power-consuming device

This section deals with the loads that conform to the last definition. Chapter 2 explained that electrical power systems are large, complex, structures consisting of power sources, transmission and sub transmission networks, distribution networks and a variety of energy consumers. As the transmission and sub-transmission networks connect the main generation and load centres, they are quite sparse, whilst as the distribution networks must reach every consumer in their service area they are very dense. This means that a typical power system may consist of several hundred nodes at the transmission and sub-transmission level, but there could be a hundred thousand nodes at the distribution level. Consequently when power systems are analysed only the transmission and sub-transmission level is considered, whilst the distribution networks are not usually modelled as such but are replaced by equivalent loads, sometimes referred to as *composite loads*. Usually each composite load represents a relatively large fragment of the system typically comprising of low and medium voltage distribution networks, small power sources operating at distribution levels, reactive power compensators, distribution voltage regulators etc. and includes a large number of different component loads such as motors, lighting, and electrical appliances. Determining a simple and valid composite load model is therefore not an easy problem and is still the subject of intensive research (IEEE Task Force, 1995). In this chapter only a simple static composite load model will be described while dynamic models will be described in Chapter 10.

In the steady-state the demand of the composite load depends on the busbar voltage V and the system frequency f. The functions describing the dependency of the active and reactive load demand on the voltage and frequency $P(V, f)$ and $Q(V, f)$ are called the *static load characteristics*. The characteristics $P(V)$, and $Q(V)$ taken at constant frequency, are called the *voltage characteristics* while the characteristics $P(f)$ and $Q(f)$, taken at constant voltage, are called the *frequency characteristics*.

The slope of the voltage or frequency characteristic is referred to as the voltage (or frequency) sensitivity of the load. Figure 3.20 illustrates this concept with respect to voltage sensitivities. Frequency sensitivities are defined in a similar way.

The voltage sensitivities k_{PV} and k_{QV} and the frequency sensitivities k_{Pf} and k_{Qf} are usually expressed in pu with respect to the given operating point:

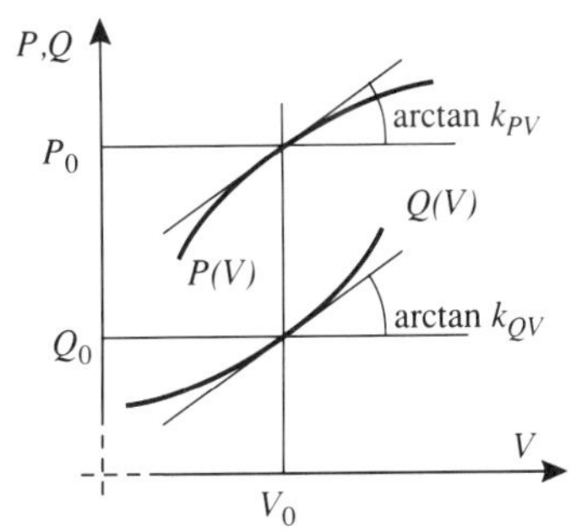

Figure 3.20 Illustration to the definition of voltage sensitivities

$$k_{PV} = \frac{\dfrac{\Delta P}{P_0}}{\dfrac{\Delta V}{V_0}}, \quad k_{QV} = \frac{\dfrac{\Delta Q}{Q_0}}{\dfrac{\Delta V}{V_0}}, \quad k_{Pf} = \frac{\dfrac{\Delta P}{P_0}}{\dfrac{\Delta f}{f_0}}, \quad k_{Qf} = \frac{\dfrac{\Delta Q}{Q_0}}{\dfrac{\Delta f}{f_0}} \tag{3.90}$$

where P_0, Q_0, V_0 and f_0 are the real power, reactive power, voltage and frequency at a given operating point.

As the characteristics of the composite load depend on the characteristics of its individual components it is first of all necessary to examine the characteristics of some of the more important individual loads. This will then be developed into a more general composite load model.

3.4.1 Lighting and heating

About one-third of electricity consumption goes to lighting and heating. Traditional bulb lighting prevails in residential areas while discharge lights (fluorescent, mercury vapour, sodium vapour) dominate in commercial and industrial premises. Traditional electric bulbs consume no reactive power and their power demand is frequency independent but, as the temperature of the filament depends on the voltage, the bulb cannot be treated as a constant-impedance. Figure 3.21a shows the relevant characteristic.

Fluorescent and discharge lighting depends heavily on the supply voltage. When the voltage drops to below 65–80% of the rated value the discharge lights extinguish and will restart with a one-two second delay only when the voltage recovers to a value above the extinguish level. When the voltage is above the extinguish level, the real and reactive power vary non-linearly with voltage as shown in Figure 3.21b.

Heating loads basically constitute a constant resistance. If a heater is equipped with a thermostat, the thermostat will maintain constant temperature and power output despite any variations in voltage. In such cases the load can be modelled as a constant power rather than a constant resistance.

3.4.2 Induction motors

About 50–70% of all electricity is consumed by electric motors with about 90% of this being used by induction motors. Generally such motors dominate industrial loads to a far greater degree than they do commercial or residential loads.

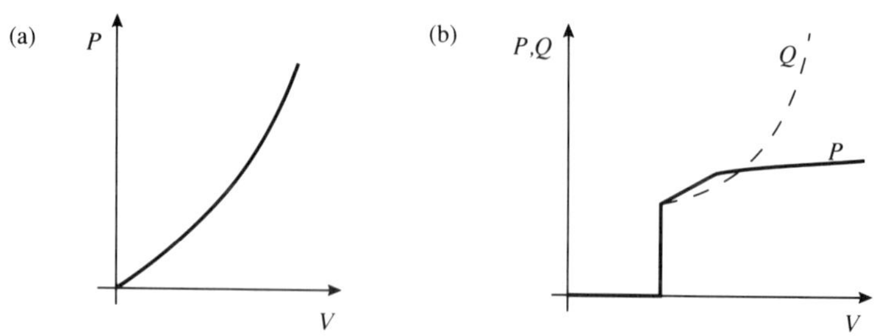

Figure 3.21 Voltage characteristics of: (a) electric bulbs; (b) discharge lighting

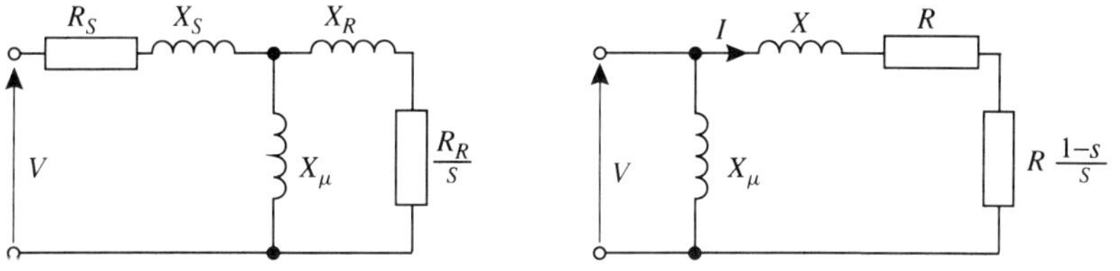

Figure 3.22 Induction motor: (a) equivalent circuit; (b) simplified equivalent circuit

Figure 3.22a shows the steady-state equivalent circuit of an induction motor with all the quantities referred to the stator. In this equivalent circuit R_S and X_S are the stator resistance and leakage reactance, R_R and X_R are the rotor resistance and leakage reactance, X_μ is the magnetising reactance and s is the *motor slip* defined as $s = (\omega_s - \omega)/\omega_s$. As R_S is small and $X_\mu \gg X_S$, the equivalent circuit can be simplified to that shown in Figure 3.22b where X is the equivalent reactance of the stator and rotor windings ($X = X_S + X_R$) and $R = R_R$ is the rotor resistance. The quantity R/s is usually separated into two components: R itself which represents the power dissipated in the rotor resistance while the resistance term $R(1 - s)/s$ represents the developed mechanical power.

Power-slip characteristic

The square of the current flowing through the series branch in Figure 3.22b is $I^2 = V^2/[X^2 + (R/s)^2]$ where V is the supply voltage. The real power demand due to this current is then

$$P_e = I^2\frac{R}{s} = V^2\frac{Rs}{R^2 + (Xs)^2} \tag{3.91}$$

The solid lines in Figure 3.23a show the variation of P_e with slip for different values of V. By differentiating the power expression it can be shown that the maximum loading

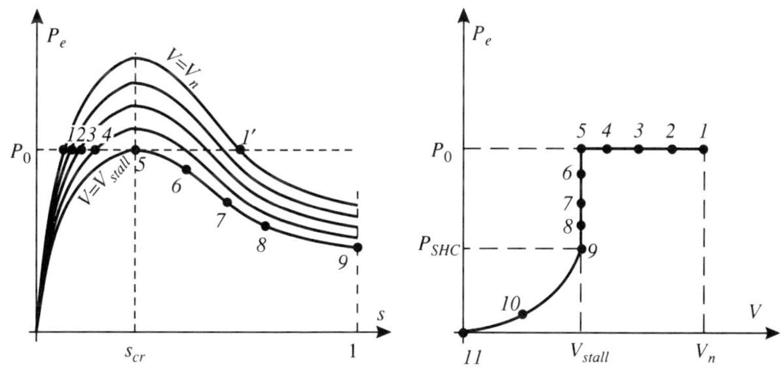

Figure 3.23 Induction motor characteristics: (a) family of $P_e(s)$ characteristics for different values of V; (b) $P_e(V)$ characteristic of the motor loaded with a constant torque

$P_{\max} = V^2/2X$ is reached at the *critical slip* $s = s_{cr} = R/X$. Note that s_{cr} does not depend on the voltage.

The stable operating part of the induction motor characteristic is to the left of the peak, where $s < s_{cr}$. To prove this consider point *1′* lying to the right of the peak on the highest characteristic. Ignoring for the moment any losses, if point *1′* is an equilibrium point then the electrical power supplied P_e is equal to the mechanical power P_m shown on the diagram as P_o. Now assume that the motor experiences a momentary disturbance causing a decrease in slip (i.e. an increase in speed). This will cause the supplied electrical power to increase. As P_m can be assumed to remain unchanged, the motor draws more power than the load can absorb ($P_e > P_m$) and the excess power accelerates the rotor so that the speed increases further. This increase in speed results in the operating point moving further to the left away from point *1′*. Now consider another momentary disturbance causing an increase in slip (drop in speed). This will cause P_e to drop resulting in a deficit of power and further slowing down of the rotor. The slip further increases and the motor finally stops at $s = 1$.

The situation is reversed at the stable equilibrium point *1*. At this point any disturbance causing a decrease in slip (i.e. an increase in speed) will result in a decrease in the power supplied to the motor. This deficit of power ($P_e < P_m$) will result in the motor slowing down (increase the slip) until it returns to point 1. In a similar way the motor will return to point 1 when subjected to a disturbance causing an increase in slip.

Real power-voltage characteristic

The motor's operating point corresponds to the intersection of the motor characteristic $P_e(s)$ with that of the mechanical load $P_m(s)$. Mechanical loads can be categorised as either *easy starting* (zero torque at starting) or *heavy starting* (a non-zero torque at starting). For further analysis the special case of a heavy starting load with a constant mechanical torque τ_m will be considered. This type of load will allow the voltage characteristics of the motor to be derived. These characteristics will then be generalised to include other types of mechanical load.

If τ_m is constant then the mechanical power developed by the motor is $P_m = \tau_m\omega = \tau_m\omega_s(1-s) = P_0(1-s)$, where $P_0 = \tau_m\omega_s$, and is represented in the equivalent circuit by the power loss in the resistance $R(1-s)/s$. The total motor power demand P_d is obtained by adding P_m to the power loss in the rotor resistance, where $P_{loss} = I^2R$, that is

$$P_d = P_m + P_{loss} = P_0(1-s) + I^2R = P_0(1-s) + \frac{V^2Rs^2}{(Xs)^2 + R^2} \tag{3.92}$$

The value of the operating slip is given by the intersection of the $P_e(s)$ and the $P_d(s)$ characteristics. Equation $P_e(s) = P_d(s)$ gives

$$\frac{V^2Rs}{(Xs)^2 + R^2} = P_0(1-s) + \frac{V^2Rs^2}{(Xs)^2 + R^2} \tag{3.93}$$

which, assuming that the motor is not stalled ($s \neq 1$) gives, after some simple algebra, the following equation

$$s^2 - 2as_{cr}s + s_{cr}^2 = 0 \tag{3.94}$$

where $a = V^2/(2P_0X)$ and $s_{cr} = R/X$. The roots of this equation are

$$s_{1,2} = s_{cr}\left[a \pm \sqrt{a^2 - 1}\right] \tag{3.95}$$

The roots are real only if $|a| \geq 1$, that is $V \geq \sqrt{2P_0X}$. Hence the minimum supply voltage for which the motor can still operate, referred to as the *stalling voltage*, is $V_{stall} = \sqrt{2P_0X}$. For $V > V_{stall}$ there are two different roots s_1 and s_2 with the smaller root corresponding to the stable operating point (i.e. that part of the characteristic to the left of the peak). Substituting this root into equation 3.91 gives

$$P_e|_{s=s_{1,2}} = \frac{V^2}{X}\frac{Rs_{cr}\left(a \pm \sqrt{a^2-1}\right)}{R^2 + X^2s_{cr}^2\left(a \pm \sqrt{a^2-1}\right)^2} = \frac{V^2}{X}\frac{a \pm \sqrt{a^2-1}}{1 + \left(a \pm \sqrt{a^2-1}\right)^2}$$
$$= \frac{V^2}{2aX} = P_0 \tag{3.96}$$

This shows that for a heavy-starting load with $\tau_m =$ constant the real power demand is independent of both the supply voltage and the slip. This is shown in Figure 3.23a by the dashed horizontal line $P = P_0$. Figure 3.23b shows the resulting $P_e(V)$ characteristic obtained in the following way.

Assume that the motor initially operates at rated voltage ($V = V_n$) so that the equilibrium point corresponds to point *1* on the top characteristic. A decrease in the supply voltage will lower the power-slip characteristic and shift the operating point to the right, as shown by points *2*, *3* and *4* in Figure 3.23a. As real power demand is constant the $P_e(V)$ characteristic corresponding to point *1*, *2*, *3*, *4* is horizontal as shown in Figure 3.23b. The lowest characteristic in Figure 3.23a corresponds to the stalling voltage $V = V_{stall}$. The operating point is at point *5* and the slip is critical $s = s_{cr}$. Any further slight decrease in voltage will result in the motor slip increasing along the lowest, unstable, characteristic (points *6*, *7*, *8*) until the motor finally stalls ($s = 1$) at point *9*. As the motor moves along this stalling characteristic, the real power decreases while the voltage is constant and slip increases. Consequently the $P_e(V)$ characteristic for points *6*, *7*, *8* is vertical as shown in Figure 3.23b. At point *9* the slip is $s = 1$ and the motor consumes its short-circuit power, $P_{SHC} = V^2R/(X^2 + R^2)$. Any further decrease in the voltage will cause a decrease in the real power demand which can be calculated from equation 3.91 as

$$P_e(s)|_{s=1} = \frac{V^2R}{X^2 + R^2} \tag{3.97}$$

This is shown by a parabola *9*, *10*, *11* in Figure 3.23b.

Reactive power-voltage characteristic

The reactive power consumed by each phase of the motor is made up of two components corresponding to each of the two parallel branches shown in Figure 3.22b:

$$Q_\mu = \frac{V^2}{X_\mu} \quad \text{and} \quad Q_s = I^2 X = \frac{P_e s}{R} X = P_e \frac{s}{s_{cr}} \tag{3.98}$$

The component Q_μ is associated with the motor magnetising reactance while the component Q_s depends on the motor load. For a heavy starting load, with $\tau_m =$ constant, the real power demand is $P_e = P_0$ and the slip corresponding to the stable operating point is $s = s_{cr}\left[a - \sqrt{a^2 - 1}\right]$, equation 3.95. Substituting these values into the expression for Q_s gives

$$Q_s = \frac{V^2}{2X} - \sqrt{\left(\frac{V^2}{2X}\right)^2 - P_0^2} \quad \text{for } V > V_{stall} \tag{3.99}$$

Figure 3.24 shows Q_μ, Q_s and $Q = Q_\mu + Q_s$ as a function of the voltage. The component $Q_\mu(V)$ is a parabola starting from the origin and increasing to infinity. The component $Q_s(V)$ tends to zero as $V \to \infty$. As V decreases, $Q_s(V)$ increases until at $V = V_{stall}$ it reaches a value corresponding to point *5*. Any decrease in voltage will cause $Q_s(V)$ to move along the unstable part defined by the vertical line *5, 6, 7, 8, 9*. At point *9* the motor stalls ($s = 1$). Any further decrease in the voltage causes $Q_s(V)$ to decrease along a parabola *9, 10, 11* describing the short-circuit characteristic (equation 3.98 for $s = 1$).

The resulting characteristic, $Q(V) = Q_s(V) + Q_\mu(V)$, is shown by the bold line in Figure 3.24. Inspection of this characteristic leads to the following conclusions:

- for voltages close to the rated voltage V_n the slope of the characteristic is positive
- as voltage decreases, the characteristic first becomes flatter (indicating a reduced voltage sensitivity) and then, with a further reduction in voltage, the reactive power increases rapidly to a large value when the motor stalls

Although a heavy starting load has been considered here the actual stalling voltage depends on the type of mechanical load that is being driven by the motor. For heavily-loaded motors, driving heavy-starting loads, it can be quite close to the rated voltage. For lightly-loaded motors, especially those driving easy-starting mechanical loads, the stalling voltage can be quite small.

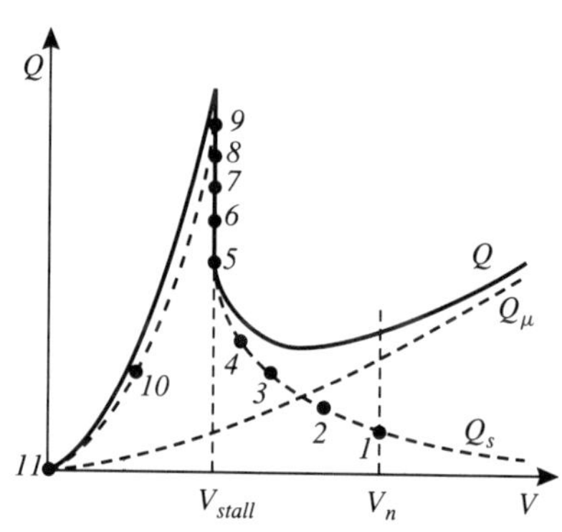

Figure 3.24 Reactive power-voltage characteristics of an induction motor

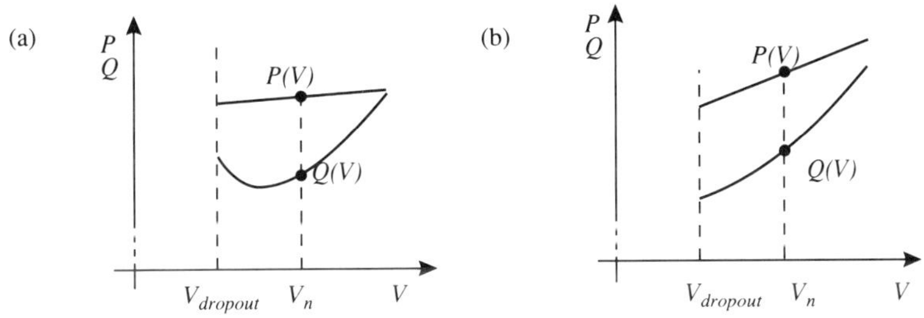

Figure 3.25 Examples of induction motor characteristics with starter control: (a) heavy-starting load; (b) easy-starting load

Influence of motor protection and starter control

The above considerations do not take into account motor protection or starter control. Many industrial motors have starter controls with electromechanically held ac contactors. If the voltage is too low, such contactors immediately drop out tripping the motor supply. The drop-out voltage ranges from about 0.3 to 0.7 per unit. The effect of this can be modelled by assuming zero real and reactive power demand for voltages lower than the drop-out voltage as shown in Figure 3.25. Only heavy-loaded motors may stall above the drop-out voltage and give a large increase in the reactive power demand, Figure 3.25a. When the stalling voltage is smaller than the drop-out voltage, the motor will not stall or exhibit the rapid increase in the reactive power demand, Figure 3.25b.

When operating at a low voltage the motor may also be tripped by the overcurrent protection. This usually involves some time delay.

Motors operating in residential and commercial loads are usually oversized, operate at less than 60% of rated power, and therefore have characteristics similar to those shown in Figure 3.25b. On the other hand large industrial motors are usually properly sized and, when driving heavy-starting mechanical loads, may have characteristics similar to those shown in Figure 3.25a.

3.4.3 Static characteristics of the load

The aggregate characteristic of the load depends on the characteristics of its individual components. A rough estimate of the aggregate characteristic, viewed from the medium voltage side (the secondary of the feeder transformer), can be obtained by summing the individual load characteristics. Figure 3.26 shows two examples of load characteristics obtained by this technique. Figure 3.26a shows an industrial load characteristic with a predominance of heavily-loaded induction motors and discharge lighting. Near the nominal operating point (voltage V_n), the $P(V)$ curve is flat while the $Q(V)$ curve is steeper with a positive slope. As the voltage decreases, the $Q(V)$ curve becomes flatter and even rises due to the increased reactive power demand of the stalled motors. When the voltage drops below about 0.7 per unit, the $P(V)$ and $Q(V)$ curves rapidly decrease due to tripping of induction motors and extinguishing of discharge lighting.

Figure 3.26b shows an example of a residential/commercial load that is dominated by traditional bulb lighting and heating. Near the nominal voltage both the $P(V)$ and $Q(V)$ curves are quite steep . Again the real and reactive power demand drops rapidly at about

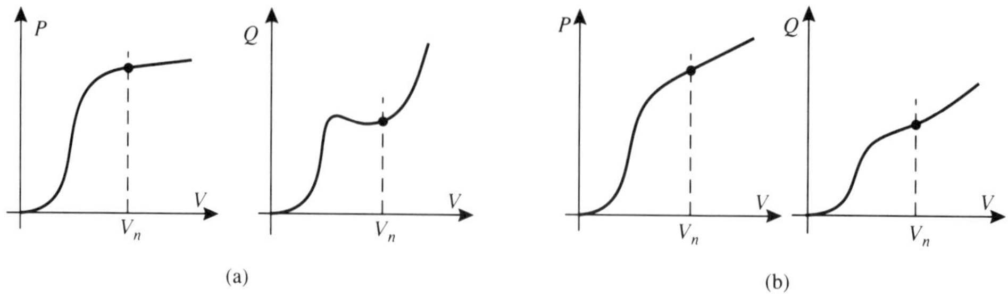

Figure 3.26 Examples of the voltage characteristics of a load: (a) dominated by large heavily-loaded induction motors; (b) dominated by lighting and heating

0.7 per unit. As the induction motors stall voltage is now below the drop-out voltage, drop-out is not preceded by an increase in the reactive power demand.

The curves shown in Figure 3.26 can only give an indication of the kind of shape a load voltage characteristic may have. They cannot be treated in a general manner as the characteristic of a particular load may be quite different. For example reactive power compensation can cause the $Q(V)$ curve to be flatter near the nominal voltage. Also relatively small, non-utility, generation embedded in the load area will significantly affect the load characteristic.

There is also a difference between the characteristic as seen from the primary and secondary side of the feeder transformer. Firstly, the real and reactive power loss in the transformer must be added to the load demand. Secondly, the feeder transformer is usually equipped with an on-load tap changer to help control the voltage in the distribution network and this also affects the characteristic as illustrated in Figure 3.27.

In Figure 3.27 the bold dashed line represents the load voltage characteristic at the nominal transformation ratio. Tap changing is controlled in discrete steps so that if the transformer tap setting is changed, the voltage characteristic moves to the left or right in discrete steps as shown by the dotted lines. The extreme left and right characteristics represent the tap-changer limits. A dead-zone is also present in the regulator in order to prevent any tap changes if the voltage variations are within limits. The resulting voltage characteristic is shown by the bold line and is quite flat within the regulation range, as can be seen by sketching an average line through the resulting characteristic.

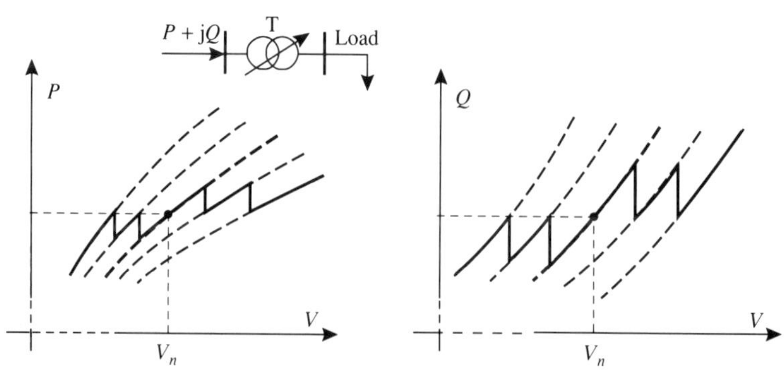

Figure 3.27 Influence of a tap-changing transformer on the voltage characteristic of a load

3.4.4 Load models

The last sub-section described how the real and reactive power of particular types of load depend on the load voltage but did not explain how these could be represented by a mathematical model. As all power system analysis programs, for example load flow or dynamic simulation, require such a load model this sub-section describes some of the most popular models currently in use.

Constant power/current/impedance

The simplest load models assume one of the following features:

- a constant power demand (P)
- a constant current demand (I)
- a constant impedance (Z)

A constant power model is voltage invariant and allows loads with a stiff voltage characteristic $k_{PV} \approx k_{QV} \approx 0$ to be represented. This model is often used in load-flow calculations, Section 3.6, but is generally unsatisfactory for other types of analysis, such as transient stability analysis, in the presence of large voltage variations. The constant current model gives a load demand that changes linearly with voltage $k_{PV} \approx 1$ and is a reasonable representation of the real power demand of a mix of resistive and motor devices. When modelling the load by a constant impedance the load power changes proportionally to the voltage squared $k_{PV} \approx k_{QV} \approx 2$ and represents some lighting loads well but does not model stiff loads at all well. To obtain a more general voltage characteristic the benefits of each of these characteristics can be combined by using the so-called *polynomial* or *ZIP model* consisting of the sum of the constant impedance (Z), constant current (I), and constant power (P) terms:

$$P = P_0\left[a_1\left(\frac{V}{V_0}\right)^2 + a_2\left(\frac{V}{V_0}\right) + a_3\right]$$

$$Q = Q_0\left[a_4\left(\frac{V}{V_0}\right)^2 + a_5\left(\frac{V}{V_0}\right) + a_6\right] \tag{3.100}$$

where V_0, P_0 and Q_0 are normally taken as the values at the initial operating conditions. The parameters of this polynomial model are the coefficients (a_1 to a_6) and the power factor of the load.

In the absence of any detailed information on the load composition, the real power is usually represented by the constant current model while the reactive power is represented by a constant impedance.

Exponential load model

In this model the power is related to the voltage by

$$P = P_0\left(\frac{V}{V_0}\right)^{n_p} \quad \text{and} \quad Q = Q_0\left(\frac{V}{V_0}\right)^{n_q} \tag{3.101}$$

where n_p and n_q are the parameters of the model. Note that by setting the parameters to 0, 1, or 2, the load can be represented by constant power, constant current, or constant impedance, respectively.

The slope of the characteristics given by equation 3.101 depends on the parameters n_p and n_q. By linearising these characteristics it can be shown that n_p and n_q are equal to the voltage sensitivities given by equation 3.90, that is $n_p = k_{PV}$ and $n_q = k_{QV}$.

Piecewise approximation

None of the models described so far will correctly model the rapid drop in load that occurs when the voltage drops below about 0.7 per unit. This can be remedied by using a two-tier representation with the exponential, or polynomial, model being used for voltages close to rated and the constant-impedance model being used at voltages below 0.3–0.7 per unit. Figure 3.28 shows an example of such an approximation which gives similar characteristics to those shown in Figure 3.26.

Frequency-dependent load model

Frequency dependence is usually represented by multiplying either a polynomial or exponential load model by a factor $[1 + a_f(f - f_0)]$ where f is the actual frequency, f_0 is the rated frequency and a_f is the model frequency sensitivity parameter. Using the exponential model this gives

$$
\begin{aligned}
P &= P(V)\left[1 + k_{Pf}(f - f_0)\right] \\
Q &= Q(V)\left[1 + k_{Qf}(f - f_0)\right]
\end{aligned}
\tag{3.102}
$$

where $P(V)$ and $Q(V)$ represent any type of the voltage characteristic and k_{Pf}, k_{Qf} are the frequency sensitivity parameters.

Test results

A number of papers have been published describing the results of field tests used to identify the parameter values for different load models; a survey of such papers can be found in Concordia & Ihara (1982) and Vaahedi *et al.* (1987). An alternative to having

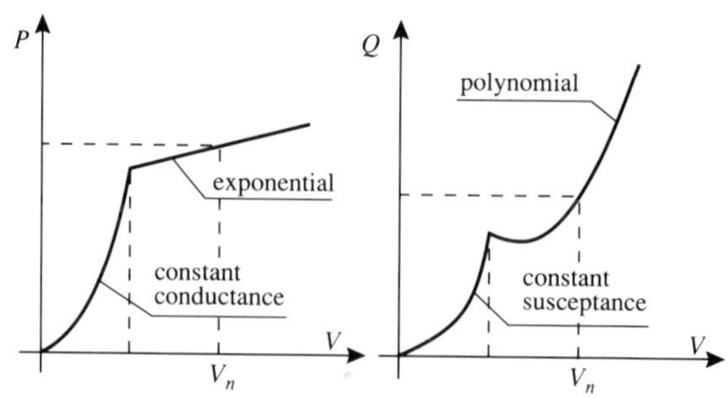

Figure 3.28 Example of a two-tier approximation of the voltage characteristics

Table 3.3 Typical load model parameters (IEEE, 1993)

Type of load	p.f.	k_{PV}	k_{QV}	k_{Pf}	k_{Qf}
Residential	0.87 to 0.99	0.9 to 1.7	2.4 to 3.1	0.7 to 1	−1.3 to −2.3
Commercial	0.85 to 0.9	0.5 to 0.8	2.4 to 2.5	1.2 to 1.7	−0.9 to −1.6
Industrial	0.8 to 0.9	0.1 to 1.8	0.6 to 2.2	−0.3 to 2.9	0.6 to 1.8

to conduct field tests is to use a *component-based approach* (IEEE, 1993) where the composite load model is constructed by aggregating individual components. With this technique the component characteristics are determined for particular classes of loads, for example residential, commercial or industrial, either by theoretical analysis or experiment. The composite load model is then constructed by aggregating the fraction of the load consumed by each particular class of load. This approach has the advantage of not requiring field measurements and of being adaptable to different systems and conditions. Table 3.3 shows some typical voltage and frequency sensitivity coefficients obtained in this way.

3.5 Network Equation

All electrical networks consist of inter-linked transmission lines and transformers, each of which can be modelled by the π-equivalent circuits described in Section 3.1 and 3.2. These individual models are combined to model the whole network by forming the *nodal network equation*:

$$\begin{bmatrix} \underline{I}_1 \\ \vdots \\ \underline{I}_i \\ \vdots \\ \underline{I}_N \end{bmatrix} = \begin{bmatrix} \underline{Y}_{11} & \cdots & \underline{Y}_{1i} & \cdots & \underline{Y}_{1N} \\ \vdots & \ddots & \vdots & & \vdots \\ \underline{Y}_{i1} & \cdots & \underline{Y}_{ii} & \cdots & \underline{Y}_{iN} \\ \vdots & & \vdots & & \vdots \\ \underline{Y}_{N1} & \cdots & \underline{Y}_{Ni} & \cdots & \underline{Y}_{NN} \end{bmatrix} \begin{bmatrix} \underline{V}_1 \\ \vdots \\ \underline{V}_i \\ \vdots \\ \underline{V}_N \end{bmatrix} \quad \text{or} \quad \underline{\mathbf{I}} = \underline{\mathbf{Y}}\underline{\mathbf{Y}} \tag{3.103}$$

The suffices i,j represent node numbers so that $\underline{V}_i$ is the voltage at node i, $\underline{I}_i$ is the *current injection* at node i and is equal to the algebraic sum of the currents in all the branches that terminate on node i, $\underline{Y}_{ij}$ is the *mutual admittance* between nodes i and j and is equal to the negative of the branch series admittance $\underline{Y}_{ij}$ that links nodes i and j, $\underline{Y}_{ii} = \Sigma_{i=1}^{N} \underline{Y}_{ij}$ is the *self-admittance* of node i and is equal to the sum of all the admittances terminating on node i (including any shunt admittance $\underline{Y}_{i0}$), and N is the number of nodes in the network.

The matrix $\underline{\mathbf{Y}}$ is called the *nodal admittance matrix*. Within this matrix, the algebraic sum of all the elements in any row i is equal to the shunt admittance Y_{i0} connecting node i to the reference node (ground), i.e. $\underline{Y}_{i0} = \Sigma_{j=1}^{N} \underline{Y}_{ij}$. The nodal admittance matrix is singular if it does not have any shunt branches. In this case the sum of the elements in all rows is zero and det $\underline{\mathbf{Y}} = 0$. If the matrix is not singular then its inverse $\underline{\mathbf{Z}} = \underline{\mathbf{Y}}^{-1}$ exists and equation 3.103 can be re-written as:

$$\begin{bmatrix} \overline{V}_1 \\ \vdots \\ \overline{V}_i \\ \vdots \\ \overline{V}_N \end{bmatrix} = \begin{bmatrix} \overline{Z}_{11} & \cdots & \overline{Z}_{1i} & \cdots & \overline{Z}_{1N} \\ \vdots & \ddots & \vdots & & \vdots \\ \overline{Z}_{i1} & \cdots & \overline{Z}_{ii} & \cdots & \overline{Z}_{iN} \\ \vdots & & \vdots & & \vdots \\ \overline{Z}_{N1} & \cdots & \overline{Z}_{Ni} & \cdots & \overline{Z}_{NN} \end{bmatrix} \begin{bmatrix} \overline{I}_1 \\ \vdots \\ \overline{I}_i \\ \vdots \\ \overline{I}_N \end{bmatrix} \quad \text{or} \quad \overline{\mathbf{V}} = \overline{\mathbf{Z}}\overline{\mathbf{I}} \tag{3.104}$$

where $\overline{\mathbf{Z}}$ is the *nodal impedance matrix.*

Any off-diagonal element $\overline{Y}_{ij}$ is non-zero only if there is a branch linking nodes i and j. If there are L branches in the system then $k = L/N$ is the ratio of the number of branches to the number of nodes and $\overline{\mathbf{Y}}$ has N^2 elements, $(2k+1)N$ of which are non-zero. The ratio of the number of non-zero elements to the total number of elements is $\alpha = (2k+1)/N$. High-voltage transmission networks are rather sparse so that typically k has a value of between 1 and 3. Thus, for example, if a network has one hundred nodes, only 3% to 7% of the elements in the admittance may be non-zero. Such a matrix is referred to as a *sparse matrix*. Power network matrices are usually sparse and, to save computer memory, only non-zero elements are stored with additional indices being used to define their position in the matrix. All matrix manipulations are performed only on the non-zero elements using *sparse matrix techniques*. The description of such techniques is beyond the scope of this book but can be found in Tewerson (1973) or Brameller *et al.* (1976). As the principal objective of the present text is to describe power system dynamics and stability from an engineering point of view the full, formal, matrix notation will be used.

For any node i, the current injection at the node can be extracted from equation 3.103 as:

$$\overline{I}_i = \overline{Y}_{ii}\overline{V}_i + \sum_{j=1;\,j\neq i}^{N} \overline{Y}_{ij}\overline{V}_j \tag{3.105}$$

where the complex voltage and admittance can be generally written as $\overline{V}_i = V_i \angle \delta_i$ and $\overline{Y}_{ij} = Y_{ij}\angle\theta_{ij}$. Using polar notation the apparent power injected at any node i is expressed as:

$$\begin{aligned} \overline{S}_i = P_i + \mathrm{j}Q_i = \overline{V}_i\overline{I}_i^* &= V_i e^{\mathrm{j}\delta_i}\left[Y_{ii}V_i e^{-\mathrm{j}(\delta_i+\theta_{ii})} + \sum_{j=1;\,j\neq i}^{N} V_j Y_{ij} e^{-\mathrm{j}(\delta_j+\theta_{ij})}\right] \\ &= V_i^2 Y_{ii} e^{-\mathrm{j}\theta_{ii}} + V_i \sum_{j=1;\,j\neq i}^{N} V_j Y_{ij} e^{\mathrm{j}(\delta_i-\delta_j-\theta_{ij})} \end{aligned} \tag{3.106}$$

Separating the real and imaginary parts gives :

$$\begin{aligned} P_i &= V_i^2 Y_{ii}\cos\theta_{ii} + \sum_{j=1;\,j\neq i}^{N} V_i V_j Y_{ij}\cos(\delta_i - \delta_j - \theta_{ij}) \\ Q_i &= -V_i^2 Y_{ii}\sin\theta_{ii} + \sum_{j=1;\,j\neq i}^{N} V_i V_j Y_{ij}\sin(\delta_i - \delta_j - \theta_{ij}) \end{aligned} \tag{3.107}$$

a,b — rectangular co-ordinates

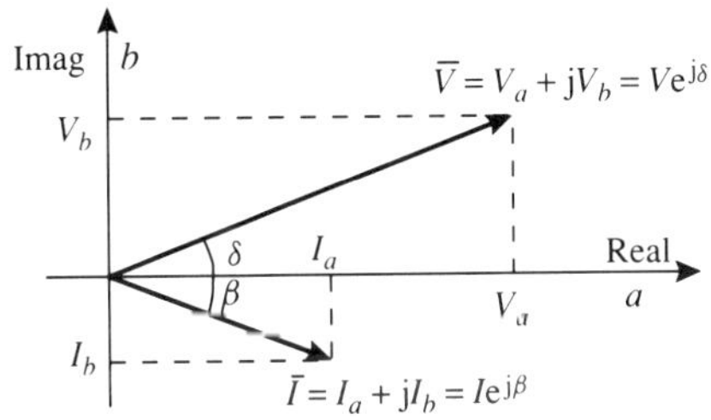

Figure 3.29 Voltage and current on the complex plane

Alternatively the rectangular co-ordinate system (a, b) shown in Figure 3.29 can be used when the apparent power injected at a node is written as

$$\overline{S}_i = P_i + jQ_i = \overline{V}_i \overline{I}_i^* = (V_{ai} + jV_{bi})(I_{ai} - jI_{bi})$$

and

$$P_i = V_{ai}I_{ai} + V_{bi}I_{bi}, \qquad Q_i = V_{bi}I_{ai} - V_{ai}I_{bi} \tag{3.108}$$

Expressing the admittance in rectangular co-ordinates as $\overline{Y}_{ij} = G_{ij} + jB_{ij}$, the current injection at each node i, given by equation 3.103, can be written as

$$\overline{I}_i = I_{ai} + jI_{bi} = \sum_{j=1}^{N} \overline{Y}_{ij}\overline{V}_j = \sum_{j=1}^{N}(G_{ij} + jB_{ij})(V_{aj} + jV_{bj}) \tag{3.109}$$

Separating real and imaginary parts gives

$$I_{ai} = \sum_{j=1}^{N}(G_{ij}V_{aj} - B_{ij}V_{bj}), \qquad I_{bi} = \sum_{j=1}^{N}(B_{ij}V_{aj} + G_{ij}V_{bj}) \tag{3.110}$$

With this notation the complex network equation 3.103 can be transformed to the real number domain to give

$$\begin{bmatrix} \mathbf{I}_1 \\ \vdots \\ \mathbf{I}_i \\ \vdots \\ \mathbf{I}_N \end{bmatrix} = \begin{bmatrix} \mathbf{Y}_{11} & \cdots & \mathbf{Y}_{1i} & \cdots & \mathbf{Y}_{1N} \\ \vdots & & \vdots & & \vdots \\ \mathbf{Y}_{i1} & \cdots & \mathbf{Y}_{ii} & \cdots & \mathbf{Y}_{iN} \\ \vdots & & \vdots & & \vdots \\ \mathbf{Y}_{N1} & \cdots & \mathbf{Y}_{Ni} & \cdots & \mathbf{Y}_{NN} \end{bmatrix} \begin{bmatrix} \mathbf{V}_1 \\ \vdots \\ \mathbf{V}_i \\ \vdots \\ \mathbf{V}_N \end{bmatrix} \quad \text{or} \quad \mathbf{I} = \mathbf{YV} \tag{3.111}$$

where all the elements are now (2 × 1) and (2 × 2) real submatrices of the form

$$\mathbf{I}_i = \begin{bmatrix} I_{ai} \\ I_{bi} \end{bmatrix}, \qquad \mathbf{V}_i = \begin{bmatrix} V_{ai} \\ V_{bi} \end{bmatrix}, \qquad \mathbf{Y}_{ij} = \begin{bmatrix} G_{ij} & -B_{ij} \\ B_{ij} & G_{ij} \end{bmatrix} \tag{3.112}$$

Quite often it is convenient to "mix" the co-ordinate systems so that the voltages are expressed in polar co-ordinates as $\overline{V}_i = V_i\angle\delta_i$ while the admittances are expressed in rectangular co-ordinates as $\overline{Y}_{ij} = G_{ij} + \text{j}B_{ij}$. Equations 3.107 then take the form

$$P_i = V_i^2 G_{ii} + \sum_{j=1;\, j\neq i}^{N} V_i V_j \left[B_{ij}\sin(\delta_i - \delta_j) + G_{ij}\cos(\delta_i - \delta_j)\right]$$
$$Q_i = -V_i^2 B_{ii} + \sum_{j=1;\, j\neq i}^{N} V_i V_j \left[G_{ij}\sin(\delta_i - \delta_j) - B_{ij}\cos(\delta_i - \delta_j)\right] \qquad (3.113)$$

Because of the mixing of complex variables in both the polar and rectangular co-ordinate systems, the above equations are called the *hybrid network equations.*

Linearisation of power network equations

The real and reactive power injection at each node is a non-linear function of the system voltages, i.e. $P = P(V, \delta)$ and $Q = Q(V, \delta)$, and it is often both convenient, and necessary, to linearise these functions in the vicinity of the operating point. This linearisation is carried out using a first order Taylor expansion and neglecting the higher-order terms. The change in the real and reactive power injection at all the system nodes can be then written, using matrix algebra and the hybrid network equations, as:

$$\begin{bmatrix} \Delta\mathbf{P} \\ \Delta\mathbf{Q} \end{bmatrix} = \begin{bmatrix} \mathbf{H} & \mathbf{M} \\ \mathbf{N} & \mathbf{K} \end{bmatrix} \begin{bmatrix} \Delta\boldsymbol{\delta} \\ \Delta\mathbf{V} \end{bmatrix} \qquad (3.114)$$

where $\Delta\mathbf{P}$ is the vector of the real power changes at all the system nodes, $\Delta\mathbf{Q}$ is the vector of reactive power changes, $\Delta\mathbf{V}$ is the vector of voltage magnitude increments and $\Delta\boldsymbol{\delta}$ is the vector of voltage angle increments. The elements of the Jacobian submatrices **H**, **M**, **N**, **K** are the partial derivatives of the functions in equation 3.113, that is:

$$H_{ij} = \frac{\partial P_i}{\partial \delta_j}, \quad M_{ij} = \frac{\partial P_i}{\partial V_j}, \quad N_{ij} = \frac{\partial Q_i}{\partial \delta_j}, \quad K_{ij} = \frac{\partial Q_i}{\partial V_j} \qquad (3.115)$$

In order to obtain simpler and more symmetrical Jacobian submatrices, equation 3.114 is often modified by multiplying the submatrices **M** and **K** by the nodal voltage magnitude. This then requires that the voltage increments $\Delta\mathbf{V}$ to be divided by the voltage magnitude and equation 3.114 takes the form

$$\begin{bmatrix} \Delta P_1 \\ \vdots \\ \Delta P_N \\ \hline \Delta Q_1 \\ \vdots \\ \Delta Q_N \end{bmatrix} = \left[\begin{array}{ccc|ccc} \dfrac{\partial P_1}{\partial \delta_1} & \cdots & \dfrac{\partial P_1}{\partial \delta_N} & V_1\dfrac{\partial P_1}{\partial V_1} & \cdots & V_N\dfrac{\partial P_1}{\partial V_N} \\ \vdots & \ddots & \vdots & \vdots & \ddots & \vdots \\ \dfrac{\partial P_N}{\partial \delta_1} & \cdots & \dfrac{\partial P_N}{\partial \delta_N} & V_1\dfrac{\partial P_N}{\partial V_1} & \cdots & V_N\dfrac{\partial P_N}{\partial V_N} \\ \hline \dfrac{\partial Q_1}{\partial \delta_1} & \cdots & \dfrac{\partial Q_1}{\partial \delta_N} & V_1\dfrac{\partial Q_1}{\partial V_1} & \cdots & V_N\dfrac{\partial Q_1}{\partial V_N} \\ \vdots & \ddots & \vdots & \vdots & \ddots & \vdots \\ \dfrac{\partial Q_N}{\partial \delta_1} & \cdots & \dfrac{\partial Q_N}{\partial \delta_N} & V_1\dfrac{\partial Q_N}{\partial V_1} & \cdots & V_N\dfrac{\partial Q_N}{\partial V_N} \end{array}\right] \begin{bmatrix} \Delta\delta_1 \\ \vdots \\ \Delta\delta_N \\ \hline \Delta V_1/V_1 \\ \vdots \\ \Delta V_N/V_N \end{bmatrix} \qquad (3.116)$$

or

$$\begin{bmatrix} \Delta\mathbf{P} \\ \Delta\mathbf{Q} \end{bmatrix} = \begin{bmatrix} \mathbf{H} & \mathbf{M}' \\ \mathbf{N} & \mathbf{K}' \end{bmatrix} \begin{bmatrix} \Delta\boldsymbol{\delta} \\ \Delta\mathbf{V}/\mathbf{V} \end{bmatrix}$$

with the elements of the Jacobian being:

$$H_{ij} = \frac{\partial P_i}{\partial \delta_j} = -V_i V_j [B_{ij}\cos(\delta_i - \delta_j) - G_{ij}\sin(\delta_i - \delta_j)] \quad \text{for} \quad i \neq j$$

$$H_{ii} = \frac{\partial P_i}{\partial \delta_i} = \sum_{\substack{j=1 \\ j\neq i}}^{N} V_i V_j [B_{ij}\cos(\delta_i - \delta_j) - G_{ij}\sin(\delta_i - \delta_j)] = -Q_i - V_i^2 B_{ii} \tag{3.117}$$

$$N_{ij} = \frac{\partial Q_i}{\partial \delta_j} = -V_i V_j [G_{ij}\cos(\delta_i - \delta_j) + B_{ij}\sin(\delta_i - \delta_j)] \quad \text{for} \quad i \neq j$$

$$N_{ii} = \frac{\partial Q_i}{\partial \delta_j} = \sum_{\substack{j=1 \\ j\neq i}}^{N} V_i V_j [G_{ij}\cos(\delta_i - \delta_j) + B_{ij}\sin(\delta_i - \delta_j)] = P_i - V_i^2 G_{ii} \tag{3.118}$$

$$M'_{ij} = V_j \frac{\partial P_i}{\partial V_j} = V_i V_j [G_{ij}\cos(\delta_i - \delta_j) + B_{ij}\sin(\delta_i - \delta_j)] \quad \text{for} \quad i \neq j$$

$$M'_{ii} = V_i \frac{\partial P_i}{\partial V_i} = 2V_i^2 G_{ij} + \sum_{\substack{j=1 \\ j\neq i}}^{N} V_i V_j \left[G_{ij}\cos(\delta_i - \delta_j) + B_{ij}\sin(\delta_i - \delta_j)\right] \tag{3.119}$$

$$= P_i + V_i^2 G_{ii}$$

$$K'_{ij} = V_j \frac{\partial Q_i}{\partial V_j} = V_i V_j [G_{ij}\sin(\delta_i - \delta_j) - B_{ij}\cos(\delta_i - \delta_j)] \quad \text{for} \quad i \neq j$$

$$K'_{ii} = V_i \frac{\partial Q_i}{\partial V_i} = -2V_i^2 B_{ij} + \sum_{\substack{j=1 \\ j\neq i}}^{N} V_i V_j \left[G_{ij}\sin(\delta_i - \delta_j) - B_{ij}\cos(\delta_i - \delta_j)\right] \tag{3.120}$$

$$= Q_i - V_i^2 B_{ii}$$

Change of reference frame

The network equations 3.103 and 3.111 are in the network complex co-ordinates (a,b) while the generator equations 3.77 are in the generator (d,q) orthogonal co-ordinates. Figure 3.30 shows the relative position of the two systems of co-ordinates. The q-axis of a given generator is shifted with respect to the network real axis by the rotor angle δ. The relationship between the two systems of co-ordinates is

$$\begin{bmatrix} E_a \\ E_b \end{bmatrix} = \begin{bmatrix} -\sin\delta & \cos\delta \\ \cos\delta & \sin\delta \end{bmatrix} \begin{bmatrix} E_d \\ E_q \end{bmatrix} \quad \text{or} \quad \mathbf{E}_{ab} = \mathbf{T}\mathbf{E}_{dq} \tag{3.121}$$

The transformation matrix $\mathbf{T}$ is unitary as $\mathbf{T}^{-1} = \mathbf{T}$. The inverse transformation is governed by the same matrix

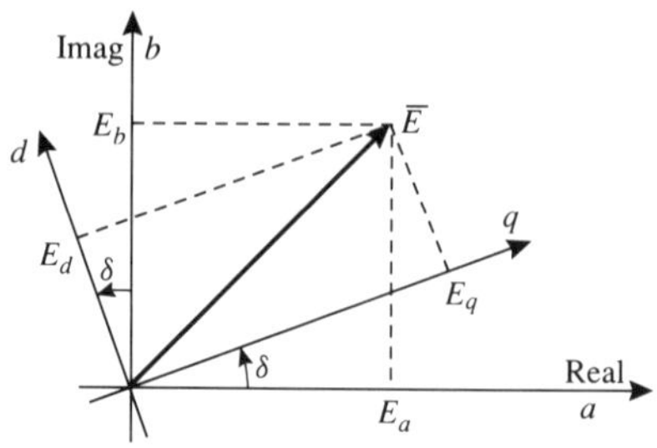

Figure 3.30 Relative position of the generator (d,q) orthogonal co-ordinates with respect to the network complex (a,b) co-ordinates

$$\begin{bmatrix} E_d \\ E_q \end{bmatrix} = \begin{bmatrix} -\sin\delta & \cos\delta \\ \cos\delta & \sin\delta \end{bmatrix} \begin{bmatrix} E_a \\ E_b \end{bmatrix} \quad \text{or} \quad \mathbf{E}_{dq} = \mathbf{T}\mathbf{E}_{ab} \tag{3.122}$$

All the other current and voltage phasors can be transformed in a similar way. Obviously, each generator may operate at a different rotor angle δ which gives as many (d,q) systems of co-ordinates as there are generators in the network.

3.6 Power Flow Problem

Power flow (or *load flow*) studies play a major role in planning the future expansion of a power system as well as in helping to run the existing system in the best possible way. Basically a power flow solution predicts what the electrical state of the network will be when it is subject to a specified loading condition. The result of the power flow is the voltage magnitude and angle at each of the system nodes. These bus voltage magnitudes and angles are defined as the system *state variables* (or *independent variables*) as they allow all the other system quantities such as the real and reactive power flows, current flows, voltage drops, power losses etc. to be computed.

The starting point for any power flow study is the set of generation and load data with the electrical network being described by the nodal admittance matrix $\overline{Y}$ of equation 3.103. Usually the generation and load data is given in terms of the *scheduled* real and reactive power generation at the generator nodes and the *predicted* real and reactive power demand at the load nodes, rather than in terms of current injections. This means that the relationship between the data inputs (real and reactive power nodal injections) and the state variables (nodal voltage magnitudes and angles) is non-linear, as indicated in equations 3.107 or 3.113.

Generator nodes have a different specification from load nodes. For the load nodes, the real and reactive power demand can be predicted and both equations 3.107 (or 3.113) can be used. However only real power generation can be specified at the generator nodes which means that only the first of the equations in 3.107 (or 3.113) is included. To understand this refer to Figure 2.2. A power station has two main controllers, the turbine governor and the automatic voltage regulator (AVR). The turbine governor allows the real power output to be specified and kept constant. On the other hand the AVR maintains the generator terminal voltage at a constant value by adjusting the generator excitation, subject to any operating limits. This means that the reactive power generation of a power station is controlled indirectly by specifying the demanded voltage at the station terminals.

Table 3.4 Node types in the power-flow problem

Node type	Number of nodes	Quantities specified	Unknown state variables	Other unknown variables
slack	usually 1	$\delta = 0, V$	—	P, Q
PV (source)	N_G	P, V	δ	Q
PQ (load)	$N - N_G - 1$	P, Q	δ, V	—
Total	N	$2N$	$2N - N_G - 2$	$N_G + 2$

The voltage is usually set high (around 1.05-1.1 p.u.) in order to support the voltages at the load nodes. As the voltage magnitude is specified there is no need for the power-flow program to calculate it, and only the voltage angle needs to be found.

The input data for all the nodes in the system is summarised in Table 3.4. For the generation nodes, referred to as *voltage-controlled* or *PV-nodes*, the required inputs are the net real power injection (scheduled generation minus predicted local demand) and the magnitude of the bus voltage, whilst the voltage angle is the unknown state variable. The unknown net reactive power injection will be determined once all the system state variables are calculated. For the load nodes, also referred to as *PQ-nodes*, the inputs are in the form of the predicted real and reactive power demand (negative injections) with the unknown state variables being the bus voltage magnitude and angle. Usually one of the generator nodes is selected as the *slack node*. At this node the voltage angle and magnitude are specified and the unknown values of the real and reactive power injection (generation or demand) determined once the state variables at all the other nodes are calculated. Any power system imbalance will then appear as the required generation from the slack generator. Removing the row and column corresponding to the slack node from the admittance matrix remedies the problem of the admittance matrix being singular for networks with no shunt branches.

The power flow problem described by equations 3.107 or 3.113 is non-linear and therefore must be solved iteratively. The first load-flow computer programs used the *Gauss-Seidel* method as this required little computer memory. Nowadays, with increased computer speed and on-chip memory, the *Newton-Raphson* method is used almost exclusively. A detailed description of these algorithms is beyond the scope of this book and can be found in a number of textbooks on power system analysis, e.g. Gross (1986), Grainger & Stevenson (1994).

4
Electromagnetic Phenomena

Chapter 1 explained how the different types of power system dynamics can be categorised according to their time scale. It also identified the fastest dynamics of interest to this book to be those associated with the electromagnetic interactions that occur within the generator immediately after a disturbance appears on the system. These dynamics lead to the generation of high currents and torques inside the generator and typically have a time scale of several milliseconds. Over this time period the inertia of the turbine and generator is sufficient to prevent any significant change in rotor speed so that the speed of the rotor can be assumed constant. Later chapters will consider the longer time scale electromechanical dynamics when the effect of rotor speed changes must be included.

To help understand how the fault currents and torques are produced this chapter will use basic physical laws to explain the electromagnetic interactions taking place within the generator. Although this approach requires some simplifications to be made it does allow a physical, qualitative, approach to be adopted. These explanations are then used to produce equations that quantify the currents and torques in a generator with a uniform air-gap. Unfortunately complications arise if the air-gap is non-uniform, for example in the salient-pole generator. However the currents, and torques, in such generators are produced by exactly the same mechanism as in a generator with a uniform air-gap only their effects are a little more difficult to quantify, the details of which are best left to the more advanced analysis techniques presented in Chapter 10. Nevertheless the way in which the current and torque expressions are modified in the salient-pole generator are discussed at the end of each section.

The chapter opens by considering the basic principles with regards flux linkage and its application to short-circuits in single-phase circuits. These principles are then used to study short-circuit effects in the synchronous generator where the interactions between the stator and rotor circuits must be included. Two types of fault are considered; firstly a three-phase short-circuit on the generator terminals and, secondly, a phase-to-phase short-circuit. Fortunately terminal three-phase short-circuits are rare but, should one occur, the generator must be able to withstand the high currents and forces produced and remain intact. From a theoretical point of view the analysis of such a fault allows a number of important terms and parameters to be introduced and quantified. A more common type of fault is the phase-to-phase fault. This type of fault introduces asymmetry into the problem and is considered later in the chapter. Finally the chapter analyses the currents and torques developed when a generator is synchronised to the grid.

4.1 Basic Principles

The currents and fluxes inside the generator after a fault can be analysed using the *law of constant flux linkages*. This law, based on the principle of *conservation of energy*, states that "the magnetic flux linking a closed winding cannot change instantaneously". As the energy stored in the magnetic field is proportional to the flux linkage this simply says that an instantaneous change in the stored energy is not possible.

The application of the law of constant flux linkages to a simple coil fed from a dc source is illustrated Figure 4.1a. At time $t = 0$ the switch disconnects the coil from the source and, at the same time, short-circuits the coil. Before the switching occurs the coil flux linkages are $\Psi_0 = N\Phi_0 = Li_0$, where L is the coil inductance, N is the number of turns and i_0 the current. The law of constant flux linkages requires the coil flux linkage just before, and just after, the short circuit to be the same, that is $\Psi(t = 0^+) = \Psi(t = 0^-)$. If the coil resistance is zero the circuit is purely inductive and the current will remain constant at a value $i(t) = i_0$, as shown by the horizontal dashed line in Figure 4.1b. Normally the coil resistance is greater than zero so that the magnetic energy stored in the circuit dissipates over a period of time and the current decays exponentially to zero with a time constant $T = L/R$. This is shown by the solid line in Figure 4.1b.

The law of constant flux linkage can now be used to analyse the way in which the generator responds to a short-circuit. A natural first step is to consider the response of the simple *RL* circuit shown in Figure 4.2a as this circuit contains a number of features that are similar to the generator equivalent circuit described in Chapter 3. In this circuit the ac driving voltage is $e = E_m \sin(\omega t + \theta_0)$ and, as the circuit is assumed to be initially open-circuit, $i(0^+) = i(0^-) = 0$. If the switch is now suddenly closed a short-circuit current $i(t)$ will flow which can be determined by solving the differential equation

$$E_m \sin(\omega t + \theta_0) = L\frac{di}{dt} + Ri \tag{4.1}$$

with the initial condition $i(0) = 0$. Solving this equation gives the current as

$$i(t) = \frac{E_m}{Z}\sin(\omega t + \theta_0 - \phi) - \frac{E_m}{Z}\sin(\theta_0 - \phi)e^{-\frac{R}{L}t} \tag{4.2}$$

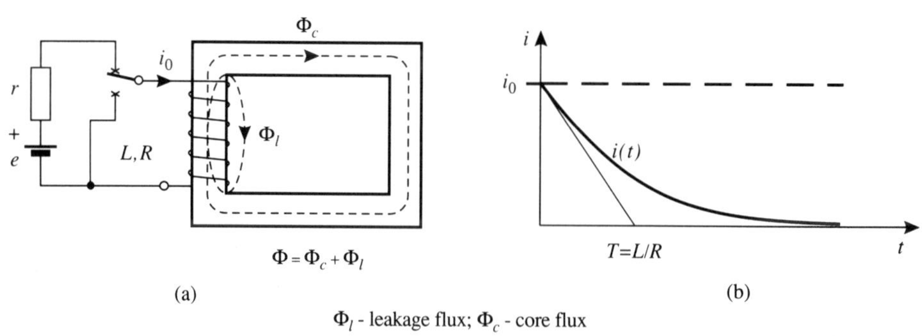

Figure 4.1 Continuity of the flux linking a coil after a change in the circuit configuration: (a) circuit diagram; (b) current as a function of time

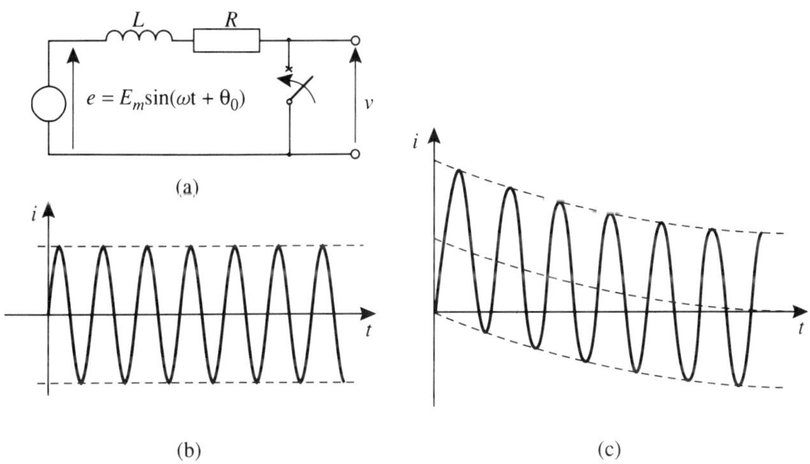

Figure 4.2 Single phase RL circuit: (a) equivalent circuit; (b) short-circuit currents with $(\theta_0 - \phi) = 0$; (c) short-circuit currents with $(\theta_0 - \phi) = -\pi/2$

where θ_0 defines the point in the ac cycle that the fault occurs, $\phi = \arctan(\omega L/R)$ is the phase angle and $Z = \sqrt{\omega^2 L^2 + R^2}$. The instant of the short-circuit is taken as the time origin.

The first term in equation 4.2 gives the forced response of the circuit. This ac current is driven by the sinusoidal emf and will be the current remaining when the circuit settles down into its final steady-state condition. The second term constitutes the natural response of the circuit and is referred to as the *dc offset*. The ac component has a constant magnitude equal to E_m/Z whilst the initial magnitude of the dc offset depends on the point in the ac cycle where the fault occurs, and decays with the time constant $T = L/R$. Figure 4.2b shows the current waveform when $(\theta_0 - \phi) = 0$ and there is no dc offset. Figure 4.2c shows the current waveform when $(\theta_0 - \phi) = -\pi/2$ and the initial dc offset is at its highest value.

The response of a synchronous generator to a short-circuit is similar to that described above in the sense that it also consists of an ac component (the forced response) and a dc component (the natural response). However equation 4.2 must be modified to account for the three-phase nature of the generator and the effect of a varying impedance in the emf source.

4.2 Three-Phase Short-Circuit on a Synchronous Generator

4.2.1 Three-phase short-circuit with the generator on no-load and winding resistance neglected

The cross-section of a basic generator is drawn in Figure 4.3 and shows the relative position of all the windings. This diagram symbolises both the round-rotor and salient-pole generator and, compared to the cross-section in Figure 3.13, shows the rotor to have one

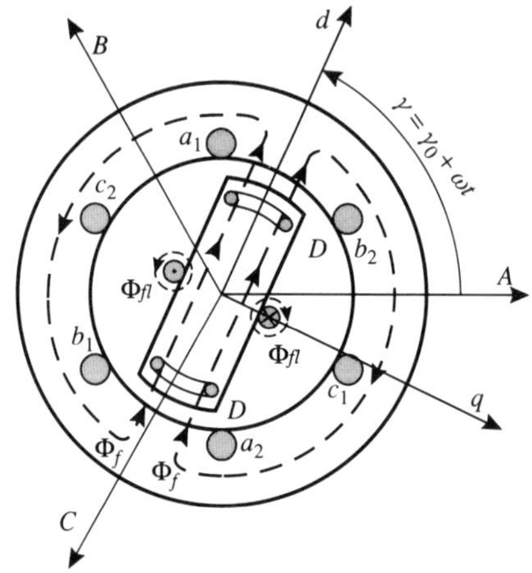

Figure 4.3 The generator and its windings

additional winding; the rotor d-axis damper winding denoted as D. At any point in time the position of the rotor is defined with reference to the axis of phase A by the angle γ. However, as a fault may occur when the rotor is at any position $\gamma = \gamma_0$, the time of the fault is taken as the time origin so that the rotor position at some time instant t after the fault is given by

$$\gamma = \gamma_0 + \omega t \tag{4.3}$$

Before the fault occurs the generator is assumed to be running at no-load with the three armature phase windings open-circuit when the only flux present in the generator is the excitation flux produced by the rotor field winding. This flux links with each phase of the armature winding producing the armature flux linkages shown in Figure 3.10a and given by equations 3.35. When the fault occurs $\gamma = \gamma_0$ and the net flux linking each of the armature phase windings is

$$\Psi_{fA0} = \Psi_{fa} \cos \gamma_0, \; \Psi_{fB0} = \Psi_{fa} \cos(\gamma_0 - 2\pi/3), \quad \Psi_{fC0} = \Psi_{fa} \cos(\gamma_0 - 4\pi/3) \tag{4.4}$$

where $\Psi_{fa} = N_\phi \Phi_{f0}$ is the amplitude of the excitation flux linkage of an armature phase winding before the fault occurs, Φ_{f0} is the pre-fault excitation flux per pole and $N_\phi = k_w N$. In this last equation N is the number of turns on each of the armature phase windings and k_w is the armature winding factor. An example of these phase flux linkages is shown by the horizontal dashed lines in Figure 4.4 which is drawn assuming that the fault occurs at $\gamma_0 = 0$.

If the resistance of the armature winding is neglected then, according to the law of constant flux linkage, the value of the flux linking each phase must remain constant after the fault at the value Ψ_{fA0}, Ψ_{fB0}, and Ψ_{fC0} defined by the equations in 4.4. After the fault the rotor continues to rotate and the linkages Ψ_{fA}, Ψ_{fB}, and Ψ_{fC} continue to change sinusoidally as shown by the solid lines in Figure 4.4 and have an equivalent effect to the emf in Figure 4.2a. To keep the total flux linkages of each of the phase windings constant additional currents i_A, i_B, and i_C must be induced in the short-circuited phase windings to produce the flux linkages Ψ_{AA}, Ψ_{BB}, and Ψ_{CC} shown by the sinusoidal dashed lines in

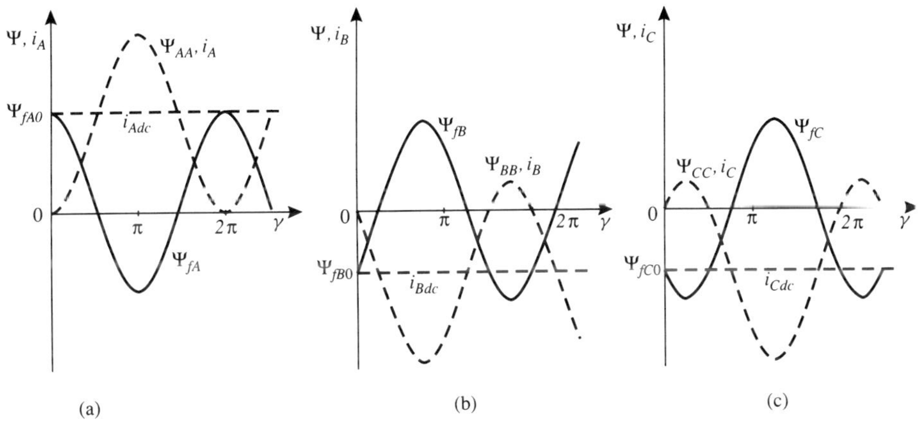

Figure 4.4 Application of the law of constant flux linkages to determine the fault currents due to a three-phase short-circuit occurring at the instant $\gamma_0 = 0$: (a) phase A; (b) phase B; (c) phase C

Figure 4.4. The total flux linkage of each phase winding can now be obtained by adding the two components of flux linkage together to give

$$\begin{aligned} \Psi_A(t) &= \Psi_{AA} + \Psi_{fA} = \Psi_{fA0} = \text{constant} \\ \Psi_B(t) &= \Psi_{BB} + \Psi_{fB} = \Psi_{fB0} = \text{constant} \\ \Psi_C(t) &= \Psi_{CC} + \Psi_{fC} = \Psi_{fC0} = \text{constant} \end{aligned} \tag{4.5}$$

which on rearranging gives the flux linkages Ψ_{AA}, Ψ_{BB}, and Ψ_{CC} as

$$\begin{aligned} \Psi_{AA} &= \Psi_{fA0} - \Psi_{fA} \\ \Psi_{BB} &= \Psi_{fB0} - \Psi_{fB} \\ \Psi_{CC} &= \Psi_{fC0} - \Psi_{fC} \end{aligned} \tag{4.6}$$

As flux linkage is equal to the product of inductance and current these flux linkages can now be used to obtain expressions for the phase current provided that the inductance value is known.† In the case of a generator with a uniform air-gap the equivalent inductance of each of the phase-windings L_{eq} is the same and does not depend on rotor position (see Section 3.3). In this case the short-circuit phase currents $i_A = \Psi_{AA}/L_{eq}$, $i_B = \Psi_{BB}/L_{eq}$, and $i_C = \Psi_{CC}/L_{eq}$ have the same shape as the flux linkages Ψ_{AA}, Ψ_{BB} and Ψ_{CC} and consist of both an ac and a dc component. The ac component currents can be expressed as

$$\begin{aligned} i_{A\,ac} &= -i_m(t)\cos(\omega t + \gamma_0) \\ i_{B\,ac} &= -i_m(t)\cos(\omega t + \gamma_0 - 2\pi/3) \\ i_{C\,ac} &= -i_m(t)\cos(\omega t + \gamma_0 - 4\pi/3) \end{aligned} \tag{4.7}$$

† Unfortunately the inductance value required is not simply the winding self-inductance but an equivalent inductance that takes into account the way the winding is coupled with all the other windings in the generator. The importance of this coupling will be examined in Sections 4.2.3 and 4.2.4 when actual values for L_{eq} are discussed.

and are proportional to the flux linkages Ψ_{fA}, Ψ_{fB}, and Ψ_{fC} whilst the dc component currents can be expressed as:

$$\begin{aligned} i_{A\,dc} &= i_m(0)\cos\gamma_0 \\ i_{B\,dc} &= i_m(0)\cos(\gamma_0 - 2\pi/3) \\ i_{C\,dc} &= i_m(0)\cos(\gamma_0 - 4\pi/3) \end{aligned} \tag{4.8}$$

and are proportional to Ψ_{fA0}, Ψ_{fB0}, and Ψ_{fC0}. In order to generalise these two sets of equations the maximum value of the current in equation 4.7 has been defined as $i_m(t)$ whilst in equation 4.8 it is $i_m(0)$, the value of $i_m(t)$ at the fault instant. When the winding resistance is neglected $i_m(t)$ is constant and equal to its initial value $i_m(0)$. The net current is obtained by adding the component currents to give

$$\begin{aligned} i_A &= -i_m(t)\cos(\omega t + \gamma_0) + i_m(0)\cos\gamma_0 \\ i_B &= -i_m(t)\cos(\omega t + \gamma_0 - 2\pi/3) + i_m(0)\cos(\gamma_0 - 2\pi/3) \\ i_C &= -i_m(t)\cos(\omega t + \gamma_0 - 4\pi/3) + i_m(0)\cos(\gamma_0 - 4\pi/3) \end{aligned} \tag{4.9}$$

where the dc offset depends on γ_0, the instant in the ac cycle the fault occurs, and is different for each phase of the generator.

The combined effect of the ac and the dc components of the three armature phase currents is to produce an armature reaction mmf that drives an armature reaction flux across the air-gap to link with the rotor windings and induce currents in them. Consider first the effect of the ac phase currents $i_{A\,ac}$, $i_{B\,ac}$, and $i_{C\,ac}$ as shown in the upper row of diagrams in Figure 4.5. These ac phase currents, Figure 4.5a, produce an ac armature reaction mmf $F_{a\,ac}$ that behaves in a similar way to the steady-state armature reaction mmf F_a discussed in Section 3.3. Recall that in the steady-state F_a rotates at the same speed as the rotor (and the excitation mmf F_f), demagnetises the machine and produces a torque proportional to the sine of the angle between F_a and F_f. During the short-circuit the developed electrical torque and power are both zero so that the angle between $F_{a\,ac}$ and F_f is 180°, $F_{a\,ac}$ directly opposes F_f, and the flux $\Phi_{a\,ac}$, produced by $F_{a\,ac}$, takes the path shown in Figure 4.5b. Consequently the rotor flux linkages $\Psi_{a\,ac\,r}$ produced by $\Phi_{a\,ac}$ are constant and negative (with respect to the excitation linkages) as shown by the solid line in Figure 4.5c. Here the subscript "r" has been added to emphasise that $\Psi_{a\,ac\,r}$ is the rotor flux linkage produced by the armature flux $\Phi_{a\,ac}$.

The lower row of diagrams in Figure 4.5 shows the combined effect of the dc phase currents $i_{A\,dc}$, $i_{B\,dc}$, and $i_{C\,dc}$. These currents, Figure 4.5a, produce a stationary dc mmf $F_{a\,dc}$ which drives the stationary armature flux $\Phi_{a\,dc}$ shown in Figure 4.5b. The space direction of $F_{a\,dc}$ can be obtained by adding the component mmfs due to i_{Adc}, i_{Bdc} and i_{Cdc} in a similar way as in equation 3.40 when the resulting mmf $F_{a\,dc}$ is found to be always directed at an angle γ_0 with respect to the A-axis. As the rotor is also at angle γ_0 at the instant of fault, $F_{a\,dc}$ is always aligned initially with the rotor d-axis and then counter-rotates with respect to it. This means that the rotor flux linkage $\Psi_{a\,dc\,r}$, produced by $\Phi_{a\,dc}$, is initially positive (magnetises the machine) and then, as the rotor rotates, $\Psi_{a\,dc\,r}$ changes cosinusoidally as shown in Figure 4.5c.

As the rotor field and damper windings are closed, their total flux linkage must remain unchanged immediately after the fault. Consequently additional currents must flow in

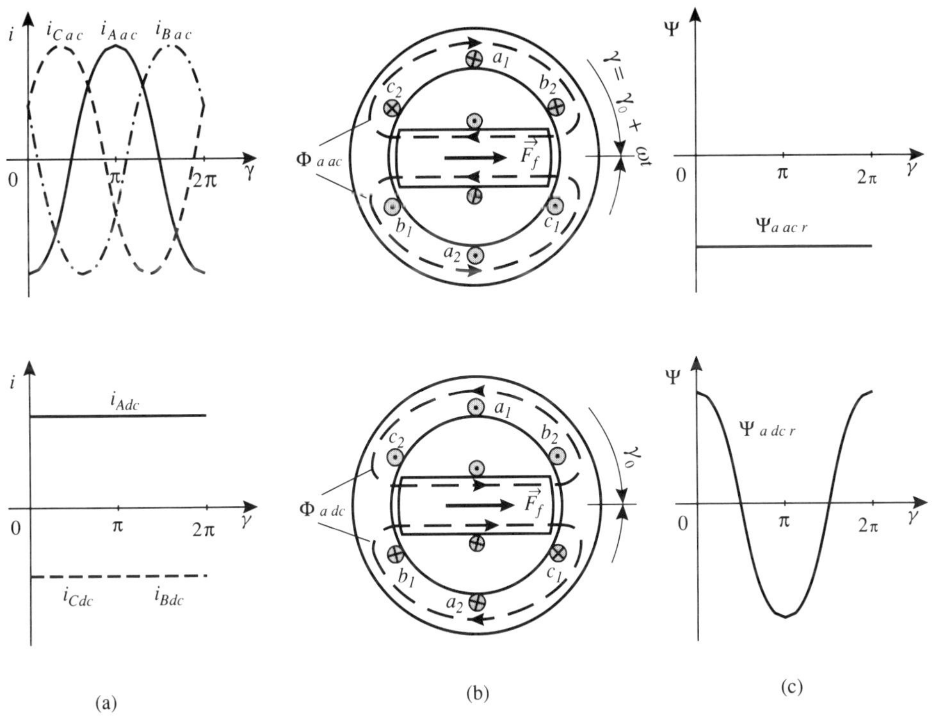

Figure 4.5 Effect of the ac component (upper row) and the dc component (lower row) of the armature currents: (a) ac and dc component of the currents; (b) path of the armature reaction flux (rotating for the ac component, stationary for the dc component); (c) linkages with the rotor windings. Position shown at the instant of fault $\gamma_0 = 0$

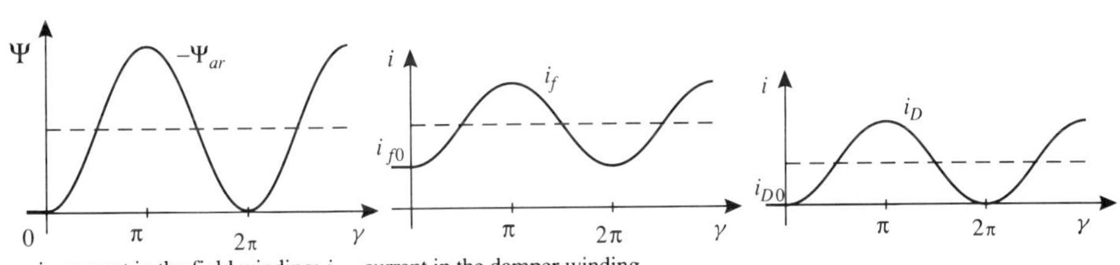

Figure 4.6 Application of the law of constant flux linkage to determine the currents flowing in the rotor windings

these windings to compensate the armature flux $\Psi_{ar} = \Psi_{a\,ac\,r} + \Psi_{a\,dc\,r}$ that links with the rotor. Figure 4.6 shows the rotor linkages equal to $-\Psi_{ar}$ necessary for this, and the field and damper currents which must flow to set up such linkages. Both currents contain dc and ac components.

To summarise, a three-phase fault causes the flow of short-circuit armature currents which have both an ac and a dc component. At the instant of fault both the ac and dc armature mmfs are directed along the rotor d-axis but then, as the rotor rotates, the ac armature mmf rotates with it inducing additional dc currents in the rotor, whilst the dc armature mmf is stationary inducing additional ac currents in the rotor. *There is always a complementary pair of stator to rotor currents of the form: ac → dc and dc → ac*. The value of the dc offset may be different for each stator phase and depends on the instant of fault.

4.2.2 Including the effect of winding resistance

A winding dissipates energy in its resistance at a rate proportional to the current squared. Consequently the stored magnetic energy decays with time and the induced dc currents maintaining the flux linkages decay exponentially to zero at a rate determined by the circuit time constant $T = L/R$.

The dc component of the armature phase current decays with the *armature time constant* T_a determined by the equivalent inductance and resistance of the phase winding. To account for this decay the dc currents in equations 4.8 must be modified to

$$\begin{aligned} i_{A\,dc} &= i_m(0)e^{-t/T_a}\cos\gamma_0 \\ i_{B\,dc} &= i_m(0)e^{-t/T_a}\cos(\gamma_0 - 2\pi/3) \\ i_{C\,dc} &= i_m(0)e^{-t/T_a}\cos(\gamma_0 - 4\pi/3) \end{aligned} \tag{4.10}$$

when the total phase currents become

$$\begin{aligned} i_A &= -i_m(t)\cos(\omega t + \gamma_0) + i_m(0)e^{-t/T_a}\cos\gamma_0 \\ i_B &= -i_m(t)\cos(\omega t + \gamma_0 - 2\pi/3) + i_m(0)e^{-t/T_a}\cos(\gamma_0 - 2\pi/3) \\ i_C &= -i_m(t)\cos(\omega t + \gamma_0 - 4\pi/3) + i_m(0)e^{-t/T_a}\cos(\gamma_0 - 4\pi/3) \end{aligned} \tag{4.11}$$

These equations are identical in form to that obtained for the single phase circuit in equation 4.2 with $\gamma_0 = \theta_0$ and a phase angle $\phi = \pi/2$. However in equation 4.2 $i_m(t)$ is constant whereas for the synchronous generator it varies with time for reasons that will now be explained.

The currents $i_{A\,ac}$, $i_{B\,ac}$, and $i_{C\,ac}$ induce dc currents in the rotor windings. These dc currents decay with a time constant determined by the particular rotor circuit in which they flow. Consequently the dc component of the damper winding current decays with the damper winding time constant, called the *d-axis subtransient short-circuit time constant* T''_d, while the dc component of the field current decays with the field winding time constant, called the *d-axis transient short-circuit time constant* T'_d. Normally the resistance of the damper winding is much higher than the resistance of the field winding so that $T''_d \ll T'_d$ and the dc component of the damper winding current decays much faster than the dc component of the field current. As the dc rotor currents induce ac stator currents, the magnitude of the ac component of the stator current $i_m(t)$ will decay to its steady-state value with these two time constants. Conversely as the dc phase currents i_{Adc}, i_{Bdc} and i_{Cdc} induce ac currents in the field and the damper windings, these ac rotor currents will also decay with the time constant T_a.

Table 4.1 summarises the relationship between the relevant components of the stator and the rotor currents and their decay time constants.

Table 4.1 Pairs of coupled stator-rotor currents

Three-phase stator winding	Rotor windings (field and damper)	Means of energy dissipation	Time constant
dc	ac	resistance of the armature winding	T_a
ac	dc	resistance of the damper winding resistance of the field winding	T''_d T'_d

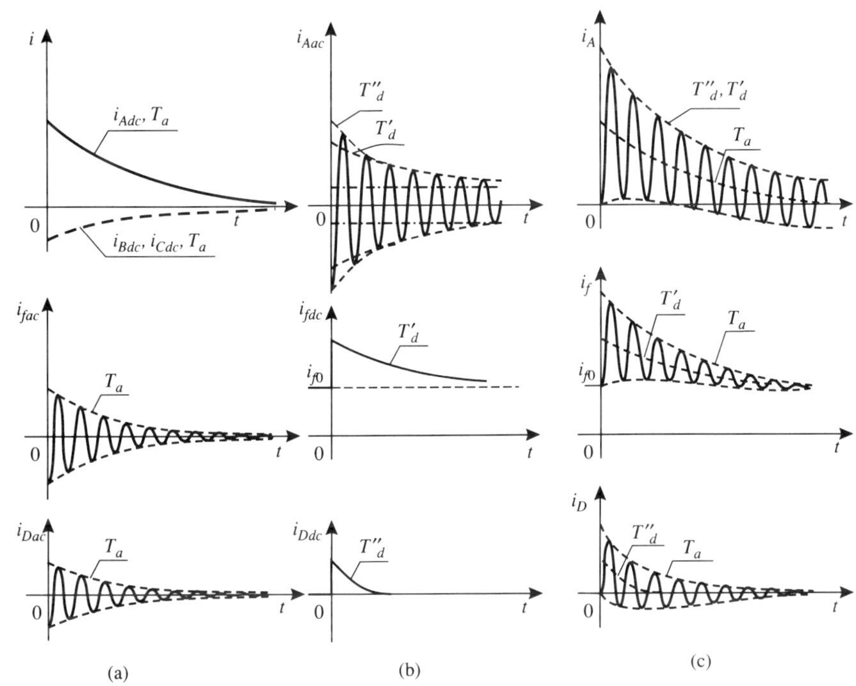

Figure 4.7 Short-circuit currents in the generator: (a) dc component of the phase current and the corresponding ac component of the field and damper winding current; (b) ac component of the current in phase A and the corresponding dc component of the field and damper winding current; (c) the resulting current in phase *A*, the field and the damper winding as the sum of the currents shown in (a) and (b)

Figure 4.7 shows a typical time variation of the generator currents resulting from a three-phase terminal short-circuit. The diagrams constituting Figure 4.7 have been organised so that the first row shows the currents in phase *A* of the stator armature winding, the second row the field winding currents and the third row the damper winding currents. Only phase *A* of the armature winding is shown as the currents in phases *B* and *C* are similar but shifted by $\pm 2\pi/3$. The first column in Figure 4.7 shows the dc component of the armature current and the corresponding induced rotor currents, the second column shows the ac component of the armature current and the corresponding induced rotor currents whilst the third column shows the resultant currents in the armature, the field and the damper winding made up from their ac and dc components. The dashed lines indicate the exponential envelopes that correspond to the different time constants.

Note that the ac component of the armature current does not decay completely to zero but to a steady-state value. The field current decays to its pre-fault value i_{f0} whilst the damper winding current decays completely to zero, its pre-fault value.

If $T_a > T_d''$ the fault current in phase A for $\gamma_0 = 0$ is positive for the first few cycles and does not cross zero until the dc offset decays. This is typical for large generators. If $T_d' > T_a$ the field current i_f will always oscillate above its steady-state value i_{f0}. If $T_d'' < T_a$, the damper current i_D may go negative.

It now remains to determine $i_m(t)$. To do this it is necessary to consider the path taken by the ac armature flux as this will determine the equivalent inductance L_{eq}.

4.2.3 Armature flux paths and the equivalent reactances

Section 3.3 explained how the effect of the steady-state ac armature flux on the generator performance could be accounted for in the equivalent circuit model by the voltage drop across the synchronous reactances X_d and X_q. A similar approach can be adopted during the fault period but now the value of the armature reactance will be different as the additional currents induced in the rotor windings force the armature flux to take a different path from that in the steady-state. As the additional rotor currents prevent the armature flux from entering the rotor windings they have the effect of *screening the rotor* from these changes in armature flux.

Figure 4.8 shows three characteristic states that correspond to three different stages of rotor screening. Immediately after the fault the current induced in both the rotor field and damper windings force the armature reaction flux completely out of the rotor to keep the rotor flux linkages constant, Figure 4.8a, and the generator is said to be in the *subtransient-state*. As energy is dissipated in the resistance of the rotor windings the currents maintaining constant rotor flux linkages decay with time allowing flux to enter the windings. As the rotor damper winding resistance is the largest the damper current is the first to decay allowing the armature flux to enter the rotor pole face. However it is still forced out of the field winding itself, Figure 4.8b, and the generator is said to be in the *transient-state*. The field current then decays with time to its steady-state value allowing

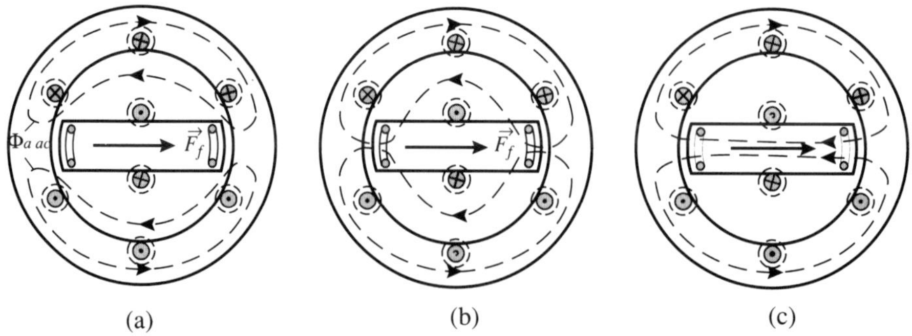

Figure 4.8 The path of the armature flux in: (a) the subtransient-state (screening effect of the damper winding and the field winding); (b) the transient-state (screening effect of the field winding only); (c) steady-state. In all three cases the rotor is shown to be in the same position but the actual rotor position corresponding to the three states will be separated by a number of rotations

the armature reaction flux to eventually enter the whole rotor and assume the minimum reluctance path. This *steady-state* is illustrated in Figure 4.8c and corresponds to the flux path shown in the top diagram of Figure 4.5b.

It is convenient to analyse the generator dynamics separately when it is in the subtransient, transient and steady-state. This is accomplished by assigning a different equivalent circuit to the generator when it is in each of the above states but in order to do this it is first necessary to consider the generator reactances in each of the characteristic states.

The inductance of a winding is defined as the ratio of the flux linkages to the current producing the flux. Thus, a low-reluctance path results in a large flux and a large inductance (or reactance) and vice versa. Normally a flux path will consist of a number of parts each of which has a different reluctance. In such circumstances it is convenient to assign a reactance to each part of the flux path when the equivalent reactance is then made up from the individual path reactances. In combining the individual reactances it must be remembered that parallel flux paths correspond to a series connection of reactances whilst series paths correspond to a parallel connection of reactances. This is illustrated in Figure 4.9 for a simple iron-cored coil with an air-gap in the iron circuit. Here the total coil flux Φ consists of the leakage flux Φ_l and the core flux Φ_c. The reactance of the leakage flux path is X_l, whilst the reactance of the core flux path has two components. The first component corresponds to the flux path across the air-gap X_{ag} and the second component corresponds to the flux path through the iron core X_{Fe}. In the equivalent circuit X_l is connected in series with the parallel combination X_{ag} and X_{Fe}. As the reluctance of a flux path in iron is very small compared to that in air $X_{Fe} \gg X_{ag}$ and the total reactance of the winding is dominated by those parts of the path that are in air; the air-gap and the flux leakage path. Consequently $X \approx X_{ag} + X_l$.

These principles are applied to the synchronous generator in Figure 4.10 for each of the three characteristic states. It also introduces a number of different reactances, each of which corresponds to a particular flux path:

- X_l corresponds to the path the armature leakage flux takes around the stator windings and is referred to as the *armature leakage reactance*
- X_a corresponds to the flux path across the air-gap and is referred to as the *armature reaction reactance*
- X_D corresponds to the flux path around the damper winding
- X_f corresponds to the flux path around the field winding

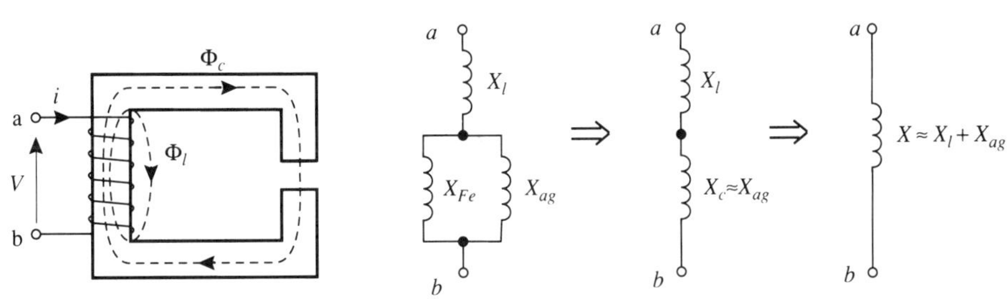

Figure 4.9 A coil with an air-gap and its equivalent circuit diagram

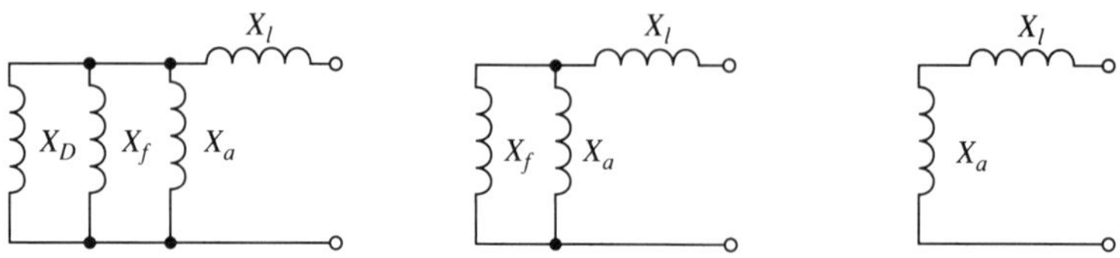

$$X''_d = X_l + \frac{1}{\frac{1}{X_a}+\frac{1}{X_D}+\frac{1}{X_f}} \qquad X'_d = X_l + \frac{1}{\frac{1}{X_a}+\frac{1}{X_f}} \qquad X_d = X_l + X_a$$

(a) (b) (c)

Figure 4.10 Equivalent reactance of a synchronous generator in: (a) subtransient-state; (b) transient-state; (c) steady-state

The reactance of the damper winding and the field winding also correspond to the flux path around the winding. Thus X_D and X_f are proportional to the actual damper and field winding reactance respectively.

In Figure 4.10 the armature reactances in each of the characteristic states are combined to give the following equivalent reactances:

X''_d — *direct-axis subtransient reactance*

X'_d — *direct-axis transient reactance*

X_d — *direct-axis synchronous reactance*

In the subtransient-state the flux path is almost entirely in air and so the reluctance of this path is very high. In contrast the flux path in the steady-state is mainly through iron with the reluctance being dominated by the length of the air-gap. Consequently the reactances, in increasing magnitude, are $X''_d < X'_d < X_d$. In the case of large synchronous generators X'_d is about twice as large as X''_d, while X_d is about ten times as large as X''_d.

Quadrature-axis reactances

Figure 4.8 shows the path taken by the ac armature flux in the three characteristic states and is drawn assuming that the armature mmf is directed along the d-axis. This is true when the generator is on no-load prior to the three-phase short-circuit but generally, for other types of disturbances or for a generator operating on-load prior to a disturbance, the armature mmf will have both a d and a q component. In this more general case it is necessary to analyse the influence of the two armature mmf components separately using the two-reaction method introduced in Section 3.3.

If a generator is in the subtransient-state, and the armature mmf is directed along the rotor d-axis, then the armature reaction flux will be forced out of the rotor by the currents induced in the field winding, the damper winding and the rotor core. This flux path corresponds to the direct axis subtransient reactance X''_d. On the other hand if the armature mmf is directed along the rotor q-axis, then the only currents forcing the armature reaction flux out of the rotor are the rotor core eddy-currents and the currents in the q-axis damper winding. If a generator only has a d-axis damper winding, then the q-axis screening effect is much weaker than that for the d-axis, and the corresponding *quadrature axis subtransient reactance* X''_q is greater than X''_d. This difference between X''_q and X''_d is called *subtransient saliency*. For a generator with a damper winding in both the d-axis and the q-axis the screening effect in both axes is similar, subtransient saliency is negligible and $X''_q \approx X''_d$.

When the generator is in the transient-state, screening is provided by the field winding which is only in the d-axis. However in the round-rotor generator some q-axis screening will be produced by eddy currents in the rotor iron with the effect that $X'_q > X'_d$. The actual value of X'_q is somewhere between X'_d and X_q with typically $X'_q \approx 2X'_d$. In the salient-pole generator the laminated rotor construction prevents eddy-currents flowing in the rotor body, there is no screening in the q-axis and $X'_q = X_q$. Because of the absence of a field winding in the q-axis there is some degree of *transient saliency* in all types of generators.

Table 4.2 summarises the different types of saliency and the reasons for it, whilst typical values for all the generator parameters are tabulated in Table 4.3.

In Table 4.3 the quadrature-axis subtransient short-circuit time constant T''_q and the transient short-circuit time constant T'_q are defined in the same way as for the direct-axis time constants T''_d, T'_d in Table 4.1. For the round-rotor generator, as there is no field winding in the q-axis, it is assumed that rotor body eddy-currents contribute to the q-axis parameters.

Values of the short-circuit time constants T'_d, T'_q, T''_d, T''_q correspond to the armature winding short-circuited. Some manufacturers may quote the time constants for the armature on open-circuit as T'_{do}, T'_{qo}, T''_{do}, T''_{qo}. Approximate relationships between the

Table 4.2 Saliency in the three characteristic states of the synchronous generator

State	Generator type	Reactance	Saliency	
			Yes/No	Reason
Subtransient	Any type but with a d-axis damper only	$X''_q > X''_d$	Yes	Weaker screening in the q-axis because of the lack of damper winding
	Any type but with both d and q-axis damper windings	$X''_q \approx X''_d$	No	Similar screening in both axes
Transient	Round-rotor	$X'_q > X'_d$	Yes	Strong screening in the d-axis due to the field winding but weak screening in the q-axis due to the rotor body currents
	Salient-pole	$X'_q = X_q$	Yes	No screening in the q-axis because of the laminated rotor core
Steady-state	Round-rotor	$X_q \approx X_d$	No	Symmetrical air-gap in both axes
	Salient-pole	$X_q < X_d$	Yes	Larger air-gap on the q-axis

Table 4.3 Typical parameter values for large generators. Reactances are in per unit to the rated MVA and time constants are in seconds

Parameter	Round-rotor			Salient-pole rotor	
	200 MVA	600 MVA	1500 MVA	150 MVA	230 MVA
X_d	1.65	2.00	2.20	0.91	0.93
X_q	1.59	1.85	2.10	0.66	0.69
X'_d	0.23	0.39	0.44	0.3	0.3
X'_q	0.38	0.52	0.64	—	—
X''_d	0.17	0.28	0.28	0.24	0.25
X''_q	0.17	0.32	0.32	0.27	0.27
T'_d	0.83	0.85	1.21	1.10	3.30
T'_q	0.42	0.58	0.47	—	—
T''_d	0.023	0.028	0.030	0.05	0.02
T''_q	0.023	0.058	0.049	0.06	0.02

open- and short-circuit time constants may be found from the equivalent circuits shown in Figure 4.10. If X_l is neglected then

$$X'_d \cong \frac{X_a X_f}{X_a + X_f}, \qquad \frac{1}{X'_d} \cong \frac{1}{X_a} + \frac{1}{X_f} \tag{4.12}$$

and

$$X''_d \cong \frac{1}{\dfrac{1}{X_D} + \dfrac{1}{X'_d}}, \qquad X_D \cong \frac{X'_d X''_d}{X'_d - X''_d} \tag{4.13}$$

In order to determine the time constants T''_d, T''_{d0} it is necessary to insert the resistance R_D into the branch with X_D in Figure 4.10a. This is illustrated in Figure 4.11. If X_l is neglected then the short-circuit in Figure 4.11a by-passes all the shunt branches and the time constant is

$$T''_d = \frac{X_D}{\omega R_D} \tag{4.14}$$

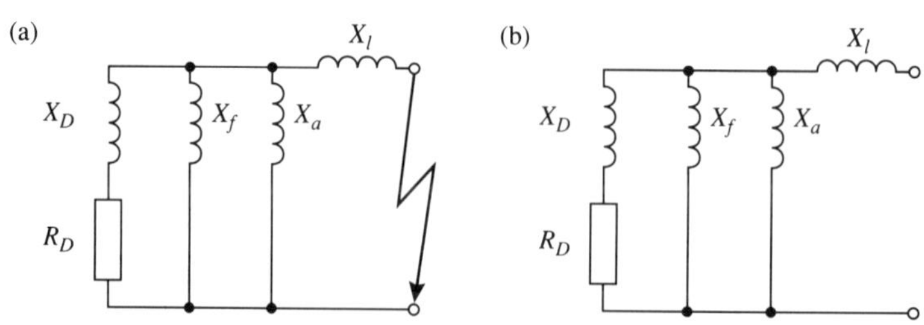

Figure 4.11 Equivalent circuits for determining the subtransient time constants: (a) short-circuit; (b) open-circuit

When the circuit is open-circuit as in Figure 4.11b the time constant is

$$T''_{d0} = \frac{X_D + \dfrac{X_a X_f}{X_a + X_f}}{\omega R_D} \approx \frac{X_D + X'_d}{\omega R_D} \tag{4.15}$$

Dividing equation 4.14 by equation 4.15, and substituting for X_D from equation 4.13, gives $T''_d/T''_{d0} \approx X''_d/X'_d$. Following a similar procedure for the remaining time constants in the d and q-axes gives the relationships:

$$T''_d \cong T''_{d0}\frac{X''_d}{X'_d}, \quad T''_q \cong T''_{q0}\frac{X''_q}{X'_q}, \quad T'_d \cong T'_{d0}\frac{X'_d}{X_d}, \quad T'_q \cong T'_{q0}\frac{X'_q}{X_q} \tag{4.16}$$

The open-circuit time constants are larger than the short-circuit ones.

4.2.4 Generator electromotive forces and equivalent circuits

Equivalent circuits were first used in Section 3.3 to analyse the steady-state behaviour of the generator and in this sub-section their use is extended to cover all three characteristic states. As these equivalent circuits are valid only for ac circuit analysis purposes they can only be used to analyse the ac component of the armature current and the dc component must be neglected. What follows is a physical justification for using such equivalent circuits to model the synchronous generator with a full mathematical derivation being left until Chapter 10.

In the steady-state the generator on-load can be modelled by the equivalent circuit shown in Figure 3.17 where the generator is represented in both the d and the q-axis by constant internal emfs E_q and E_d acting behind the synchronous reactances X_d and X_q. The emf E_q is proportional to the excitation flux and, as the excitation flux is proportional to the field current, is also given the symbol E_f. As there is no excitation in the q-axis the corresponding emf E_d is zero. Consequently the total internal emf is $\overline{E} = \overline{E}_q + \overline{E}_d = \overline{E}_q = \overline{E}_f$.

A similar representation can be used when the generator is in the subtransient and transient state. However as each of the three generator characteristic states is characterised by a different pair of reactances there are now three different equivalent circuits, each valid at the beginning of a corresponding state, in which the generator is represented by a pair of constant internal emfs behind the appropriate reactances. It is important to realise that in each of the states the internal emfs will be different and correspond to that part of the rotor flux linkages that are assumed to remain constant in that particular characteristic state.

In the steady-state the three emfs E, E_f and E_q are all equal but during a disturbance they have a different interpretation and it is important to differentiate between them. E_q is proportional to the field current (which produces a d-axis mmf) and will vary proportionally to the changes in the field current. During a disturbance there are also currents induced in the q-axis rotor body so that E_d will no longer be zero but will vary proportionally to these rotor body currents. Consequently neither E_q or E_d are constant during either the subtransient or transient state.

The emf E_f is defined as being proportional to the voltage V_f applied to the field winding and it is only in the steady-state when the field current $i_f = i_{f0} = V_f/R_f$ that $E_f = E_q$ (where R_f is the field winding resistance). In all other characteristic states $i_f \neq V_f/R_f$ and $E_f \neq E_q$. The importance of E_f is that it reflects the effect of 'excitation control'.

Subtransient-state

During the subtransient period the armature flux is forced into high reluctance paths outside the rotor circuits by currents induced in the field and damper winding. This is shown in Figure 4.8a for the flux acting along the d-axis only but in general the armature flux will have both d and q components. In this more general case the flux path is distorted not only by the d-axis field and damper currents but also by the q-axis damper current (if there is a q-axis damper winding) and q-axis rotor body currents. The reactances associated with the flux path are X_d'' and X_q''. As the rotor flux linkages in both axes are assumed to remain constant during the subtransient state, the internal emf corresponding to these linkages may also be assumed to remain constant and the generator may be represented by a constant *subtransient internal emf* $\overline{E}'' = \overline{E}_q'' + \overline{E}_d''$ acting behind the subtransient reactances X_d'' and X_q''. The generator circuit equation is then

$$\overline{E}'' = \overline{V}_g + R\overline{I} + \mathrm{j}\overline{I}_d X_d'' + \mathrm{j}\overline{I}_q X_q'' \tag{4.17}$$

where $\overline{I} = \overline{I}_d + \overline{I}_q$ is the subtransient ac armature current immediately after the fault. Its value can be calculated once $\overline{E}''$ is known. The emf E_q'' is proportional to the rotor flux linkages along the d-axis, that is the sum of the field winding and the d-axis damper winding flux linkages. Similarly E_d'' is proportional to q-axis rotor flux linkages, that is the sum of the q-axis rotor body and the q-axis damper winding flux linkages.

It is important to understand the difference between the steady-state and subtransient emfs. Consider first E_q and E_q''. E_q is proportional to the field current and therefore to the field winding self-flux linkages. In order to screen the field winding from the change in armature flux the field current varies following the fault and E_q varies accordingly. The emf E_q'' is proportional to the total rotor d-axis flux linkages (field and damper windings), and includes the linkages due to the armature flux. This total flux linkage must remain constant and equal to its pre-fault value following the fault. If the generator is on no-load prior to the fault then the pre-fault armature current and flux are zero. In this case the total pre-fault rotor flux linkages are equal to the linkages due to the excitation flux only, and the pre-fault values of E_q and E_q'' are the same and equal to the terminal voltage. If, however, the generator is on load prior to the fault then the armature current is not zero and the total rotor flux linkages will include the effect of the pre-fault linkages due to the armature flux. In this case the pre-fault values of E_q and E_q'' will not be the same. A similar argument is valid for the q-axis flux linkages and emfs. Thus the pre-fault value of E_d'' is equal to zero only when the generator is on no-load prior to the fault.

The emf $\overline{E} = \overline{E}_q + \overline{E}_d$ can be found by observing that

$$\begin{aligned}\overline{E} &= \overline{V}_g + R\overline{I} + \mathrm{j}\overline{I}_d X_d + \mathrm{j}\overline{I}_q X_q \\ &= \overline{V}_g + R\overline{I} + \mathrm{j}\overline{I}_d (X_d - X_d'') + \mathrm{j}\overline{I}_d X_d'' + \mathrm{j}\overline{I}_d (X_q - X_q'') + \mathrm{j}\overline{I}_q X_q'' \\ &= \overline{E}' + \mathrm{j}\overline{I}_d (X_d - X_d'') + \mathrm{j}\overline{I}_q (X_q - X_q'')\end{aligned} \tag{4.18}$$

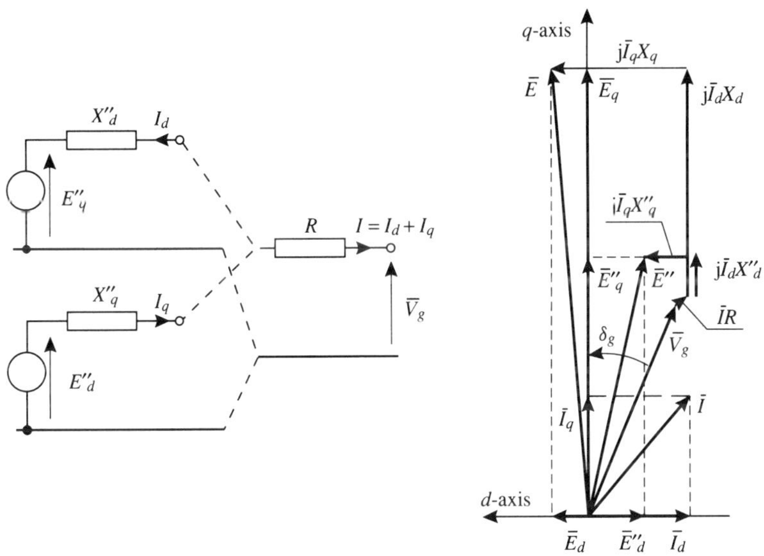

Figure 4.12 Equivalent circuit and phasor diagram of the generator in the subtransient state

Figure 4.12 shows the equivalent circuit and the phasor diagram of the generator in the subtransient state. $\overline{E}$ is no longer along the q-axis and has both E_d and E_q components.

Transient-state

During the transient period the armature flux is forced into high reluctance paths outside the field winding by currents induced in the field winding. This is shown in Figure 4.8b for the flux acting along the d-axis only but in general the armature flux will have both d and q components. In this more general case the flux path is distorted not only by the d-axis field current but also by the q-axis rotor body currents. The reactances associated with the flux path are X'_d and X'_q. As the rotor flux linkages in both axes are assumed to remain constant during the transient state the internal emf corresponding to these linkages may also be assumed to remain constant and equal to their pre-fault values. Thus the generator may be represented in the transient state by constant *transient internal emfs* E'_q and E'_d acting behind the transient reactances X'_d and X'_q. The circuit equation of the generator is then

$$\overline{E}' = \overline{E}'_q + \overline{E}'_d = \overline{V}_g + R\overline{I} + \mathrm{j}\overline{I}_d X'_d + \mathrm{j}\overline{I}_q X'_q \tag{4.19}$$

where $\overline{I} = \overline{I}_d + \overline{I}_q$ is the armature current at the beginning of the transient period. Its value is different than that given by equation (4.17) and can be calculated once $\overline{E}'$ is known. E'_q is proportional to the field winding flux linkages Ψ_f whilst E'_d is proportional to the flux linkage of the q-axis rotor body. Both components include the effect of the pre-fault armature current and are assumed to remain constant during transient state. Following a similar argument as for the subtransient state the pre-fault values of the emf's are $E'_{q0} = E_{q0} = V_g$ and $E'_{d0} = E_{d0} = 0$ only if prior to the fault the generator is on no-load

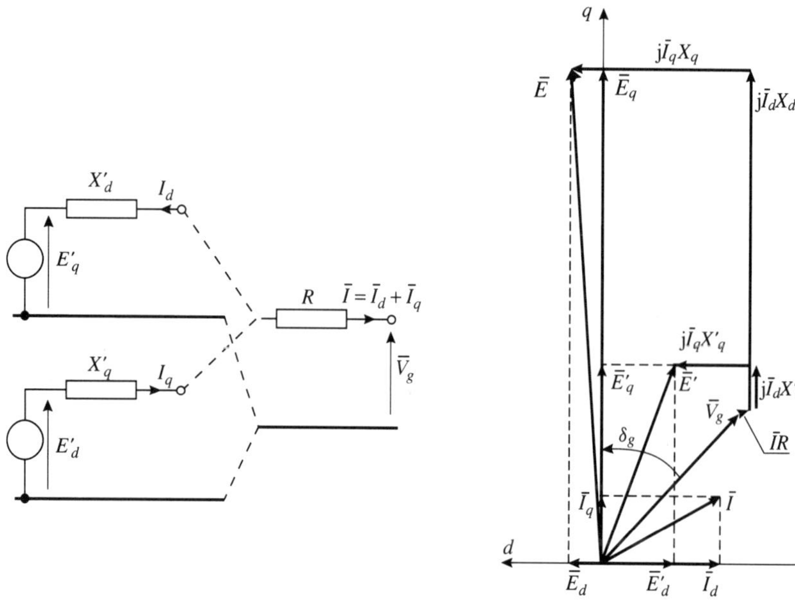

Figure 4.13 Equivalent circuit and phasor diagram of the round-rotor generator in the transient state (The length of IR has been exaggerated for clarity)

(zero armature reaction flux). If prior to the fault the generator is on-load then E'_q and E'_d include the effect of the load current and $\overline{E}'_0 \neq \overline{E}''_0 \neq \overline{E}_0$.

Similarly as in the subtransient state the emf $\overline{E}$ can be found from:

$$\begin{aligned}\overline{E} &= \overline{V}_g + R\overline{I} + \mathrm{j}\overline{I}_d X_d + \mathrm{j}\overline{I}_q X_q \\ &= \overline{E}' + \mathrm{j}\overline{I}_d (X_d - X'_d) + \mathrm{j}\overline{I}_q (X_q - X'_q)\end{aligned} \tag{4.20}$$

Figure 4.13 shows the equivalent circuit and the phasor diagram of the round-rotor generator in the transient state. The transient emf of such a generator has both d and q components. For the salient-pole generator with laminated rotor there is no screening in the q-axis and $X'_q = X_q$ so that $E'_d = 0$ and $E' = E'_q$. A summary of the different emfs and their associated flux linkage can be found in the List of Symbols.

Establishing initial emf values

The previous discussion explained why the subtransient and transient emfs do not change during their respective periods (neglecting the influence of resistances) and are equal to their pre-fault values.

The simplest situation occurs when the generator is on no-load prior to the fault when the initial values of the transient and subtransient emfs are equal to the pre-fault steady-state emf E (and the generator pre-fault terminal voltage). When the generator is on-load the influence of the pre-fault armature current $\overline{I}_0$ on the internal emfs must be taken into account. Note that the three equivalent circuits shown in Figures 3.17, 4.12 and 4.13 are valid for different states of the generator and therefore for different values of

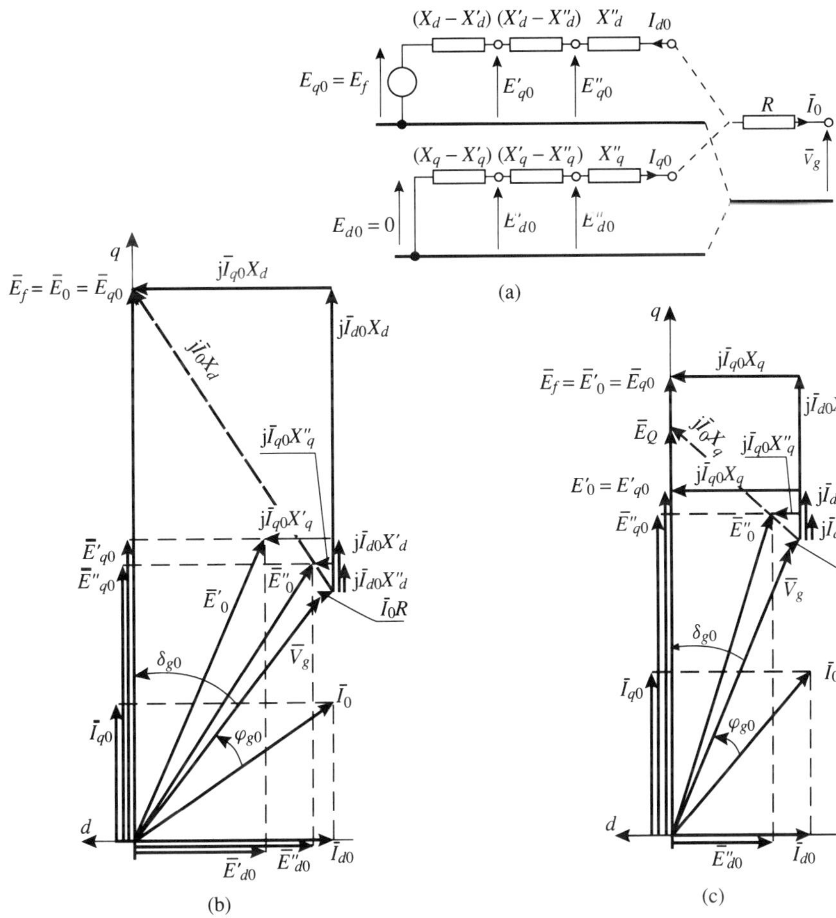

Figure 4.14 Finding the initial values of the emfs: (a) equivalent circuits; (b) phasor diagram of the round-rotor generator; (c) phasor diagram of the salient-pole generator (The length of IR has been exaggerated for clarity)

current $\overline{I}$. However, as the subtransient and transient emfs are equal to their pre-fault values, the three equivalent circuits must also be valid for the pre-fault current $\overline{I}_0$. Thus, according to Kirchhoff's law, the three equivalent circuits can be combined for the same current $\overline{I}_0$, as shown in Figure 4.14a. By following a similar procedure to that described in Section 3.3 the initial values of $\overline{E}_0$, $\overline{E}'_0$ and $\overline{E}''_0$ can then be found from the phasor diagram, Figure 4.14b, assuming that $\overline{I}_0$ and its phase angle ϕ_g are known. As excitation is only in the d-axis E_0 acts along the q-axis while $\overline{E}''_0$ and $\overline{E}'_0$ usually have non-zero direct and quadrature components. For the salient pole-generator, Figure 4.14c, $E_{d0} = 0$ and $\overline{E}'_0 = E'_{q0}$. In the phasor diagram the value of the voltage drop across the generator resistance has been exaggerated for clarity but, in practice, the resistance is very small and can often be neglected. Once the initial values of the transient and subtransient emfs have been found, equations 4.17 and 4.19 can be used to calculate the magnitude of the ac current component at the start of the subtransient and the transient period respectively.

Example 4.1

A 200 MVA round-rotor generator with the parameters given in Table 4.3 is loaded with 1 pu of real power and 0.5 pu of reactive power (lagging). The voltage at the generator terminals is 1.1 pu. Find the pre-fault values of the steady-state, transient and subtransient emfs. Assume $X_d = X_q = 1.6$ and neglect the armature resistance.

Solution

Assuming the generator voltage to be the reference, the load current is

$$\underline{I}_0 = \left(\frac{\underline{S}}{\underline{V}_g}\right)^* = \frac{P - \mathrm{j}Q}{V_g} = \frac{1 - \mathrm{j}0.5}{1.1} = 1.016\angle -26.6^\circ$$

so that $\phi_{g0} = 26.6^\circ$. The steady-state internal emf is

$$E_{q0} = \underline{V}_g + \mathrm{j}X_d\underline{I}_0 = 1.1 + \mathrm{j}1.6 \times 1.016\angle -26.6^\circ = 2.336\angle 38.5^\circ$$

Thus $E_{q0} = 2.336$ and $\delta_{g0} = 38.5^\circ$. The (d,q) components of the current and voltage are:

$$I_{d0} = -I_0 \sin(\phi_{g0} + \delta_{g0}) = -1.016 \sin(26.6^\circ + 38.5^\circ) = -0.922$$

$$I_{q0} = -I_0 \cos(\phi_{g0} + \delta_{g0}) = 0.428$$

$$V_{gd} = -V_g \sin \delta_{g0} = -1.1 \sin 38.5^\circ = -0.685$$

$$V_{gq} = V_g \cos \delta_{g0} = 0.861$$

Now the (d, q) components of the transient and subtransient emfs can be calculated from the phasor diagram in Figure 4.14 as

$$E'_{d0} = V_{gd} + X'_q I_{q0} = -0.685 + 0.38 \times 0.428 = -0.522$$

$$E'_{q0} = V_{gq} + X'_d I_{d0} = -0.861 + 0.23 \times (-0.922) = 1.073$$

$$E''_{d0} = V_{gd} + X''_q I_{q0} = -0.612$$

$$E''_{q0} = V_{gq} + X''_d I_{d0} = 1.018$$

Example 4.2

Solve a similar problem to that in Example 4.1 but for the 230 MVA salient-pole generator in Table 4.3.

Solution

The main problem with the salient-pole generator is in finding the direction of the q-axis. Equation 3.64 gives $\underline{E}_Q = \underline{V}_g + \mathrm{j}X_q\underline{I}_0 = 1.1 + \mathrm{j}0.69 \times 1.016\angle -26.6^\circ = 1.546\angle 23.9^\circ$.

Thus $\delta_{g0} = 23.9^\circ$ and

$$I_{d0} = -1.016\sin(26.6^\circ + 23.9^\circ) = -0.784$$

$$I_{q0} = -1.016\cos(26.6^\circ + 23.9^\circ) = 0.647$$

$$V_{gd} = -1.1\sin 23.9^\circ = -0.446$$

$$V_{gq} = 1.1\cos 23.9^\circ = 1.006$$

$$E_{q0} = V_{gq} - I_{d0}X_d = 1.006 - (-0.784) \times 0.93 = 1.735$$

$$E'_{d0} = -0.446 + 0.69 \times 0.647 = 0$$

$$E'_{q0} = 1.006 - 0.3 \times (-0.784) = 1.241$$

$$E''_{d0} = -0.446 + 0.27 \times 0.647 = -0.271$$

$$E''_{q0} = 1.006 - 0.25 \times (-0.784) = 1.202$$

Flux decrement effects

Although the subtransient and transient emfs remain constant immediately after a fault their values change with time as the armature flux penetrates the rotor circuits. The time scale of these changes is such that changes during the subtransient period affect the short-circuit currents and torques, as will be discussed later in this chapter, whilst changes during the transient period can affect generator stability and are discussed at length in Chapters 5 and 6. Although a full mathematical treatment of these *flux decrement* effects must be left until Chapter 10 a basic understanding of the mechanisms behind them can be obtained by considering the simple d-axis coupled circuit shown in Figure 4.15. This circuit models the field winding and the d-axis armature coil, and the coupling that exists between them, and gives the field flux linkage as

$$\Psi_f = L_f i_f + M_{fd} i_d \tag{4.21}$$

where L_f is the self-inductance of the field-winding and M_{fd} is the mutual inductance between the two windings. Chapter 10 will show that $M_{fd} = \sqrt{3/2}M_f$.

According to the theorem of constant flux linkage Ψ_f will remain constant immediately following any change in i_f or i_d and, as $E'_q \propto \Psi_f$, E'_q will also remain constant. However as the field winding has a resistance greater than zero this will allow some of the magnetic stored energy to be dissipated and the winding flux linkage to change. This change in

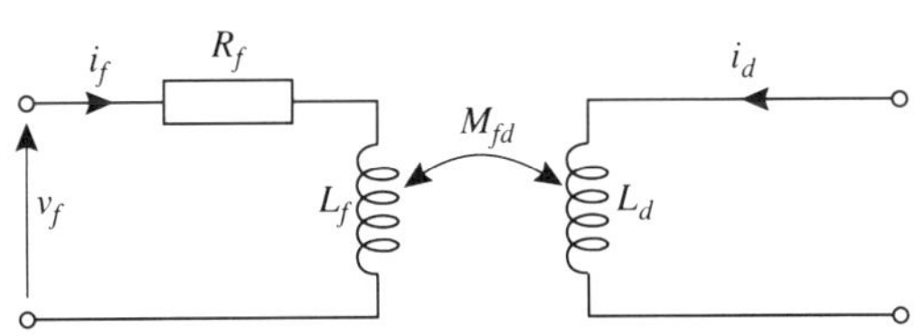

Figure 4.15 Coupling between the field winding and the d-axis armature coil

flux linkage is governed by the differential equation

$$\frac{d\Psi_f}{dt} = v_f - R_f i_f \tag{4.22}$$

which, when rearranging and substituting equation 4.21, gives

$$v_f = L_f \frac{di_f}{dt} + R_f i_f + M_{fd} \frac{di_d}{dt} \tag{4.23}$$

At this point it is convenient to consider the changes Δi_f, Δi_d and Δv_f rather than absolute values. Introducing the Laplace operator and writing equation 4.23 in terms of these changes gives

$$\Delta v_f(s) = L_f s \Delta i_f(s) + R_f \Delta i_f(s) + M_{fd} s \Delta i_d(s) \tag{4.24}$$

This equation can then be written in transfer function form as

$$\Delta i_f(s) = \frac{1/R_f}{(1 + T'_{do}s)} \Delta v_f(s) - \frac{M_{fd}/R_f}{(1 + T'_{d0}s)} s \Delta i_d(s) \tag{4.25}$$

where $T'_{d0} = L_f/R_f$. Substituting equation 4.25 into equation 4.21 gives

$$\Delta \Psi_f(s) = \frac{L_f/R_f}{(1 + T'_{d0}s)} \Delta v_f(s) + \frac{M_{fd}}{(1 + T'_{d0}s)} \Delta i_d(s) \tag{4.26}$$

As $\Delta E'_q \propto \Delta \Psi_f$ and $\Delta E_f \propto \Delta v_f$ this equation can be written in terms of $\Delta E'_q$ and ΔE_f. Figure 4.13 shows that for a generator I_d flows into the winding while I flows out of the winding. Thus, assuming no change in the rotor angle, Δi_d can be replaced by a term proportional to $-\Delta I$ when a more rigorous analysis similar to the above but using the relationships established in Chapter 10 to evaluate the constants of proportionality would show that

$$\Delta E'_q(s) = \frac{1}{(1 + T'_{d0}s)} \Delta E_f(s) - \frac{K}{(1 + T'_{d0}s)} \Delta I(s) \tag{4.27}$$

where K is a constant. As can be seen a change in the field winding flux linkage, and hence a change in E'_q, can be produced by either a change in the excitation voltage or by a change in the armature current. The rate at which these changes occur is determined by the transient time constant T'_{d0} with the field winding filtering out high frequency changes in ΔI and ΔE_f. Assuming for the moment that $\Delta I = 0$ and that a step change is made on E_f, E'_q will change exponentially as shown in Figure 4.16a. Similarly a step increase in the armature current I, such as occurs during a short circuit, would lead to an exponential reduction in E'_q as shown in Figure 4.16b.

In practice the armature will be connected to the system (a voltage source) by a transmission line of finite reactance when the actual time constants involved will depend on the impedance in the armature circuit. For a generator connected to an infinite busbar through a transmission link Anderson & Fouad (1977) show that this modifies equation 4.27 to

$$\Delta E'_q(s) = \frac{B}{(1 + BT'_{d0}s)} \Delta E_f(s) - \frac{AB}{(1 + BT'_{d0}s)} \Delta \delta(s) \tag{4.28}$$

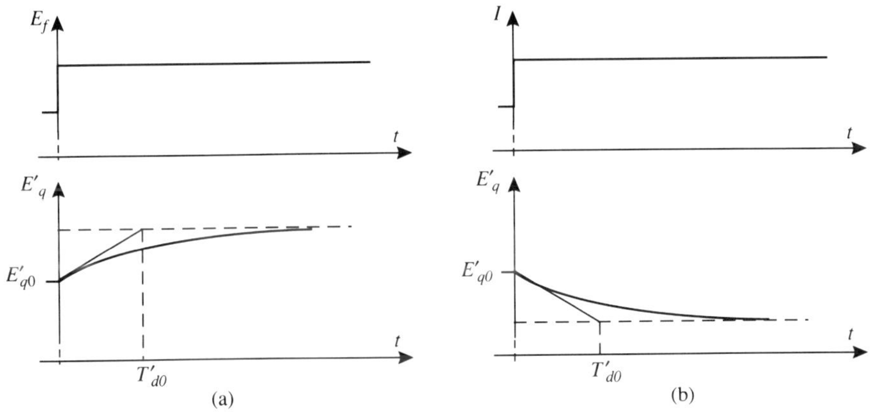

Figure 4.16 Changes in E'_q (a) to a step change in excitation ΔE_f; (b) to a step change in load ΔI

where a change in the generator loading is now reflected by a change in the power angle $\Delta\delta$ (measured with respect to the system) and B is a constant that takes into account the effect of the impedance in the armature circuit. If both the generator armature resistance and the resistance of the transmission link are neglected then $B = (X'_d + X_s)/(X_d + X_s) = x'_d/x_d$ and $A = [(1 - B)/B]V_s \sin\delta_0$. The importance of equations 4.27 and 4.28 cannot be overemphasised as they show how changes in the excitation and generator loading alters E'_q.

4.2.5 Short-circuit currents with the generator initially on no-load

To consider the time variation of the fault current assume the generator to be on no-load prior to the fault. In this case the pre-fault values of the three internal emfs E, E', and E'' are all equal to the terminal voltage, that is $E'' = E' = E = E_f = V_g$, and all lie along the q-axis. As the d-axis component of all these emfs is equal to zero the d and q-axis equivalent circuits shown in Figures 3.17, 4.12 and 4.13 can be simplified to those shown in Figure 4.17. The amplitude of the ac fault current component in each of the three characteristic states can be obtained directly from the equivalent circuits in Figure 4.17 by short-circuiting the generator terminals, when:

$$\text{for the subtransient state: } i''_m = \frac{E_{fm}}{X''_d} \tag{4.29}$$

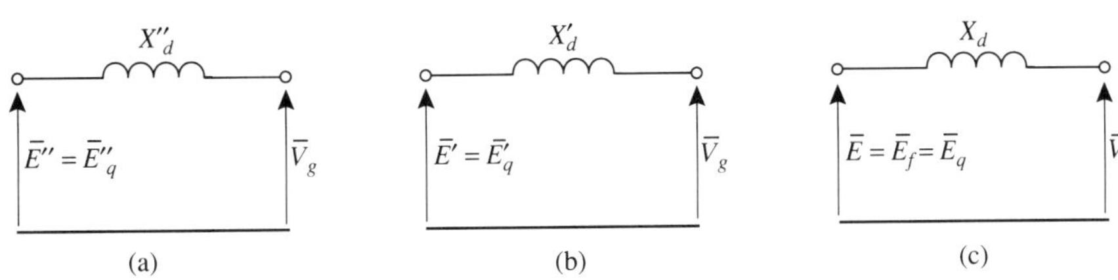

Figure 4.17 Equivalent circuit for a synchronous generator in; (a) subtransient-state; (b) transient-state; (c) steady-state (The generator is assumed to be on no-load prior to the fault)

$$\text{for the transient state: } i'_m = \frac{E_{fm}}{X'_d} \tag{4.30}$$

$$\text{for the steady-state: } i_m^\infty = \frac{E_{fm}}{X_d} \tag{4.31}$$

where $E_{fm} = \sqrt{2}E_f$.

As the ac component of the phase current, and its decay time constant, in all three characteristic states is known it is now possible to derive formulae for the short-circuit currents as a function of time. Figure 4.18 shows the envelopes of the maximum value of the ac component of the short-circuit current corresponding to those shown in the top diagram of Figure 4.7b. The resultant envelope is defined by $i_m(t)$ and is obtained by adding the three components, each of which decays with a different time constant, to give

$$i_m(t) = \Delta i'' e^{-t/T''_d} + \Delta i' e^{-t/T'_d} + \Delta i \tag{4.32}$$

where

$\Delta i = i_m^\infty = E_{fm}/X_d$ is the maximum value of the ac component neglecting the screening effect of all the rotor windings - equation 4.31

$\Delta i + \Delta i' = E_{fm}/X'_d$ is the maximum value of the ac component including the screening effect of the field winding but neglecting the screening effect of the damper winding — equation 4.30

$\Delta i + \Delta i' + \Delta i'' = E_{fm}/X''_d$ is the maximum value of the ac component when the screening effect of both the damper winding and the field winding are included — equation 4.29

Some simple algebra applied to Figure 4.18 then gives the equations for the three components of the short-circuit current as:

$$\Delta i = E_{fm}\frac{1}{X_d}, \qquad \Delta i' = E_{fm}\left(\frac{1}{X'_d} - \frac{1}{X_d}\right), \qquad \Delta i'' = E_{fm}\left(\frac{1}{X''_d} - \frac{1}{X'_d}\right) \tag{4.33}$$

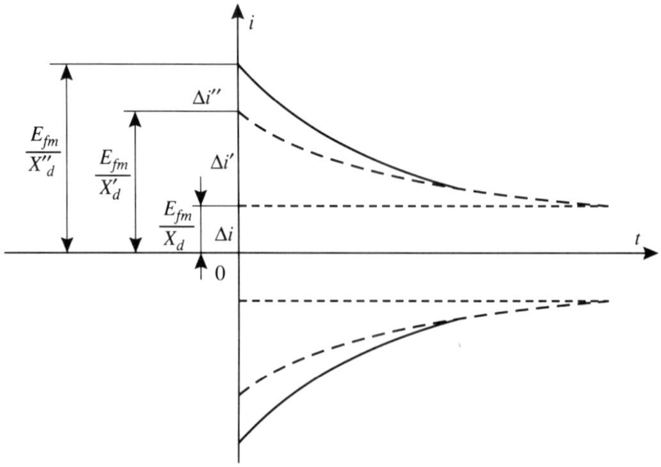

Figure 4.18 Envelopes of the three characteristic ac components of the short-circuit current

Substituting equation 4.33 into equation 4.32 gives the following expression for the envelope of the ac component of the fault current:

$$i_m(t) = E_{fm}\left[\left(\frac{1}{X''_d} - \frac{1}{X'_d}\right)e^{-t/T''_d} + \left(\frac{1}{X'_d} - \frac{1}{X_d}\right)e^{-t/T'_d} + \frac{1}{X_d}\right] \tag{4.34}$$

The initial value of equation 4.34 at $t = 0$ is

$$i_m(0) = \frac{E_{fm}}{X''_d} \tag{4.35}$$

Substituting equations 4.34 and 4.35 into equation 4.11 gives the expression for the short-circuit currents as

$$\begin{aligned}
i_A &= -\frac{E_{fm}}{X''_d}[g_3(t)\cos(\omega t + \gamma_0) - e^{-t/T_a}\cos\gamma_0] \\
i_B &= -\frac{E_{fm}}{X''_d}[g_3(t)\cos(\omega t + \gamma_0 - 2\pi/3) - e^{-t/T_a}\cos(\gamma_0 - 2\pi/3)] \\
i_C &= -\frac{E_{fm}}{X''_d}[g_3(t)\cos(\omega t + \gamma_0 - 4\pi/3) - e^{-t/T_a}\cos(\gamma_0 - 4\pi/3)]
\end{aligned} \tag{4.36}$$

The function $g_3(t)$ is defined as

$$g_3(t) = X''_d\left[\left(\frac{1}{X''_d} - \frac{1}{X'_d}\right)e^{-t/T''_d} + \left(\frac{1}{X'_d} - \frac{1}{X_d}\right)e^{-t/T'_d} + \frac{1}{X_d}\right] \tag{4.37}$$

and accounts for the decay of the ac component from the subtransient-state to the transient-state and then to the steady-state. The subscript "3" relates to the three-phase short-circuit.

The maximum, instantaneous, value of the current in each phase depends on the instant in the ac cycle that the fault occurs. For example, the current in phase A reaches its maximum value when the fault is applied at $\gamma_0 = 0$, that is when the excitation flux Ψ_{fA} linking phase A reaches its maximum value and the voltage E_{fA} is zero. Figure 4.7 showed currents flowing in a large generator after a three-phase fault assuming the armature time constant T_a to be much greater than the ac cycle time. With such a long armature time constant the current i_A may reach a value almost equal to $2E_{fm}/X''_d$. On the other hand, when the fault occurs at $\gamma_0 = \pi/2$, the excitation flux linkage Ψ_{fA} is zero and the current i_A is minimal as it does not contain a dc component. However the dc component will be evident in currents i_B and i_C.

Influence of the rotor subtransient saliency

When saliency effects are included, the different reluctance on the d and q axes requires the effect of the armature mmf on the short-circuit current to be analysed using the two reaction theory (Jones, 1967). However an intuitive understanding of the effects of saliency can be obtained by considering the ac and dc armature mmf components separately.

Section 4.2.1 explained that the armature mmf due to the stator fault currents consists of a rotating component $F_{a\,ac}$ directed along the d-axis and a stationary component $F_{a\,dc}$ which is also initially directed along the d-axis but then counter-rotates with respect to the rotor. As $F_{a\,ac}$ is always directed along the d-axis, it is associated with a reactance equal to X''_d and the ac term in equation 4.36 remains unaffected by rotor saliency.

On the other hand the stationary mmf component $F_{a\,dc}$ drives a flux across the air-gap the width of which appears to be continuously changing. This has two effects. Firstly the maximum value of the dc component of the fault current, given by the second component in equation 4.36, has to be modified by making it dependent on the mean of the subtransient reactances and equal to $E_{fm}(1/X''_d + 1/X''_q)/2$. Secondly an additional double-frequency component of magnitude $E_{fm}(1/X''_d - 1/X''_q)/2$ is introduced into the fault current due to $F_{a\,dc}$ lying twice every cycle along the d-axis (q-axis). Both the dc and the double-frequency components decay with the same time constant equal to the mean of the d and q-axis time constants $T_a = (X''_d + X''_d)/(2\omega R)$ whilst their values depend on the fault instant.

The short-circuit current in phase A described by equation 4.36 can now be adapted to include the modified dc component and the double-frequency ac component

$$\begin{aligned} i_A = &-\frac{E_{fm}}{X''_d}[g_3(t)\cos(\omega t + \gamma_0)] \\ &+\frac{E_{fm}}{2}e^{-t/T_a}\left[\left(\frac{1}{X''_d}+\frac{1}{X''_d}\right)\cos\gamma_0 + \left(\frac{1}{X''_d}-\frac{1}{X''_q}\right)\cos(2\omega t+\gamma_0)\right] \end{aligned} \tag{4.38}$$

At time $t = 0$ the second component in this equation is equal to $E_{fm}\cos\gamma_0/X''_d$ and corresponds to the dc component of the armature mmf always being in phase with the rotor d-axis when the fault is applied. In the case of a generator with damper windings in both axes $X''_q \approx X''_d$ and equation 4.38 reduces to equation 4.36.

4.2.6 Short-circuit currents in the loaded generator

When the generator is on load prior to the fault, the initial values of the emfs E''_{q0}, E'_{q0}, E''_{d0} and E'_{d0} are functions of the load current and, knowing the initial loading conditions, their value may be determined from the phasor diagram of Figure 4.14. These emfs now further modify equation 4.38 to account for the pre-fault load current. Both d and q-axis ac emfs and currents have now to be considered. The q-axis emfs will force the flow of d-axis ac currents (changing as cosine with time) while d-axis emfs will force the flow of q-axis ac currents (changing as sine with time). Denoting the angle between the q-axis and the terminal voltage V_g by δ_g, the following expression for the short-circuit current flowing in phase A is obtained:

$$\begin{aligned} i_A = &-\left[\left(\frac{E''_{qm0}}{X''_d}-\frac{E'_{qm0}}{X'_d}\right)e^{-t/T''_d} + \left(\frac{E'_{qm0}}{X'_d}-\frac{E_{qm0}}{X_d}\right)e^{-t/T'_d} + \frac{E_{qm0}}{X_d}\right]\cos(\omega t+\gamma_0) \\ &+\left[\left(\frac{E''_{dm0}}{X''_q}-\frac{E'_{dm0}}{X'_q}\right)e^{-t/T''_q} + \frac{E'_{dm0}}{X'_q}e^{-t/T'_q}\right]\sin(\omega t+\gamma_0) \\ &+\frac{V_{gm0}}{2}e^{-t/T_a}\left[\left(\frac{1}{X''_d}+\frac{1}{X''_q}\right)\cos(\gamma_0+\delta_g) + \left(\frac{1}{X''_d}-\frac{1}{X''_q}\right)\cos(2\omega t+\gamma_0+\delta_g)\right] \end{aligned} \tag{4.39}$$

All the emfs, and the terminal voltage, include an additional subscript "0" to emphasise the fact that they are pre-fault quantities. The first two components represent the fundamental-frequency ac currents due to the direct and quadrature components of the emfs. The last component consists of a dc term and a double-frequency term resulting from the subtransient saliency.

Influence of the automatic voltage regulator (AVR)

Depending on the type of exciter used the action of the AVR can have a considerable effect on the shape of the short-circuit current. In the case of a rotating exciter, Figure 2.3a, the fault does not influence the generating capabilities of the excitation unit and the large voltage error, equal to the difference between the reference voltage and the actual value of the terminal voltage, quickly drives up the excitation voltage. This increases the value of the short-circuit current compared to its value without the AVR active, as shown in Figure 4.19a. As E_q follows the field current, it will not now decay to E_{q0} but to a higher value corresponding to the new field current. Consequently E_q' will not decay to the same extent as when the AVR was inactive. If a static exciter is used, fed solely from the generator terminal voltage as in Figure 2.3e, the three-phase short-circuit reduces the excitation voltage to zero, the unit completely loses its excitation capabilities and all the emfs will decay to zero, Figure 4.19b. If the static exciter is fed by compounding both the generator voltage and current, as shown in Figure 2.3f, then excitation will not be lost during the short-circuit as the exciter input voltage is augmented by the fault currents. The short-circuit current will not now vanish to zero but its shape will depend on the strength of the current component in the compounded excitation signal. An example of this is shown in Figure 4.19c.

4.2.7 Subtransient torque

The torque creation mechanism was discussed in sub-Section 3.2.2, equation 3.71, with regards a two-pole generator in the steady state. This mechanism is equally applicable here except that now the d-axis and q-axis damper flux Φ_D and Φ_Q must be added into equation 3.71 to give

$$\tau = \tau_d - \tau_q = \frac{\pi}{2}(\Phi_f + \Phi_D + \Phi_{ad})F_{aq} - \frac{\pi}{2}(\Phi_Q + \Phi_{aq})F_{ad} \tag{4.40}$$

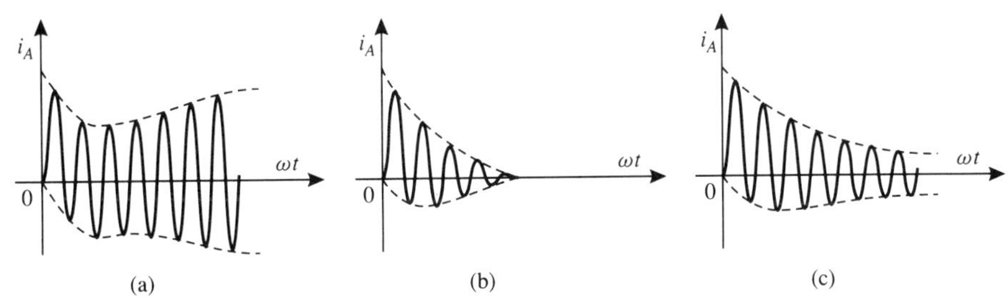

Figure 4.19 Influence of the AVR on the short-circuit current: (a) rotating exciter; (b) static exciter fed from the generator terminal voltage only; (c) static exciter fed from the generator terminal voltage and current

In this equation the field flux, Φ_f, is made up of two components Φ_{f0} and $\Delta\Phi_f$ where Φ_{f0} is flux due to the initial field current i_{f0} and $\Delta\Phi_f$ is the change in the flux due to the change in the field current. If the generator is assumed to be on no-load prior to the fault and if resistance effects are neglected, then the rotor flux produced by the damper windings Φ_D and Φ_Q and the change in the field flux $\Delta\Phi_f$ completely cancels the armature reaction flux so that $\Phi_D + \Delta\Phi_f = -\Phi_{ad}$ and $\Phi_Q = -\Phi_{aq}$. Substituting these values into equation 4.40 shows that the electromagnetic torque is produced solely by the interaction of the armature q-axis mmf with the original field flux Φ_{f0}. Denoting this torque as τ_ω gives

$$\tau_\omega(t) = \frac{\pi}{2}\Phi_{f0}F_{aq} \tag{4.41}$$

As explained earlier, the ac component of the short-circuit armature mmf acts along the d-axis and does not contribute to the q-axis armature mmf. In contrast the dc component of armature flux is directed along the d-axis the moment that the fault occurs but then rotates relative to the rotor, with the angle λ between the field and the armature mmfs changing as $\lambda = \omega t$. The amplitude of the dc armature mmf can be evaluated as $F_{a\,dc} = 1.5N_a i_m(0)$ by using a similar technique to that used to obtain equation 3.42. The q-axis armature mmf with respect to the rotor is then given by

$$F_{aq} = F_{a\,dc}\sin\lambda = \frac{3}{2}N_a i_m(0)\sin\omega t \tag{4.42}$$

As F_{aq} is solely dependent on the dc component of armature fault current, the torque is created by the initial field flux interacting with the dc component of the armature reaction mmf. This can be compared to two magnets, one rotating and the other stationary, as shown diagrammatically in Figure 4.20. It is important to note that for the three phase fault the angle λ is independent of the time the fault is applied and simply varies as ωt. The torque variation is obtained by substituting 4.42 into equation 4.41 to give

$$\tau_\omega(t) = \frac{\pi}{2}\Phi_{f0}N_a\frac{3}{2}i_m(0)\sin\omega t \tag{4.43}$$

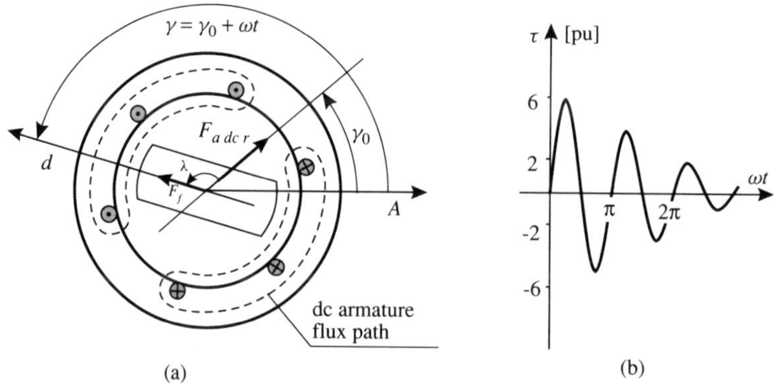

Figure 4.20 Subtransient electromagnetic torque following a three-phase fault: (a) torque creation mechanism; (b) example of torque variation with time

For the two pole generator considered here equation 3.39 gives $N_\phi = \pi N_a/2$, when substituting for $E_f = \omega N_\phi \Phi_{f0}/\sqrt{2}$ from equation 3.37 and for $i_m(0)$ from equation 4.35 finally gives the short circuit torque as

$$\tau_\omega(t) = \frac{3}{\omega}\frac{E_f^2}{X_d''}\sin\omega t \quad \text{Nm} \tag{4.44}$$

This equation shows that the short-circuit electromagnetic torque is independent of the fault instant but varies as $\sin\omega t$. During the first half-cycle the electromagnetic torque opposes the mechanical driving torque, whilst during the second half-cycle it assists the driving torque. The average value of the torque is zero. Equation 4.44 expresses the torque in MKSA units but, as explained in Appendix A, this can easily be changed to per unit by multiplying by $\omega/3$ to give the per unit form $\tau_\omega(t) = E_f^2 \sin\omega t/X_d''$ and is valid for any generator regardless of pole number.

The effect of including winding resistance is twofold. Firstly the dc phase currents that sustain the constant stator flux linkages $\Psi_{a\,dc}$ decay with a time constant T_a. Secondly, the rotor currents sustaining the constant rotor linkages decay with time constants T_d'' and T_d' allowing the armature reaction flux to penetrate the rotor. The net effect is that the torque τ_ω will decay to zero with the armature time constant T_a and the subtransient and transient time constants T_d'' and T_d' which modify equation 4.44 to give

$$\tau_\omega(t) = \frac{3}{\omega}\frac{E_f^2}{X_d''}g_3(t)e^{-t/T_a}\sin\omega t \quad \text{Nm} \tag{4.45}$$

where $g_3(t)$ is defined by equation 4.37. As the electromagnetic torque is equal to the product of two decaying functions, it vanishes very quickly, Figure 4.20b.

Strictly the torque expression in equation 4.45 should be modified to take into account the torque developed due to the power loss in the armature resistance. As this power loss is due to the flow of ac armature currents, which decay with the subtransient and transient time constants defined by the function $g_3(t)$ in equation 4.37, the torque corresponding to the power loss in all three-phases $\tau_R(t)$ may be expressed as

$$\tau_R(t) = \frac{3}{\omega}\left[\frac{E_f}{X_d''}g_3(t)\right]^2 R \quad \text{Nm} \tag{4.46}$$

Similarly, the ac currents induced in the rotor windings (damper and field) also cause a power loss. These currents decay with time constant T_a and the torque corresponding to the rotor losses may be approximately expressed as

$$\tau_r(t) = \frac{3}{\omega}\left(\frac{i_m(0)}{\sqrt{2}}e^{-\frac{t}{T_a}}\right)^2 r = \frac{3}{\omega}\left(\frac{E_f}{X_d''}\right)^2 re^{-\frac{2t}{T_a}} \quad \text{Nm} \tag{4.47}$$

where r is an equivalent resistance of all the rotor windings, referred to the stator (similarly as for a transformer). The torque τ_r is initially high but rapidly decays to zero.

As the armature resistance of a large generator is usually very small, the torque τ_R, corresponding to the stator losses, is several times smaller than the torque τ_r corresponding to the rotor losses. However the former decays at a much slower rate.

The resultant torque acting on the rotor during the fault is equal to the sum of the three components τ_ω, τ_R, τ_r. None of these components depends on the fault instant γ_0. The function $g_3(t)$, given by equation 4.37, does not decay to zero so that the component τ_R decays to a steady-state value corresponding to the losses that would be incurred by the steady-state short-circuit currents. If the generator is on-load prior to the fault then the initial value of the torque would correspond to the pre-fault load torque rather than zero as shown in Figure 4.20.

Equation 4.38 suggests that there should also be an additional periodic torque component corresponding to the double-frequency currents caused by the rotor subtransient saliency. Assuming that the resistance of the armature windings is small, the double-frequency torque $\tau_{2\omega}$ may be expressed as

$$\tau_{2\omega}(t) = -\frac{3}{2}\frac{E_f^2}{\omega}\left(\frac{1}{X_d''} - \frac{1}{X_q''}\right) e^{-\frac{2t}{T_a}} \sin 2\omega t \quad \text{Nm} \tag{4.48}$$

Typically $X_d'' \approx X_q''$ and the torque $\tau_{2\omega}$ only slightly distorts the single-frequency torque τ_ω. In some rare cases of high subtransient saliency the double-frequency component may produce a substantial increase in the maximum value of the ac torque.

4.3 Phase-to-phase Short-circuit

The phase-to-phase fault will be analysed in a similar way as the three-phase fault with the generator initially considered to be on no-load and with both winding resistance and subtransient saliency neglected. The effect of saliency and resistance on the short-circuit current will then be considered before finally developing expressions for the short-circuit torque. The short-circuit will be assumed to occur across phases B and C of the stator winding.

4.3.1 Short-circuit current and flux with winding resistance neglected

Figure 4.21 shows the armature coils with a short-circuit across phases B and C. The short-circuit connects the armature coils such that the current flows in opposite directions in the two coils and $i_B = -i_C$. In the case of the three-phase fault there were three closed armature circuits, one for each of the phase windings, and the flux linkage of

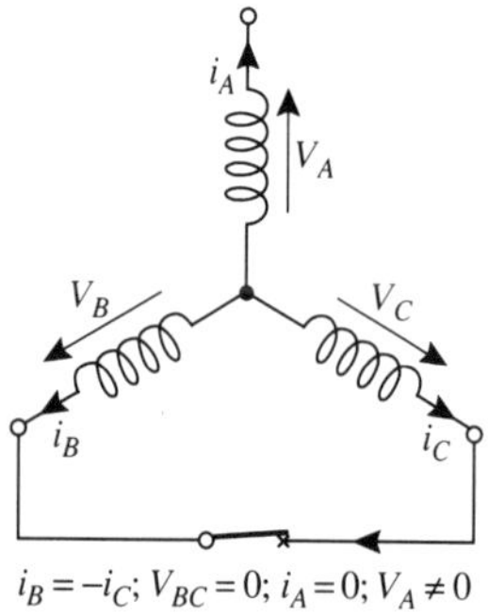

Figure 4.21 Phase-to-phase short-circuit

each of these windings could be considered separately. In the case of the phase-to-phase short-circuit there is only one closed circuit, the series connection of the two phase windings. When the fault occurs the total energy stored in this closed circuit cannot change instantaneously and it is the net flux linkage $\Psi_B - \Psi_C$ of this closed circuit that must remain constant.

The flux linkages Ψ_B and Ψ_C of the two short-circuited windings consist of two components: the winding self flux linkages Ψ_{BB} and Ψ_{CC} produced by the fault currents and the flux linkages produced by the excitation flux Ψ_{fB} and Ψ_{fC}. As the fault current in the two short-circuited phases is equal and opposite $\Psi_{BB} = -\Psi_{CC}$ and the net flux linkage $\Psi_B - \Psi_C$ can be expressed as

$$\Psi_B - \Psi_C = (\Psi_{BB} + \Psi_{fB}) - (\Psi_{CC} + \Psi_{fC}) = \Psi_{fB} - \Psi_{fC} - 2\Psi_{CC} \tag{4.49}$$

With the generator on no-load prior to the fault, $\Psi_{CC}(0^-) \propto i_C(0^-) = i_B(0^-) = 0$ and the only flux linking the windings is the excitation flux. Thus

$$\Psi_B(0^-) - \Psi_C(0^-) = \Psi_{fB0} - \Psi_{fC0} \tag{4.50}$$

The law of constant flux linkage requires the net flux linkage $(\Psi_B - \Psi_C)$ just before and just after the fault to be equal so that

$$\Psi_{fB} - \Psi_{fC} - 2\Psi_{CC} = \Psi_{fB0} - \Psi_{fC0} \tag{4.51}$$

Rearranging gives

$$\Psi_{CC} = \tfrac{1}{2}[(\Psi_{fB} - \Psi_{fC}) - (\Psi_{fB0} - \Psi_{fC0})] \tag{4.52}$$

The variation of the flux linkages Ψ_{fB} and Ψ_{fC} are shown in Figure 4.4, with the initial values defined by equations 4.4, so that

$$\begin{aligned} \Psi_{fB} &= \Psi_{fa}\cos(\gamma - 2\pi/3), \quad & \Psi_{fC} &= \Psi_{fa}\cos(\gamma - 4\pi/3) \\ \Psi_{fB0} &= \Psi_{fa}\cos(\gamma_0 - 2\pi/3), \quad & \Psi_{fC0} &= \Psi_{fa}\cos(\gamma_0 - 4\pi/3) \end{aligned} \tag{4.53}$$

which, when substituted into equation 4.52, gives

$$\Psi_{CC} = \frac{\sqrt{3}}{2}\Psi_{fa}(\sin\gamma - \sin\gamma_0) \tag{4.54}$$

Recall from equation 3.37 that $E_{fm} = \omega\Psi_{fa}$ whilst immediately after the fault the subtransient current $i_C = \Psi_{CC}/L''_d$. This allows the formula for the phase-to-phase short-circuit current to be written as

$$i_C = -i_B = \frac{\sqrt{3}}{2}\frac{E_{fm}}{X''_d}(\sin\gamma - \sin\gamma_0) \tag{4.55}$$

This short-circuit current consists of two components: an ac component

$$i_{C\,ac} = -i_{B\,ac} = \frac{\sqrt{3}E_{fm}}{2X''_d}\sin\gamma \tag{4.56}$$

whose magnitude is independent of the fault instant, and a dc component

$$i_{C\,dc} = -i_{B\,dc} = \frac{\sqrt{3}E_{fm}}{2X''_d}\sin\gamma_0 \tag{4.57}$$

the value of which depends on the fault instant. If the fault occurs at $\gamma_0 = 0$, the voltage of phase A is zero, and the fault current is purely sinusoidal with no dc component present. If, on the other hand, the fault occurs at $\gamma_0 = -\pi/2$, the voltage in phase A is at its negative peak, and the dc component of the fault current will be at its maximum value.

The flux linkages for the case when $\gamma_0 = 0$ and for $\gamma_0 = -\pi/2$ are shown in Figure 4.22a and 4.22b respectively. When $\gamma_0 = 0$ the flux linkages Ψ_{fB0} and Ψ_{fC0} are identical, thc net flux linkage $(\Psi_{fB0} - \Psi_{fC0})$ is zero, and no dc current component is required to maintain the flux linkages. The flux linkages Ψ_{BB} and Ψ_{CC}, and hence i_B and i_C, simply vary sinusoidally with opposite sign, Figure 4.22a. However when the fault occurs at $\gamma_0 = -\pi/2$ the voltage in phase A is a maximum and the flux linkages Ψ_{fB0} and Ψ_{fC0} now have the same magnitude but opposite sign. The net flux linkage is $\sqrt{3}\Psi_{fa}$ and a large dc circulating current $i_{C\,dc} = -i_{B\,dc}$ is necessary in order to maintain constant flux linkages, Figure 4.22b.

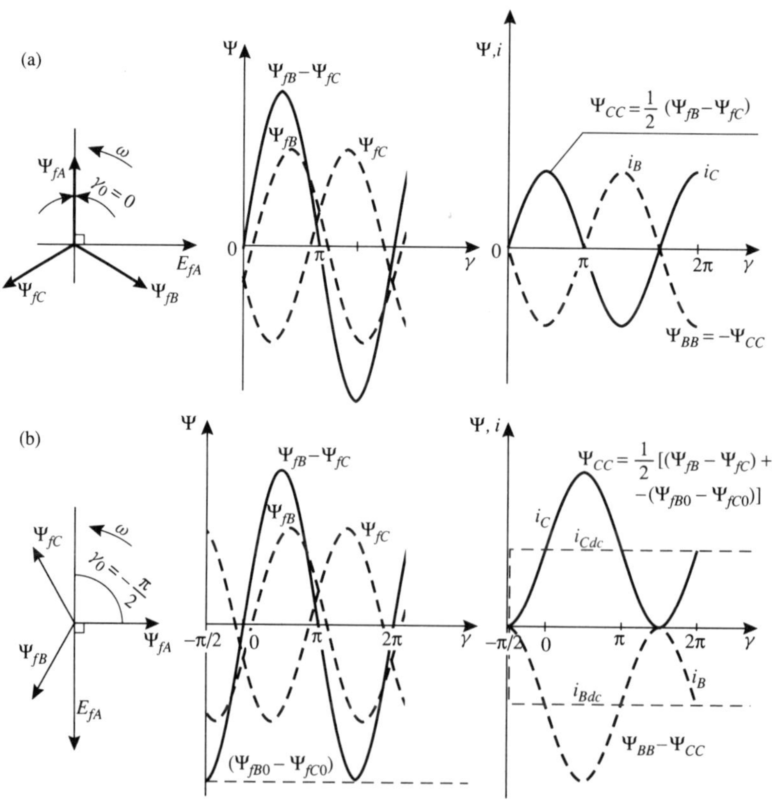

Figure 4.22 Application of the law of constant flux linkages to determine the phase-to-phase fault currents when the fault instant is: (a) $\gamma_0 = 0$; (b) $\gamma_0 = -\pi/2$

The resultant armature mmf $\vec{F}_a$ can be found by the vector addition of the individual mmfs $\vec{F}_B$ and $\vec{F}_C$ produced by the stator fault currents i_B and i_C. Taking phase A as a reference (real axis on the complex plane) gives

$$\begin{aligned}\vec{F}_a &= \vec{F}_B + \vec{F}_C = F_B e^{j2\pi/3} + F_C e^{j4\pi/3} = F_C(-e^{j2\pi/3} + e^{j4\pi/3}) \\ &= -j\sqrt{3}F_C = -j\sqrt{3}N_a i_C = j\sqrt{3}N_a \frac{\sqrt{3}}{2}\frac{E_{fm}}{X''_d}(\sin\gamma - \sin\gamma_0)\end{aligned} \tag{4.58}$$

The result of this phasor addition is shown in Figure 4.23 where the resultant armature mmf $\vec{F}_a$ is at all times perpendicular to the axis of phase A and proportional to the short-circuit current i_C. Consequently $\vec{F}_a$ is stationary in space but pulsates at frequency ω. As the rotor position at any instant in time is determined by the angle γ measured with respect to the A axis, this stationary armature mmf counter-rotates with respect to the rotor with the angle between the armature and the field mmf changing as $\lambda = \gamma + \pi/2$. This produces mmf components along the d-axis and q-axis

$$\begin{aligned}F_{ad} &= F_a \cos\lambda = -\sqrt{3}F_C \sin\gamma \\ F_{aq} &= F_a \sin\lambda = \sqrt{3}F_C \cos\gamma\end{aligned} \tag{4.59}$$

which, when substituting for i_C from equation 4.55, gives

$$\begin{aligned}F_{ad} &= -\frac{3}{2}N_a\frac{E_{fm}}{X''_d}(\sin\gamma - \sin\gamma_0)\sin\gamma = -\frac{3}{4}N_a\frac{E_{fm}}{X''_d}(1 - 2\sin\gamma_0\sin\gamma - \cos 2\gamma) \\ F_{aq} &= \frac{3}{2}N_a\frac{E_{fm}}{X''_d}(\sin\gamma - \sin\gamma_0)\cos\gamma\end{aligned} \tag{4.60}$$

The d-axis armature mmf has three components: a dc component, an ac component of fundamental frequency and an ac component of double frequency. The magnitude of both the dc component and the double frequency component are independent of the fault instant whilst the value of the fundamental ac component depends on γ_0.

This variation in the d-axis armature mmf also explains the changes that occur in the rotor field current and the d-axis damper current following the fault. The d-axis armature mmf forces an armature flux across the air-gap that produces flux linkages Ψ_{ar} with the rotor windings. These flux linkages are proportional to F_{ad} and are shown in

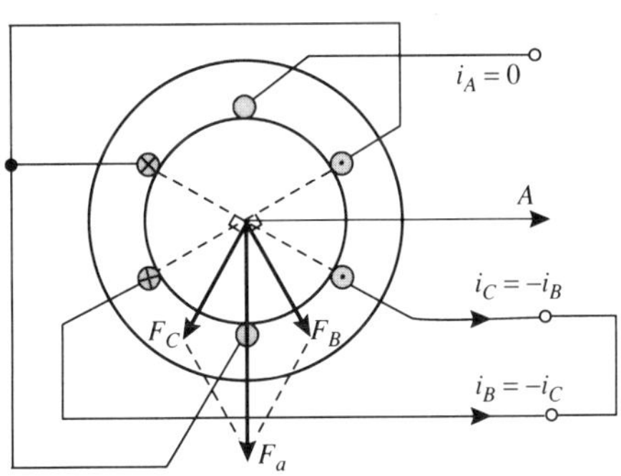

Figure 4.23 Phase-to-phase fault currents creating a stationary armature mmf F_a perpendicular to the axis of phase A (Generator windings are star-connected)

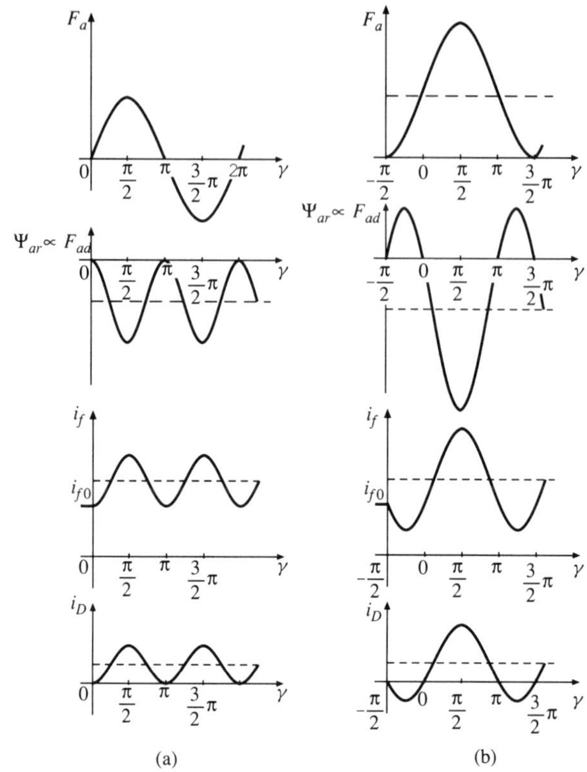

Figure 4.24 Illustration of the influence of the armature flux on the rotor currents when the fault occurs at: (a) $\gamma_0 = 0$; (b) $\gamma_0 = -\pi/2$

Figure 4.24. As the flux linkage of the rotor circuits must remain constant following the fault, additional currents flow in both the field winding and the d-axis damper winding to produce a flux linkage that is equal and opposite to that produced by the armature flux. These additional field and damper currents therefore have the same form as F_{ad} but are of opposite sign. The actual magnitude of the field current will depend on how well the field winding is screened by the d-axis damper winding. If perfect screening is obtained then the field current will remain constant with the armature flux being totally compensated by the d-axis damper current. If the fault occurs at $\gamma_0 = 0$ then the armature current and the resultant armature mmf do not contain a dc component, Figure 4.24a, and the flux linkages in the field and damper windings produced by the armature mmf simply pulsate at twice the system frequency. The induced rotor currents are then similar to those flowing during a three-phase fault, except they are now at twice system frequency. If, on the other hand, the fault occurs at $\gamma_0 = -\pi/2$, Figure 4.24b, then the armature mmf pulsates with a large dc component. The rotor flux linkages now contain both fundamental and double frequency components and the induced rotor currents have a different shape from before in that both the field and the damper currents initially reduce their value as shown in Figure 4.24b.

The q-axis mmf F_{aq} will be compensated by damper currents in a similar manner as F_{ad} and its influence will be discussed when the torque is considered.

4.3.2 Influence of the subtransient saliency

The armature current equation 4.55 is valid for a magnetically symmetrical rotor where the magnetic reluctance encountered by the armature flux is constant and does not depend on the rotor position. However if the machine has substantial subtransient saliency then the reluctance of the d and q axes will be different and the solution must be obtained using two-reaction theory (Ching & Adkins, 1954), (Gopal & Jones, 1971). The result of such an analysis shows that in equation 4.55 X''_d is replaced by $(X''_d \sin^2\gamma + X''_q \cos^2\gamma)$ to give

$$i_C = -i_B = \frac{\sqrt{3}E_{fm}(\sin\gamma - \sin\gamma_0)}{2(X''_d \sin^2\gamma + X''_q \cos^2\gamma)} = \frac{\sqrt{3}E_{fm}(\sin\gamma - \sin\gamma_0)}{X''_d + X''_q - (X''_d - X''_q)\cos 2\gamma} \tag{4.61}$$

Generators without a q-axis damper winding may have a large value of subtransient saliency which will distinctly distort the fault currents. Figure 4.25 shows these distorted currents for the same γ_0 as the sinusoidal waveforms shown previously in Figure 4.22 when X''_d was equal to X''_q.

The harmonic content of the distorted fault current can be quantified by expanding equation 4.61 into a Fourier series†

$$\begin{aligned} i_C = -i_B = {} & \frac{\sqrt{3}E_{fm}}{X''_d + \sqrt{X''_d X''_q}}(\sin\gamma - b\sin 3\gamma + b^2\sin 5\gamma - b^3\sin 7\gamma + \ldots) \\ & - \frac{\sqrt{3}E_{fm}\sin\gamma_0}{\sqrt{X''_d X''_q}}\left(\frac{1}{2} - b\cos 2\gamma + b^2\cos 4\gamma - b^3\cos 6\gamma + \ldots\right) \end{aligned} \tag{4.62}$$

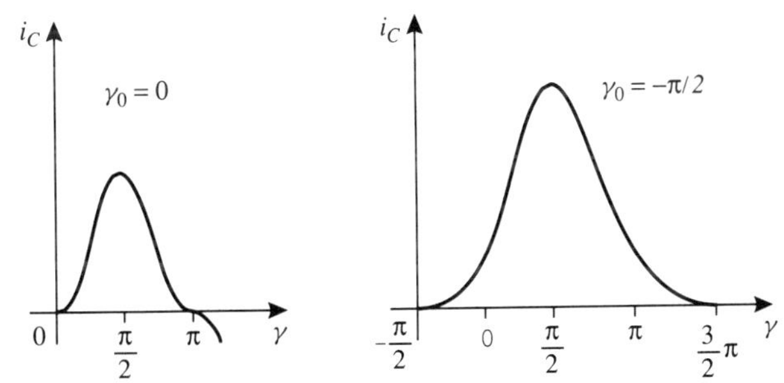

Figure 4.25 Phase-to-phase fault current waveforms strongly distorted due to the subtransient saliency

† Fourier series expansion of the two components in equation 4.61 gives

$$\frac{\sin\gamma}{A + B - (A - B)\cos 2\gamma} = \frac{1}{A + \sqrt{AB}}(\sin\gamma - b\sin 3\gamma + b^2\sin 5\gamma - b^3\sin 7\gamma + \ldots)$$

$$\frac{1}{A + B - (A - B)\cos 2\gamma} = \frac{1}{\sqrt{AB}}(\frac{1}{2} - b\cos 2\gamma + b^2\cos 4\gamma - b^3\cos 6\gamma + \ldots)$$

where $b = \dfrac{\sqrt{B} - \sqrt{A}}{\sqrt{B} + \sqrt{A}}$

where the asymmetry coefficient, b, is

$$b = \frac{\sqrt{X''_q} - \sqrt{X''_d}}{\sqrt{X''_q} + \sqrt{X''_d}} = \frac{\sqrt{X''_d X''_q} - X''_d}{\sqrt{X''_d X''_q} + X''_d} \tag{4.63}$$

In the example shown in Figure 4.25 the generator possessed a high degree of subtransient saliency with $X''_q = 2X''_d$, giving an asymmetry coefficient $b \approx 0.17$, whilst the magnitude of the third, fifth and seventh harmonics are respectively 17%, 3%, and 0.5% of the fundamental.

Even-numbered harmonics appear only when the fault current contains a dc component, that is when $\sin\gamma_0 \neq 0$ and subtransient saliency is present. However the odd-numbered harmonics appear whenever the generator possesses subtransient saliency, $b \neq 0$, regardless of the fault instant. The source of the odd-numbered harmonics is the negative-sequence component of the fault current as explained below.

Symmetrical component analysis of the phase-to-phase fault

Symmetrical components allow any unsymmetrical set of three-phase currents to be expressed as the phasor sum of three symmetrical ac current components: the positive sequence component $\bar{i}_1$, the negative sequence component $\bar{i}_2$, and the zero sequence component $\bar{i}_0$; the detailed derivation of which can be found in any standard textbook on power system analysis such as Grainger & Stevenson (1994).

The phase-to-phase fault currents shown in Figure 4.22a are given by $i_A = 0$ and $i_B = -i_C = i$. Applying the symmetrical component transformation gives

$$\begin{bmatrix} \bar{i}_0 \\ \bar{i}_1 \\ \bar{i}_2 \end{bmatrix} = \frac{1}{3}\begin{bmatrix} 1 & 1 & 1 \\ 1 & a & a^2 \\ 1 & a^2 & a \end{bmatrix}\begin{bmatrix} 0 \\ \bar{i} \\ -\bar{i} \end{bmatrix} = \frac{i}{3}\begin{bmatrix} 0 \\ a - a^2 \\ a^2 - a \end{bmatrix} = \frac{\bar{i}}{\sqrt{3}}\begin{bmatrix} 0 \\ \mathrm{j} \\ -\mathrm{j} \end{bmatrix} \tag{4.64}$$

where $a = e^{\mathrm{j}2\pi/3}$. The positive sequence component $\bar{i}_1$ leads the current in the B-phase by $\pi/2$, while the negative sequence component $\bar{i}_2$ lags the current in the B-phase by $\pi/2$. The zero sequence component is equal to zero. The reverse transformation from the symmetric components to phase components gives the actual phase currents which can be conveniently written as the sum of two three-phase systems representing the positive sequence and negative sequence phase currents:

$$\begin{bmatrix} \bar{i}_A \\ \bar{i}_B \\ \bar{i}_C \end{bmatrix} = \begin{bmatrix} 1 & 1 & 1 \\ 1 & a^2 & a \\ 1 & a & a^2 \end{bmatrix}\begin{bmatrix} 0 \\ \bar{i}_1 \\ \bar{i}_2 \end{bmatrix} = \begin{bmatrix} \bar{i}_{A1} \\ \bar{i}_{B1} \\ \bar{i}_{C1} \end{bmatrix} + \begin{bmatrix} \bar{i}_{A2} \\ \bar{i}_{B2} \\ \bar{i}_{C2} \end{bmatrix} \tag{4.65}$$

where $\bar{i}_{A1} = \bar{i}_1$, $\bar{i}_{B1} = a^2\bar{i}_1$, $\bar{i}_{C1} = a\bar{i}_1$, $\bar{i}_{A2} = \bar{i}_2$, $\bar{i}_{B2} = a\bar{i}_2$, and $\bar{i}_{C2} = a^2\bar{i}_2$.

Figure 4.26 shows the phase and symmetrical components of the phase-to-phase fault currents.

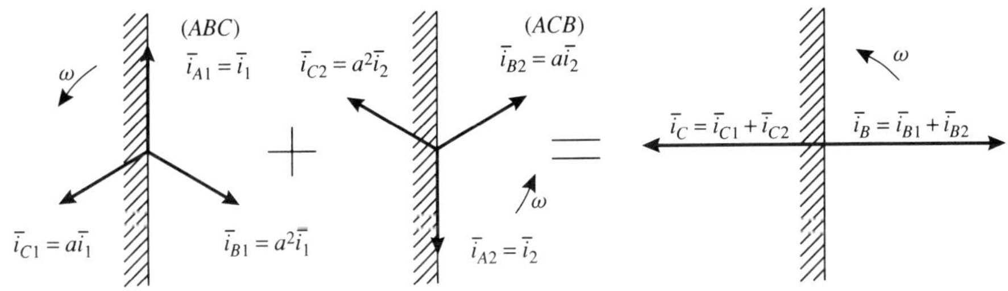

Figure 4.26 The phase-to-phase fault current transformed into its positive and negative sequence components rotating in the opposite direction

Phase-to-phase fault current harmonics explained by symmetrical components

The phase-to-phase fault current defined by equation 4.61 can be decomposed into two fundamental parts, an ac part containing terms in $\sin\gamma$ and a dc part containing terms in $\sin\gamma_0$, each part of which may be analysed separately using symmetrical components.

The ac part of the fault current produces an armature mmf $F_{a\,ac}$ which is stationary with respect to the stator but pulsates with time. By representing the ac component of the fault current by a positive sequence and negative sequence current, the effect of this pulsating mmf can be replicated by two counter-rotating mmf waves produced by the sequence currents. The positive-sequence currents (i_{A1}, i_{B1}, i_{C1}) produce an mmf $F_{a\,ac1}$ which rotates with the rotor and is stationary with respect to it. Changes in the flux linkages produced by this mmf are opposed by dc currents flowing in the rotor windings, just as in the generator with no subtransient saliency shown in Figure 4.24. The negative sequence currents (i_{A2}, i_{B2}, i_{C2}) produce mmf $F_{a\,ac2}$ which rotates in the opposite direction to the rotor, and induces double-frequency currents in the rotor windings again as shown in Figure 4.24 for a generator with no subtransient saliency. In turn these double frequency rotor currents produce a pulsating mmf which is stationary with respect to the rotor. This pulsating mmf can be represented by two mmf waves one rotating with respect to the rotor with velocity (-2ω) and the other with velocity 2ω, with the relative magnitudes of these two mmf waves depending on the degree of subtransient saliency. As the rotor itself rotates with angular velocity ω, these two rotor mmf components rotate with respect to the stator with velocities $(-\omega)$ and 3ω and introduce a third harmonic component into the fault current. Analysing the effect of this third harmonic on the rotor currents, and the effect of the rotor currents on the armature, leads to an explanation for the presence of the fifth harmonic in the fault current. This analysis may be continuously repeated to account for all the odd-numbered harmonics in the armature fault current. For a generator with no subtransient saliency the magnitude of the rotor mmf wave that rotates at 3ω relative to the armature is zero and so no third or higher order harmonics are present.

When $\sin\gamma_0 \neq 0$ a dc component of fault current appears and produces a stationary armature mmf $F_{a\,dc}$ which counter-rotates with respect to the rotor. The rotor windings oppose this mmf by inducing fundamental frequency ac currents. These rotor currents produce a sinusoidally-pulsating mmf which is stationary with respect to the rotor. As before, this mmf can be represented as two counter-rotating mmfs, but in this case with angular velocities ω and $(-\omega)$ with respect to the rotor. Again the relative magnitudes of these two mmfs depends on the degree of subtransient saliency present. As the rotational velocity of the rotor is ω, one of these mmfs is stationary with respect to the armature whilst the other rotates with velocity 2ω and induces a second harmonic component into the fault current. Analysing the influence of this harmonic explains the presence of the fourth, and the other even-numbered, harmonics in the fault current when subtransient saliency is present. The phase-to-phase fault current may then be expressed as a sum of the harmonics

$$i_C = -i_B = i_{(\omega)} + \underbrace{i_{(3\omega)} + i_{(5\omega)} + \dots}_{\text{induced by the negative sequence currents}} + i_{dc} + \underbrace{i_{(2\omega)} + i_{(4\omega)} + i_{(6\omega)} + \dots}_{\text{induced by the dc currents depending on the fault instant}} \tag{4.66}$$

Positive and negative sequence reactances

Equation 4.62 can now be used to determine the reactance with which the generator opposes the flow of the negative sequence currents. Figure 4.27a shows the representation of the generator for the positive and negative sequence currents, while Figure 4.27b shows the connection of the sequence equivalent circuits required to represent the phase-to-phase fault (Grainger & Stevenson, 1994).

Applying Ohm's law to the circuit of Figure 4.27b gives

$$\bar{I}_1'' = -\bar{I}_2'' = \frac{\bar{E}''}{j(X_1'' + X_2'')} \tag{4.67}$$

where X_1'' is the positive sequence subtransient reactance and X_2'' the negative sequence subtransient reactance. Using the inverse symmetrical component transformation gives the phase currents

$$-\bar{I}_B = \bar{I}_C = \frac{\sqrt{3}\,\bar{E}''}{X_1'' + X_2''} \tag{4.68}$$

The positive sequence component of the fault current produces a rotating flux whose interaction with the rotor is similar to that produced by the ac component of the three-phase short-circuit current. As the generator opposed the flow of currents produced by this flux

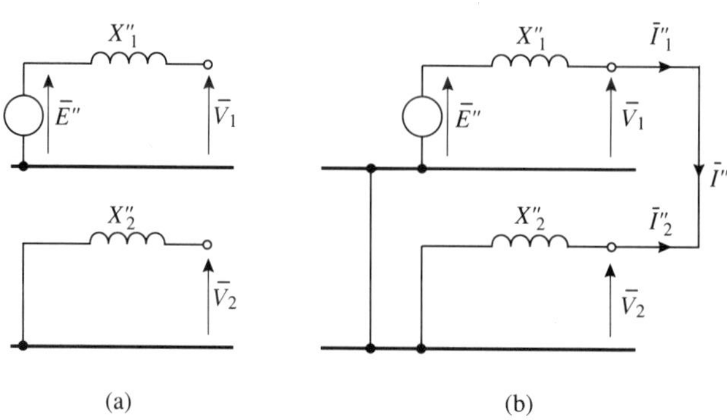

Figure 4.27 Equivalent circuits of the generator: (a) positive and negative sequence circuits; (b) connection of the circuits to account for the phase-to-phase fault

with the reactance X''_d, the reactance with which the generator opposes the flow of the positive sequence currents is also equal to X''_d and

$$X''_1 = X''_d. \tag{4.69}$$

Comparing equation 4.68 with the first part of equation 4.62 allows the negative sequence reactance X''_2 to be evaluated as

$$X''_2 = \sqrt{X''_d X''_q} \tag{4.70}$$

and the generator opposes the flow of the negative sequence currents with a reactance equal to the geometric average of the two subtransient reactances. Finally the second part of equation 4.62 shows that the generator opposes the flow of the dc component of the fault current with a reactance that has the same value as the negative sequence reactance i.e. $\sqrt{X''_d X''_q}$.

4.3.3 Influence of winding resistance

The winding resistance dissipates the magnetic energy stored in the winding and causes a decay in the dc component of the current flowing in that winding. As explained above the generator opposes the flow of the dc component of the phase-to-phase fault current with the reactance $X''_2 = \sqrt{X''_d X''_q}$. The resistance of the phase winding will then produce a decay in this dc component of the fault current with a time constant

$$T_\alpha = \frac{X''_2}{\omega R} = \frac{\sqrt{L''_d L''_q}}{R} \tag{4.71}$$

Equation 4.61 can now be modified by multiplying the $\sin\gamma_0$ term by the exponential function $\exp(-t/T_\alpha)$.

The decay of the ac component of fault current can best be explained by considering its positive and negative sequence components as derived in the previous section. The positive sequence component produces a rotating flux whose penetration into the rotor is opposed by dc rotor currents. The resistance of the rotor windings causes this component of the current to decay, first in the damper windings and then in the field winding, so that the armature reaction flux penetrates deeper and deeper into the rotor. As each of these flux conditions corresponds to the subtransient, transient, and steady-state condition the generator effectively opposes the flow of the positive sequence component of the fault current with the same reactances as for the three-phase fault:

$X''_1 = X''_d$ in the subtransient-state

$X'_1 = X'_d$ in the transient-state

$X_1 = X_d$ in the steady-state

The armature mmf produced by the negative sequence current rotates in the opposite direction from the rotor. The penetration of the flux produced by this mmf into the rotor is prevented by the flow of double-frequency rotor currents so that the same reactance opposes the flow of the negative sequence fault currents in all three characteristic generator states. The value of this negative sequence reactance is

$$X_2'' = X_2' = X_2 = \sqrt{X_d'' X_q''} \tag{4.72}$$

As the positive and negative sequence reactances of the generator are connected in the same way for the transient-state and the steady-state as for the subtransient-state (Figure 4.27) the ac phase-to-phase transient and steady-state fault currents are respectively $(X_d'' + X_2)/(X_d' + X_2)$ and $(X_d'' + X_2)/(X_d + X_2)$ times smaller than the subtransient fault currents. Equation 4.61 can be used to express the ac component of the fault current in the three characteristic states as

$$\begin{aligned} i''_{C\,ac}(t) &= \frac{\sqrt{3}E_{fm} \sin\gamma}{2(X_d'' \sin^2\gamma + X_q'' \cos^2\gamma)} \\ i'_{C\,ac}(t) &= i''_{C\,ac}(t) \frac{X_d'' + X_2}{X_d' + X_2} \\ i_{C\,ac}(t) &= i''_{C\,ac}(t) \frac{X_d'' + X_2}{X_d + X_2} \end{aligned} \tag{4.73}$$

Similarly as during the three-phase fault, the difference $(i''_{C\,ac} - i'_{C\,ac})$ decays at a rate determined by the subtransient time constant, while the difference $(i'_{C\,ac} - i_{C\,ac})$ decays with the transient time constant. Following similar arguments to those used to produce equations 4.33 and 4.36, the following expression is obtained for the phase-to-phase fault current with the effect of winding resistance included

$$i_C(t) = \frac{\sqrt{3}E_{fm}}{2(X_d'' \sin^2\gamma + X_q'' \cos^2\gamma)} [g_2(t) \sin\gamma - e^{-t/T_\alpha} \sin\gamma_0] \tag{4.74}$$

where the function

$$\begin{aligned} g_2(t) = (X_d'' + X_2) \Bigg[&\left(\frac{1}{X_d'' + X_2} - \frac{1}{X_d' + X_2} \right) e^{-t/T_\beta''} \\ &+ \left(\frac{1}{X_d' + X_2} - \frac{1}{X_d + X_2} \right) e^{-t/T_\beta'} + \frac{1}{X_d + X_2} \Bigg] \end{aligned} \tag{4.75}$$

is responsible for describing the decay of the ac component of the fault current (subscript "2" denotes the phase-to-phase fault).

As the reactances with which the generator opposes the flow of the phase-to-phase fault current are different from those opposing the flow of the three-phase fault current, the subtransient and transient time constants must be changed in proportion to the ratio of the reactances

$$T_\beta'' = T_d'' \left(\frac{X_d'}{X_d''} \right) \left(\frac{X_d'' + X_2}{X_d' + X_2} \right), \quad T_\beta' = T_d' \left(\frac{X_d}{X_d'} \right) \left(\frac{X_d' + X_2}{X_d + X_2} \right) \tag{4.76}$$

4.3.4 Subtransient torque

Similarly as in the three-phase fault, the flux produced by the damper windings during the phase-to-phase fault is equal and opposite to the d and q axis armature flux, and the electromagnetic torque is due solely to the interaction between the field flux and the quadrature component of armature reaction mmf. Neglecting subtransient saliency and substituting equation 4.60 into equation 4.41, similarly as for the three-phase fault, gives

$$\tau_{ac} = \frac{\pi}{2}\Phi_{f0}\frac{3}{2}N_a\frac{E_{fm}}{X''_d}(\sin\gamma - \sin\gamma_0)\cos\gamma \tag{4.77}$$

but, as $E_f = \frac{1}{\sqrt{2}}\omega\left(\frac{\pi}{2}N_a\right)\Phi_{f0}$, the subtransient torque can be written as

$$\tau_{ac} = \frac{3}{\omega}\frac{E_f^2}{X''_d}(\sin\gamma - \sin\gamma_0)\cos\gamma \quad \text{Nm} \tag{4.78}$$

This equation neglects winding resistance, the effect of which is to cause the rotor currents to decay with time as determined by the function $g_2(t)$, while the decay of the stator currents is determined by the functions $g_2(t)$ and $\exp(-t/T_\beta)$. Multiplying the corresponding components of equation 4.78 by these functions gives

$$\begin{aligned}\tau_{ac} &= \frac{3}{\omega}\frac{E_f^2}{X''_d}[g_2(t)\sin\gamma - e^{-t/T_\beta}\sin\gamma_0]g_2(t)\cos\gamma \\ &= \frac{3}{\omega}\frac{E_f^2}{X''_d}\left[\frac{1}{2}g_2^2(t)\sin 2\gamma - \sin\gamma_0 e^{-t/T_\beta}g_2(t)\cos\gamma\right] \quad \text{Nm}\end{aligned} \tag{4.79}$$

Unlike the torque due to the three-phase fault, the electromagnetic torque produced during a phase-to-phase fault depends on the fault instant. The component of the torque that depends on γ_0 has a periodic variation at fundamental frequency and a maximum value when the fault occurs at the instant the voltage in phase A reaches its negative maximum, that is when $\gamma_0 = -\pi/2$. In this situation the fundamental-frequency component is initially the highest and has a value twice the double-frequency component. As time progresses the magnitude of the fundamental frequency torque component reduces until, in the steady-state, it vanishes completely (Figure 4.28) and the torque varies at double-frequency.

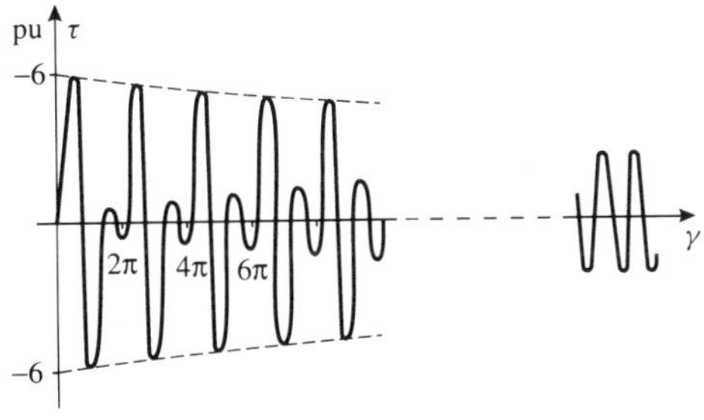

Figure 4.28 Example of the changes in the electromagnetic torque during phase-to-phase fault when $\gamma_0 = -\pi/2$

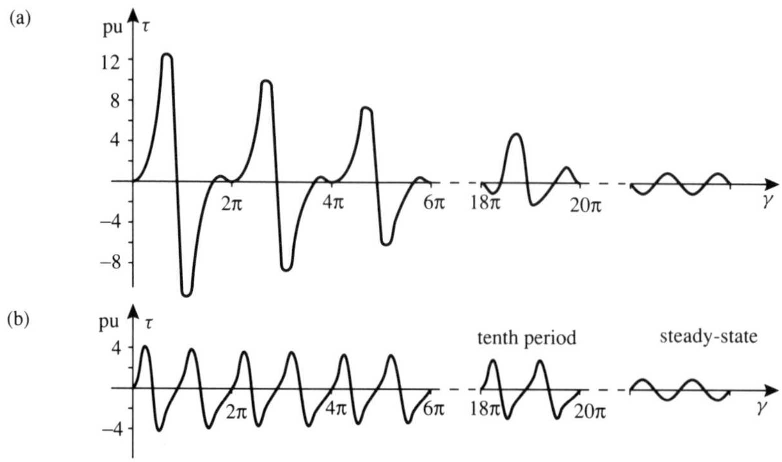

Figure 4.29 Example of torque variation during phase-to-phase fault in a case of high subtransient saliency of $X''_q \approx 2X''_d$ when the fault instant is : (a) $\gamma_0 = -\pi/2$; (b) $\gamma_0 = 0$

Including the effect of subtransient saliency introduces considerable complications to the formulae. Firstly the application of two-reaction theory leads to a term $X''_d \sin^2\gamma + X''_q \cos^2\gamma$ appearing in the denominator, as in the expression for the current. Secondly, as in the case of the three-phase fault, an ac component at double-frequency appears that is dependent on $(X''_q - X''_d)$. The torque is then expressed as

$$\tau_{ac} = \frac{3}{\omega}\frac{E_f^2}{X''_d \sin^2\gamma + X''_q \cos^2\gamma}\left\{[g_2(t)\sin\gamma - e^{-t/T_\beta}\sin\gamma_0]g_2(t)\cos\omega t \right.$$
$$\left. + \frac{1}{2}\frac{(X''_q - X''_d)[g_2(t)\sin\gamma - e^{-t/T_\beta}\sin\gamma_0]^2 \sin 2\gamma}{X''_d \sin^2\gamma + X''_q \cos^2\gamma}\right\} \quad \text{Nm} \tag{4.80}$$

When the difference $(X''_q - X''_d)$ is small the torque given by equation 4.80 is similar in both shape and maximum value to that obtained from equation 4.79. If subtransient saliency is high then the second part of equation 4.80 will distort the torque profile and may also increase the instantaneous value of the torque. Figure 4.29 shows the variation of the torque in a case of high subtransient saliency when $X''_q \approx 2X''_d$.

Equation 4.80 defines one part of the electromagnetic torque occurring during a phase-to-phase short-circuit and, just as in the case of the three-phase fault, there will be other torque components due to the power losses in the armature phase windings and the rotor field and damper windings. However these are usually small compared to the torque defined in equation 4.80.

4.4 Synchronisation

Synchronisation is the name given to the process of connecting a generator, with its field winding excited, to the power system. Such a synchronisation process is shown

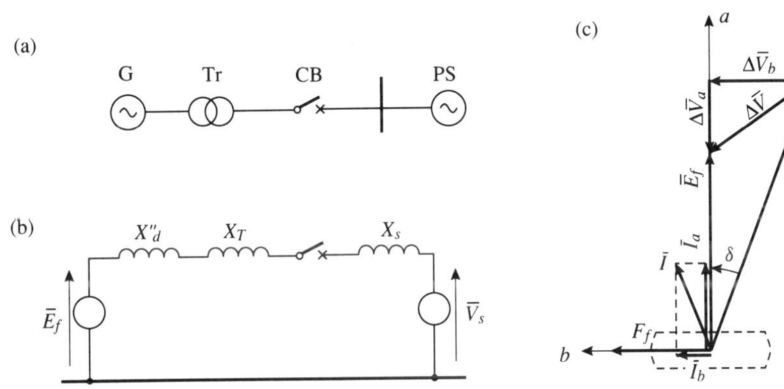

G-generator; Tr-transformer; PS-power system; CB-circuit-breaker (synchronising switch)

Figure 4.30 Synchronisation: (a) schematic diagram; (b) equivalent circuit in the subtransient state; (c) phasor diagram

schematically in Figure 4.30a,b where the power system is replaced by an infinite busbar of voltage $\overline{V}_s$ behind an equivalent reactance X_s. It is assumed that just before the switch is closed the generator rotates at a speed ω that is close to synchronous speed ω_s and that the excitation produces a no-load terminal voltage E_f that is close to the system voltage V_s.

Ideal synchronisation occurs when $\omega = \omega_s$, $\overline{E}_f = \overline{V}_s$, and $\delta = 0$ so that when the synchronising switch is closed no circulating current will flow. Normally a generator is connected to a power system when these conditions are almost, but not exactly, satisfied, when closing the synchronising switch will result in the flow of a circulating current that contains both an ac and a dc component. If the generator is assumed to feed no local load prior to synchronisation then it can be represented in the subtransient state by the excitation emf E_f behind the subtransient reactance X''_d. The circulating current can then be evaluated using Thevenin's Theorem based on the voltage $\Delta\overline{V}$ across the synchronising switch and the impedance as seen from the switch terminals. For simplicity armature resistance and rotor subtransient saliency will be neglected by assuming $R = 0$ and $X''_d = X''_q$.

Currents and torques

The subtransient armature current can be found by resolving the voltage across the switch $\Delta\overline{V}$ into two orthogonal components directed along the axes a and b as shown in Figure 4.30c. It is important to stress that although the a and b axes are directed along the d and q axes the resulting (a, b) voltage and current components have a different meaning from the (d, q) components. The (a, b) components are simply the components of the phase quantities resolved along two axes in the complex plane while the (d, q) voltage and current components are associated with fictitious, rotating, orthogonal armature windings as explained in Chapter 3. As the circuit is assumed to be purely reactive the current must lag the forcing voltage by $\pi/2$. This means that the a-component voltage ΔV_a will force the flow of the b-component of current I_b and vice versa.

Applying Thevenin's theorem gives the maximum value of the b-axis current i_{bm} due to ΔV_a as

$$i_{bm} = \frac{\Delta V_{am}}{x''_d} = \frac{V_{sm}\cos\delta - E_{fm}}{x''_d} \tag{4.81}$$

where $x''_d = X''_d + X_T + X_s$, $E_{fm} = \sqrt{2}E_f$ and $V_{sm} = \sqrt{2}V_s$. As the voltage component ΔV_a acts along the q-axis its effect in the Thevenin equivalent circuit, when looking from the terminals of the switch, is similar to that of the emf E_f during the three-phase short-circuit on an unloaded generator. As a result ac and dc currents will be induced in the armature windings similar to those expressed by equations 4.7 and 4.8 but with i_m replaced by i_{bm}.

Similarly, as in the case of the three-phase fault, the ac component of the phase currents due to I_b produces an armature mmf which rotates with the rotor and is in phase with the excitation mmf F_f, Figure 4.30c. As there is no angular displacement between these two mmfs they cannot produce an electromagnetic torque. On the other hand the dc component of the phase currents produces a stationary mmf which counter-rotates with respect to the rotor. Consequently the angular displacement between the field flux and the armature mmf varies with time producing a periodic torque given by equation 4.43 as

$$\tau_I = -\frac{3}{2}\frac{1}{\omega}E_{fm}i_{bm}\sin\omega t = -\frac{3}{\omega}\frac{E_f}{x''_d}(V_s\cos\delta - E_f)\sin\omega t \quad \text{Nm} \tag{4.82}$$

When compared with equation 4.43 the minus sign is due to the voltage $\Delta\overline{V}_a$ directly opposing E_f.

The b-axis voltage $\Delta\overline{V}_b$ produces the a-axis circulating current $\overline{I}_a$. The amplitude of this current is

$$i_{am} = \frac{\Delta V_{bm}}{x''_d} = \frac{V_{sm}\sin\delta}{x''_d} \tag{4.83}$$

$\overline{I}_a$ produces an mmf which rotates with the rotor and is directed along its q-axis, Figure 4.30c. The interaction between this component of armature mmf and the excitation flux produces a constant driving torque given by

$$\tau_{II} = \frac{3}{2}\frac{1}{\omega}E_{fm}i_{am} = \frac{3}{\omega}\frac{E_f V_s}{x''_d}\sin\delta \quad \text{Nm} \tag{4.84}$$

The dc component of the phase current driven by $\Delta\overline{V}_b$ produces a stationary flux shifted by $\pi/2$ with respect to the stationary flux driven by ΔV_a, so that the resulting periodic torque is equal to

$$\tau_{III} = \frac{3}{2}\frac{1}{\omega}E_{fm}i_{am}\sin\left(\omega t - \frac{\pi}{2}\right) = -\frac{3}{\omega}\frac{E_f V_s}{x''_d}\sin\delta\cos\omega t \quad \text{Nm} \tag{4.85}$$

At this point it is worth noting that although the circulating current resulting from synchronisation has a similar shape to the three-phase fault current, the direct component can be larger, or smaller, than the three-phase fault current depending on the actual values of X_T, X_s and δ. In addition the synchronisation torque may be quite different from the three-phase fault torque due to the influence of the voltage component $\Delta\overline{V}_b$. Adding the three torque components in equations 4.82, 4.84 and 4.85, gives

$$\tau = \tau_I + \tau_{II} + \tau_{III} = \frac{3}{\omega}\left[\frac{E_f V_s}{x_d''}\sin\delta(1-\cos\omega t) - \frac{E_f}{x_d''}(V_s\cos\delta - E_f)\sin\omega t\right] \quad \text{Nm} \tag{4.86}$$

which can be rewritten by substituting $\cos\delta = 1 - \sin\delta\tan(\delta/2)$ as

$$\tau = \frac{3}{\omega}\left[\frac{E_f V_s}{x_d''}\sin\delta\left(1-\cos\omega t + \tan\frac{\delta}{2}\sin\omega t\right) + \frac{E_f(E_f - V_s)}{x_d''}\sin\omega t\right] \quad \text{Nm} \tag{4.87}$$

The maximum value of the torque produced on synchronisation is highly dependent on the angle δ at which the switch is closed. When $E_f \approx V_s$, the second component in the above equation is zero and the torque is determined by the trigonometric expression

$$T_\delta(t) = \sin\delta\left(1-\cos\omega t + \tan\frac{\delta}{2}\sin\omega t\right) \tag{4.88}$$

The variation of this expression with time for three different values of δ is shown in Figure 4.31a whilst Figure 4.31b shows the maximum value of $T_\delta(t)$ as a function of δ. From these diagrams it can be seen that when δ is large the maximum torque is large and occurs early in the cycle with the highest torque occurring when $\delta = 2\pi/3$.

If $E_f \neq V_s$ then according to equation 4.87 the torque is increased by a periodic component dependent on the difference $(E_f - V_s)$.

Example 4.3

The per unit network data, referred to the generator unit, is: $X_d'' = X_q'' = 0.18$; $X_T = 0.11$; $X_s = 0.01$; $E_f = V_s = 1.1$. The per unit subtransient torque produced on synchronisation is obtained from equation 4.87 by multiplying by $\omega/3$ to obtain $\tau = T_\delta(1.1)^2/(0.18 + 0.11 + 0.01) = 4T_\delta$. The torque may be evaluated by taking $T_\delta(t)$ from Figure 4.31b

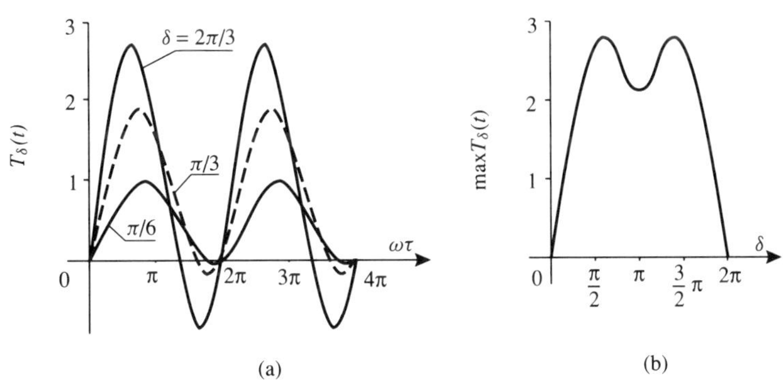

Figure 4.31 Equation 4.88: (a) as a function of time for various values of the synchronisation angle δ; (b) its maximum value as a function of the synchronisation angle δ

Table 4.4 Results for Example 4.3

synchronisation angle	$\pi/6$	$\pi/3$	$\pi/2$	$2\pi/3$	$5\pi/6$	π
torque in pu	4.08	7.48	9.64	10.4	9.64	8

for various values of the synchronisation angle δ. The results in Table 4.4 show that the torque may exceed the rated torque.

The equations derived in this section are valid immediately after synchronisation, that is during the subtransient period. As time progresses magnetic energy will be dissipated in the winding resistances, the circulating current will decay, and the generator will move from the subtransient state, through the transient state, until it eventually reaches the steady-state. As the circulating current decays so do the corresponding torque components.

4.5 Short Circuit in a Network and its Clearing

Section 4.2 discussed in some detail the effect of a three-phase short-circuit across the generator terminals. Fortunately this type of fault does not occur very often and much more common are faults some distance from the generator elsewhere in the power system. Figure 4.32 shows a general case where the generator is connected to the power system via a transformer and two parallel lines.

When a three-phase-to-earth fault occurs at the beginning of one of the lines, point $F2$, then the fault can be treated as a generator short-circuit but with some of the reactances and time constants modified to include the effect of the transformer. Firstly, the transformer reactance X_T must be added in series to the reactances X''_d, X'_d and X_d to form the resultant reactances

$$x''_d = X''_d + X_T, \quad x'_d = X'_d + X_T, \quad x_d = X_d + X_T \tag{4.89}$$

This has the effect of suppressing the magnitude of the currents given by equations 4.29 to 4.31 and the electromagnetic torque containing both periodic and aperiodic components as given by equations 4.45 to 4.48. Secondly the transformer resistance R_T increases the rate at which the stored magnetic energy can be dissipated so that the dc component of the short-circuit current, determined by T'_a, decays more rapidly. Thirdly the time constants T'_d and T''_d have to be modified to include the effect of the transformer reactance

$$T''_{d(network)} = T''_d \left(\frac{X'_d}{X''_d}\right)\left(\frac{X''_d + X_T}{X'_d + X_T}\right), \quad T'_{d(network)} = T'_d \left(\frac{X_d}{X'_d}\right)\left(\frac{X'_d + X_T}{X_d + X_T}\right) \tag{4.90}$$

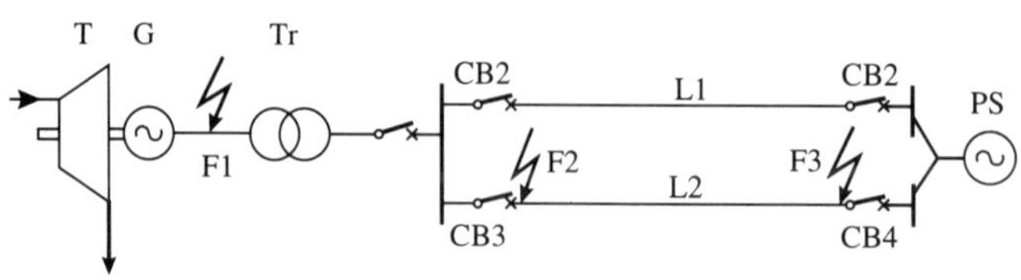

Figure 4.32 Schematic diagram of a fragment of a power system with examples of the short-circuit points $F1$, $F2$, $F3$

This leads to an increase in the value of these time constants. As a consequence of these changes the instantaneous value of the short-circuit current may pass through zero even on the first cycle.

The shape of the currents and torques during the short circuit and following fault clearing are shown in Figure 4.33. During the short-circuit high currents flow in all three phases and the torque oscillates around near-zero average values. As the clearing time is usually small the rotor angle, shown by the dashed line on the torque diagram, increases only slightly. When the fault is cleared the changes in the torque and the currents are similar to those occurring during synchronisation when the rotor angle first increases then decreases, but the torque oscillates around an average value.

In the case of an unsymmetrical fault in the network the generator currents are distorted by the step-up transformer. This effect is illustrated in Figure 4.34 assuming the transformer to be delta-star connected when a phase-to-phase fault on the secondary of the transformer, Figure 4.34a, is seen as a three-phase unsymmetrical current on the

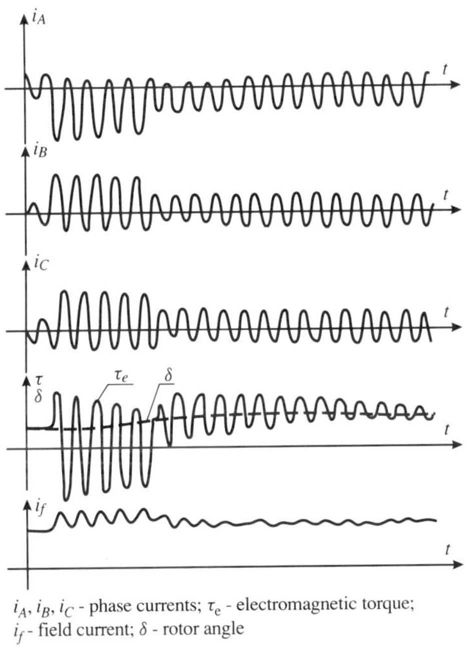

Figure 4.33 Three-phase fault and its clearing in the line connecting a generator to the system (Adapted from Kulicke & Webs (1975))

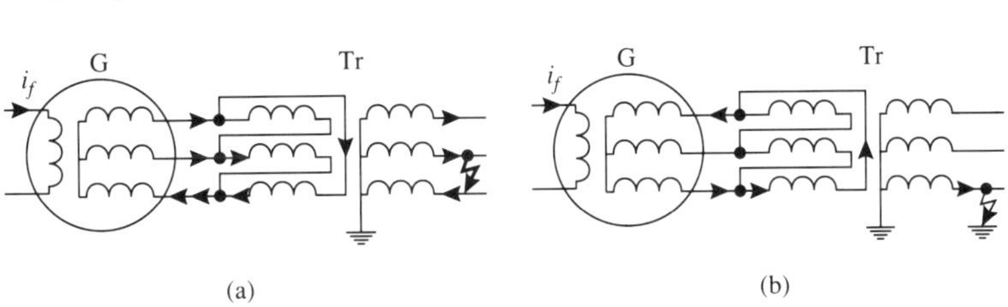

Figure 4.34 Transformation of the short-circuit current in the delta-star connection of the transformer: (a) phase-to-phase fault; (b) single-phase fault

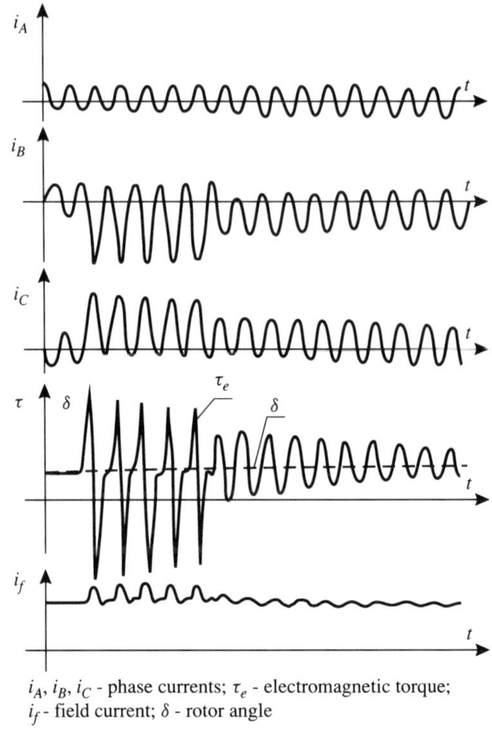

Figure 4.35 Phase-to-phase fault and its clearing in the line connecting a generator to the system (Adapted from Kulicke & Webs (1975))

primary. In the case of a single-phase fault on the secondary, Figure 4.34b, this is seen on the primary side as a phase-to-phase fault.

Figure 4.35 shows the time variation of the phase currents, air-gap torque and field current during, and after clearing, a short-circuit between phases B and C in the line connecting the generator to a system. Such an unsymmetrical fault produces negative sequence currents and consequently the electromagnetic torque contains a double-frequency component.

The dynamic behaviour of the generator during the short-circuit period and the importance of the clearing time is considered in more detail in Chapter 5.

5

Electromechanical Dynamics — Small Disturbances

In the previous chapter the currents and torques produced in a synchronous generator as the result of a system disturbance were discussed and, as the duration of the disturbance was very short, the generator rotational speed could be considered constant. In this chapter a longer time scale is considered during which the rotor speed will vary and interact with the electromagnetic changes to produce *electromechanical dynamic* effects. The time scale associated with these dynamics is sufficiently long for them to be influenced by the turbine and the generator control systems.

The aim of this chapter is to provide both an explanation of how, and why, mechanical movement of the generator rotor is influenced by electromagnetic effects and to examine how this movement varies depending on the generator operating state. During the course of this discussion some important stability concepts will be introduced together with their basic mathematical description and explanation of the physical implications.

5.1 Swing Equation

Section 2.2 described the constructional features of a turbine and explained how a multi-stage turbine drives the generator rotor through a common drive shaft. A diagram of a multi-stage turbine consisting of high pressure, intermediate pressure and low pressure stages is shown in Figure 5.1 where each turbine stage contributes a proportion of the total mechanical driving torque. The drive system can be modelled by a series of rotational masses, to represent the inertia of each turbine stage, connected together by springs, to represent the torsional stiffness of the drive shaft and coupling between the stages. Such a model can be used to compute the torsional natural frequencies of the drive system and can also be used in a detailed computer simulation to obtain information on the actual shaft torques occurring following a major fault or disturbance. This is discussed further in Section 6.6. One of the natural frequencies of the turbine/generator drive system will be at 0 Hz and represents free body rotation where the turbine and generator inertias move together with no relative displacement of the individual rotor

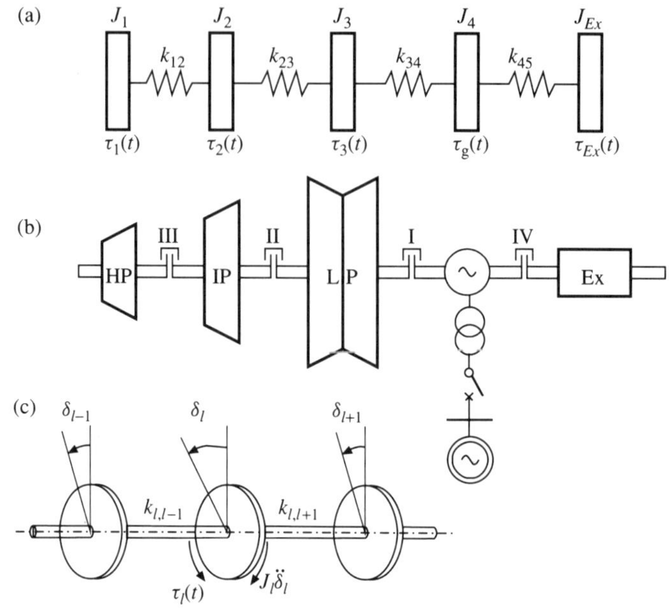

HP, IP, LP - high-pressure, intermediate-pressure, and low-pressure sections of the turbine; G - generator; Ex-rotating exciter; J - moment of inertia of individual sections; τ-external torque acting on a mass; k - stiffness of a shaft section; δ - angular displacement of a mass; I, II, III, IV - shaft couplers

Figure 5.1 Generating unit as an oscillating system: (a) division of the rotor mass into individual sections; (b) schematic diagram; (c) torsional displacement

masses. When connected to the power system this free-body rotation will appear as a low frequency oscillation of typically 1 to 2 Hz. It is this free-body rotation that is addressed in this section.

When considering free-body rotation the shaft can be assumed to be rigid when the total inertia of the rotor J is simply the sum of the individual inertias. Any unbalanced torque acting on the rotor will result in the acceleration or deceleration of the rotor as a complete unit according to Newton's second law

$$J\frac{\mathrm{d}\omega_m}{\mathrm{d}t} + D_d\omega_m = \tau_t - \tau_e \tag{5.1}$$

where J is the total moment of inertia of the turbine and generator rotor (kg m^2), ω_m is the rotor shaft velocity (mechanical rad/s), τ_t is the torque produced by the turbine (Nm), τ_e is the counter-acting electromagnetic torque and D_d is the damping-torque coefficient (Nms) and accounts for the mechanical rotational loss due to windage and friction.

Although the turbine torque τ_t changes relatively slowly, due to the long thermal time constants associated with the boiler and turbine, the electromagnetic torque τ_e may change its value almost instantaneously. In the steady-state the rotor angular speed is synchronous speed ω_{sm} whilst the turbine torque τ_t is equal to the sum of the electromagnetic torque τ_e and the damping (or rotational loss) torque $D_d\omega_{sm}$

$$\tau_t = \tau_e + D_d\omega_{sm} \quad \text{or} \quad \tau_m = \tau_t - D_d\omega_{sm} = \tau_e \tag{5.2}$$

where τ_m is the net mechanical shaft torque i.e. the turbine torque less the rotational losses at $\omega_m = \omega_{sm}$. It is this torque that is converted into electromagnetic torque. If, due to some disturbance, $\tau_m > \tau_e$ then the rotor accelerates, if $\tau_m < \tau_e$ then it decelerates.

In Section 3.3 the rotor position with respect to a synchronously rotating reference axis was defined by the rotor, or power, angle δ. The rotor velocity can therefore be expressed as

$$\omega_m = \omega_{sm} + \Delta\omega_m = \omega_{sm} + \frac{d\delta_m}{dt} \tag{5.3}$$

where δ_m is the rotor angle expressed in mechanical radians and $\Delta\omega_m = d\delta_m/dt$ is the *speed deviation* in mechanical radians per second.

Substituting equation 5.3 into equation 5.1 gives:

$$J\frac{d^2\delta_m}{dt^2} + D_d\left(\omega_{sm} + \frac{d\delta_m}{dt}\right) = \tau_t - \tau_e \quad \text{or} \quad J\frac{d^2\delta_m}{dt^2} + D_d\frac{d\delta_m}{dt} = \tau_m - \tau_e \tag{5.4}$$

Multiplying through by the rotor synchronous speed ω_{sm} gives

$$J\omega_{sm}\frac{d^2\delta_m}{dt^2} + \omega_{sm}D_d\frac{d\delta_m}{dt} = \omega_{sm}\tau_m - \omega_{sm}\tau_e \tag{5.5}$$

As power is the product of angular velocity and torque, the terms on the right hand side of this equation can be expressed in power to give

$$J\omega_{sm}\frac{d^2\delta_m}{dt^2} + \omega_{sm}D_d\frac{d\delta_m}{dt} = \frac{\omega_{sm}}{\omega_m}P_m - \frac{\omega_{sm}}{\omega_m}P_e \tag{5.6}$$

where P_m is the net shaft power input to the generator and P_e is the electrical air-gap power, both expressed in watts. During a disturbance the speed of a synchronous machine is normally quite close to synchronous speed so that $\omega_m \approx \omega_{sm}$ and equation 5.6 becomes

$$J\omega_{sm}\frac{d^2\delta_m}{dt^2} + \omega_{sm}D_d\frac{d\delta_m}{dt} = P_m - P_e \tag{5.7}$$

The coefficient $J\omega_{sm}$ is the *angular momentum* of the rotor at synchronous speed and, when given the symbol M_m, allows equation 5.7 to be written as

$$M_m\frac{d^2\delta_m}{dt^2} = P_m - P_e - D_m\frac{d\delta_m}{dt} \tag{5.8}$$

where $D_m = \omega_{sm}D_d$ is the damping coefficient. Equation 5.8 is called the *swing equation* and is the fundamental equation governing the rotor dynamics.

It is common practice to express the angular momentum of the rotor in terms of a normalised *inertia constant* when all generators of a particular type will have similar "inertia" values regardless of their rating. The inertia constant is given the symbol H defined as the stored kinetic energy in megajoules at synchronous speed divided by the machine rating S_n in MVA so that

$$H = \frac{0.5J\omega_{sm}^2}{S_n} \quad \text{and} \quad M_m = \frac{2HS_n}{\omega_{sm}} \tag{5.9}$$

The units of H are seconds. In effect H simply quantifies the kinetic energy of the rotor at synchronous speed in terms of the number of seconds it would take the generator to provide an equivalent amount of electrical energy when operating at a power output equal to its MVA rating. In Continental Europe the symbol T_m is used for *mechanical time constant* where

$$T_m = \frac{J\omega_{sm}^2}{S_n} = 2H \quad \text{and} \quad M_m = \frac{T_m S_n}{\omega_{sm}} \tag{5.10}$$

Again the units are seconds but the physical interpretation is different. In this case, if the generator is at rest and a mechanical torque equal to S_n/ω_{sm} is suddenly applied to the turbine shaft, the rotor will accelerate, its velocity will increase linearly and it will take T_m seconds to reach synchronous speed ω_{sm}.

Section 3.3 showed that the power angle and angular speed can be expressed in electrical radians and electrical radians per second respectively, rather than their mechanical equivalent, by substituting

$$\delta = \frac{\delta_m}{p/2} \quad \text{and} \quad \omega_s = \frac{\omega_{sm}}{p/2} \tag{5.11}$$

where p is the number of poles. Introducing the inertia constant and substituting equations 5.11 into equation 5.8 allows the swing equation to be written as

$$\frac{2HS_n}{\omega_s}\frac{d^2\delta}{dt^2} + D\frac{d\delta}{dt} = P_m - P_e \quad \text{or} \quad \frac{T_m S_n}{\omega_s}\frac{d^2\delta}{dt^2} + D\frac{d\delta}{dt} = P_m - P_e \tag{5.12}$$

where D, the damping coefficient, is $D = 2D_m/p$. The equations 5.12 can be rationalised by defining an *inertia coefficient* M and *damping power* P_D such that

$$M = \frac{2HS_n}{\omega_s} = \frac{T_m S_n}{\omega_s}, \qquad P_D = D\frac{d\delta}{dt} \tag{5.13}$$

when the swing equation takes the common form

$$M\frac{d^2\delta}{dt^2} = P_m - P_e - P_D = P_{acc} \tag{5.14}$$

where P_{acc} is the net accelerating power. The time derivative of the rotor angle $d\delta/dt = \Delta\omega = \omega - \omega_s$ is the *rotor speed deviation* in electrical radians per second. Often it is more convenient to replace the second-order differential equation 5.14 by two first-order equations

$$M\frac{d\Delta\omega}{dt} = P_m - P_e - P_D = P_{acc}$$

$$\frac{d\delta}{dt} = \Delta\omega \tag{5.15}$$

It is also common power system practice to express the swing equation in a *per unit* form. This simply means normalising equation 5.14 to a common MVA base. Assuming that this base is the MVA rating of the generator then dividing both sides of equation 5.14 by

S_n does not modify the equation structure but all parameters are now normalised to the three-phase MVA base.

5.2 Damping Power

Damping of the rotor motion by mechanical losses is small and can be neglected for all practical considerations. The main source of damping in the synchronous generator is provided by the *damper*, or *amortisseur*, windings described in Section 2.2.1. The damper windings have a high resistance/reactance ratio and act in a similar way to the short-circuited squirrel-cage rotor windings in an induction motor. In the subtransient state these windings act as a perfect screen and the changes in the armature flux cannot penetrate them. In the transient state the air-gap flux, which rotates at the synchronous speed, penetrates the damper windings and induces an emf and current in them whenever the rotor speed ω is different from the synchronous speed ω_s. This induced current produces a damping torque which, according to Lenz's Law, tries to restore the synchronous speed of the rotor. As this additional torque only appears when $\omega \neq \omega_s$ it is proportional to $\Delta\omega = \mathrm{d}\delta/\mathrm{d}t$ and is referred to as the *asynchronous torque*.

Damper windings can be on both rotor axes, or on the d-axis only. In the round-rotor generator the solid steel rotor body provides paths for eddy currents which have the same effect as damper windings. Machines with laminated salient poles require explicit damper windings for effective damping.

Rigorous derivation of an expression for damping power is long and complicated but an approximate equation for the generator-infinite busbar system can be derived if the following assumptions are made:

(i) that the resistance of both the armature and the field winding are neglected
(ii) that damping is produced only by the damper windings
(iii) that the leakage reactance of the armature winding can be neglected
(iv) that excitation does not affect the damping torque

With these assumptions an equation for damping power can be derived using the induction motor equivalent circuit shown in Figure 3.22.

Figure 4.10a showed that in subtransient state the equivalent reactance of the generator, as seen from the network, consists of the armature winding leakage reactance X_l connected in series with the parallel connected damper winding leakage reactance X_D, the field winding leakage reactance X_f and the armature reaction reactance X_a. Figure 5.2a shows

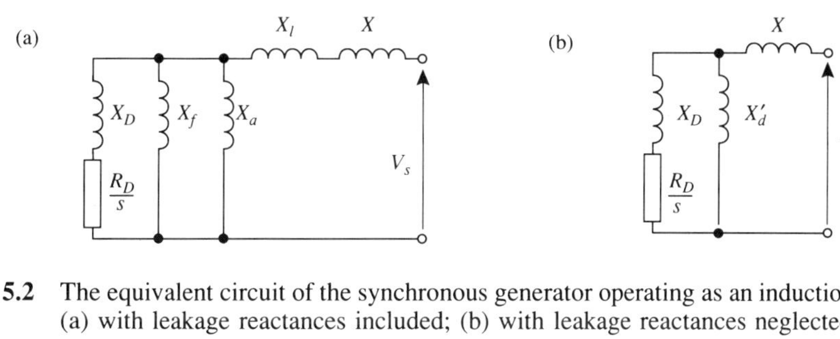

Figure 5.2 The equivalent circuit of the synchronous generator operating as an induction machine: (a) with leakage reactances included; (b) with leakage reactances neglected

a similar equivalent circuit of the generator-infinite busbar system with the generator operating as an induction machine. The reactance X represents the combined reactance of the step-up transformer and the network while V_s is the infinite busbar voltage. Due to assumption (iv) the field winding is closed but not excited. To take the speed deviation into account the equivalent resistance in the damper branch is divided by $s = \Delta\omega/\omega_s$ in a similar way as for the induction machine. Initially rotor saliency is neglected.

Assumption (iii) allows the reactance X_l to be neglected when the subtransient and transient reactances are given approximately by the approximate expressions in equations 4.12 and 4.13

$$X'_d \cong \frac{1}{\frac{1}{X_f} + \frac{1}{X_a}}, \qquad X''_d \cong \frac{1}{\frac{1}{X_f} + \frac{1}{X_a} + \frac{1}{X_D}} \tag{5.16}$$

The first of these equations allows the parallel connection of X_f and X_a in Figure 5.1a to be replaced by X'_d, as shown in Figure 5.1b.

Equation 5.16 also allows the damper winding leakage reactance to be approximately expressed as a function of the subtransient and transient reactances as

$$X_D \cong \frac{X'_d X''_d}{X'_d - X''_d} \tag{5.17}$$

The subtransient short-circuit time constant $T''_d = X_D/\omega_s R_D$ can now be calculated as

$$T''_d = \frac{X_D}{\omega_s R_D} \cong \frac{X'_d X''_d}{\omega_s R_D (X'_d - X''_d)} \tag{5.18}$$

allowing the damper winding resistance (R_D/s) as seen from the armature to be written

$$\frac{R_D}{s} = \frac{X'_d X''_d}{X'_d - X''_d} \frac{1}{T''_d \omega_s s} = \frac{X'_d X''_d}{X'_d - X''_d} \frac{1}{T''_d \Delta\omega} \tag{5.19}$$

For small values of speed deviation the term R_D/s is large and the current flows mostly through X'_d allowing the series connection of X and X'_d to be treated as a voltage divider. With this assumption the voltage drop across X'_d is equal to $V_s X'_d/(X + X'_d)$. Ohm's law applied to the damper equivalent branch then gives

$$I_D^2 \cong V_s^2 \left(\frac{X'_d}{X + X'_d}\right)^2 \frac{1}{\left(\frac{R_D}{s}\right)^2 + X_D^2} \tag{5.20}$$

and the damping power is

$$P_D = I_D^2 \frac{R_D}{s} \cong V_s^2 \frac{(X'_d)^2}{(X + X'_d)^2} \frac{\frac{R_D}{s}}{\left(\frac{R_D}{s}\right)^2 + \frac{(X'_d X''_d)^2}{(X'_d - X''_d)^2}} \tag{5.21}$$

Substituting (R_D/s) from equation 5.19 gives

$$P_D \cong V_s^2 \frac{X'_d - X''_d}{(X + X'_d)^2} \frac{X'_d}{X''_d} \frac{T''_d \Delta\omega}{1 + (T''_d \Delta\omega)^2} \tag{5.22}$$

When rotor saliency is taken into account a similar formula can be derived for the quadrature axis. The resultant damping power can be found by replacing the driving voltage V_s in both d and q equivalent circuits by the respective voltage components, $V_d = -V_s \sin\delta$ and $V_q = V_s \cos\delta$ to give

$$P_D = V_s^2 \left[\frac{X'_d - X''_d}{(X + X'_d)^2} \frac{X'_d}{X''_d} \frac{T''_d \Delta\omega}{1 + (T''_d \Delta\omega)^2} \sin^2\delta + \frac{X'_q - X''_q}{(X + X'_q)^2} \frac{X'_q}{X''_q} \frac{T''_q \Delta\omega}{1 + (T''_q \Delta\omega)^2} \cos^2\delta \right] \tag{5.23}$$

The damper power depends on the rotor angle δ and fluctuates with the rotor speed deviation $\Delta\omega = \mathrm{d}\delta/\mathrm{d}t$. For small speed deviations the damping power is proportional to speed deviation while for larger speed deviations it is a non-linear function of speed deviation and resembles the power-slip characteristic of the induction motor, Figure 3.23.

For a small speed deviation $s = \Delta\omega/\omega_s \ll 1$ and the $(T''_d \Delta\omega)^2$ term in the denominator of equation 5.23 can be neglected. Equation 5.23 then simplifies to

$$P_D = V_s^2 \left[\frac{X'_d - X''_d}{(X + X'_d)^2} \frac{X'_d}{X''_d} T''_d \sin^2\delta + \frac{X'_q - X''_q}{(X + X'_q)^2} \frac{X'_q}{X''_q} T''_q \cos^2\delta \right] \Delta\omega \tag{5.24}$$

This equation is identical to the expression derived by Dahl (1938) and quoted by Kimbark (1956). In both equations the time constants T''_d and T''_q are the subtransient short-circuit time constants.

When δ is large damping is strongest in the d-axis while when δ is small the q-axis damper winding produces the stronger damping . Equation 5.24 can usefully be re-written as

$$P_D = [D_d \sin^2\delta + D_q \cos^2\delta]\Delta\omega = D(\delta)\Delta\omega \tag{5.25}$$

where $D(\delta) = D_d \sin^2\delta + D_q \cos^2\delta$ and D_d, D_q are damping coefficients in both axes. Figure 5.3 shows the variation of the damping coefficient D with δ as expressed by equation 5.25. The function reaches extrema for $\delta = 0, \pi$ or $\delta = \pi/2, 3\pi/2$ with the corresponding extremal values of damping coefficient being equal to D_d and D_q respectively. The average value of the damping coefficient is $D_{av} = (D_d + D_q)/2$.

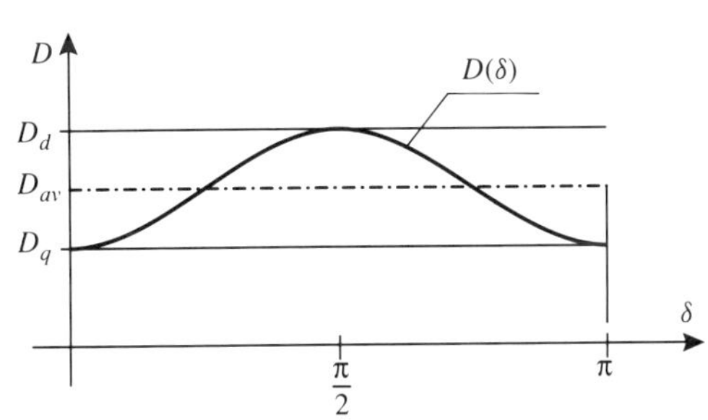

Figure 5.3 Damping coefficient as a function of rotor angle

The network equivalent reactance X has a significant influence on the damping (asynchronous) power as its squared value appears in the denominator of equations 5.23 and 5.24. In comparison its influence on the synchronous power, equation 3.84, is much less where it appears in the denominator and it is not squared.

Damping power at large speed deviations

When analysing the damping power over a wide range of speed deviation values it is convenient to re-write equation 5.23 as

$$P_D = P_{D(d)} \sin^2 \delta + P_{D(q)} \cos^2 \delta \tag{5.26}$$

where $P_{D(d)}$ and $P_{D(q)}$ both depend on the speed deviation. Both components are proportional to non-linear functions of the speed deviation of the form $\alpha/(1+\alpha^2)$, where $\alpha = T''_d \Delta\omega$ for the first component and $\alpha = T''_q \Delta\omega$ for the second component. This function reaches a maximum for $\alpha = 1$ so that each component of the damping power will normally reach a maximum, critical, value at a different *critical speed deviation* given by

$$s_{cr(d)} = \frac{\Delta\omega_{cr(d)}}{\omega_s} = \frac{1}{T''_d \omega_s}, \qquad s_{cr(q)} = \frac{\Delta\omega_{cr(q)}}{\omega_s} = \frac{1}{T''_q \omega_s} \tag{5.27}$$

with

$$P_{D(d)cr} = \frac{V_s^2}{2} \frac{X'_d - X''_d}{(X + X'_d)^2} \frac{X'_d}{X''_d}, \qquad P_{D(q)cr} = \frac{V_s^2}{2} \frac{X'_q - X''_q}{(X + X'_q)^2} \frac{X'_q}{X''_q} \tag{5.28}$$

Figure 5.4 shows the variation of both $P_{D(d)}$ and $P_{D(q)}$ as a function of the speed deviation. Both factors are equal to zero when $\Delta\omega = 0$ and increase their value as the speed deviation increases until the critical value is reached, after which they decrease. With δ changing the damping power P_D will assume values between $P_{D(d)}$ and $P_{D(q)}$ with the average value lying between the axes characteristics.

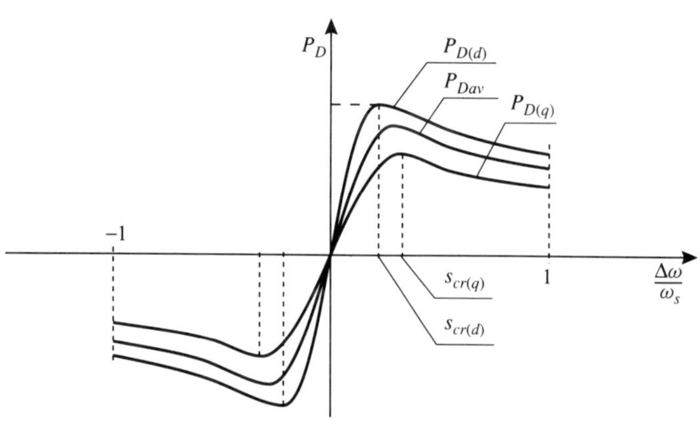

Figure 5.4 Average value of the damping power as a function of speed deviation

5.3 Equilibrium Points

Section 5.1 showed how the accelerating power is dependent on the difference between the turbine power P_m and the electrical air-gap power P_e (minus the damping power P_D). The mechanical power is supplied by the turbine and its value is controlled by the turbine governor. The electrical air-gap power depends on the generator loading and varies depending on the generator paramcters and the power angle. It depends also on the generator operating state but in this section only the steady-state model of the generator-infinite busbar system will be considered. The infinite busbar voltage V_s, introduced in Section 3.3 and shown in Figure 3.18, will be assumed as the datum so that the rotor of the fictitious system equivalent generator will provide a reference axis rotating at a constant speed. Expressions will be derived for P_e assuming that the combined generator and system resistance r is small and can be neglected. However it is important to realise that the air-gap power is the power supplied to the system P_s plus the power loss in the equivalent resistance I^2r.

Section 3.3 showed that in the steady-state the generator can be represented by a constant emf E_q behind the synchronous reactances X_d and X_q. If all the resistances and shunt admittances of the generator-infinite busbar system shown in Figure 3.18 are neglected then the air-gap power P_e is equal to the power delivered to the system P_{sEq} and is given by Equation 3.89:

$$P_e = P_{E_q} = \frac{E_q V_s}{x_d} \sin\delta + \frac{V_s^2}{2}\frac{x_d - x_q}{x_q x_d} \sin 2\delta \tag{5.29}$$

where $x_d = X_d + X$, $x_q = X_q + X$ and $X = X_T + X_s$ is the combined reactance of the step-up transformer and the equivalent network. The suffix E_q indicates that the air-gap power is calculated with E_q assumed to be constant.

Section 3.3 showed that the angle δ is the angle between the $\overline{E}_q$ and $\overline{V}_s$ phasors (referred to as the *power angle*) and, at the same time, it is the spatial angle between the generator rotor and the rotor of the fictitious system generator (referred to as the *rotor angle*). This is of vital importance as it allows the swing equation 5.14, describing the dynamics of the rotor, to be linked with equation 5.29, describing the electrical state of the generator.

Equation 5.29 describes the *steady-state*, or *static*, power-angle characteristic of the generator. For constant E_q and V_s the characteristic becomes a function of the power/rotor angle δ only, that is $P_e = P_e(\delta)$, and equation 5.14 can be re-written as

$$M\frac{\mathrm{d}^2\delta}{\mathrm{d}t^2} = P_m - P_e(\delta) - D\frac{\mathrm{d}\delta}{\mathrm{d}t} \tag{5.30}$$

in order to emphasise that P_e is a function of δ.

When in equilibrium the generator operates at synchronous speed $\omega = \omega_s$, so that

$$\left.\frac{\mathrm{d}\delta}{\mathrm{d}t}\right|_{\delta=\hat{\delta}} = 0 \quad \text{and} \quad \left.\frac{\mathrm{d}^2\delta}{\mathrm{d}t^2}\right|_{\delta=\hat{\delta}} = 0 \tag{5.31}$$

where $\hat{\delta}$ is the rotor angle at the equilibrium point. Substituting the above conditions into equation 5.30 shows that at equilibrium $P_m = P_e(\hat{\delta})$. To simplify considerations the round rotor generator with $x_d = x_q$ is assumed, when the expression for the air-gap power

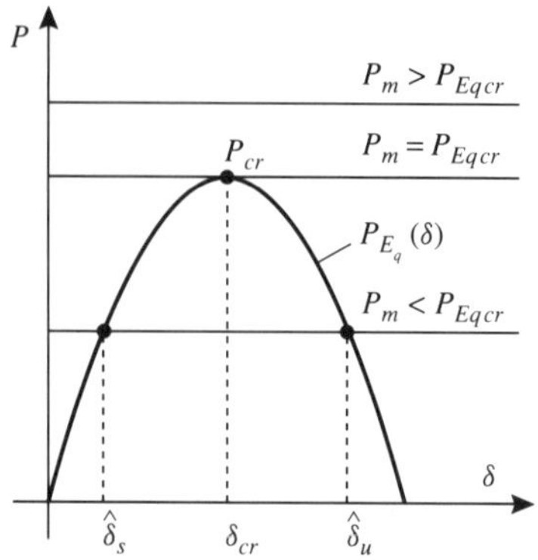

Figure 5.5 Equilibrium points for various values of mechanical power

simplifies to

$$P_e(\delta) = P_{E_q}(\delta) = \frac{E_q V_s}{x_d} \sin\delta \tag{5.32}$$

This characteristic is shown in Figure 5.5. The maximum value of $P_{E_q}(\delta)$ is referred to as the *critical power* $P_{E_q cr}$ while the corresponding value of the rotor angle is referred to as the *critical angle* δ_{cr}. For the round-rotor generator described by equation 5.32 $P_{E_q cr} = E_q V_s / x_d$ and $\delta_{cr} = \pi/2$.

As the mechanical power depends only on the flow of the working fluid through the turbine, and not on δ, the turbine mechanical power characteristic can be treated as a horizontal line P_m = constant on the (δ, P) plane. The intersection between the horizontal P_m characteristic and the sine-like $P_e(\delta)$ characteristic gives the equilibrium points of the generator. Three situations, shown in Figure 5.5, are now possible:

(i) $P_m > P_{E_q cr}$. Clearly no equilibrium points exist and the generator cannot operate at such a condition

(ii) $P_m = P_{E_q cr}$. There is only one equilibrium point at δ_{cr}

(iii) $P_m < P_{E_q cr}$. There are two equilibrium points at $\hat{\delta}_s$ and $\hat{\delta}_u$. This condition corresponds to normal operation and will be discussed in the next section

5.4 Steady-State Stability of the Unregulated System

Classically a system is said to be steady-state stable for a particular operating condition if, following any small disturbance, it reaches a steady-state operating point which is identical, or close to, the pre-disturbance condition. This is also known as *small disturbance*, or *small signal, stability*. A small disturbance is a disturbance for which the equations that describe the dynamics of the power system may be linearised for analysis purposes.

The dynamics of the generator, and its stability, are generally affected by automatic control of the generator and the turbine (Figure 2.2). To simplify considerations analysis of the generator dynamics will be considered in two sections. In this section the *unregulated system* will be considered when the mechanical power and the excitation voltage are assumed to be constant. This corresponds to analysing the system *steady-state inherent* or *natural stability*. The influence of the AVR will be considered in Section 5.5.

5.4.1 Pull-out power

Section 4.4. showed that when a generator is about to be synchronised to the system it must rotate at synchronous speed and its terminal voltage must be equal to, and in phase with, the busbar voltage. When the synchronising switch is closed the steady-state equilibrium point is reached at $\delta = 0$ and $P_m = 0$ and corresponds to the origin of the power-angle characteristic shown in Figure 5.5. If now the mechanical power P_m is slowly increased by a small amount the electrical power P_e must follow the changes so that a new equilibrium point of $P_m = P_e$ is reached. In other words the system is steady-state stable if *an increase (decrease) in mechanical power causes a corresponding increase (decrease) in electrical power*. If the system reaction is opposite to this, that is an increase in mechanical power is accompanied by a decrease in electrical power, then no equilibrium point can be reached. These stability considerations are illustrated in Figure 5.6.

For a certain value of mechanical power, marked as "old", there are two equilibrium points: *1* and *2*. If mechanical power is increased to a "new" value, Figure 5.6a, then this gives rise to a surplus of power at point *1*. This surplus acceleration power is equal to segment *1–3* and will, according to equation 5.14, accelerate the rotor and so increase the power angle and the electrical power. The resulting motion is shown by the arrow and is towards the new equilibrium point *5*.

The opposite situation occurs at equilibrium point *2*. Here the acceleration power, equal to segment *2–4*, again accelerates the rotor and leads to an increase in the power angle but this now results in a reduction in the electrical power. The motion, as shown by the arrow, is away from the new equilibrium point *6*.

A similar response is obtained if the mechanical power is reduced, Figure 5.6b. For equilibrium points on the left-hand-side of the power-angle characteristic the rotor motion is from point *1* towards the new equilibrium point *5*. On the other hand when starting from equilibrium point *2*, on the right-hand-side of the characteristic, it is not possible

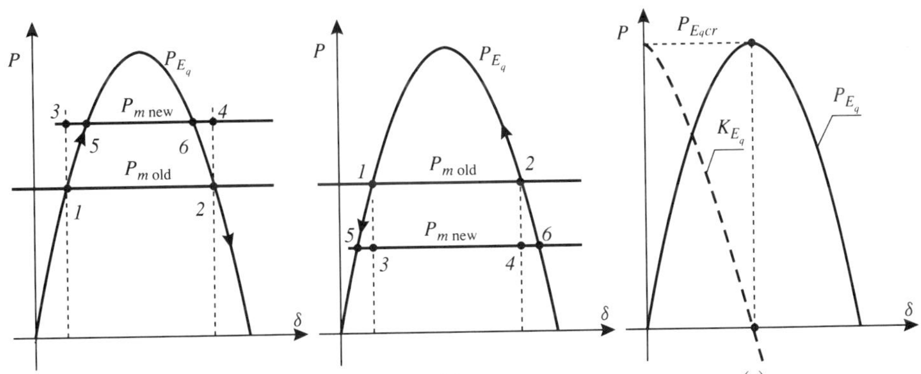

Figure 5.6 Illustration of the conditions for steady-state stability: (a) increase in mechanical power; (b) decrease in mechanical power; (c) the generator steady-state power and synchronising power coefficient

to reach the new equilibrium point *6* as the rotor motion is in the opposite direction. Obviously an increase in the mechanical power to a value exceeding P_{cr} results in the loss of synchronism due to the lack of any equilibrium point.

From this discussion it is apparent that the generator-infinite busbar system with constant excitation emf E_q is steady-state stable only on the left-hand-side of the power-angle characteristic, that is when the slope K_{E_q} of the characteristic is positive:

$$K_{E_q} = \frac{\partial P_{E_q}}{\partial \delta}\bigg|_{\delta=\hat{\delta}_s} > 0 \tag{5.33}$$

K_{E_q} is referred to as the *steady-state synchronising power coefficient* whilst the critical power $P_{E_q\,cr}$ is often referred to as the *pull-out power* to emphasise the fact that a larger mechanical power will result in the unregulated generator losing synchronism with the rest of the system. Figure 5.6c shows the plot of $K_{E_q}(\delta)$ and $P_{E_q\,cr}$. The value of $P_{E_q\,cr}$ is also referred to as the *steady-state stability limit* and can be used to determine the *steady-state stability margin* as

$$c_{E_q} = \frac{P_{E_q\,cr} - P_m}{P_{E_q\,cr}} \tag{5.34}$$

where P_m is the actual loading of the generator. The stability margin varies between $c_{E_q} = 1$ (when the generator is unloaded) and $c_{E_q} = 0$ (when the generator is critically loaded).

It should be emphasised that the pull-out power is determined by the steady-state characteristic $P_{E_q}(\delta)$ whilst the dynamic response of the generator to a disturbance is determined by the transient power-angle characteristic described in the next sub-section.

5.4.2 Transient power-angle characteristics

Chapter 4 explained how any disturbance acting on a generator will produce a sudden change in the armature currents and flux. This flux change induces additional currents in the rotor windings (field and damper) that expel the armature flux into high reluctance paths around the rotor so as to screen the rotor and keep the rotor flux linkage constant. As the emf E_q is proportional to the field current the additional induced field current will cause changes in E_q so that the assumption of constant E_q used to derive the static power-angle characteristic (equation 5.32) is invalid in the analysis of post-disturbance rotor dynamics.

Chapter 4 also explained how the induced rotor currents decay with time as the armature flux penetrates first the damper windings (subtransient period) and then the field winding (transient period). Usually the frequency of rotor oscillations is about 1–2 Hz which corresponds to an electromechanical swing period of about 1–0.5 seconds. This period can be usefully compared with the generator subtransient open-circuit time constants T''_{d0} and T''_{q0} which are in the region of a few hundredths of a second. On the other hand the d-axis transient time constant T'_{d0} is in the region of a few seconds while the q-axis time constant T'_{q0} is about one second (see Table 4.3 and equations 4.16). Consequently, on the time scale associated with rotor oscillations, it may be assumed that changes in the armature flux can penetrate the damper windings but that the field winding and the round-rotor body act as perfect screens maintaining constant flux linkages. This corresponds to assuming that the transient emfs E'_d and E'_q are constant. This assumption will modify the generator power-angle characteristic as described below.

Constant flux linkage model

Assume that the generator is connected to the infinite busbar as shown in Figure 3.18 and that all the resistances and shunt impedances associated with the transformer and network can be neglected. The corresponding equivalent circuit and phasor diagram of the round-rotor and salient-pole generator in the transient state are then as shown in Figure 5.7. The fictitious rotor of the infinite busbar serves as the synchronously rotating reference axis. The reactances of the step-up transformer and the connecting network can be combined with that of the generator to give

$$x'_d = X'_d + X, \qquad x'_q = X'_q + X \tag{5.35}$$

where X'_d and X'_q are the d- and q-axis transient reactances of the generator, and $X = X_T + X_s$. The voltage equations can now be constructed using Figure 5.7a to give

$$E'_d = V_{sd} + x'_q I_q, \qquad E'_q = V_{sq} - x'_d I_d \tag{5.36}$$

where V_{sd} and V_{sq} are the d and q components of the infinite busbar voltage V_s, given by $V_{sd} = -V_s \sin\delta$ and $V_{sq} = V_s \cos\delta$. As all resistances are neglected the air-gap power is $P_e = V_{sd}I_d + V_{sq}I_q$. Substituting the values for V_{sd} and V_{sq} and the currents I_d and I_q

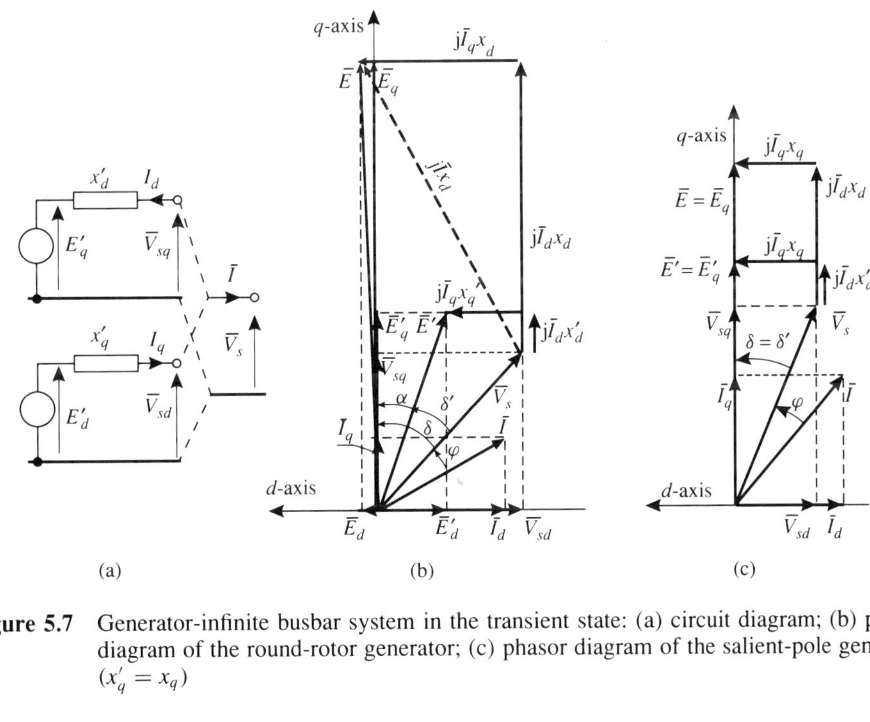

(a) (b) (c)

Figure 5.7 Generator-infinite busbar system in the transient state: (a) circuit diagram; (b) phasor diagram of the round-rotor generator; (c) phasor diagram of the salient-pole generator ($x'_q = x_q$)

calculated from equations 5.36 gives

$$P_e = P_s = V_{sd}I_d + V_{sq}I_q = -\frac{E'_qV_{sd}}{x'_d} + \frac{V_{sd}V_{sq}}{x'_d} + \frac{E'_dV_{sq}}{x'q} - \frac{V_{sd}V_{sq}}{x'_q}$$

$$P_e = P_{E'}(\delta) = \frac{E'_qV_s}{x'_d}\sin\delta + \frac{E'_dV_s}{x'_q}\cos\delta - \frac{V_s^2}{2}\frac{x'_q - x'_d}{x'_qx'_d}\sin 2\delta \tag{5.37}$$

Equation 5.37 defines the air-gap power as a function of δ and the (d,q)-components of the transient emf, and is valid for any generator (with or without transient saliency). The phasor diagram in Figure 5.7b shows that $E'_d = -E'\sin\alpha$, $E'_q = E'\cos\alpha$ and $\delta = \delta' + \alpha$. Substituting these relationships into equation 5.37 gives, after some simple but tedious algebra,

$$P_e = P_{E'}(\delta') = \frac{E'V_s}{x'_d}\left[\sin\delta'\left(\cos^2\alpha + \frac{x'_d}{x'_q}\sin^2\alpha\right) + \frac{1}{2}\left(\frac{x'_q - x'_d}{x'_q}\right)\cos\delta'\sin 2\alpha\right] - \frac{V_s^2}{2}\frac{x'_q - x'_d}{x'_dx'_q}\sin 2(\delta' + \alpha) \tag{5.38}$$

Assuming constant rotor flux linkages, the values of the emfs E'_d and E'_q are also constant implying that both $E' =$ constant and $\alpha =$ constant. Equation 5.38 describes the generator power-angle characteristic $P_e(E', \delta')$ in terms of the transient emf and the transient power angle and is valid for any type of generator.

A generator with a laminated, salient-pole, rotor cannot produce effective screening in the q-axis with the effect that $x'_q = x_q$. Inspection of the phasor diagram in Figure 5.7c shows that in this case E' lies along the q-axis so that $\alpha = 0$ and $\delta' = \delta$. Consequently, in this special case, equation 5.38 simplifies to

$$P_e = P_{E'_q}(\delta')\Big|_{x'_q = x_q} = \frac{E'_qV_s}{x'_d}\sin\delta' - \frac{V_s^2}{2}\frac{x_q - x'_d}{x_qx'_d}\sin 2\delta' \tag{5.39}$$

Classical model

The constant flux linkage model expressed by equation 5.38 can be simplified by ignoring transient saliency, that is assuming $x'_d \approx x'_q$. With this assumption equation 5.38 simplifies to

$$P_e = P_{E'}(\delta')\Big|_{x'_d \approx x'_q} \approx \frac{E'V_s}{x'_d}\sin\delta' \tag{5.40}$$

The assumption of $x'_d = x'_q$ allows the separate d- and q-axis circuits shown in Figure 5.7b to be replaced by one simple equivalent circuit shown in Figure 5.8. As this figure is drawn assuming $x'_d = x'_q$, α is also calculated under this assumption and is designated α'. Normally $x'_q > x'_d$ and $\alpha' > \alpha$. In this *classical model* all the voltages, emfs and currents are phasors in the network reference frame rather than their components resolved along the (d,q) axes.

Generally, as Table 4.3 shows, there is always some degree of transient saliency and $x'_d \neq x'_q$. However it should be noted that the generator is connected to the infinite busbar

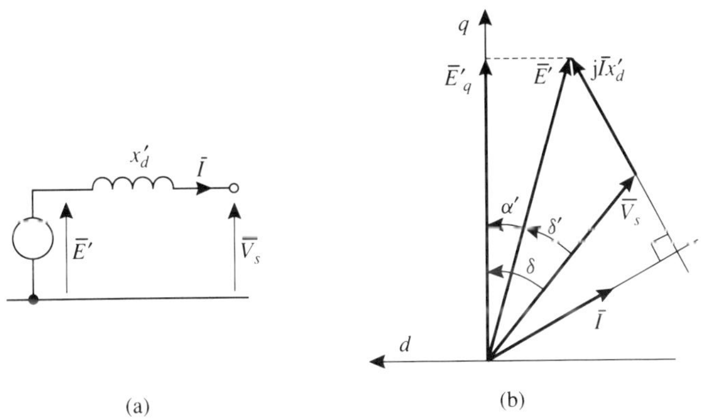

Figure 5.8 Classical model of the generator in the transient state: (a) circuit diagram; (b) phasor diagram

through the reactance $X = X_T + X_s$ so that $x'_d = X'_d + X$ and $x'_q = X'_q + X$. The influence of X is such that as its magnitude increases the closer the term x'_d/x'_q approaches unity whilst the term $(x'_q - x'_d)/x'_d x'_q$ approaches zero. Consequently when the network reactance is large the classical model, and the constant flux linkage model defined by equation 5.38, give very similar results even for a generator with a laminated, salient-pole, rotor.

It is important to note that δ' is the angle between V_s and E' and not the angle between V_s and the q-axis. However during the transient period the emfs E'_d and E'_q are assumed to be constant (with respect to the rotor axes) and α (or α') is also constant with

$$\delta = \delta' + \alpha, \qquad \frac{d\delta}{dt} = \frac{d\delta'}{dt} \quad \text{and} \quad \frac{d^2\delta}{dt^2} = \frac{d^2\delta'}{dt^2} \tag{5.41}$$

This allows δ' to be used in the swing equation instead of δ when equation 5.14 becomes

$$M\frac{d^2\delta'}{dt^2} = P_m - \frac{E'V_s}{x'_d}\sin\delta' - D\frac{d\delta'}{dt} \tag{5.42}$$

An important advantage of the classical model is that the generator reactance may be treated in a similar way to the reactance of the transmission lines and other network elements. This has particular importance for multi-machine systems when combining the algebraic equations describing the generator and the network is not as easy as for the generator-infinite busbar system. Due to its simplicity, the classical model will be used extensively throughout this book to analyse and explain rotor dynamics.

Steady-state and transient characteristics on the power-angle diagram

It is now important to understand how the dynamic characteristic of a generator is located with respect to the static characteristic on the power-angle diagram. For a given stable equilibrium point, when $P_e = P_m$, the balance of power must be held whichever characteristic is considered so that both the static and the dynamic characteristic must intersect at the stable equilibrium point. Generally the angle α between δ_0 and δ'_0 is not equal to zero and the transient characteristic is shifted to the right. For salient-pole generators, $\alpha = 0$, both characteristics originate at the same point, and the intersection between them at the equilibrium point is due to a distortion in the sine shape of the transient characteristic.

Example 5.1

The round-rotor generator considered in Example 4.1 is connected to the power system (infinite busbar) via a transformer with series reactance $X_T = 0.13$ pu and a transmission line with series reactance $X_L = 0.17$ pu. Find, and plot, the steady-state and the transient characteristics using both the constant flux linkage and thc classical generator model. As in Example 4.1 the generator real power output is 1 pu, the reactive power output is 0.5 pu and the terminal voltage is 1.1 pu.

From Example 4.1, $I_0 = 1.016$, $\varphi_{g0} = 26.6°$, $E_{q0} = 2.336$, $\delta_{g0} = 38.5°$, $I_{d0} = -0.922$, $I_{q0} = 0.428$, $E'_q = E'_{q0} = 1.073$, $E'_d = E'_{d0} = -0.522$. Angle α can be found from $\alpha = \arctan|E'_d/E'_q| = 26°$. The total reactances are

$$x_d = x_q = X_d + X_T + X_L = 1.9,$$

$$x'_d = X'_d + X_T + X_L = 0.53,$$

$$x'_q = X'_q + X_T + X_L = 0.68$$

Taking V_g as the reference the phasor of the transient emf is $\overline{E}' = 1.193\angle 12.5°$. It is now necessary to calculate the system voltage V_s and calculate the position of E_q and E' with respect to V_s. The system voltage can be calculated from

$$\overline{V}_s = \overline{V}_g - \mathrm{j}(X_T + X_L)\overline{I} = 1.1 - \mathrm{j}0.3 \times 1.016\angle 26.6° = 1.0\angle -15.8°$$

The angle δ_0 is therefore equal to $38.5 + 15.8 = 54.3°$, δ'_0 is $12.5 + 15.8 = 28.3°$ and φ_0 is $26.6 - 15.8 = 10.8°$. The (d, q) components of the system voltage are:

$$V_{sd} = -1\sin 54.3° = -0.814,\ V_{sq} = 1\cos 54.3° = 0.584$$

The steady-state power-angle characteristic is

$$P_{E_q}(\delta) = \frac{E_q V_s}{x_d}\sin\delta = \frac{2.336 \times 1}{1.9}\sin\delta = 1.23\sin\delta.$$

The transient characteristic (constant flux linkage model) can be calculated from equation 5.37 as

$$P_{E'}(\delta) = \frac{1.07 \times 1}{0.53}\sin\delta + \frac{-0.522 \times 1}{0.68}\cos\delta - \frac{1^2}{2}\frac{0.68 - 0.53}{0.68 \times 0.53}\sin 2\delta$$
$$= 2.02\sin\delta - 0.768\cos\delta - 0.208\sin 2\delta$$

The approximated transient characteristic for the classical model can be calculated from equation 5.40. The transient emf can be calculated with respect to V_s as $\overline{E}' = V_s + \mathrm{j}x'_d\overline{I}_0 = 1 + \mathrm{j}0.53 \times 1.016\angle -10.8° = 1.223\angle 25.7°$. Thus $E' = 1.233$, $\delta'_0 = 25.7°$

and $\alpha' = \delta - \delta' = 54.3 - 25.7 = 28.6°$. The approximated transient characteristic can now be calculated from equation 5.40 as

$$P_{E'}(\delta') \approx \frac{1.223 \times 1}{0.53} \sin\delta' = 2.31 \sin\delta'$$

This characteristic is shifted with respect to $P_{E_q}(\delta)$ by 28.6°. Figure 5.9a shows all three characteristics. It can be seen that the classical model gives a good approximation of the constant flux linkage model.

Example 5.2

Re-calculate all the characteristics from Example 5.1 for the salient-pole generator considered previously in Example 4.2.
From Example 4.2, $I_0 = 1.016$, $\varphi_{g0} = 26.6°$, $E_q = E_{q0} = 1.735$, $I_{d0} = -0.784$, $I_{q0} = 0.647$, $E'_q = E'_{q0} = 1.241$, $E'_d = 0$, $\delta_{g0} = 23.9°$, $\alpha = 0$. The total reactances are

$$x_d = X_d + X_T + X_L = 1.23,$$

$$x_q = x'_q = X_q + X_T + X_L = 0.99,$$

$$x'_d = X'_d + X_T + X_L = 0.6$$

Taking V_g as the reference the phasor of the transient emf is $\overline{E'} = 1.241\angle 23.9°$. The system voltage was calculated in Example 5.1 and is $V_s = 1.0\angle 15.8°$. Thus $\delta_0 = \delta'_0 = 23.9 + 15.8 = 39.7°$ and $\delta_0 + \varphi_0 = 39.7 + 10.8 = 50.5°$. The (d,q) components of the system voltage are:

$$V_{sd} = -1\sin 39.7° = -0.64, \; V_{sq} = 1\cos 39.7° = 0.77$$

The steady-state power-angle characteristic is

$$P_{E_q}(\delta) = \frac{E_q V_s}{x_d}\sin\delta + \frac{V_s^2}{2}\frac{x_d - x_q}{x_d x_q}\sin 2\delta = \frac{1.735 \times 1}{1.23}\sin\delta + \frac{1}{2}\frac{1.23 - 0.99}{1.23 \times 0.99}\sin 2\delta$$

$$= 1.41\sin\delta + 0.099\sin 2\delta \cong 1.41\sin\delta$$

The transient characteristic (constant flux linkage model) can be calculated from equation 5.37 as

$$P_{E'}(\delta) = \frac{1.241 \times 1}{0.6}\sin\delta - \frac{1^2}{2}\frac{0.99 - 0.6}{0.99 \times 0.6}\sin 2\delta = 2.07\sin\delta - 0.322\sin 2\delta$$

The approximated transient characteristic for the classical model can be calculated assuming $x'_d = x'_q$. The transient emf, calculated with respect to V_s, is then $\overline{E'} = V_s + \mathrm{j}x'_d\overline{I} = 1 + \mathrm{j}0.6 \times 1.016\angle -10.8° = 1.265\angle 28.3°$. Thus $E' = 1.265$, $\delta' = 28.3°$ and $\alpha' = \delta - \delta' = 39.7 - 28.3 = 11.4°$. The approximated transient characteristic can now be calculated from equation 5.40 as

$$P_{E'}(\delta') \approx \frac{1.265 \times 1}{0.6}\sin\delta' = 2.108\sin\delta'$$

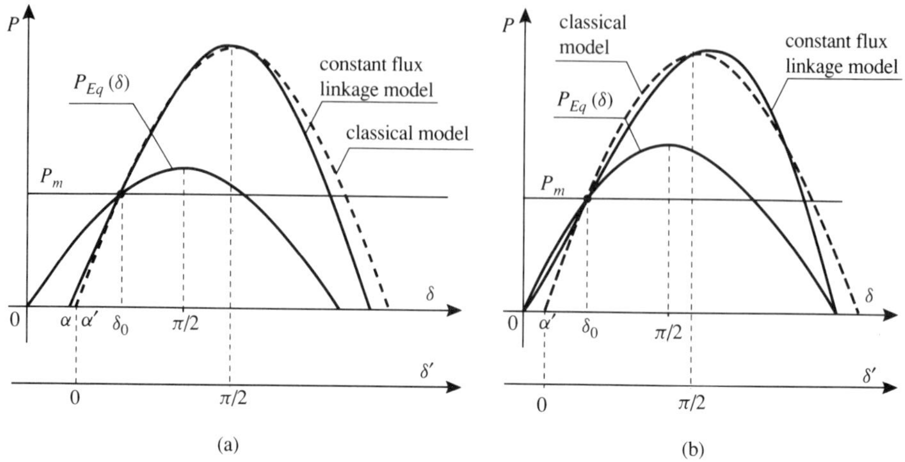

Figure 5.9 Steady-state and transient characteristics: (a) round-rotor generator ($\alpha > 0$); (b) laminated salient-pole machine ($\alpha = 0$)

This characteristic is shifted with respect to $P_{E_q}(\delta)$ by $\alpha' = 11.4°$. Figure 5.9b shows all three characteristics. It can be seen that the classical model gives a good approximation of the constant flux linkage model.

Figure 5.9 shows that by neglecting transient saliency the classical model does not generally significantly distort the transient characteristics. As $x'_d < x_d$, the amplitude of the transient characteristic is greater than the amplitude of the steady-state characteristic. Consequently the slope of the transient characteristic at the stable equilibrium point, referred to as the *transient synchronising power coefficient*

$$K_{E'} = \left.\frac{\partial P_{E'}}{\partial \delta'}\right|_{\delta'=\hat{\delta}'_s} \tag{5.43}$$

is steeper than the slope of the steady-state characteristic, K_{E_q}, defined in equation 5.33. For the classical model $K_{E'} = E'V_s \cos\hat{\delta}'_s / x'_d$.

When the generator loading is changed P_m changes and the stable equilibrium point is shifted to a new position on the steady-state characteristic $P_{E_q}(\delta)$, provided that P_m does not exceed the pull-out power. This increased load modifies the transient characteristic as illustrated in Figure 5.10a which shows three transient characteristics corresponding to three different mechanical powers $P_{m(1)}$, $P_{m(2)}$, $P_{m(3)}$. The generator excitation (and therefore amplitude of the steady-state characteristic) is assumed to be unchanged. Each new steady-state equilibrium point corresponds to a different value of E' and a different transient characteristic $P_{E'}(\delta)$ crossing the equilibrium point. Note that increased loading results in a smaller transient emf E' so that the amplitude of the transient characteristic is reduced. This can be verified by inspecting the pre-disturbance d-axis phasor diagram, Figure 4.14, assuming E_f constant. This shows that an increase in the armature current I_0 results in a larger voltage drop across the reactances $(x_d - x'_d)$ and $(x_q - x'_q)$ and therefore a smaller $E'_0 = E'$ value. Figure 5.10b illustrates the same effect shown in Figure 5.10a but in a slightly different way as it shows the steady-state electrical power P_{E_q}, the steady-state synchronising power coefficient K_{E_q} and the transient synchronising power

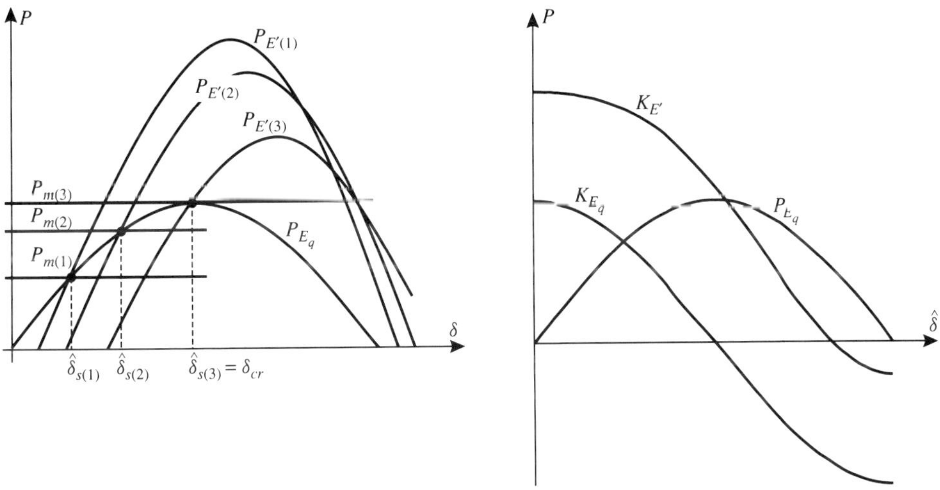

Figure 5.10 Effect of increased loading on the round-rotor generator: (a) the steady-state and transient characteristics for three different loads; (b) the steady-state electrical power and the steady-state and transient synchronising power coefficients as a function of the steady-state equilibrium rotor angle

coefficient $K_{E'}$ as a function of the generation loading as expressed by the steady-state equilibrium angle $\hat{\delta}$. The case of the round-rotor generator is shown. An increase in the generator loading results in smaller steady-state and transient synchronising power coefficients K_{E_q} and $K_{E'}$ but nevertheless $K_{E'}$ is always larger than K_{E_q}.

5.4.3 Rotor swings and equal area criterion

The generator models derived in the previous sub-section can now be used to describe and analyse the rotor dynamics assuming that the generator is subjected to a sudden disturbance. Such a disturbance will result in additional currents being induced in the rotor windings so as to maintain constant rotor flux linkages and therefore constant E'. As the synchronous emfs E_d and E_q follow the changes in the field winding current and the rotor body current respectively, they cannot be assumed to be constant and any rotor swings must therefore follow the transient power-angle curve $P_e = P_{E'}(\delta)$.

Disturbances in the generator-infinite busbar system may arise due to a change in the turbine mechanical power or a change in the equivalent system reactance. The effect of such practical disturbances will be considered in detail later and here it suffices to consider the effect of disturbing the rotor angle δ from its equilibrium value $\hat{\delta}_s$ to a new value $(\hat{\delta}_s + \Delta\delta_0)$. Although such a disturbance is unlikely from a technical point of view it does allow the introduction of a number of important concepts that are fundamental to understanding the effects of other, more practical, types of disturbance. The initial disturbed conditions for the solution of the system differential equations are

$$\Delta\delta(t = 0^+) = \Delta\delta_0 \neq 0, \qquad \Delta\omega(t = 0^+) = \Delta\omega_0 = 0 \tag{5.44}$$

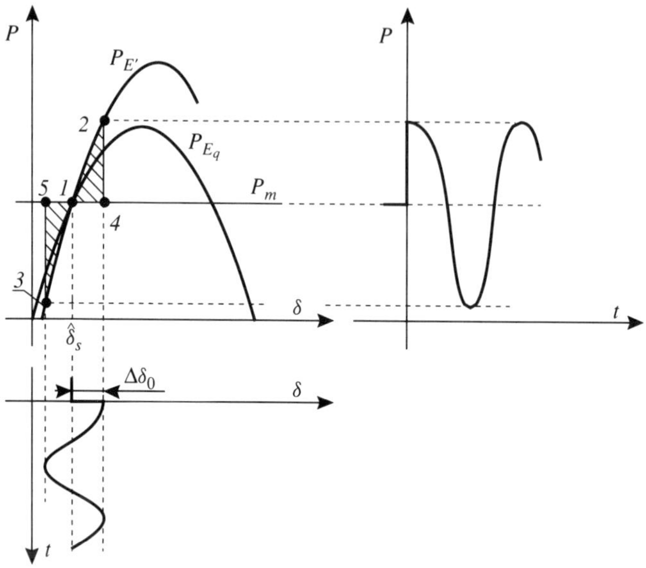

Figure 5.11 Rotor and power swings following a disturbance

The reaction of the system to this disturbance is illustrated in Figure 5.11 which shows both the steady-state and the transient power characteristics as a function of the rotor angle δ. As the disturbance cannot change the rotor flux linkage the initial (disturbed) generator operating point will be at point *2* on the transient characteristic $P_{E'}(\delta)$ which crosses the pre-disturbance stable equilibrium point *1*.

Any movement requires work to be done so that by increasing the rotor angle from $\hat{\delta}_s$ to $(\hat{\delta}_s + \Delta\delta_0)$ the disturbance performs work on the rotor. In rotational motion, work is equal to the integral of the torque acting over the angular displacement and in this case the net torque is equal to the difference between the electrical (transient) torque and the mechanical torque. As power is equal to the product of torque and angular velocity, and assuming $\omega \approx \omega_s$, the work done by the disturbance is proportional to the integral of the net power acting over the angular displacement, that is:

$$W_{1-2} = \int_{\hat{\delta}_s}^{\hat{\delta}_s+\Delta\delta_0} [P_{E'}(\delta) - P_m]\,\mathrm{d}\delta = \text{ area } (1\text{-}2\text{-}4) \tag{5.45}$$

Throughout this book the quantity W will be treated as energy (or work) although, strictly speaking, it should be divided by the synchronous speed ω_s. As the rotor speed deviation $\Delta\omega$ at point *2* is assumed to be zero (the speed is synchronous) the kinetic energy of the rotor is the same as that at equilibrium point *1*. This means that the work done by the disturbance W_{1-2} increases the system potential energy (with respect to equilibrium point *1*) by

$$E_p = W_{1-2} = \text{ area } (1\text{-}2\text{-}4) \tag{5.46}$$

This initial potential energy provides the impetus necessary to move the rotor back towards its equilibrium point *1*. At the disturbed rotor position *2* the mechanical driving torque (and power) is less than the opposing electrical torque (and power), the resulting net deceleration power (equal to segment *4–2*) starts to reduce the rotor speed (with respect to synchronous speed) and the rotor angle will decrease. At the equilibrium point *1* all the potential energy in equation 5.46 will be converted into kinetic energy (relative to synchronous speed) and the deceleration work done is equal to

$$E_k = W_{1-2} = \text{area } (1-2-4) - \tfrac{1}{2}M\Delta\omega^2 \tag{5.47}$$

The kinetic energy will now push the rotor past the equilibrium point $\hat{\delta}_s$ so that it continues to move along curve *1–3*. On this part of the characteristic the mechanical driving torque is greater than the opposing electrical torque and the rotor begins to accelerate. Acceleration will continue until the work performed by the acceleration torque (proportional to the integral of the accelerating power) becomes equal to the work performed previously by the deceleration torque. This happens at point *3* when

$$\text{area } (1-3-5) = \text{area } (1\text{-}2\text{-}4) \tag{5.48}$$

At this point the generator speed is again equal to synchronous speed but, as $P_m > P_{E'}$, the rotor will continue to accelerate increasing its speed above synchronous and will swing back towards $\hat{\delta}_s$. In the absence of any damping the rotor will continually oscillate between points *2* and *3*, as described by the swing equation 5.30. The resulting swing curve is shown in the lower part of Figure 5.11 with the corresponding power swings being shown to the right.

Equation 5.48 defines the maximum deflection of the rotor in either direction on the basis of equalising the work done during deceleration and acceleration. This concept will be used later in Chapter 6 to define the *equal area criterion* of stability.

5.4.4 Effect of damper windings

Equation 5.25 showed that for small deviations in rotor speed the damper windings produce a damping power $P_D = D\Delta\omega$ that is proportional to the rotor speed deviation. To help explain the effect of the damper windings on the system behaviour it is convenient to re-write the swing equation 5.30 as

$$M\frac{d^2\delta}{dt^2} = P_m - [P_e(\delta) + P_D] \tag{5.49}$$

when the damping power is seen to either add to, or subtract from, the electrical air-gap power $P_e(\delta)$ depending on the sign of the speed deviation. If $\Delta\omega < 0$ then P_D is negative effectively opposing the air-gap power and shifting the resulting $(P_{E'} + P_D)$ characteristic downwards. If $\Delta\omega > 0$ then P_D is positive effectively assisting the air-gap power and shifting the resultant characteristic upwards. The rotor will therefore move along a modified power-angle trajectory such as that shown in Figure 5.12. To help increase clarity, this diagram shows an enlarged part of the power-angle diagram in the vicinity of the equilibrium point.

As before, the rotor is initially disturbed from the equilibrium point *1* to point *2*. At point 2 the driving mechanical power is less than the opposing electrical power and the

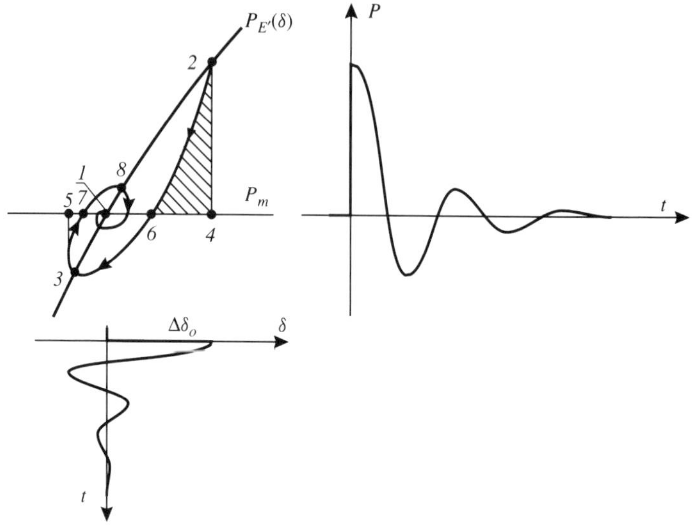

Figure 5.12 Rotor and power oscillations with damping included

decelerating torque will force the rotor back towards the equilibrium point. On deceleration the rotor speed drops and P_D becomes negative decreasing the resulting decelerating torque. The rotor therefore moves along the line *2–6* when the work done by the decelerating torque is equal to the area *2-4-6*. This is less than the area *2-4-1* in Figure 5.11 which represents the work that would have been done if no damping was present. At point *6* the rotor speed reaches a minimum and, as it continues to move along the curve *6–3*, the accelerating torque counteracts further movement of the rotor and is assisted by the negative damping term. The rotor again reaches synchronous speed when the area *6-3-5* is equal to the area *2-4-6* which is achieved earlier than in the case without damping. The rotor then starts to swing back, still accelerating, so that the speed increases above synchronous speed. The damping term changes sign, becoming positive, and decreases the resulting accelerating torque. The rotor moves along the curve *3–7* and the work performed during the acceleration is equal to the small area *3-5-7*. As a result the rotor reaches synchronous speed at point *8*, much earlier than in the case without damping. The rotor oscillations are damped and the system quickly reaches the equilibrium point *1*.

5.4.5 Effect of rotor flux linkage variation

The discussion so far has assumed that the total flux linking the field winding and rotor body remains constant during rotor oscillations so that $E' = \text{constant}$. However as the armature flux enters the rotor windings (Chapter 4) the rotor flux linkages change with time and the effect this has on the rotor swings must now be considered.

Linearised form of the generator equations

For small disturbance analysis, suitable for steady-state stability purposes, the equations describing the generator behaviour can be linearised in the vicinity of the pre-disturbance operating point. Assume that this operating point is defined by the transient rotor angle $\hat{\delta}'_s$ and the transient emf E'_0. If the angle $\alpha = \delta - \delta'$ is assumed to be constant an approximate expression for the change in power $\Delta P = P_e(\delta') - P_m$ with respect to the steady-state equilibrium point can be obtained as a function of the change in the transient rotor angle $\Delta\delta' = \delta'(t) - \hat{\delta}'_s$ and a change in the transient emf $\Delta E' = E'(t) - E'_0$ as

$$\Delta P_e = \left.\frac{\partial P_e(\delta', E')}{\partial \delta'}\right|_{E'=E'_0} \Delta\delta' + \left.\frac{\partial P_e(\delta', E')}{\partial E'}\right|_{\delta'=\hat{\delta}'_s} \Delta E' = K_{E'}\Delta\delta' + D_{\delta'}\Delta E' \tag{5.50}$$

where $P_e(\delta', E')$ is the air-gap power given by equation 5.38, 5.39, or 5.40, $K_{E'} = \partial P_e/\partial\delta'$ is the transient synchronising power coefficient and $D_{\delta'} = \partial P_e/\partial E'$ is a constant coefficient all of which depend on the generator load. As the angle $\alpha = \delta - \delta'$ is assumed to be constant $\Delta\delta' = \Delta\delta$ and the swing equation 5.30 can be linearised in the vicinity of the operating point as

$$M\frac{\mathrm{d}^2\Delta\delta}{\mathrm{d}t^2} + D\frac{\mathrm{d}\Delta\delta}{\mathrm{d}t} + K_{E'}\Delta\delta + D_{\delta'}\Delta E' = 0 \tag{5.51}$$

The transient emf can be resolved into two components E'_d and E'_q along the d and q rotor axes. Analysing the time changes of both components is complicated as the variations take place with different time constants. To simplify the discussion the case of the salient-pole machine will be considered when $\delta' = \delta$, $E' = E'_q$ and $E'_d = 0$ and only the flux linkages of the field winding need be considered.

Now assume that the salient-pole generator operates at a steady-state equilibrium point defined by $\delta' = \delta = \hat{\delta}_s$ and $E'_q = E'_{q0}$. The incremental swing equation 5.51 is

$$M\frac{\mathrm{d}^2\Delta\delta}{\mathrm{d}t^2} + D\Delta\omega + K_{E'_q}\Delta\delta + D_{\delta'}\Delta E'_q = 0 \tag{5.52}$$

where $\Delta\omega = \mathrm{d}\Delta\delta/\mathrm{d}t$ and the values of the coefficients $K_{E'_q} = \partial P_e/\partial\delta'$ and $D_{\delta'} = \partial P_e/\partial E'_q$ can be obtained by differentiating the expression for the air-gap power of the salient-pole generator, equation 5.39. Analysis of equation 5.52 suggests that if the changes $\Delta E'_q$ are in phase with the rotor speed deviation $\Delta\omega$ then, just as with the damper winding, additional positive damping power will be introduced into the system.

Chapter 4 showed that the rate at which flux can penetrate the field winding is primarily determined by the field winding transient time constant T'_{d0}, although the generator and system reactances and the operating point will also have some influence. Assuming that the excitation emf E_f and the infinite busbar voltage V_s are constant, equation 4.28 can be re-written as

$$\Delta E'_q(s) = -\frac{AB}{1 + BT'_{d0}s}\Delta\delta(s) \tag{5.53}$$

where A and B are constants that depend on the operating conditions (angle δ_0) and the generator and network reactances. If both the generator armature resistance and the network resistance are neglected then $A = \left(\frac{1-B}{B}\right) V_s \sin\delta_0$ and $B = (X'_d + X)/(X_d + X) = x'_d/x_d$. The time constant BT'_{d0} is often referred to as the *effective field winding time constant*.

If sinusoidal variations of $\Delta\delta$ are considered then the frequency response of equation 5.53 can be obtained by setting $s = j\omega_d$, where ω_d is the frequency of the rotor swings. This allows the phase of $\Delta E'_q$ to be compared with that of $\Delta\delta$. As T'_{d0} is typically much longer than the swing period $T'_{d0} >> 2\pi/\omega_d$, $BT'_{d0}\omega_d > 1$ and

$$\Delta E'_q(j\omega_d) \cong -\frac{AB}{BT'_{d0}j\omega_d}\Delta\delta(j\omega_d) = j\frac{A}{T'_{d0}\omega_d}\Delta\delta(j\omega_d) \tag{5.54}$$

Thus, the changes $\Delta E'_q$ are seen to lead the changes $\Delta\delta$ by $\pi/2$, that is they are in-phase with the rotor speed deviation $\Delta\omega = d\Delta\delta/dt$ thereby providing some additional positive damping torque.

Equation 5.54 also shows that the magnitude of the variation in $\Delta E'_q$ depends, via the coefficient $A = \left(\frac{1-B}{B}\right) V_s \sin\delta_0$, on the power angle δ_0 at the linearisation (operating) point. Thus the same variations $\Delta\delta$ cause deeper variations in $\Delta E'_q$ if the system is heavily loaded (higher value of δ_0) whilst the assumption of constant flux linkage (E' = constant) is more accurate when the generator is lightly loaded (small δ_0) and the variations in $\Delta E'_q$ are small.

The effect of variations in the rotor flux linkages will be examined for two cases using the equal area method. In the first case the generator mechanical power $P_m < P_{E_q cr}$ while in the second case $P_m = P_{E_q cr}$. In both cases the effect of the damper winding will be neglected from the discussion in order to see more clearly the effect the variation in rotor flux linkage has on the damping.

Equilibrium point for $P_m < P_{E_q cr}$

Figure 5.13a shows how Figure 5.11 has to be modified to include the effect of the variation in rotor flux linkage. Again only that part of the power-angle characteristic in the vicinity of the operating point is shown. As before the disturbance causes the electrical power to move from point *1* to point *2* on the pre-disturbance transient $P_{E'_{q0}}(\delta)$ curve as the field winding acts initially like a perfect screen maintaining constant rotor flux linkages. As the rotor starts to decelerate, its speed drops so that the speed deviation $\Delta\omega$ becomes negative. The resistance of the field winding dissipates magnetic energy and the rotor linkages start to decay reducing the value of transient emf E'_q as shown in Figure 5.13b. Consequently the electrical power is less than would be the case with E'_q = constant and the rotor motion is along the curve *2–6* rather than *2–1*. The resulting deceleration area *2-4-6* is smaller than area *2-4-1* thereby reducing the kinetic energy that is responsible for providing the impetus for the backswing. As the rotor passes point *6* it starts to accelerate, the emf E'_q starts to recover and reaches its pre-disturbance value when $\Delta\omega$ is zero at point *3* such that area *6-3-5* is equal to area *2-4-6*. The rotor then continues to accelerate and swings back towards the equilibrium point. The speed deviation $\Delta\omega$ increases and E'_q continues to increase so modifying the dynamic characteristic which now lies above $P_{E'_{q0}}(\delta)$. This reduces the accelerating area and the kinetic energy providing the impetus for the forward swing. The accelerating area is *3-5-7*, the forward swing ends at point *8* and the whole cycle is repeated with a reduced amplitude of rotor swings.

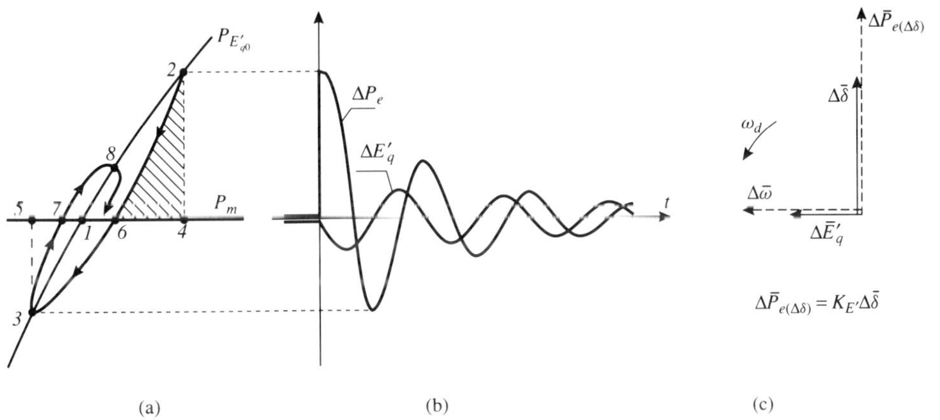

Figure 5.13 Including the effect of variation in rotor flux linkage: (a) trajectory of the operating point; (b) time variation of electrical power and transient emf; (c) rotating phasors of $\Delta\delta$, $\Delta E'_q$ and $\Delta\omega = \mathrm{d}\delta/\mathrm{d}t$

Figure 5.13b shows that the E'_q variations are in phase with $\Delta\omega$ and lead the $\Delta\delta$ variations by $\pi/2$. Figure 5.13c illustrates the same effect using phasor representation of the time-varying sinusoidal quantities. Strictly speaking the phasors represent the phase and initial values of the quantities as their rms values decay with time. The synchronous power $\Delta\overline{P}_{e(\Delta\delta)}$ (i.e. the first component in equation 5.50) is in phase with the rotor angle $\Delta\bar{\delta}$ while $\Delta\overline{E'}_q$ is in phase with the rotor speed deviation $\Delta\overline{\omega}$. Obviously $\Delta\omega$ is a derivative of $\Delta\delta$ and leads it by $\pi/2$. It should be emphasised that the changes in E'_q are very small, in the range of a few percent.

Figure 5.13 shows that the field winding provides a similar damping action to the damper windings, but much weaker. The combined effect of the two damping mechanisms helps to return the rotor quickly to the equilibrium point. If the damping effect is large then the rotor motion may be aperiodic (without oscillations).

The constants A and B in equations 5.53 and 5.54 were stated assuming that the effect of the network resistance could be neglected. A fuller analysis would show that A and B depend on the network resistance (Anderson and Fouad, 1977) and that at large values of this resistance A can change sign when, according to equation 5.54, the term with $\Delta E'_q$ changes sign to negative (with respect to $\Delta\omega$) and negative damping is introduced to the system. This may happen when a generator operates in a medium-voltage distribution network (Venikov, 1978) and such a case is illustrated in Figure 5.14. If $\Delta E'_q$ lags $\Delta\delta$ then, during the backswing, the rotor will move from point *2* along a characteristic that is higher than the initial one thereby performing deceleration work equal to the area *2-3-2'*. This area is larger than if the rotor followed the initial characteristic *2–1* (compare with Figure 5.11). To balance this deceleration work the rotor must swing back to point *4* such that area *3-4-4'* is equal to area *2-3-2'*. Consequently the amplitude of the rotor swings increases and, if this negative damping is larger than the positive damping introducing by the damper windings, the generator may lose stability.

By analysing rotor swings similar to those shown in Figure 5.13 and 5.14 it can be shown that the positive damping due to the flux linkage variation described above can only take place when the slope of the transient characteristic is steeper than the slope of the steady-state characteristic, $K_{E'} > K_{E_q}$ (Machowski & Bernas, 1989). Consequently

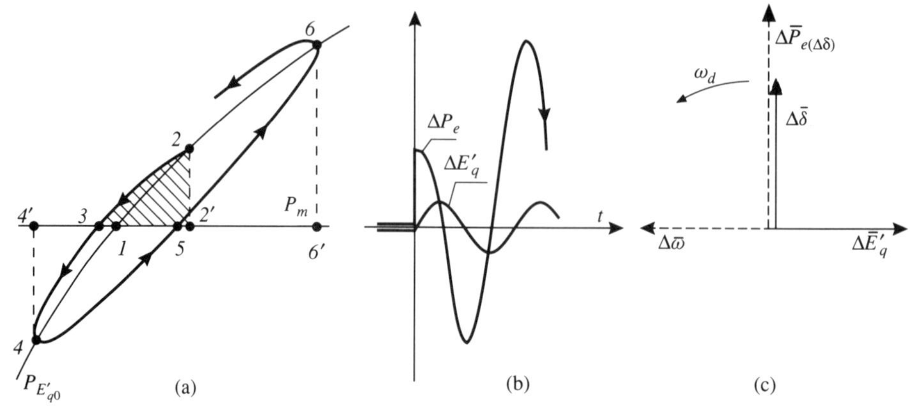

Figure 5.14 Negative damping: (a) trajectory of the operating point; (b) time variation of electrical power and transient emf; (c) relative position of phasors

the steady-state stability condition defined in equation 5.33 must be modified to take into account the rotor dynamics as

$$K_{E_q} = \frac{\partial P_{E_q}}{\partial \delta} > 0 \quad \text{and} \quad K_{E'} = \frac{\partial P_{E'}}{\partial \delta} > K_{E_q} \tag{5.55}$$

Another type of disturbance, consisting of a small increase in the mechanical power to $(P_m + \Delta P_m)$, is shown in Figure 5.15. Again the characteristics are shown only in the vicinity of the equilibrium point. Initially the system operates at point *1*, the intersection between the characteristics P_m, P_{E_q} and $P_{E'_{q0}}$. The increased mechanical loading produces

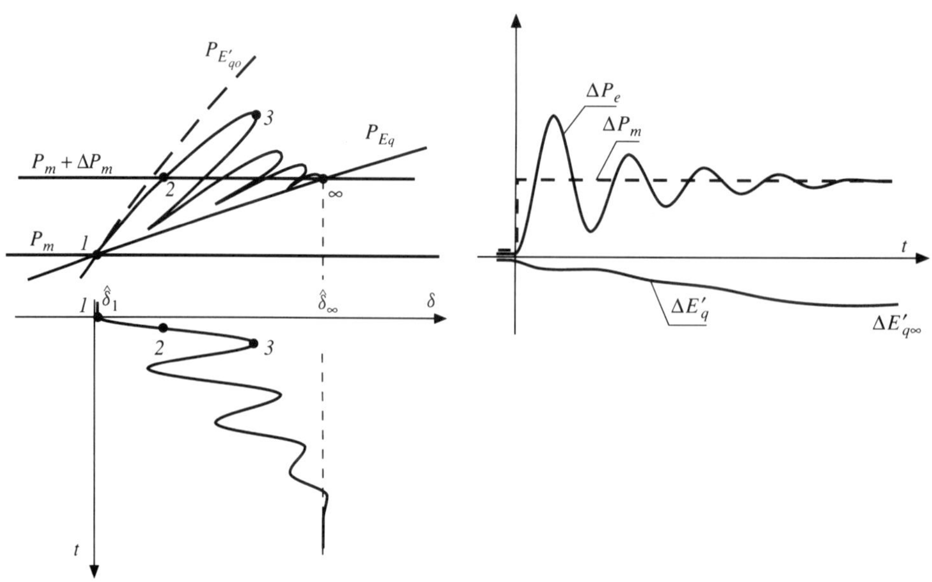

Figure 5.15 Rotor and power swings due to a disturbance in mechanical power

a new, final, equilibrium point ∞. The final value of the transient emf $E'_{q\infty}$ is smaller than the initial value E'_{q0} as indicated in Figure 5.10. Consequently E'_q must reduce as the rotor moves from point 1 towards point ∞. This means that the rotor oscillations will be similar to those shown in Figure 5.13, but occurring along a series of dynamic characteristics of decreasing amplitude, all of them lying below the initial characteristic $P_{E'_{q0}}$. Small oscillations of E'_q around its average decaying value cause some damping similar to that shown in Figure 5.13. This damping again adds to the main damping coming from the damper windings.

Equilibrium point at $P_m = P_{E_q\,cr}$

If $P_m = P_{E_q\,cr}$ then there is only one steady-state equilibrium point. As shown in Figure 5.10a the transient characteristic $P_{E'_q}(\delta)$ crossing point $(\delta_{cr}, P_{E_q\,cr})$ has a positive slope, and a momentary disturbance in the angle δ will produce oscillations around the equilibrium point that are damped by the damper and field windings. Although this would suggest a stable operation, the system is in practice unstable as will be explained below.

Any generator will be subjected to minor disturbances resulting from vibrations, slight changes in supplied power, switching in the network etc. When the system operates at $P_m = P_{E_q\,cr}$ any such disturbance may cause a relative shift between the P_m and $P_{E_q}(\delta)$ characteristics. This is shown in Figure 5.16 where it is assumed that P_m is increased by ΔP_m such that the new mechanical $(P_m + \Delta P_m)$ characteristic lies above $P_{E_q}(\delta)$. Initially, as the transient $P_{E'_q}(\delta)$ characteristic has a positive slope, the rotor starts to oscillate around $(P_m + \Delta P_m)$ in a similar way to that shown in Figure 5.15. Gradually the armature flux penetrates the field winding, the emf E'_q decays and the rotor follows $P_{E'_q}(\delta)$ characteristics of declining amplitude. Eventually the decay in E'_q is such that the $P_{E'_q}(\delta)$ curve lies below the new mechanical $(P_m + \Delta P_m)$ characteristic and the generator loses synchronism. The behaviour of the generator during asynchronous operation will be described in Section 6.4.

The oscillatory loss of stability shown in Figure 5.16 takes place when the influence of the damper windings is neglected. In practice, damping at high loading can be significant

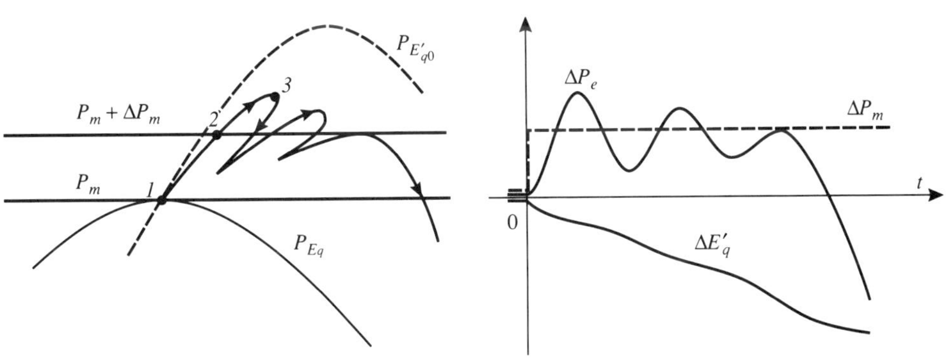

Figure 5.16 Rotor, power and transient emf oscillations after a small increase in P_m at the critical equilibrium point

when the loss of stability may occur in an aperiodic way for the reasons discussed in the next sub-section.

5.4.6 Analysis of rotor swings around the equilibrium point

In this sub-section a quantitative analysis of rotor dynamics around the equilibrium point is presented. Assuming the constant flux linkage generator model with constant E', the incremental swing equation 5.51 becomes

$$M\frac{d^2\Delta\delta}{dt^2} + D\frac{d\Delta\delta}{dt} + K_{E'}\Delta\delta = 0 \tag{5.56}$$

with the initial disturbed conditions being

$$\Delta\delta\left(t = 0^+\right) = \Delta\delta_0 \neq 0 \quad \text{and} \quad \Delta\omega = \Delta\dot{\delta}\left(t = 0^+\right) = 0 \tag{5.57}$$

Equation 5.56 is a second-order linear differential equation whose solution is determined by the roots of the *characteristic equation*

$$\lambda^2 + \frac{D}{M}\lambda + \frac{K_{E'}}{M} = 0 \tag{5.58}$$

where the two roots λ_1 and λ_2 are

$$\lambda_{1,2} = -\frac{D}{2M} \pm \sqrt{\left(\frac{D}{2M}\right)^2 - \frac{K_{E'}}{M}} \tag{5.59}$$

Three cases are now possible:

(i) the roots are real and distinct and the solution is of the form $\Delta\delta(t) = A_1 e^{\lambda_1 t} + A_2 e^{\lambda_2 t}$, where A_1 and A_2 are the integration constants. Substitution of the initial conditions defined in equation 5.57 gives the aperiodic response

$$\Delta\delta(t) = \frac{\Delta\delta_0}{\lambda_2 - \lambda_1}\left[\lambda_2 e^{\lambda_1 t} - \lambda_1 e^{\lambda_2 t}\right] \tag{5.60}$$

(ii) the roots are real and equal, $\lambda_1 = \lambda_2 = \lambda$, and the solution is of the form $\Delta\delta(t) = e^{\lambda t}(A_1 + A_2 t)$. Substitution of the initial conditions in equation 5.57 gives the aperiodic response

$$\Delta\delta(t) = \Delta\delta_0 e^{\lambda t}(1 - \lambda t) \tag{5.61}$$

(iii) the roots form a complex conjugate pair $\lambda_{1,2} = a \pm jb$. In this case the solution of equation 5.56 can be better understood if it is compared with the standard second order differential equation

$$\frac{d^2\Delta\delta}{dt^2} + 2\zeta\omega_{nat}\frac{d\Delta\delta}{dt} + \omega_{nat}^2 = 0 \tag{5.62}$$

whose roots are

$$\lambda_{1,2} = -\zeta\omega_{nat} \pm j\omega_d \tag{5.63}$$

where ω_{nat} is the *undamped natural frequency* of rotor small oscillations, ζ is the *damping ratio* and $\omega_d = \omega_{nat}\sqrt{1-\zeta^2}$ is the *damped natural frequency*. Comparing equation 5.62 with equation 5.56 gives $\omega_{nat} = \sqrt{K_{E'}/M}$ and $\zeta = D/2\sqrt{K_{E'}M}$. The solution for $\Delta\delta(t)$ is now given by

$$\Delta\delta(t) = \frac{\Delta\delta_0}{\sqrt{1-\zeta^2}} e^{-\zeta\omega_{nat}t} \cos\left[\omega_d t - \phi\right] \tag{5.64}$$

where $\phi = \arcsin\zeta/\sqrt{1-\zeta^2}$. The damping ratio ζ determines the amount of damping present in the system response.

As the values of the roots depend on the actual values of $K_{E'}$, D and M so too does the type of response. The inertia coefficient M is constant while both D and $K_{E'}$ depend on the generator loading. Figure 5.3 shows that the damping coefficient D increases with load while Figure 5.10b shows that the transient synchronising power coefficient $K_{E'}$ decreases with load.

Equation 5.59 shows that if $K_{E'} > 0$ then, depending on the actual values of $K_{E'}$ and D, the roots of the characteristic equation can be either real or complex. For small initial values of power angle $\hat{\delta}_s$ the damping coefficient is small while the transient synchronising power coefficient $K_{E'}$ is large so that $(K_{E'}/M) > (D/2M)^2$ and the two roots form a complex conjugate pair. In this case the solution of the differential equation is given by equation 5.64. The system response oscillates along the transient characteristic $P_{E'}(\delta)$ with an amplitude that decays with time as shown schematically in the "bubble"near point A in Figure 5.17. The frequency of oscillation is $\omega_d = \omega_{nat}\sqrt{1-\zeta^2}$ and is slightly smaller than the undamped natural frequency, ω_{nat}.

As the initial value of the power angle $\hat{\delta}_s$ increases, the synchronising power coefficient $K_{E'}$ decreases while damping increases. Consequently ω_{nat} decreases but the damping ratio ζ increases with the result that the rotor oscillations become slower and more heavily damped as shown in the "bubble"near point B in Figure 5.17. At some point, when $(K_{E'}/M) = \left(D/2M^2\right)$, the damping ratio ζ is unity and the oscillations are said to be *critically damped*. In this case the roots are real and equal, $\lambda_1 = \lambda_2 = -\omega_{nat}$, giving the aperiodic response expressed by equation 5.61 and shown in the "bubble" near point C in Figure 5.17.

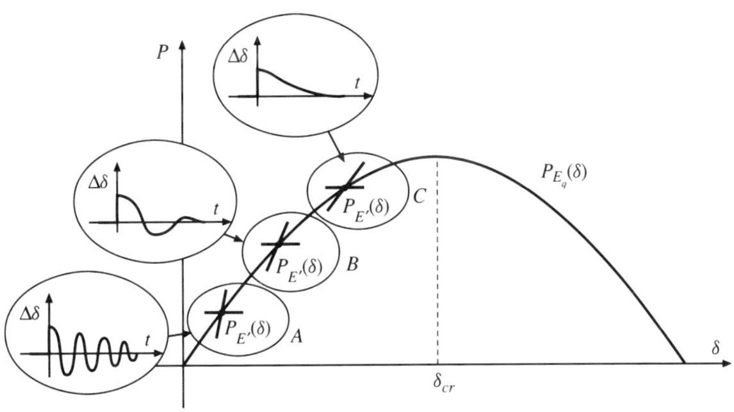

Figure 5.17 Examples of rotor swings for different stable equilibrium points

As the initial power angle $\hat{\delta}_s$ is further increased, at some point it holds that $(K_{E'}/M) < (D/2M)^2$, the two roots are real and negative and the rotor swings are expressed by equation 5.60. In this case the damping ratio $\zeta > 1$, the rotor swings are overdamped, and the response is sluggish and vanishes aperiodically with time. Typically this condition may occur when the initial operating point is near the peak of the steady-state characteristic in Figure 5.17.

When the power angle $\hat{\delta}_s$ is equal to its critical value δ_{cr}, it is not possible to analyse the system dynamics using the constant flux linkage model ($E' =$ constant) as the effect of flux decrement must be included. This was explained in Figure 5.16.

5.4.7 Mechanical analogues of the generator-infinite busbar system

Further insight into the generator response to a small disturbance can be obtained by comparing the generator-infinite busbar system with the standard mass/spring/damper system shown in Figure 5.18a and described by the equation

$$m\frac{d^2\Delta x}{dt^2} + c\frac{d\Delta x}{dt} + k\Delta x = 0 \tag{5.65}$$

Comparing this equation with equation 5.56 suggests that the generator-infinite busbar system can be treated as an "electromagnetic spring" where any increase in the spring extension Δx is equivalent to an increase in the rotor angle $\Delta\delta$. In this analogous system the mass m is equivalent to the inertia coefficient M, the spring damping coefficient c is equivalent to the generator damping coefficient D and the spring stiffness k is equivalent to the synchronising power coefficient $K_{E'}$. This analogy allows the generator response, as expressed by equations 5.60 to 5.64, to be directly related to the response of the standard mass/spring/damper system. Unlike the mechanical spring the "electromagnetic spring" is non-linear as $K_{E'}$ depends strongly on the initial value of the power angle $\hat{\delta}_s$, Figure 5.10b.

Another useful mechanical analogue is that of the pendulum of mass m and length l shown in Figure 5.18b. The pendulum has an upper and lower equilibrium point as shown where the weight of the ball is counterbalanced by the arm. When the pendulum

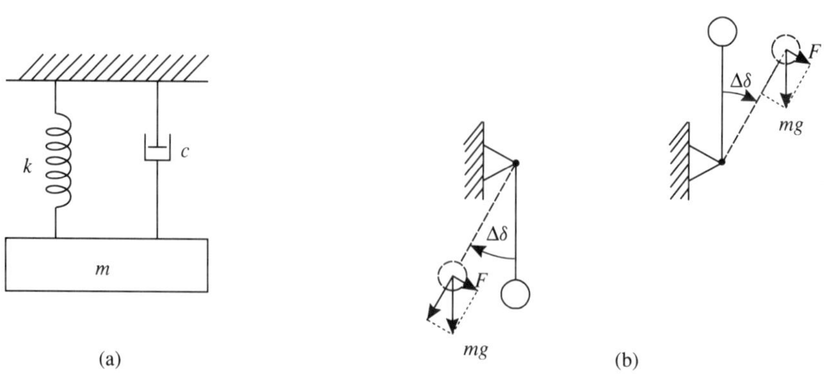

Figure 5.18 Mechanical analogues of the generator-infinite busbar system: (a) mass-spring-damper system; (b) a pendulum at stable and unstable equilibrium point

is disturbed from its equilibrium position a force equal to $F = -mg \sin \delta$ is developed and, neglecting damping, the pendulum motion is described by the differential equation

$$m \frac{d^2\delta}{dt^2} = \frac{mg}{l} \sin \delta \tag{5.66}$$

Near the lower (stable) equilibrium point the force F always acts towards the equilibrium point so that the pendulum oscillates around the equilibrium point. Near the upper (unstable) equilibrium point the force F pushes the pendulum away from the equilibrium point causing instability. The behaviour of the pendulum at these two equilibrium points can be directly compared with the behaviour of the generator at the stable and unstable equilibrium points shown in Figure 5.6.

The pendulum can also provide a useful analogue when analysing the generator dynamics. In this case equation 5.66 has to be compared with the classical generator model in equation 5.42. Again, both equations are of identical form.

5.5 Steady-State Stability of the Regulated System

The previous section considered the power-angle characteristics of a simple generator-infinite busbar system and the resulting steady-state, or small signal, stability conditions when the excitation voltage (and therefore the excitation emf E_f) was assumed to be constant. This section considers the steady-state stability when the action of the Automatic Voltage Regulator (AVR) is included. The influence of the AVR will be considered in three stages. First the modified steady-state power-angle characteristic will be derived. Secondly the possibility of operation beyond the critical point (as defined by the pull-out power) will be discussed and finally the influence of the AVR on rotor swings will be analysed.

5.5.1 *Steady-state power-angle characteristic of the regulated generator*

The static $P_{E_q}(\delta)$ power-angle characteristic, equation 5.29, was derived assuming that in steady-state the excitation emf $E_f = E_q =$ constant. In practice every generator is equipped with an AVR which tries to maintain constant voltage at the generator terminals (or at some point behind the terminals) by adjusting the value of the excitation voltage and, consequently, E_f. As the resulting formulae for the active and reactive power are more complicated than when $E_f =$ constant the following discussion will be restricted to the case of a round-rotor generator ($x_d = x_q$) with resistance neglected ($r = 0$). In this case the steady-state equivalent circuit and phasor diagrams are shown in Figure 5.19. The formulae for the active and reactive power will be derived by resolving the voltages and currents along the a and b axes, where the a-axis is located along the system voltage V_s. The co-ordinates of $\overline{E}_q$ in the (a, b) reference frame are

$$E_{qa} = E_q \cos \delta, \qquad E_{qb} = E_q \sin \delta \tag{5.67}$$

Inspection of the phasor diagram gives

$$I_a = \frac{E_{qb}}{X_d + X} \qquad I_b = \frac{E_{qa} - V_s}{X_d + X} \tag{5.68}$$

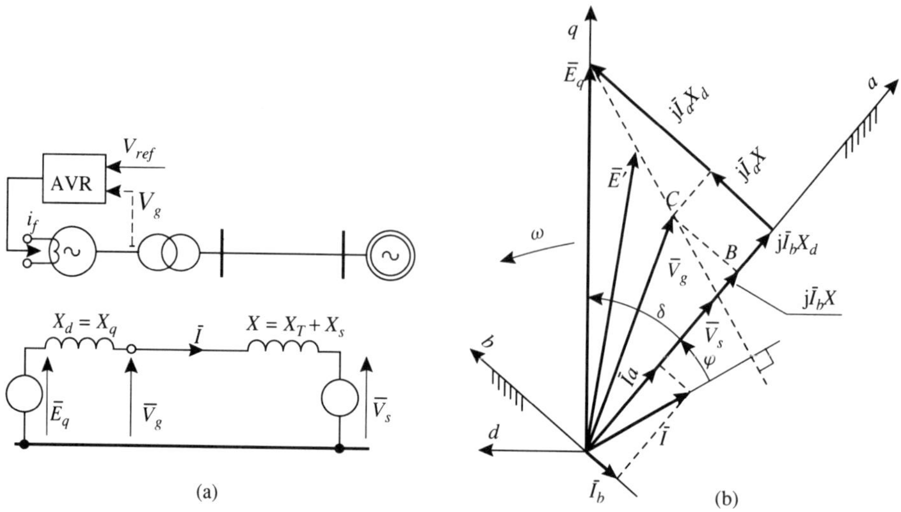

Figure 5.19 Generator operating on the infinite busbars: (a) schematic and equivalent circuit; (b) phasor diagram in the (d, q) and (a, b) reference frames

while the co-ordinates of the current are $I_a = I\cos\varphi$ and $I_b = -I\sin\varphi$. Pythagoras' theorem applied to the triangle OBC yields $(V_s + I_b X)^2 + (I_a X)^2 = V_g^2$ which, after taking into account equation 5.68, gives

$$\left(E_{qa} + \frac{X_d}{X}V_s\right)^2 + E_{qb}^2 = \left[\frac{X_d + X}{X}V_g\right]^2 \tag{5.69}$$

This equation describes a circle of radius $\rho = (X_d/X + 1)V_g$ with the centre lying on the a-axis at a distance $A = -X_d V_s/X$ from the origin. This means that with $V_g =$ constant and $V_s =$ constant, the tip of E_q moves on this circle. Figure 5.20 shows the circular locus centred on the origin made by the phasor $V_g =$ constant, and another circular locus (shifted to the left) made by phasor E_q.

The circle defined by equation 5.69 can be transformed into polar co-ordinates by substituting equation 5.67 to give

$$E_q^2 + 2\frac{X_d}{X}E_q V_s\cos\delta + \left(\frac{X_d}{X}V_s\right)^2 = \left[\frac{X_d + X}{X}V_g\right]^2 \tag{5.70}$$

One of the roots of this equation is

$$E_q = \sqrt{\left(\frac{X_d + X}{X}V_g\right)^2 - \left(\frac{X_d}{X}V_s\sin\delta\right)^2} - \frac{X_d}{X}V_s\cos\delta \tag{5.71}$$

and corresponds to the $E_f = E_q$ points that lie on the upper part of the circle. Substituting equation 5.71 into the round-rotor power-angle equation 5.32 gives the generated

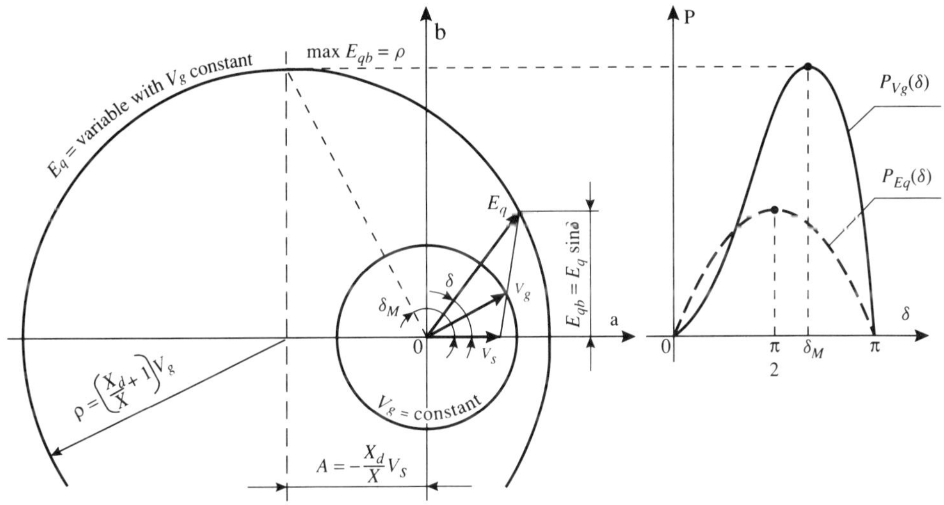

Figure 5.20 The circle diagrams and the power-angle characteristics for the round-rotor generator operating on the infinite busbars

power as

$$P_{V_g}(\delta) = \frac{V_s}{X_d + X}\sin\delta\sqrt{\left(\frac{X_d + X}{X}V_g\right)^2 - \left(\frac{X_d}{X}V_s\sin\delta\right)^2} - \frac{1}{2}\frac{X_d}{X}\frac{V_s^2}{X_d + X}\sin 2\delta \tag{5.72}$$

Equation 5.72 describes the power-angle characteristic $P_{V_g}(\delta)$ with V_g = constant and is shown, together with $P_{E_q}(\delta)$, in Figure 5.20 Comparison between the two shows that AVR can significantly increase the amplitude of the steady-state power-angle characteristic.

The maximum value of the power given by equation 5.72 can be easily found by examining Figure 5.20. Substituting the second equation 5.67 into equation 5.32 gives

$$P_{V_g}(\delta) = \frac{V_s}{X_d + X}E_{qb} \tag{5.73}$$

indicating that the generator power is proportional to the projection of E_q on the b-axis. The function defined by equation 5.73 reaches its maximum value when E_{qb} is a maximum. As can be seen from Figure 5.20 this occurs at the point on the E_q locus that corresponds to the centre of the circle. At this point E_q has the following co-ordinates

$$E_{qb} = \rho = \left(\frac{X_d}{X} + 1\right)V_g \quad \text{at} \quad \delta_M = \arctan\left(\frac{\rho}{A}\right) = \arctan\left(-\frac{X_d + X}{X_d}\frac{V_g}{V_s}\right) \tag{5.74}$$

The angle δ_M at which $P_{V_g}(\delta)$ reaches maximum is always greater than $\pi/2$ irrespective of the voltages V_g and V_s. This is typical of systems with active AVRs. Substituting the first part of equation 5.74 into equation 5.73 gives gives

$$P_{V_gM} = P_{V_g}(\delta)|_{\delta=\delta_M} = \frac{V_g V_s}{X} \tag{5.75}$$

showing that the amplitude of the power-angle characteristic of the regulated system is independent of the generator reactance. It does, however, depend on the equivalent reactance of the transmission system. The steady-state synchronising power coefficient of the regulated system is $K_{V_g} = \partial P_{V_g}(\delta)/\partial\delta$ and $K_{V_g} > 0$ when $\delta < \delta_M$.

Physical interpretation

The $\sin 2\delta$ component in equation 5.72 has negative sign, making the maximum of the $P_{V_g}(\delta)$ characteristic shown in Figure 5.20 occur at $\delta_M > \pi/2$. For small rotor angles $\delta << \pi/2$ the characteristic is concave while for $\delta > \pi/2$ the characteristic is very steep. The $\sin 2\delta$ component has nothing to do with the reluctance power (as was the case with $P_{E_q}(\delta)$) as equation 5.72 has been derived assuming $x_d = x_q$. The distortion of the characteristic is entirely due to the influence of AVR.

Physically the shape of the $P_{V_g}(\delta)$ characteristic can be explained using Figure 5.21. Assume that initially the generator operates at point *1* corresponding to the characteristic $P_{E_{q1}} = P_{E_q}(\delta)\big|_{E_q=E_{q1}}$ shown by the dashed curve *1*. An increase in the generator load causes an increase in the armature current, an increased voltage drop in the equivalent network reactance X, Figure 5.19, and therefore a decrease in the generator voltage V_g. The resulting voltage error forces the AVR to increase the excitation voltage so that E_q is increased to a value $E_{q2} > E_{q1}$ and a new operating point is established on a higher characteristic $P_{E_{q2}} = P_{E_q}(\delta)\big|_{E_q=E_{q2}}$ denoted by *2*. Subsequent increases in load will cause the resulting $P_{V_g}(\delta)$ characteristic to cross at the points *2,3,4,5* and *6* lying on consecutive $P_{E_q}(\delta)$ characteristics of increased amplitude. Note that starting from point *5* (for $\delta > \pi/2$) the synchronising power coefficient $K_{E_q} = \partial P_{E_q}(\delta)/\partial\delta$ is negative while $K_{V_g} = \partial P_{V_g}(\delta)/\partial\delta$ is still positive.

Stability

If the AVR is very slow acting (that is it has a large time constant) then it may be assumed that following a small disturbance the AVR will not react during the transient

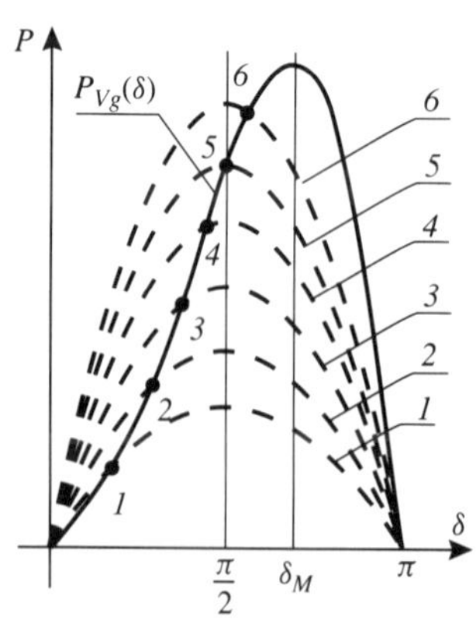

Figure 5.21 Creation of the $P_{V_g}(\delta)$ characteristic from a family of $P_{E_q}(\delta)$ characteristics

state and the regulated and unregulated systems behave in a similar manner. The stability limit then corresponds to point *5* when $\delta = \pi/2$ (for a round-rotor generator) and the stability condition is given by equation 5.33. If the AVR is fast acting so that it is able to react during the transient state then the stability limit can be moved beyond $\delta = \pi/2$ to a point lying below the top of $P_{V_g}(\delta)$ curve. In this case stability depends on the parameters of the system and the AVR when the system stability is referred to as *conditional stability*.

A fast acting AVR may also reverse the situation when the stability limit is lowered (with respect to the unregulated system) to a point $\delta < \pi/2$, for example to point *4*, or even *3*, in Figure 5.21. In this situation the system may lose stability in an oscillatory manner because of the detrimental effect of the AVR. Such a situation, and the conditional stability condition, will be discussed later in this section.

Effect of the field current limiter

Equation 5.72 was derived, and Figure 5.20 drawn, under the assumption that the AVR may change $E_q = E_f$ in order to keep the terminal voltage constant without any limit being placed on the maximum value of $E_q = E_f$. In practice the AVR is equipped with various limiters, described in Chapter 2, one of which limits the field current and hence E_f. This limiter operates with a long time delay. If the exciter reaches the maximum field current value during slow changes in operating conditions, any further increase in the load will not increase the field current despite a drop in the terminal voltage V_g. Any further changes will take place at $E_q = E_{fMAX}$ = constant and the operating point will follow a $P_{E_q}(\delta)\big|_{E_q=E_{qMAX}}$ characteristic.

Whether or not the field current limiter will act before the top of the $P_{V_g}(\delta)$ characteristic is reached depends not only on the field current limit set but also on the equivalent network reactance X. Equation 5.75 shows that the amplitude of the $P_{V_g}(\delta)$ characteristic depends on the reactance X. If X is large then the amplitude is small and the field current limit may not be reached. If X is small then the amplitude is very large and the field current limit is encountered before the peak of the characteristic is reached. Such a situation is illustrated in Figure 5.22. The limit is reached at point *Lim*. Below this point

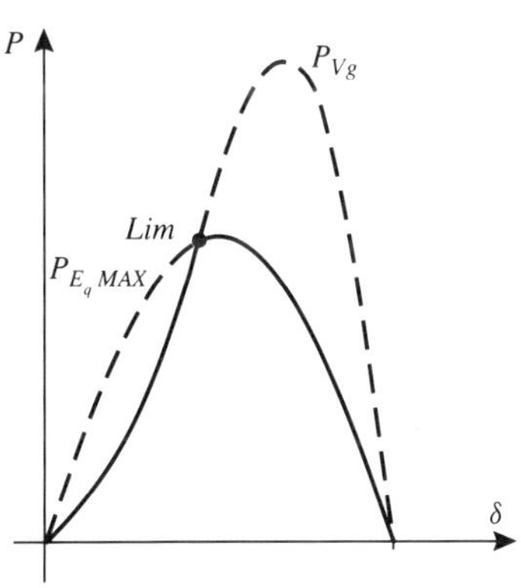

Figure 5.22 Example of the influence of the field current limiter on the steady-state power-angle characteristic

the steady-state characteristic is $P_{V_g}(\delta)$ while above this point the generator follows the $P_{E_q}(\delta)\big|_{E_q=E_{qMAX}}$ curve. The resulting characteristic is shown in bold.

From the stability point of view a more interesting case is when the field current limit is not reached. However it is important to remember that the thermal field current limit may be reached before the stability limit.

5.5.2 Transient power-angle characteristic of the regulated generator

If the AVR or the exciter have a large time constant, the regulation process is slow and the rotor swings follow the transient power-angle characteristic as discussed in Section 5.4. The transient characteristics of both the regulated and unregulated system are the same with the only difference being that the increased loading in the regulated system will cause an increase in the steady-state field current and therefore a higher value of E'_q and amplitude of the $P_{E'}(\delta')$ characteristic. Moreover the angle δ' will reach the critical value $\pi/2$ before δ reaches its critical value δ_M. This is illustrated in Figure 5.23 which is a repetition of a fragment of the phasor diagram in Figure 5.20. For the critical angle $\delta = \delta_M$, the phasor of V_g lies on the vertical axis, Figure 5.23a, and $\delta' > \pi/2$ because the emf E' leads V_g. This means that at the critical point δ_M the transient synchronising power coefficient $K_{E'} = \partial P_{E'}/\partial\delta$ is negative. When $\delta' = \pi/2$, Figure 5.23b, the emf E' lies on the vertical axis and $\delta < \delta_M$. The question now arises at which point on the $P_{V_g}(\delta)$ curve the transient synchronising power coefficient becomes zero?

This question is most easily answered if the classical generator model is used with transient saliency neglected. The following equation, similar to equation 5.69, can be obtained for the transient emf when the phasor diagram of Figure 5.19b is used:

$$\left(E'_a + \frac{X'_d}{X}V_s\right)^2 + (E'_b)^2 = \left[\frac{X'_d + X}{X}V_g\right]^2 \tag{5.76}$$

This equation describes a circle on which the tip of $\overline{E'}$ moves when the power angle and excitation are increased. This circle is similar to the circle shown in Figure 5.20 but has a radius and horizontal shift dependent on X'_d. Substituting $E'_a = E'\cos\delta'$ and $E'_b = E'\sin\delta'$

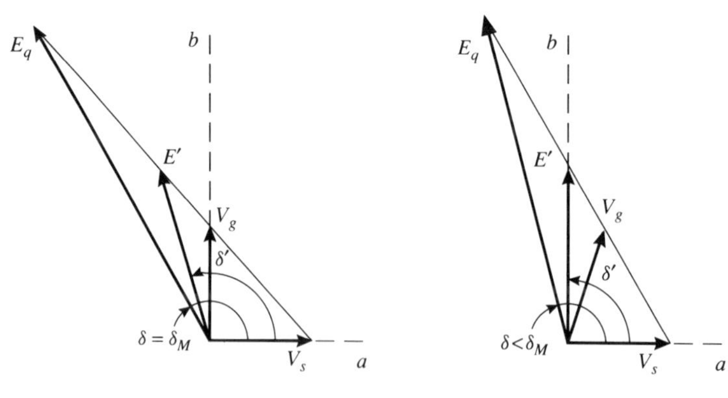

Figure 5.23 Phasor diagram of the regulated system with; (a) $\delta = \delta_M$; and (b) $\delta' = \pi/2$

into equation 5.76 gives

$$(E')^2 + 2\frac{X'_d}{X}E'V_s\cos\delta' + \left(\frac{X'_d}{X}V_s\right)^2 = \left[\frac{X'_d + X}{X}V_g\right]^2 \tag{5.77}$$

Solving this equation with respect to E' yields

$$E' = \sqrt{\left(\frac{X'_d + X}{X}V_g\right)^2 - \left(\frac{X'_d}{X}V_s\sin\delta'\right)^2} - \frac{X'_d}{X}V_s\cos\delta' \tag{5.78}$$

For $\delta' = \pi/2$ the second component in this equation is zero and the transient emf is

$$E'\big|_{\delta'=\pi/2} = \frac{V_g}{X}\sqrt{(X'_d + X)^2 - \left(X'_d\frac{V_s}{V_g}\right)^2} \tag{5.79}$$

Substituting this value into equation 5.40 and noting that $x'_d = X'_d + X$ gives

$$P_{V_{gcr}} = P_{V_g}(\delta')\big|_{\delta'=\pi/2} = \frac{V_sV_g}{X}\sqrt{1 - \left(\frac{X'_d}{X'_d + X}\right)^2\left(\frac{V_s}{V_g}\right)^2} \tag{5.80}$$

According to equation 5.75 the factor V_sV_g/X in this equation corresponds to the amplitude of the $P_{V_g}(\delta)$ curve. This means that the ratio of power at which $K_{E'} = 0$ to the power at which $K_{V_g} = 0$ corresponds to the square-rooted expression in equation 5.80 and is equal to

$$\alpha = \frac{P_{V_g(\delta'=\pi/2)}}{P_{V_g}(\delta = \delta_M)} = \frac{P_{V_{gcr}}}{P_{V_gM}} = \sqrt{1 - \left(\frac{X'_d}{X'_d + X}\right)^2\left(\frac{V_s}{V_g}\right)^2} \tag{5.81}$$

This coefficient depends strongly on the equivalent network reactance X. If X is large then α is close to unity and the transient synchronising power coefficient $K_{E'} = 0$. This takes place at a point close to the peak of $P_{V_g}(\delta)$ at which point $K_{V_g} = 0$. On the other hand if X is small then $K_{E'}$ reaches zero at a point lying well below the peak of $P_{V_g}(\delta)$. This is illustrated in Figure 5.24.

Figure 5.24a shows the case when the power angle is less than $\pi/2$, all the characteristics have positive slope and all the synchronising power coefficients are positive. Figure 5.24b shows two operating points. The lower point corresponds to the natural stability limit ($\delta = \pi/2$) when $K_{E_q} = 0$ but $K_{E'} > 0$. Further increasing the load (and the power angle) causes $K_{E_q} < 0$ whilst $K_{E'} > 0$ until the top point is reached where the $P_{E'}(\delta')$ curve, shown as a dashed line, intersects at its peak with $P_{V_g}(\delta)$. Above this point $K_{E'}$ becomes negative despite $K_{V_g} > 0$. The ratio of both peaks is given by the coefficient α defined in equation 5.81.

Stability

After a disturbance the rotor swings follow the transient power-angle characteristic $P_{E'}(\delta')$. The system is unstable above the point $P_{V_g cr}$ where $K_{E'} = \partial P_{E'}/\partial\delta < 0$ and no deceleration area is available. Therefore the necessary stability condition is

$$K_{E'} = \frac{\partial P_{E'}}{\partial\delta} > 0 \tag{5.82}$$

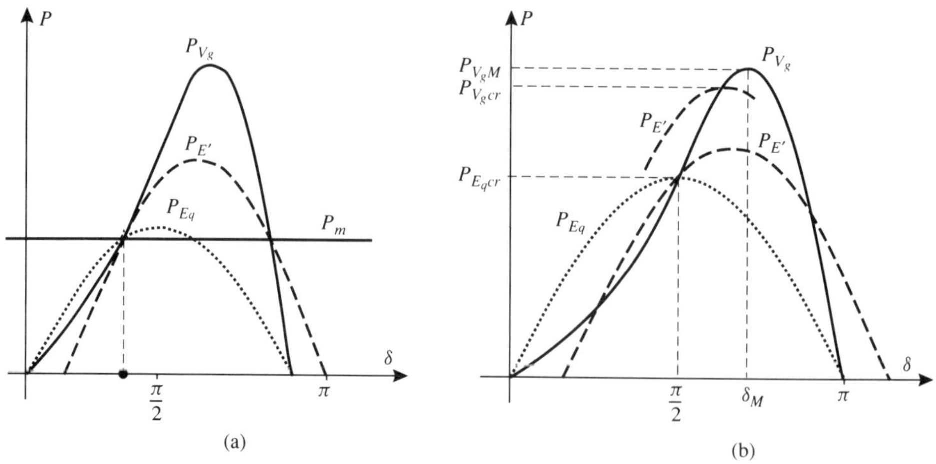

Figure 5.24 Power-angle characteristics of the regulated generator: (a) all synchronising power coefficients are positive; (b) points above the limit of natural stability ($\delta = \pi/2$)

At the point $P_{V_g cr}$ any small increase in the mechanical power will cause asynchronous operation as the mechanical power is greater than the electrical. Whether or not the generator can operate on the $P_{V_g}(\delta)$ characteristic below this point, that is when $P_m < P_{V_g cr}$, depends on the generator, network and AVR parameters. Two factors are decisive: (i) the influence the regulator has on the variations in E'_q due to changes in the excitation flux linkage during rotor swings; and (ii) the influence the regulator has on damping torques due to additional currents induced in the damper windings.

5.5.3 Effect of rotor flux linkage variation

The influence of changes in the excitation emf E_f on the transient emf E' is given by equation 4.28. For the salient-pole machine when $E' = E'_q$ this equation can be re-written as

$$\Delta E'_q = \Delta E'_{q(\Delta\delta)} + \Delta E'_{q(\Delta E_f)} \tag{5.83}$$

where
$$\Delta E'_{q(\Delta\delta)} = -\frac{AB}{1 + BT'_{d0}s}\Delta\delta, \quad \Delta E'_{q(\Delta E_f)} = +\frac{B}{1 + BT'_{d0}s}\Delta E_f \tag{5.84}$$

and $\Delta\delta = \Delta\delta'$. These two components are due to the rotor swings and voltage regulation respectively. The influence of $\Delta E'_{q(\Delta\delta)}$ was described in sub-Section 5.4.5 following equation 5.53 where it was shown in Figure 5.13 that this component is in phase with the speed deviation $\Delta\omega$ and introduces an additional damping torque in equation 5.52. A question arises as to what influence the voltage control component $\Delta E'_{q(\Delta E_f)}$ has on the problem. To answer this question it is necessary to determine the phase shift of $\Delta E'_{q(\Delta E_f)}$ with respect to $\Delta\omega$ (or $\Delta\delta$). This problem can be better understood with the help of Figure 5.25 which shows how a change in the rotor angle influences $\Delta E'_{q(\Delta E_f)}$.

The first block reflects the fact that, assuming constant infinite busbar voltage, a change in $\Delta\delta$ causes a voltage regulation error ΔV. The second block is the transfer function of

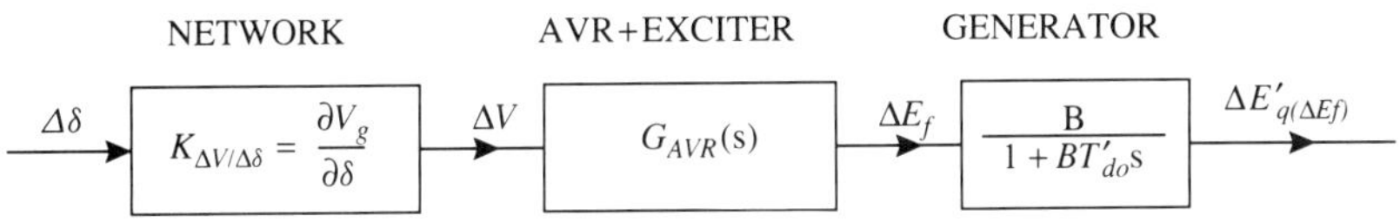

Figure 5.25 Components determining the phase shift between $\Delta\delta$ and $\Delta E'_{q(\Delta E_f)}$

the AVR and the exciter. Its effect is to convert the regulation error ΔV into a change in the excitation emf ΔE_f. The third block reflects the changes in $\Delta E'_q$ due to excitation changes and corresponds to equation 5.84.

The first block in Figure 5.25 constitutes a proportional element, as an increase in the rotor angle by $\Delta\delta$ causes a decrease in the generator voltage by $\Delta V_g \cong (\partial V_g/\partial\delta)\Delta\delta$ so that the following voltage error is produced

$$\Delta V = V_{ref} - V_g = -\frac{\partial V_g}{\partial \delta}\Delta\delta = K_{\Delta V/\Delta\delta}\Delta\delta \tag{5.85}$$

An expression for the proportionality coefficient $K_{\Delta V/\Delta\delta}$ can be obtained from equation 5.77. Solving this equation with respect to V_g gives

$$V_g = \frac{\sqrt{E'^2 + 2\frac{X'_d}{X}E'V_s\cos\delta' + \left(\frac{X'_d}{X}V_s\right)^2}}{\frac{X'_d}{X} + 1} \tag{5.86}$$

Differentiation of this expression at the linearisation point defined by δ'_0, E'_0 and V_{g0} gives

$$K_{\Delta V/\Delta\delta} = -\frac{\partial V_g}{\partial \delta'} = \frac{X'_d X}{(X'_d + X)^2}\frac{E'_0}{V_{g0}}V_s\sin\delta'_0 \tag{5.87}$$

This coefficient is positive over a wide range of angle changes which means that the voltage regulation error given by equation 5.85 is always in phase with the angle changes $\Delta\delta$. The amplitude of ΔV depends on the generator load. For a small load (and δ'_0) the coefficient $K_{\Delta V/\Delta\delta}$ is small and the resulting voltage error is small. As the load is increased, changes in ΔV caused by the changes in $\Delta\delta$ become bigger.

The second block in Figure 5.25 introduces a phase shift between ΔE_f and ΔV dependent on transfer functions of the AVR and the exciter. In the case of a static exciter, Figure 2.3d, e and f, with proportional regulator the phase shift is small and it may be assumed that ΔE_f is in phase with ΔV. In comparison a dc cascade exciter, or ac exciter with rectifier, Figure 2.3a, b and c, behaves like an inertia element which introduces a phase shift of a few tens of degrees for a frequency of oscillation of about 1Hz.

The generator block in Figure 5.25 introduces a phase shift which, as in equation 5.54, is equal to $\pi/2$. However the minus sign in equations 5.53, 5.54 and the first part of equation 5.84 will cause $\Delta E'_{q(\Delta\delta)}$ to lead the variations in $\Delta\delta$ by $\pi/2$ whilst the plus sign in the second part of equations 5.84 will cause $\Delta E'_{q(\Delta E_f)}$ to lag the changes in $\Delta\delta$ by $\pi/2$.

A knowledge of all these phase shifts allows a phasor diagram similar to that shown in Figure 5.13c to be drawn but with both of the components of equation 5.83 taken into

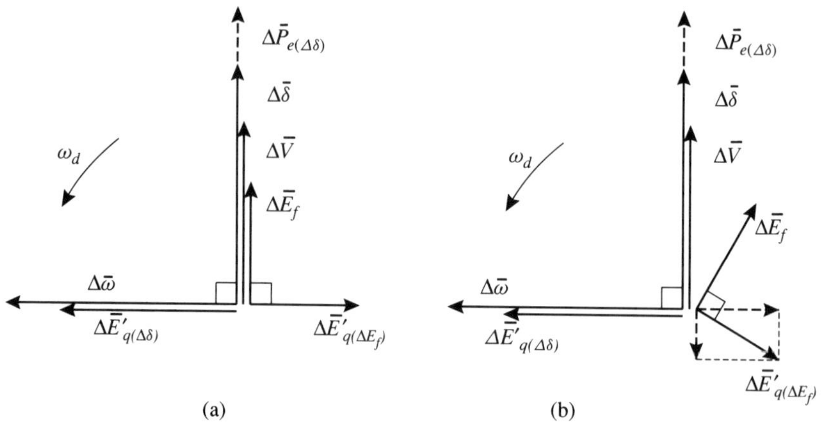

Figure 5.26 Phasors of increments on the complex plane for: (a) AVR proportional system; (b) AVR proportional system with inertia

account. This is shown in Figure 5.26 which contains two phasor diagrams drawn for two general types of AVR systems. In both diagrams the phasors $\Delta\bar{\delta}$ and $\Delta\overline{V}$ are in phase, equation 5.87. The component $\Delta\overline{E}'_{q(\Delta\delta)}$ leads $\Delta\bar{\delta}$, similarly as in Figure 5.13c.

The phasor diagram shown in Figure 5.26a is valid for a proportional AVR system when $\Delta\overline{E}_f$ and $\Delta\overline{V}$ are almost in phase. The component $\Delta\overline{E}'_{q(\Delta E_f)}$ lags $\Delta\overline{E}_f$ by $\pi/2$ and directly opposes $\Delta\overline{E}'_{q(\Delta\delta)}$. This diagram clearly shows that voltage regulation, represented by $\Delta\overline{E}'_{q(\Delta E_f)}$, weakens the damping introduced by the field winding and represented by $\Delta\overline{E}'_{q(\Delta E_f)}$. If the magnitude of $\Delta\overline{E}'_{q(\Delta E_f)}$ is greater than that of $\Delta\overline{E}'_{q(\Delta\delta)}$ then the voltage regulation will introduce a net negative damping into the system. This negative damping is enhanced by:

- large generator load (large value of δ'_0) resulting in a large value of the coefficient $K_{\Delta V/\Delta\delta}$, equation 5.87
- large gain $|G_{AVR}(s)|$ in the AVR controller determining the magnitude of $\Delta\overline{E}_f$
- large network reactance X, determining the value of the coefficient $K_{\Delta V/\Delta\delta}$ (the maximum value of $K_{\Delta V/\Delta\delta}$ is obtained for $X = X'_d$ which, for a typical network, is a large value)

Figure 5.26b shows the phasors for an AVR system with cascaded dc exciter or an ac exciter with a rectifier when $\Delta\overline{E}_f$ lags $\Delta\overline{V}$ typically by a few tens of degrees. The phasor $\Delta\overline{E}'_{q(\Delta E_f)}$ lags $\Delta\overline{E}_f$ by $\pi/2$ giving rise to two components with respect to the direction of $\Delta\bar{\delta}$: (i) the quadrature component which introduces negative damping, as in Figure 5.26a; and (ii) the in-phase component which is in phase with $\Delta\overline{P}_{e(\Delta\delta)}$ and does not influence damping. The effect of the latter component is to reduce the synchronising power coefficient $K_{E'_q}$, equation 5.52, and therefore change the frequency of oscillations.

The above analysis of the influence of an AVR system on the generator damping is of qualitative nature only with the aim of helping the understanding of these complicated phenomena. Detailed quantitative analysis can be found in De Mello & Concordia (1969), later enhanced in Anderson & Fouad (1977), Yao-nan Yu (1983) and Kundur (1994).

5.5.4 Effect of AVR action on the damper windings

Sub-Section 5.5.3 described how the AVR system can influence the damping torque due to the field winding, that is the last component in the swing equation 5.52. The second component in this equation $P_D = D\Delta\omega$ corresponds to the damping power introduced by the damper windings. Assuming constant excitation voltage, that is E_f = constant, the damping power is given by equation 5.24. Recall that the mechanism by which this power is developed is similar to that on which operation of the induction machine is based. A change in the rotor angle δ results in the speed deviation $\Delta\omega$. According to Faraday's Law an emf is induced proportional to the speed deviation. The current driven by this emf interacts with the air-gap flux to produce a torque referred to as the *natural damping torque*. To simplify considerations only the d-axis damper winding will be analysed.

Figure 5.27a shows a phasor diagram for the d-axis damper winding, similar to that shown in Figure 5.26. The emf induced in the winding $\overline{e}_{D(\Delta\omega)}$ is shown to be in phase with $\Delta\overline{\omega}$. The damper winding has a large resistance which means that the current flowing in it, $\overline{i}_{D(\Delta\omega)}$, lags the emf by an angle less than $\pi/2$. The component of this current which is in-phase with $\Delta\overline{\omega}$ gives rise to the natural damping torque. The quadrature component, which is in phase with $\Delta\overline{\delta}$, enhances the synchronising power coefficient.

Now consider the influence of the AVR on the damper windings. The d-axis damper winding lies along the path of the excitation flux produced by the field winding, Figure 4.3. This means that the two windings are magnetically coupled and may be treated as a transformer, Figure 5.27b, supplied by $\Delta\overline{E}_f$ and loaded with the resistance R_D of the damper winding. Consequently the additional current $\overline{i}_{D(\Delta E_f)}$ induced in the damper winding must lag $\Delta\overline{E}_f$. Figure 5.27c shows the position of the phasors. The horizontal component of $\overline{i}_{D(\Delta E_f)}$ directly opposes the horizontal component of $\overline{i}_{D(\Delta\omega)}$. As the former is due to AVR negative action while the latter is due to speed deviation and is responsible for the natural damping, it may be concluded that voltage regulation weakens the natural damping. This weakening effect is referred to as *artificial negative damping*.

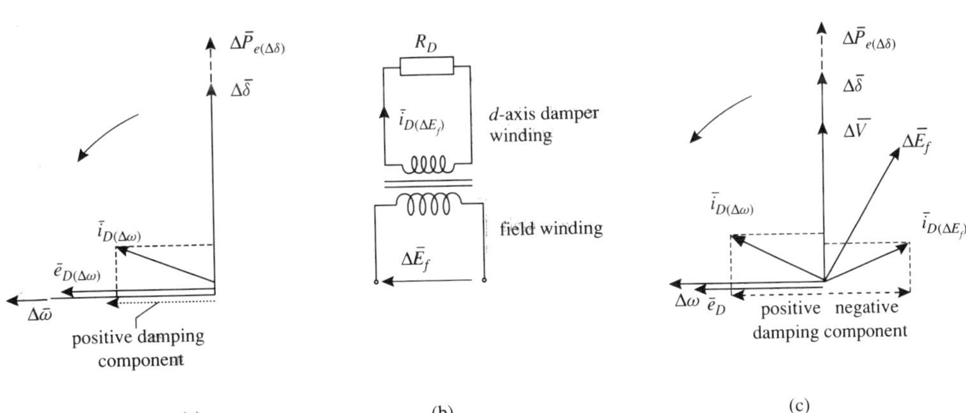

Figure 5.27 Phasor diagram for the damper windings: (a) natural damping only; (b) field and damper windings as a transformer; (b) natural and artificial damping components

Artificial damping is stronger for larger $\bar{i}_{D(\Delta E_f)}$ currents. This current is, in turn, proportional to the variations in $\Delta \overline{E}_f$ and $\Delta \overline{V}$ caused by $\Delta \overline{\delta}$. Some of the factors influencing this effect were described in the previous sub-section and are: generator load, reactance of the transmission network, and gain of the voltage controller.

5.5.5 Compensating the negative damping components

The main conclusion from the previous two sub-sections is that a voltage controller, which reacts only to the voltage error, weakens the damping introduced by the damper and field windings. In the extreme case of a heavily loaded generator operating on a long transmission link, a large gain in the voltage controller may result in net negative damping leading to an oscillatory loss of stability. This detrimental effect of the AVR can be compensated using supplementary control loop referred to as *power system stabiliser* (PSS) discussed in more detail in Section 9.1.

PSS is widely used in the USA, Canada and Western Europe. Another solution, preferred in the former Soviet Union, consists of voltage controllers with in-built feedback loops reacting to the time derivative of the voltage error and other quantities (Глебов 1970).

6

Electromechanical Dynamics — Large Disturbances

The previous chapter explained how a power system responds to a small disturbance, and determined the conditions necessary for the system to remain stable when subjected to such a disturbance. Much more dramatic from a stability point of view is the way in which a system responds to a large disturbance such as a short-circuit or line tripping. When such a fault occurs large currents and torques are produced and often action must be taken quickly if system stability is to be maintained. It is this problem of large disturbance stability, and the effect such a disturbance has on the system behaviour, that is addressed in this chapter.

6.1 Transient Stability

Assume that before the fault occurs the power system is operating at some stable steady-state operating condition. The power system transient stability problem is then defined as that of assessing whether or not the system will reach an acceptable steady-state operating point following the fault.

As the subtransient period is normally very short compared to the period of the rotor swings the effect of the subtransient phenomena on the electromechanical dynamics can be neglected. This allows the generator classical model to be used to study the transient stability problem when the swing equation is expressed by equation 5.15 and the air-gap power by equation 5.40. During a major fault, such as a short-circuit, the equivalent reactance x_d' appearing in equation 5.40 will be subject to change so that the air-gap power $P_e = P_{E'}(\delta')$ will also change and the power balance within the system will be disturbed. This will result in energy transfers between the generators producing corresponding rotor oscillations. Usually there are three states accompanying a disturbance with three, generally different, values of x_d': (i) the pre-fault state when the reactance $x_d' = x_{dPRE}'$; (ii) the fault state when $x_d' = x_{dF}'$; and (iii) the post-fault state when $x_d' = x_{dPOST}'$. This section will start by considering a fault that is cleared without any change in the network configuration being required. In this case $x_{dPOST}' = x_{dPRE}'$.

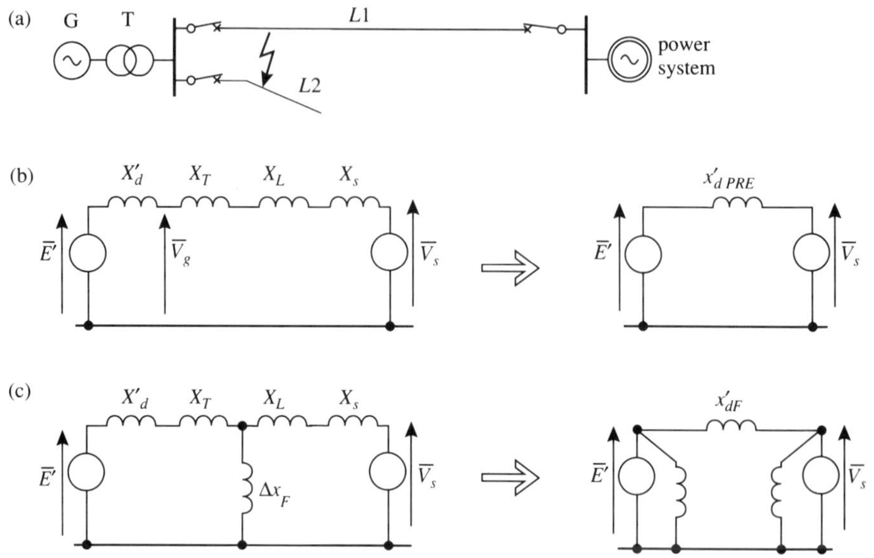

Figure 6.1 Example of a fault with the same pre- and post-fault impedance: (a) schematic diagram; (b) equivalent circuit for the pre- and post-fault state; (c) equivalent circuit during the fault

6.1.1 Fault cleared without a change in the equivalent network impedance

Figure 6.1a shows an example of a fault which is cleared by tripping the faulty element but without changing the equivalent network impedance. It is assumed that only line $L1$ of a double circuit connection is in use, with line $L2$ energised but not connected at the system end. If a fault occurs on the unconnected line $L2$, and is then cleared by opening the circuit breaker at the generator end of the line, the pre-fault and the post-fault impedance between the generator and the system are the same.

Reactance values during the pre-fault, fault and post-fault period

The equivalent circuit for the system is shown in Figure 6.1b. The generator is represented by the classical model with a constant transient emf E' behind the transient reactance X'_d whilst the system is represented by a constant voltage V_s behind the equivalent reactance X_s. The reactance of the transformer and line $L1$ are X_T and X_L respectively. The pre-fault equivalent reactance x'_{dPRE} of the whole transmission link is

$$x'_{dPRE} = X'_d + X_T + X_L + X_s \tag{6.1}$$

The use of symmetrical components allows any type of fault to be represented in the positive sequence network by a fault shunt reactance Δx_F connected between the point of the fault and the neutral, as shown in Figure 6.1c (Gross, 1986). The value of the fault shunt depends on the type of fault and is given in Table 6.1 where X_1, X_2 and X_0 are respectively the positive, negative and zero sequence Thevenin equivalent reactances as seen from the fault terminals.

Table 6.1 Shunt reactances representing different types of fault

Fault type	three-phase (3ph)	double-phase-to-ground (2ph-g)	phase-to-phase (2ph)	single phase (1ph)
Δx_F:	0	$\frac{X_2 X_0}{X_2 + X_0}$	X_2	$X_2 + X_0$

Using the star-delta transformation the fault network can be transformed as shown in Figure 6.1c so that the voltages E' and V_s are directly connected by the equivalent fault reactance

$$x'_{dF} = X'_d + X_T + X_L + X_s + \frac{(X'_d + X_T)(X_L + X_s)}{\Delta x_F} \tag{6.2}$$

The value of this reactance is heavily dependent on the value of the fault shunt Δx_F given in Table 6.1. When the fault is cleared, by opening the circuit-breaker in line $L2$, the equivalent circuit is the same as that in the pre-fault period so that $X'_{dPOST} = X'_{dPRE}$.

The circuit diagram of Figure 6.1c corresponds to the positive-sequence network so that when the reactance given in equation 6.2 is used in the power-angle equation 5.40 only the torque and power due to the flow of positive-sequence currents is accounted for. The influence of negative- and zero-sequence fault currents and torques are neglected from further considerations in this chapter.

Three-phase fault

Figure 6.2 shows how the equal area criterion, described in sub-Section 5.4.3, can be used to analyse the effect of a three-phase fault on the system stability. To simplify the discussion damping will be neglected ($P_D = 0$) and the changes in the rotor speed will be assumed to be too small to activate the turbine governor system. In this case the mechanical power input P_m from the turbine can be assumed to be constant.

For a three-phase fault $\Delta x_F = 0$ and, according to equation 6.2, $x'_{dF} = \infty$. Thus power transfer from the generator to the system is completely blocked by the fault with the fault current being purely inductive. During the fault the electrical power drops from its pre-fault value to zero as illustrated by line *1–2* in Figure 6.2 and remains at zero until the fault is cleared by opening the circuit-breaker. During this time the rotor acceleration ε can be obtained from the swing equation 5.15 by dividing both sides by M, substituting $P_e = 0$, $P_D = 0$ and writing in terms of δ' to give

$$\varepsilon = \frac{d^2\delta'}{dt^2} = \frac{P_m}{M} = \text{constant} \tag{6.3}$$

Integrating equation 6.3 twice with the initial conditions $\delta'(t = 0) = \delta'_0$ and $\Delta\omega(t = 0) = 0$ gives the power angle trajectory as

$$\delta' = \delta'_0 + \frac{\varepsilon t^2}{2} \text{ or } \Delta\delta' = \delta' - \delta'_0 = \frac{\varepsilon t^2}{2} \tag{6.4}$$

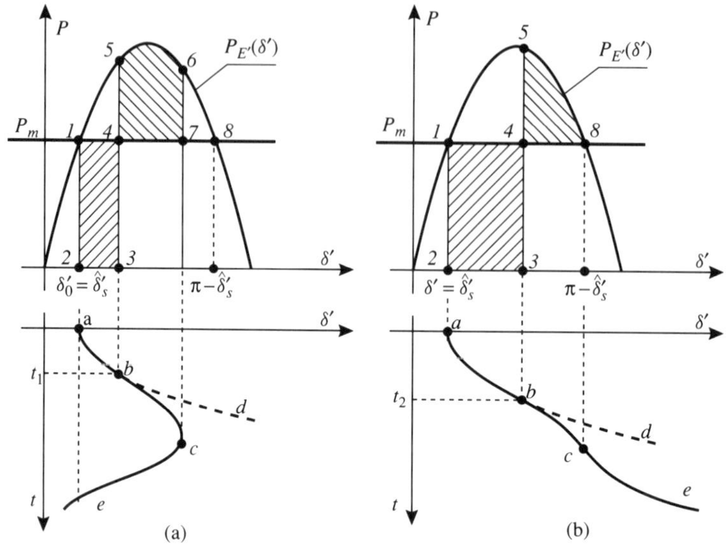

Figure 6.2 The acceleration and deceleration areas: (a) short clearing time; (b) long clearing time

This corresponds to the parabola *a-b-d* in Figure 6.2a. Before the fault is cleared the rotor moves from point *2* to point *3* on the power-angle diagram and acquires a kinetic energy proportional to the shaded area *1-2-3-4*.

When the fault is cleared at $t = t_1$ by opening the circuit-breaker, the rotor again follows the power angle characteristic $P_{E'}(\delta')$ corresponding to the reactance given by equation 6.1 so that the operating point 'jumps' from point *3* to point *5*. The rotor now experiences a deceleration torque, with magnitude proportional to the length of the line *4–5*, and starts to decelerate. However, due to its momentum, the rotor continues to increase its angle until the work done during deceleration, area *4-5-6-7*, equals the kinetic energy acquired during acceleration, area *1-2-3-4*. The rotor again reaches synchronous speed at point *6* when

$$\text{area } (4\text{-}5\text{-}6\text{-}7) = \text{area } (1\text{-}2\text{-}3\text{-}4) \tag{6.5}$$

In the absence of damping the cycle repeats and the rotor swings back and forth around point *1* performing *synchronous swings*. The generator does not lose synchronism and the system is stable.

Figure 6.2b shows a similar situation but with a substantially longer fault clearing time $t = t_2$ when the kinetic energy acquired during acceleration, proportional to the area *1-2-3-4*, is much larger than in Figure 6.2a. As a result the work performed during deceleration, proportional to the area *4-5-8*, cannot absorb the kinetic energy acquired during acceleration and the speed deviation does not become equal to zero before the rotor reaches point *8*. After passing point *8* the electrical power $P_{E'}(\delta')$ is less than the mechanical power P_m and the rotor experiences a net acceleration torque which further increases its angle. The rotor makes an *asynchronous rotation* and loses synchronism with the system. Asynchronous operation is analysed in more detail in Section 6.4.

Two important points arise from this discussion. The first is that the generator loses stability if, during one of the swings, the operating point passes point *8* on the characteristic. This point corresponds to the transient rotor angle being equal to $(\pi - \hat{\delta}'_s)$, where

$\hat{\delta}'_s$ is the stable equilibrium value of the transient rotor angle. Area *4-5-8* is therefore the *available deceleration area* with which to stop the swinging generator rotor. The corresponding transient stability condition states that the available deceleration area must be larger than the acceleration area forced by the fault. For the case shown in Figure 6.2a this criterion is

$$\text{area } (1\text{-}2\text{-}3\text{-}4) < \text{area } (4\text{-}5\text{-}8) \tag{6.6}$$

As the generator did not use the whole available decelerating area the remaining area *6-7-8*, divided by the available deceleration area, can be used to define the *transient stability margin*

$$K_{area} = \frac{\text{area } (6\text{-}7\text{-}8)}{\text{area } (4\text{-}5\text{-}8)} \tag{6.7}$$

The second important observation is that fault clearing time is a major factor in determining the stability of the generator. This is born out by equation 6.4 where the accelerating area *1-2-3-4* is seen to be proportional to the clearing time squared. The longest clearing time for which the generator will remain in synchronism is referred to as the *critical clearing time*. The relative difference between the critical clearing time and the actual clearing time can be used to give another measure of the transient stability margin

$$K_{time} = \frac{t_{cr} - t_f}{t_{cr}} \tag{6.8}$$

where t_{cr} and t_f are the critical and actual clearing times.

Unbalanced faults

During an unbalanced fault at least one of the phases is unaffected allowing some power to be transmitted to the system. The equivalent fault reactance x'_{dF} does not now increase to infinity, as for the three-phase fault, but to a finite value defined by equation 6.2. The increase in the reactance is inversely proportional to Δx_F and depends on the type of fault, as shown in Table 6.1. This allows faults to be listed in order of decreasing severity as: (i) a three-phase fault (3ph); (ii) a phase-to-phase-to-ground fault (2ph-g); (iii) a phase-to-phase fault (2ph); and (iv) a single-phase fault (1ph).

The corresponding power-angle characteristics during the fault are illustrated in Figure 6.3a.

The effect of an unbalanced fault on system stability is examined by considering the least severe, single-phase, fault. A fault clearing time is assumed slightly longer than the critical clearing time for the three-phase fault. The acceleration and the deceleration areas are shown in Figure 6.3b. In the case of the three-phase fault the acceleration area *1-2-3-4* is larger than the deceleration area *4-5-8* and the system is unstable, as in Figure 6.2a.

For the single-phase fault the power transfer is not completely blocked and the air-gap power drops from point *1* on the pre-fault characteristic to point *9* on the fault characteristic. The accelerating torque, corresponding to line *1–9*, is smaller than that for the three-phase fault (line *1–2*), the rotor accelerates less rapidly and, by the time that the fault is cleared, the rotor has reached point *10*. At this point the rotor angle is smaller than in the case of the three-phase fault. The acceleration area *1-9-10-12* is now

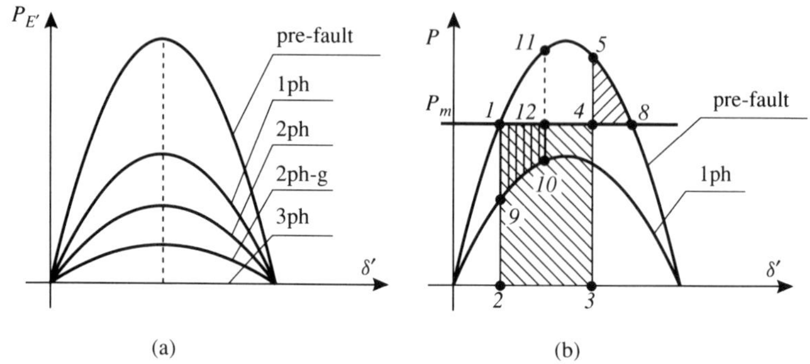

Figure 6.3 Effect of unbalanced faults: (a) comparison of power-angle characteristics; (b) acceleration and deceleration areas during a three-phase fault and a single-phase fault

much smaller than the maximum available deceleration area *11-8-12* and the system is stable with a large stability margin. Obviously a longer clearing time would result in the generator losing stability but the critical clearing time for the single-phase fault is significantly longer than that for the three-phase fault. Critical clearing times for other types of faults are in the same order as the Δx_F value given in Table 6.1.

Effect of the pre-fault load

Figure 6.4 shows a generator operating at load P_{m1} prior to a three-phase fault. The fault is cleared when the acceleration area *1-2-3-4* is smaller than the available deceleration area *4-5-8*. The system is stable with stability margin *6-7-8*. Increasing the pre-fault load by 50% to $P_{m2} = 1.5P_{m1}$ increases the acceleration power $P_{acc} = P_m - P_{E'}(\delta') = P_m$ one-and-half times so that, according to equations (6.3) and (6.4), the change in the power angle $\Delta\delta'$ also increases by a factor of 1.5. Consequently as each side of the accelerating area rectangle *1-2-3-4* has increased 1.5 times the acceleration area *1-2-3-4* is now much larger than the available deceleration area *4-5-8* and the system is unstable.

The pre-fault load is an important factor with regard to determining the critical clearing time and generator stability. The higher the load the lower the critical clearing time.

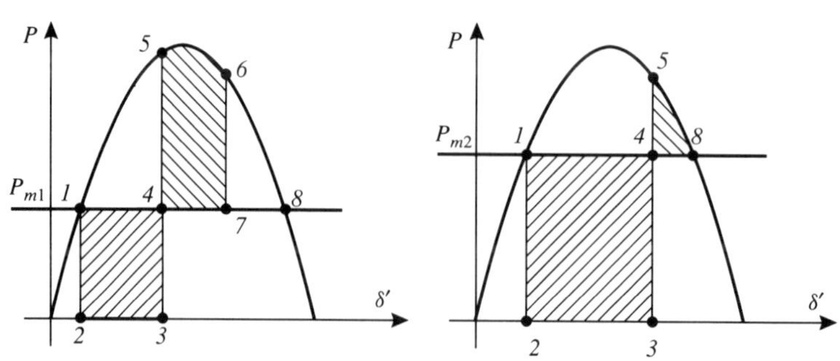

Figure 6.4 Acceleration and deceleration areas for two different pre-fault loads P_{m1} and $P_{m2} = 1.5P_{m1}$. The fault clearing time is the same for both cases

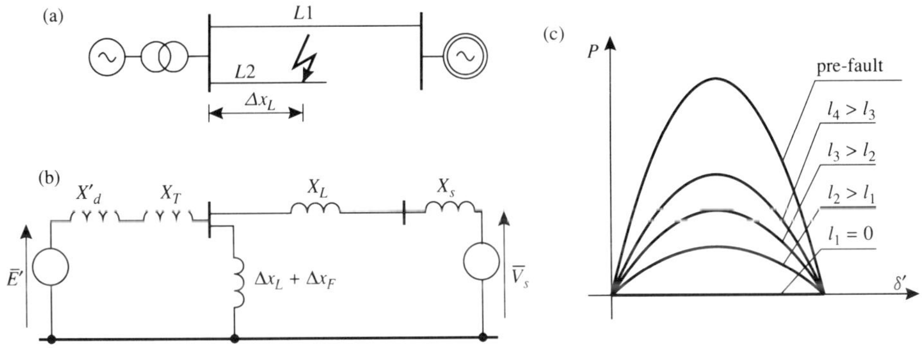

Figure 6.5 Influence of the fault distance: (a) schematic diagram; (b) equivalent circuit diagram; (c) power-angle characteristics before the fault and for various fault distances

Influence of fault distance

So far it has been assumed that in Figure 6.1 the fault occurs close to the busbar. If the point of fault is further along the line, as shown in Figure 6.5a, then the impedance of the faulted line Δx_L is proportional to the fault distance and the per-unit length reactance of the line. The resulting equivalent circuit during the fault period is shown in Figure 6.5b. The equivalent series reactance x'_{dF} can again be obtained from equation 6.2 but with Δx_F replaced by $\Delta x = \Delta x_F + \Delta x_L$.

Figure 6.5c shows a family of power-angle characteristics for a three-phase fault ($\Delta x_F = 0$) occurring at increasing distances along the line. By comparison with the discussion on unbalanced faults it can be seen that the further the distance to the fault the less severe the fault, and the longer the critical clearing time.

In the case of unbalanced faults $\Delta x_F \neq 0$ and the magnitude of the power-angle characteristic during the fault is further increased compared to the three-phase fault case. As a result the effect of the fault is less severe. In the case of a remote single-phase fault the disturbance to the generator may be very small.

6.1.2 Short-circuit cleared with/without auto-reclosing

The previous section described a particular situation where the network equivalent reactance does not change when the fault is cleared. In most situations the events surrounding the fault are more complex. Firstly, the fault itself would normally be on a loaded element, for example on line $L2$ when it was connected at both ends. Secondly, the fault will not usually clear itself and the faulted element must be switched out of circuit.

The majority of faults on transmission lines are intermittent so that, after clearing the fault by opening the necessary circuit breakers, the faulty line can be switched on again after allowing sufficient time for the arc across the breaker points to extinguish. This process is known as *auto-reclosing*. The sequence of events in a *successful auto-reclosing* cycle, shown in Figure 6.6, would be:

- both lines operate (before the fault) - Figure 6.6b
- short-circuit - Figure 6.6c

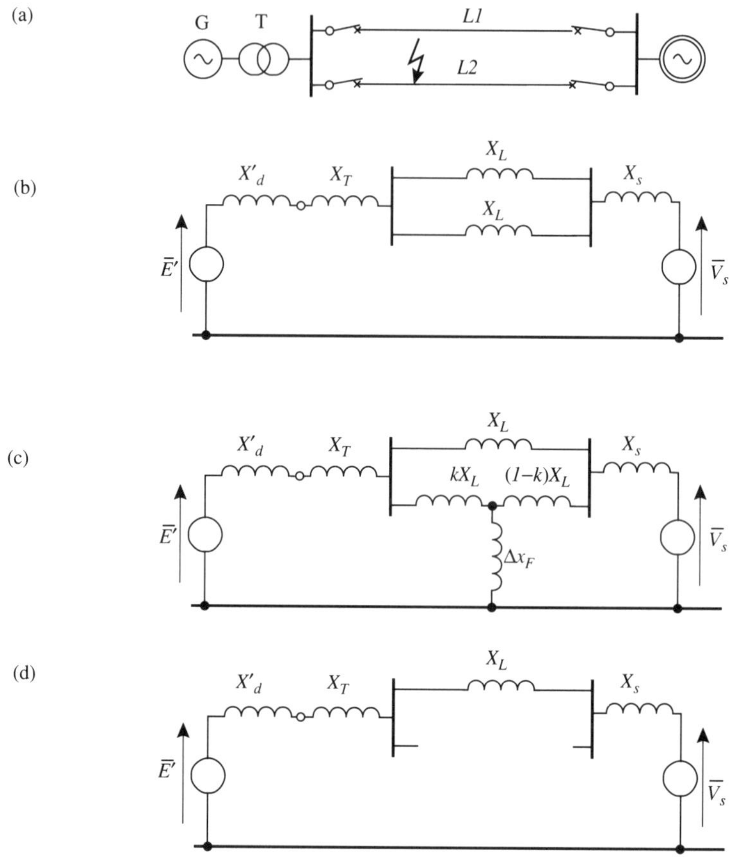

Figure 6.6 Auto-reclosing cycle: (a) schematic diagram; (b) equivalent circuit with both lines operating; (c) short-circuit on one of the lines; (d) one line operating

- the faulted line is tripped and only one line operates - Figure 6.6d
- the faulted line is automatically reclosed and both lines again operate - Figure 6.6b.

In Figure 6.6 the fault is assumed to occur on line *L2* at some distance *k* from the circuit-breaker. Each state is characterised by a different equivalent reactance x_d' in equation 5.40 and a different power-angle characteristic corresponding to that reactance. Figure 6.7 shows the effect of a three-phase fault with two fault clearing times; one producing a stable response and the other an unstable one. In both cases the accelerating power *1–2* accelerates the rotor from point *2* to point *3* during the fault. When line *L2* is tripped the operating point moves to point *5* and, because of the acquired kinetic energy, moves further along the characteristic *c*. After a certain auto-reclose time required to extinguish the arc, the auto-recloser reconnects line *L2* and the system moves from point *6* to point *7*. The power angle moves further along the characteristic *a* to point *8* when, in the stable case, the decelerating area *4-5-6-7-8-10* is equal to the accelerating area *1-2-3-4*. The system is stable with a stability margin corresponding to the area *8-9-10*. In the unstable case, Figure 6.7b, the increased clearing time enlarges the accelerating area *1-2-3-4* and

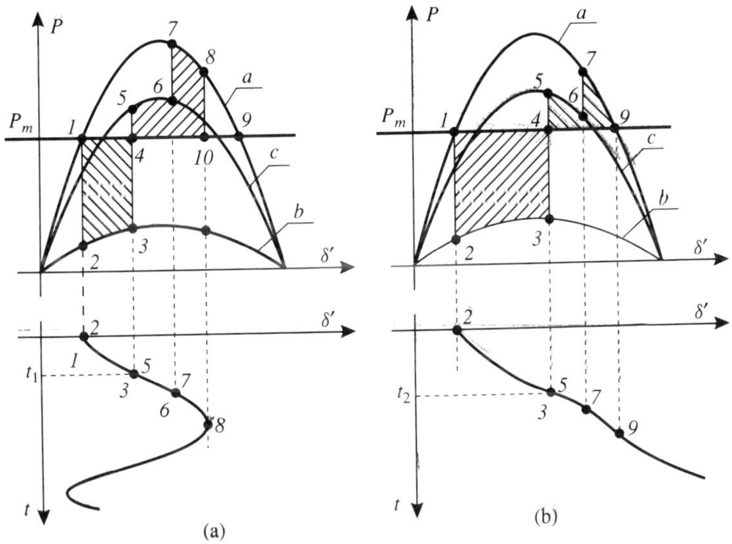

Figure 6.7 The accelerating and the decelerating areas for successful auto-reclosing: (a) stable case; (b) unstable case

the available decelerating area is too small to absorb this energy and stop the rotor. The generator rotor makes an asynchronous rotation and loses stability with the system.

In the case of a solid fault the reclosed line is again tripped and the cycle is referred to as *unsuccessful auto-reclosing*. Such a cycle poses a much greater threat to system stability than successful auto-reclosing. In the case of unsuccessful auto-reclosing the following sequence of events occurs:

- both lines operate (before the fault)
- short-circuit
- the faulted line is tripped and one line operates
- short-circuit (an attempt to auto-reclose onto the solidly faulted line)
- the faulted line is permanently tripped so that only one line remains in operation

An illustration of unstable and stable unsuccessful auto-reclosure is shown in Figure 6.8. In the first case the rotor acquires kinetic energy during the short-circuit proportional to the area *1-2-3-4*. Then, during the break in the cycle, the rotor is decelerated and loses energy proportional to the area *4-5-6-7*. An attempt to switch onto the solidly faulted line results in an increase of the kinetic energy proportional to the area *7-8-9-11*. When the faulted line is permanently tripped, the decelerating area left is *10-13-11*. As the sum of the accelerating areas *1-2-3-4* and *7-8-9-11* is greater than the sum of the decelerating areas *4-5-6-7* and *10-13-11*, the rotor passes point *13* and makes an asynchronous rotation.

If the clearing time and the auto-reclose time are reduced and, in addition, the pre-fault load on the generator is reduced from P_{m1} to P_{m2}, the system may remain stable as shown in Figure 6.8b. The sum of the accelerating areas *1-2-3-4* and *7-8-9-11* are now equal to the sum of the decelerating areas *4-5-6-7* and *10-11-14-12* and the system is stable with a stability margin corresponding to the area *12-14-13*. Assuming that the oscillations are

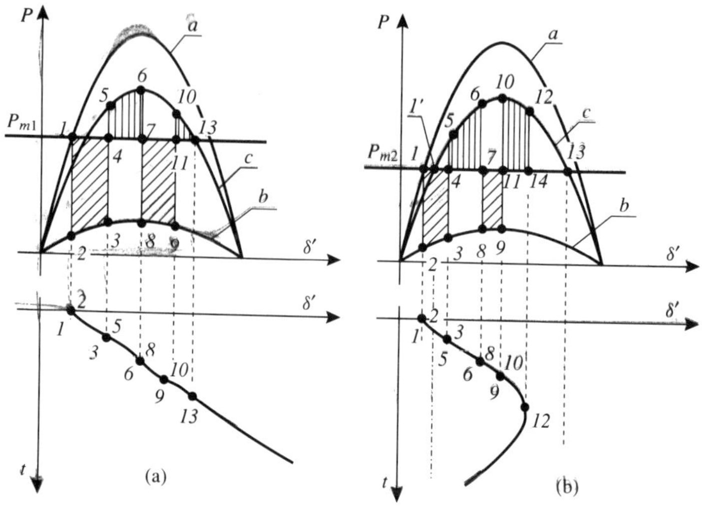

Figure 6.8 The accelerating and the decelerating areas for the unsuccessful auto-reclosing: (a) unstable case; (b) stable case

damped the new equilibrium point *1'* corresponds to the power characteristic with one line switched off.

6.1.3 Power swings

The rotor oscillations accompanying a fault also produce oscillations in the generated power. The shape of power variations can be a source of useful, although approximate, information on the transient stability margin. Again consider the system shown in Figure 6.5a and assume that the fault on line $L2$ is cleared by tripping the circuit breakers at each end of the line without auto-reclosing. If the stability margin *6-7-8*, shown in Figure 6.9a, is small, the power angle oscillations will be large and may exceed a value of $\pi/2$. The corresponding power oscillations will increase until δ' passes over the peak of the power angle characteristic when they will start to decrease. During the rotor back-swing the power will initially increase, as the rotor angle passes over the peak of the characteristic, before again decreasing. As a result the power waveform $P_e(t)$ exhibits characteristic "humps" which disappear as the oscillations are damped out. If the transient stability margin *6-7-8* is large, as in Figure 6.9b, then the humps do not appear, as the maximum value of the power angle oscillation is less than $\pi/2$ and the oscillations are only on one side of the power-angle characteristic.

It should be emphasised that power swings follow the transient power-angle characteristic $P_{E'}(\delta')$ rather than the static characteristic $P_{E_q}(\delta)$. This means that the value of power at which humps appear is usually much higher than the critical steady-state power $P_{E_q cr}$, see Figure 5.9.

6.1.4 Effect of flux decrement

The transient characteristic $P_{E'}(\delta')$ used to analyse transient stability is valid assuming that the flux linkage with the field winding is constant so that $E' = \text{constant}$. In fact, as

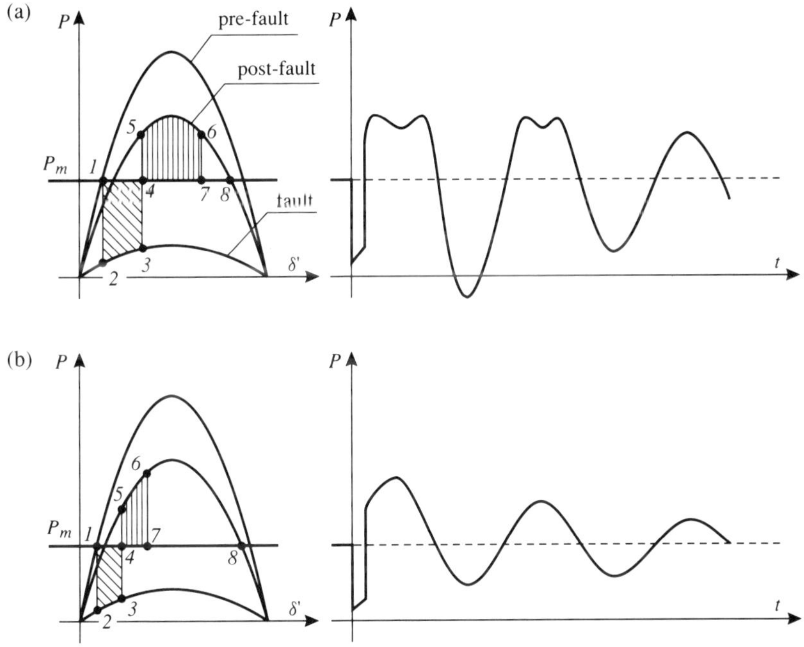

Figure 6.9 Power oscillations in the case of (a) a low stability margin (b) a large stability margin

magnetic energy is dissipated in the field winding resistance, flux decrement effects will cause E' to decrease with time. If the fault clearing time is short then flux decrement effects can be neglected for transient stability considerations but if the clearing time is long then the decay in E' may have a considerable effect. To understand this refer to Figure 6.2 which illustrates the case of $E' =$ constant. If flux decrement effects are now included, then following the fault, the amplitude of the $P_{E'}(\delta')$ characteristic will reduce leading to a decrease in the available deceleration area *4-5-8* and a deterioration in the transient stability. Consequently the use of the classical model may lead to an optimistic assessment of the critical clearing time.

6.1.5 Effect of the AVR

Section 5.5 explained how AVR action may reduce the damping of rotor swings following a small disturbance. In the case of large disturbances the influence of AVR is similar. However, immediately after a fault occurs and is cleared, a strong acting AVR may prevent a loss of synchronism. This may be explained as follows.

When a fault occurs, the generator terminal voltage drops and the large regulation error ΔV forces the AVR to increase the generator field current. However the field current will not change immediately due to a delay dependent on the gain and time constants of the AVR, and on the time constant of the generator field winding. To examine the effect of AVR action on transient stability the system shown in Figure 6.5a will be considered assuming that a three-phase short-circuit occurs some distance along the $L2$ so that $\Delta x_L \neq 0$ and $\Delta x_F = 0$.

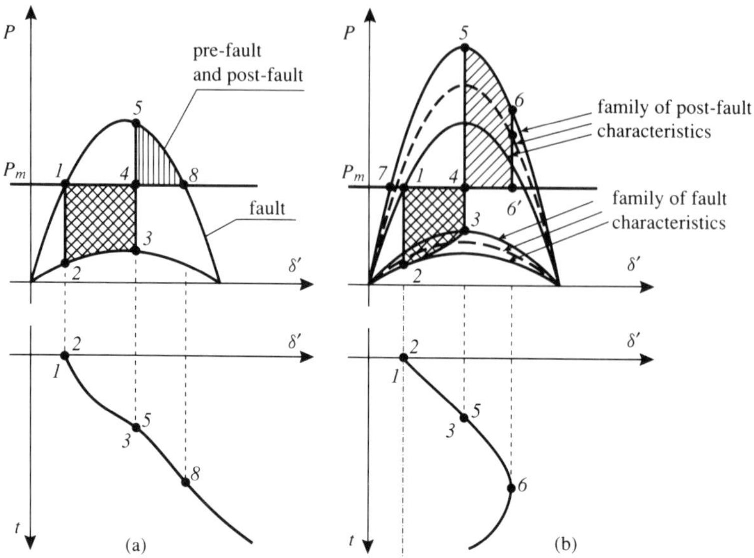

Figure 6.10 The acceleration area and the deceleration area when the influence of the voltage regulator is: (a) neglected; (b) included

When no AVR is present this system may lose stability as shown in Figure 6.10a. The effect of the AVR, shown in Figure 6.10b, is to increase the field current leading to an increase in the transient emf E' as explained in sub-Section 4.2.4 and illustrated in Figure 4.16a. This increase in E' can be accounted for by drawing a family of power-angle characteristics $P_{E'}(\delta')$ for different values of E'. A fast acting AVR and exciter can increase the excitation voltage up to its ceiling before the fault is cleared, although the change in the field current, and E', will lag behind this due to the time constant of the generator field winding. This increase in field current, and hence E', has two positive effects. Firstly, as E' increases, the accelerating power decreases and the accelerating area *1-2-3-4* is slightly reduced. Secondly, when the fault is cleared, the system will follow a higher power-angle characteristic resulting from the new E' so that a larger decelerating area is available. In this example the rotor reaches a maximum power angle at point *6*, when the decelerating area *4-5-6-6'* equals the accelerating area *1-2-3-4*, before starting to move back towards the equilibrium point.

Although a fast acting AVR reduces the first rotor swing, it can increase the second and following swings depending on the system parameters, the dynamic properties of AVR and the time constant of the field winding. Consider the voltage regulation error $\Delta V = V_{ref} - V_g$, created when the fault is cleared. Equation 5.86 is now important as it shows how V_g depends on δ' and the ratio X'_d/X. An example of such dependence is shown in Figure 6.11.

The terminal voltage V_g reaches a minimum when $\delta' = \pi$. The actual value of this minimum depends on the ratio X'_d/X with the smallest value of $V_g = (E' - V_s)/2$ being reached when $X'_d/X = 1$ (curve 1). As the generator reactance usually dominates over other reactances in the transmission link the case $X'_d/X = 1$ corresponds to a long transmission link. For a short transmission link $X'_d/X > 1$ and the minimum value of V_g is larger (curve 2).

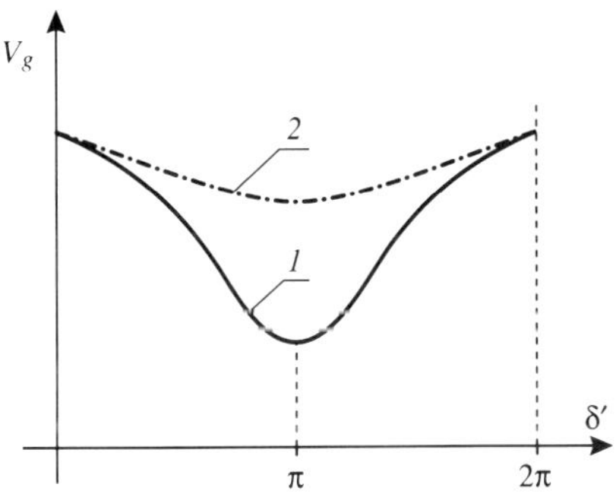

Figure 6.11 Generator terminal voltage as a function of δ' for a long transmission link (curve 1) and a short transmission link (curve 2)

Consider first the case of a long transmission link and assume that when the fault is cleared the angle δ' is large. On clearing the fault, the terminal voltage will recover from a small fault value to a somewhat small post-fault value (curve *1* in Figure 6.11). Consequently the AVR will continue to increase the excitation current in order to try to recover the terminal voltage. In this case the rotor back swing will occur on the highest possible transient characteristic $P_{E'}(\delta')$. This situation is illustrated in Figure 6.12a which shows the system trajectory following fault clearance. The AVR continues to increase the field current during the back swing increasing the deceleration area *6-8-7-6′*. This results in an increase in the amplitude of consecutive rotor swings such that, in this case, the AVR may have a detrimental effect on the generator transient stability.

Now consider the case of a short transmission link illustrated in Figure 6.12b. In this case the terminal voltage V_g recovers well when the fault is cleared, despite the large value of δ'. As a small increase in the transient emf E' will be forced by the increase in the excitation current during the fault period the terminal voltage may recover to a value that is slightly higher than the reference value. Subsequently this high terminal voltage will force the AVR to reduce the field current during the rotor back swing; the amplitude of the transient power-angle characteristic will decrease and, as a result, the deceleration area *6-8-7-6′* will be reduced. This reduction in the deceleration area will

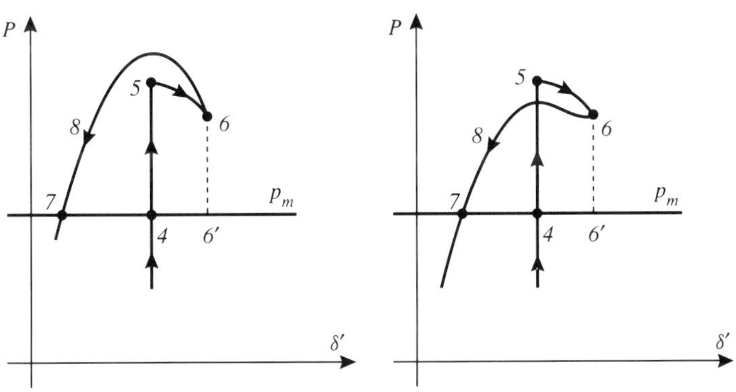

Figure 6.12 Rotor swing after the fault clearance in the case of: (a) a long transmission link; (b) a short transmission link

lead to a reduction in the amplitude of subsequent rotor swings. In this case the AVR enhances the transient stability in both the forward and the backward swing.

Figure 6.12 is also closely related to the damping produced by the AVR when a small disturbance occurs. Section 5.5 concluded that the amount of negative damping produced by the AVR increases with the length of the line as this increases the proportionality coefficient between $\Delta\delta'$ and ΔV Figure 5.25. The example shown in Figure 6.12a corresponds to the situation when this negative damping is greater than the system positive damping, while Figure 6.12b illustrates the reverse case.

As the influence of the AVR strongly depends on the post-fault network reactance, the dynamic system response depends on the fault location and its clearance. This is illustrated in Figure 6.13. If fault $F1$ appears on line $L1$ and is cleared by tripping the faulted line then the generator will operate in the post-fault state via a long transmission link that consists of line $L2$ and $L3$. Simulation results for this example are presented in Figure 6.14. Figure 6.14a shows the trajectory of the operating point while Figure 6.14b shows that the terminal voltage drops during the fault and does not recover very well. This low value of terminal voltage forces the AVR to continue to increase the excitation

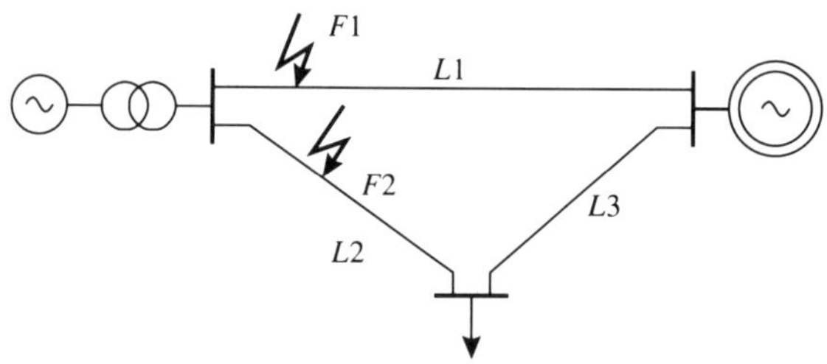

Figure 6.13 Example of simple power system

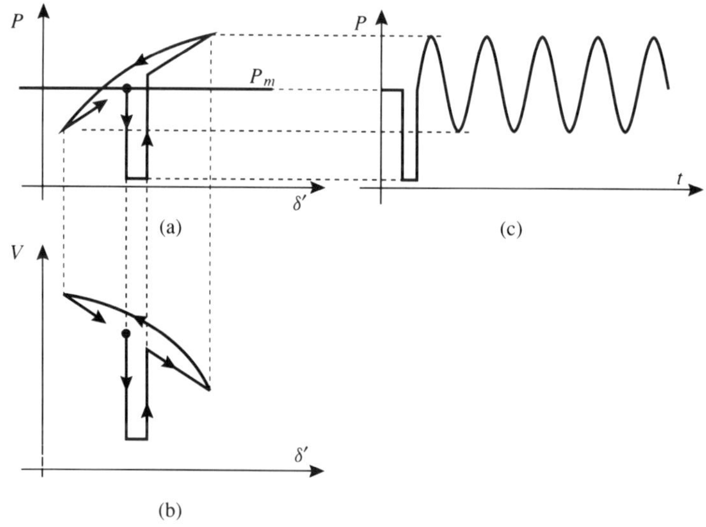

Figure 6.14 Simulation results for a fault on line $L1$ cleared by tripping the faulted line: (a) equal-area method; (b) variation in the generator voltage; (c) power swings

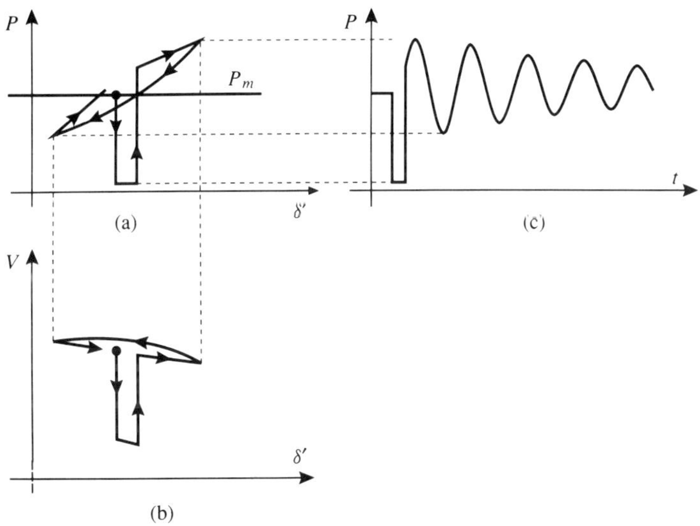

Figure 6.15 Simulation results for a fault on line $L2$ cleared by tripping the faulted line: (a) equal-area method; (b) variation in the generator voltage; (c) power swings

current so that the rotor back swing occurs along the upper characteristic in Figure 6.14a. This repeats from cycle to cycle with the effect that the rotor swings are poorly damped, Figure 6.14c.

If the fault occurs on line $L2$ and is again cleared by tripping the faulted line, the generator operates in the post-fault period via a short transmission link that consists solely of line $L1$. Simulation results for this case are shown in Figure 6.15. Now, when the fault is cleared, the generator voltage recovers well and stays at a high level, Figure 6.15b. Consequently the AVR decreases the excitation current and the rotor swings back along the lower power-angle characteristic in Figure 6.15a and the deceleration area is small. This repeats during each cycle of the rotor swings with the result that they are well damped, Figure 6.15c.

These examples show how the rotor oscillations may increase if the transmission link is long and the AVR keeps the excitation voltage at too high a value during the rotor back swing. Ideally the regulator should increase the excitation when the power angle δ' increases and lower it when δ' decreases no matter what the value of the regulation error ΔV is. Section 9.1 will show how the AVR can be provided with supplementary control loops to provide a regulation error that depends on the rotor speed deviation $\Delta\omega$ or the rate of change of real power. Such supplementary loops co-ordinate the regulation process with the rotor swings in order to ensure correct damping.

6.2 Direct Method for Stability Assessment

Throughout this chapter use has been made of the equal area criterion to explain and assess system stability. This method will now be formalised using *Lyapunov's direct method* and an *energy-type Lyapunov function*. In this section the basic concepts on which Lyapunov's direct method are based will be described and applied to the simple generator-infinite busbar system. Lyapunov's direct method is also known as *Lyapunov's second method*.

Due to its potential in assessing power system stability, without the need to solve the system differential equations, Lyapunov's direct method has been the subject of much intensive research. However the practical application of the direct method for real-time security assessment is still some time ahead due to modelling limitations and the unreliability of computational techniques. When applied to a multi-machine system, especially one operating close to its stability limits, the direct method is vulnerable to numerical problems and may give unreliable results. Interested readers looking for a detailed treatment of this topic are referred to Pai (1981 and 1989), Fouad & Vittal (1992) and Pavella & Murthy (1994).

6.2.1 Mathematical background

Dynamic systems are generally described by a set of non-linear differential equations of the form

$$\dot{\mathbf{x}} = \mathbf{F}(\mathbf{x}) \tag{6.9}$$

where $\mathbf{x}$ is the vector of state variables. The Euclidean space determined by $\mathbf{x}$ is referred to as the *state-space* whilst the point $\hat{\mathbf{x}}$, for which $\mathbf{F}(\hat{\mathbf{x}}) = \mathbf{0}$, is referred to as the *equilibrium point*. For an initial point $\mathbf{x}_0 \neq \hat{\mathbf{x}}$, $\dot{\mathbf{x}}(t = 0) = \mathbf{0}$, equation 6.9 has a solution $\mathbf{x}(t)$ in the state-space and is referred to as the *system trajectory*. The system is said to be *asymptotically stable* if the trajectory returns to the equilibrium point as $t \to \infty$. If the trajectory remains in the vicinity of the equilibrium point as $t \to \infty$ then the system is said to be *stable*.

Lyapunov's stability theory, used in the direct method of stability assessment, is based on a scalar function $V(\mathbf{x})$ defined in the state-space of the dynamic system. At a given point the direction of the largest increase in the value of $V(\mathbf{x})$, is given by the gradient of the function $\operatorname{grad} V(\mathbf{x}) = [\partial V/\partial x_i]$. The points $\tilde{\mathbf{x}}$ for which $\operatorname{grad} V(\tilde{\mathbf{x}}) = \mathbf{0}$ are referred to as the *stationary points*. Each stationary point may correspond to a *minimum*, a *maximum* or a *saddle point* as illustrated in Figure 6.16. The stationary point $\tilde{\mathbf{x}}$ corresponds to a minimum, Figure 6.16a, if any small disturbance $\Delta\mathbf{x} \neq \mathbf{0}$ causes an increase in the function, i.e. $V(\tilde{\mathbf{x}} + \Delta\mathbf{x}) > V(\tilde{\mathbf{x}})$. Similarly a given stationary point $\tilde{\mathbf{x}}$ corresponds to

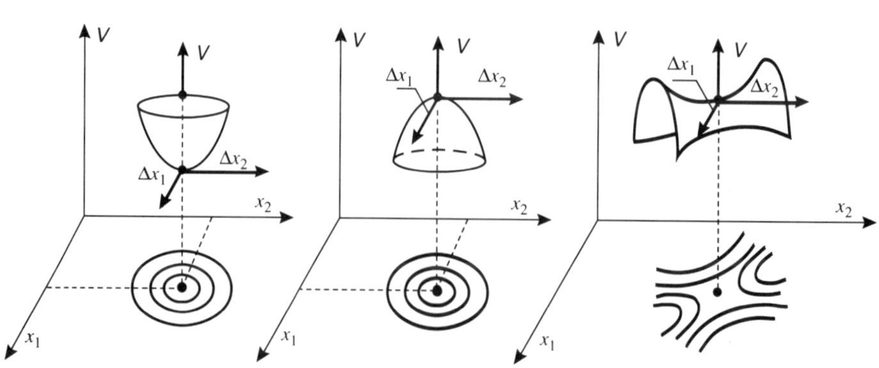

Figure 6.16 A scalar function of two variables with three types of stationary points: (a) minimum; (b) maximum; (c) saddle point

a maximum, Figure 6.16b, if any small disturbance $\Delta\mathbf{x} \neq \mathbf{0}$ causes a decrease in the function, i.e. $V(\tilde{\mathbf{x}} + \Delta\mathbf{x}) < V(\tilde{\mathbf{x}})$. The mathematical condition for checking whether a function has a maximum or minimum at a stationary point can be derived by expanding $V(\mathbf{x})$ by a Taylor series to give:

$$V(\tilde{\mathbf{x}} + \Delta\mathbf{x}) \cong V(\tilde{\mathbf{x}}) + \Delta\mathbf{x}^{\mathrm{T}}[\mathrm{grad}V] + \tfrac{1}{2}\Delta\mathbf{x}^{\mathrm{T}}\mathbf{H}\Delta\mathbf{x} + \ldots \tag{6.10}$$

where $\mathbf{H} = [\partial^2 V/\partial x_i \partial x_j]$ is the *Hessian matrix*. At the stationary points $\mathrm{grad}V(\tilde{\mathbf{x}}) = \mathbf{0}$ and the increment in $V(\mathbf{x})$ caused by the disturbance can be found as

$$\Delta V = V(\tilde{\mathbf{x}} + \Delta\mathbf{x}) - V(\tilde{\mathbf{x}}) \cong \tfrac{1}{2}\Delta\mathbf{x}^{\mathrm{T}}\mathbf{H}\Delta\mathbf{x} = \tfrac{1}{2}\sum_{i=1}^{N}\sum_{j=1}^{N} h_{ij}\Delta x_i \Delta x_j \tag{6.11}$$

where h_{ij} is the (i, j) element of the matrix $\mathbf{H}$. This equation shows that the increment in V is equal to the quadratic form of the state variables constructed using the Hessian matrix. Sylvester's theorem (Bellman, 1970) states that such a quadratic form has a minimum at a given stationary point if, and only if, all leading principal minors of the matrix $\mathbf{H}$ are positive, Figure 6.16a. The matrix is then referred to as *positive definite*. If all leading principal minors of $\mathbf{H}$ are negative then the matrix is *negative definite* and the quadratic form has a maximum at a given stationary point, Figure 6.16b. If some leading principal minors are positive and some negative then the matrix is *nondefinite* and the quadratic form has a saddle point at the stationary point, Figure 6.16c.

As the scalar function $V(\mathbf{x})$ is defined in the state space each point on the system trajectory $\mathbf{x}(t)$ corresponds to a value of $V(\mathbf{x}(t))$. The rate of change of $V(\mathbf{x})$ along the system trajectory (that is the derivative $\mathrm{d}V/\mathrm{d}t$) can be expressed as:

$$\begin{aligned}\dot{V} = \frac{\mathrm{d}V}{\mathrm{d}t} &= \frac{\partial V}{\partial x_1}\frac{\mathrm{d}x_1}{\mathrm{d}t} + \frac{\partial V}{\partial x_2}\frac{\mathrm{d}x_2}{\mathrm{d}t} + \ldots + \frac{\partial V}{\partial x_n}\frac{\mathrm{d}x_n}{\mathrm{d}t} = [\mathrm{grad}V(\mathbf{x})]^{\mathrm{T}}\dot{\mathbf{x}} \\ &= [\mathrm{grad}V(\mathbf{x})]^{\mathrm{T}}\mathbf{F}(\mathbf{x})\end{aligned} \tag{6.12}$$

To introduce the *direct* (or *second*) *Lyapunov method* assume that a positive definite scalar function $V(\mathbf{x})$ has a stationary point (minimum) at the equilibrium point $\tilde{\mathbf{x}} = \hat{\mathbf{x}}$. Any disturbance $\Delta\mathbf{x} \neq \mathbf{0}$ will move the system trajectory to an initial point $\mathbf{x}_0 \neq \hat{\mathbf{x}}$ as illustrated in Figure 6.17. If the system is asymptotically stable, as in Figure 6.17a, then the trajectory $\mathbf{x}(t)$ will tend towards the equilibrium point and $V(\mathbf{x})$ will decrease along the trajectory until $\mathbf{x}(t)$ settles at the minimum point $\tilde{\mathbf{x}} = \hat{\mathbf{x}}$. If the system is unstable, Figure 6.17b, then the trajectory will move away from the equilibrium point and $V(\mathbf{x})$ will increase along the trajectory. The essence of these considerations is summarised in *Lyapunov's stability theorem*:

> Let $\hat{\mathbf{x}}$ be an equilibrium point of a dynamic system $\dot{\mathbf{x}} = \mathbf{F}(\mathbf{x})$. Point $\hat{\mathbf{x}}$ is stable if there is a continuously differentiable positive definite function $V(\mathbf{x})$ such that $\dot{V}(\mathbf{x}) \leq 0$. Point $\hat{\mathbf{x}}$ is asymptotically stable if $\dot{V}(\mathbf{x}) < 0$.

One of the main attractions of Lyapunov's theorem is that it can be used to obtain an assessment of system stability without the need to solve the differential equation 6.9. The main problem is how to find a suitable positive definite Lyapunov function $V(\mathbf{x})$ for which the sign of the derivative $\dot{V}(\mathbf{x})$ can be determined without actually determining the system trajectory.

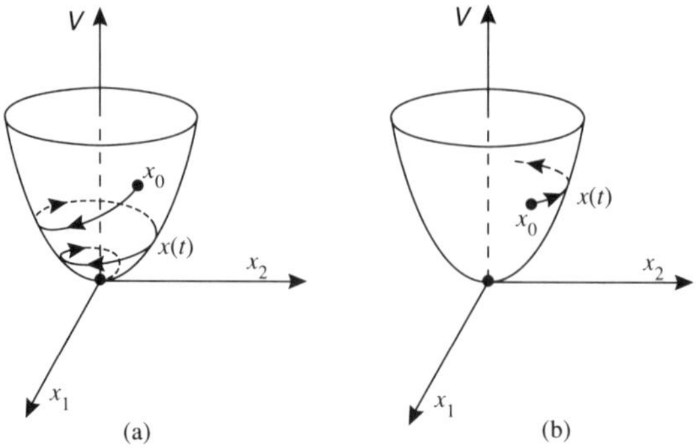

Figure 6.17 Illustration to Lyapunov's theorems on stability: (a) asymptotic stability; (b) instability

Generally the Lyapunov function $V(\mathbf{x})$ will be non-linear and may have more than one stationary point. Figure 6.18 illustrates this using two functions of one variable; one which has three stationary points, Figure 6.18a, and one which has two stationary points, Figure 6.18b. Assume that in both cases the first of the stationary points corresponds to the equilibrium point $\tilde{\mathbf{x}}_1 = \hat{\mathbf{x}}$. The Lyapunov function determines the stability only if there is no other stationary point $\tilde{\mathbf{x}}_2$ between the initial point $\mathbf{x}_0 = \mathbf{x}(t = 0)$ and the equilibrium point $\tilde{\mathbf{x}}_1 = \hat{\mathbf{x}}$. If the initial point $\mathbf{x}_0$ is beyond the second stationary point $\tilde{\mathbf{x}}_2$ then the condition $\dot{V}(\mathbf{x}) < 0$ may mean that the system trajectory will tend towards another stationary point, Figure 6.18a, or run away, as in Figure 6.18b. The value of $V(\mathbf{x})$ at the nearest stationary point $\tilde{\mathbf{x}}_2 \neq \hat{\mathbf{x}}$ is referred to as the *critical value* of the Lyapunov function. These considerations lead to the following theorem:

> If there is a positive-definite scalar function $V(\mathbf{x})$ in the vicinity of the equilibrium point $\hat{\mathbf{x}}$ of a dynamic system $\dot{\mathbf{x}} = \mathbf{F}(\mathbf{x})$ and the time derivative of this function is negative ($\dot{V}(\mathbf{x}) < 0$) then the system is asymptotically stable at $\hat{\mathbf{x}}$ for any initial conditions satisfying
>
> $$V(\mathbf{x}_0) < V_{cr} \tag{6.13}$$
>
> where $V_{cr} = V(\tilde{\mathbf{x}} \neq \hat{\mathbf{x}})$ is the value of Lyapunov function at the nearest stationary point.

It is important to realise that the stability conditions due to Lyapunov's theorems are only sufficient for assessing stability. Failure of the candidate Lyapunov function to satisfy the

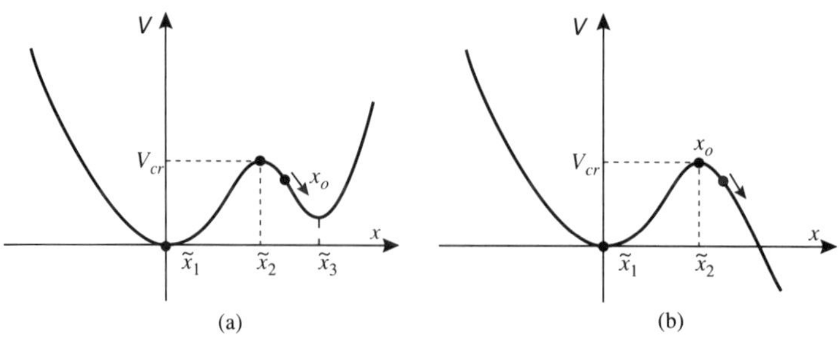

Figure 6.18 Examples of non-linear functions with more than one stationary point

stability conditions does not mean that the equilibrium point is not stable. Moreover for any given dynamic system there are usually many possible Lyapunov functions each of which gives a larger, or smaller, area of initial conditions that satisfies the stability theorems. This means that a particular Lyapunov function usually gives a pessimistic assessment of stability as it covers only a portion of the actual stability area. The function giving the largest area, and at the same time being closest to the actual stability area, is referred to as the *good Lyapunov function*. Usually good Lyapunov functions have a physical meaning.

6.2.2 Energy-type Lyapunov function

In section 5.4, equation 5.45, showed that the integral of the acceleration power is proportional to the work done by the acceleration torque. The proportionality constant ω_s was neglected and the integral of power was treated as "work" or "energy". A similar approach will be adopted here.

Energy function

Assume that the generator can be represented by the classical model defined by equations 5.15 and 5.40, when the equation describing the generator-infinite busbar system is

$$M\frac{\mathrm{d}\Delta\omega}{\mathrm{d}t} = P_m - b\sin\delta' - D\frac{\mathrm{d}\delta'}{\mathrm{d}t} \tag{6.14}$$

where $b = E'V_s/x'_d$ is the amplitude of the transient power-angle curve $P_{E'}(\delta')$ and $\Delta\omega = \mathrm{d}\delta'/\mathrm{d}t = \mathrm{d}\delta/\mathrm{d}t$ is the speed deviation. This equation has two equilibrium points:

$$(\hat{\delta}'_s;\ \Delta\hat{\omega} = 0); \qquad (\hat{\delta}'_u = \pi - \hat{\delta}'_s;\ \Delta\hat{\omega} = 0) \tag{6.15}$$

Multiplying equation 6.14 by $\Delta\omega$, neglecting the damping term and moving the right-hand-side to the left gives

$$M\Delta\omega\frac{\mathrm{d}\Delta\omega}{\mathrm{d}t} - (P_m - b\sin\delta')\frac{\mathrm{d}\delta'}{\mathrm{d}t} = 0 \tag{6.16}$$

As the function on the left-hand-side of this equation is equal to zero its integral must be constant. Integrating the function from the first of the equilibrium points defined in equation 6.15 to any point on the system transient trajectory gives

$$V = \int_0^{\Delta\omega}(M\Delta\omega)\mathrm{d}\Delta\omega - \int_{\delta'_s}^{\delta'}(P_m - b\sin\delta')\mathrm{d}\delta' = \text{constant} \tag{6.17}$$

Evaluating the integrals gives the following form of the function

$$V = \tfrac{1}{2}M\Delta\omega^2 - [P_m(\delta' - \hat{\delta}'_s) + b(\cos\delta' - \cos\hat{\delta}'_s)] = E_k + E_p = E \tag{6.18}$$

where

$$E_k = \tfrac{1}{2}M\Delta\omega^2,\ E_p = -[P_m(\delta' - \hat{\delta}'_s) + b(\cos\delta' - \cos\hat{\delta}'_s)] \tag{6.19}$$

E_k is a measure of the system kinetic energy while E_p is a measure of potential energy, both taken with respect to the first equilibrium point ($\hat{\delta}'_s$, $\Delta\hat{\omega} = 0$) and, for the remainder of this chapter, will be treated as energy. By neglecting damping, equation 6.17 shows the sum of potential and kinetic energy $V = E_k + E_p$ to be constant.

It is now necessary to check whether the function V, defined by equation 6.18, satisfies the definition of a Lyapunov function, that is whether: (i) it has stationary points at the equilibrium points defined in equation 6.15; (ii) it is positive definite in the vicinity of one of the equilibrium points; (iii) its derivative is not positive ($\dot{V} \leq 0$).

The first of the conditions can be checked by calculating the gradient of V. Differentiation of equation 6.18 gives

$$\mathrm{grad}V = \begin{bmatrix} \dfrac{\partial V}{\partial \Delta\omega} \\ \dfrac{\partial V}{\partial \delta'} \end{bmatrix} = \begin{bmatrix} \dfrac{\partial E_k}{\partial \Delta\omega} \\ \dfrac{\partial E_p}{\partial \delta'} \end{bmatrix} = \begin{bmatrix} M\Delta\omega \\ -(P_m - b\sin\delta') \end{bmatrix} \tag{6.20}$$

This gradient is equal to zero at the stationary points where $\Delta\tilde{\omega} = 0$ and the electrical power is equal to mechanical power given by

$$\tilde{\delta}'_1 = \hat{\delta}'_s, \quad \tilde{\delta}'_2 = \pi - \hat{\delta}'_s \tag{6.21}$$

Both these points are the equilibrium points of equation 6.14.

The second condition can be checked by determining the Hessian matrix given by

$$\mathbf{H} = \begin{bmatrix} \dfrac{\partial^2 V}{\partial \Delta\omega^2} & \dfrac{\partial^2 V}{\partial \Delta\omega^2 \partial \delta'} \\ \dfrac{\partial^2 V}{\partial \delta' \partial \Delta\omega} & \dfrac{\partial^2 V}{\partial \delta'^2} \end{bmatrix} = \begin{bmatrix} M & 0 \\ 0 & b\cos\delta' \end{bmatrix} \tag{6.22}$$

Sylvester's theorem says that this matrix is positive definite when $M > 0$ (which is always true) and when $b\cos\delta' > 0$, which is true for $|\delta'| < \pi/2$ and holds for the first stationary point $\tilde{\delta}'_1 = \hat{\delta}'_s$. Thus the function V is positive definite at the first equilibrium point $\hat{\delta}'_s$.

The third condition can be checked by determining $\dot{V} = \mathrm{d}V/\mathrm{d}t$ along the trajectory of equation 6.14. As V represents the total system energy its derivative $\dot{V} = \mathrm{d}V/\mathrm{d}t$ corresponds to the rate at which energy is dissipated by the damping. This can be proved by expressing the derivative of V as

$$\dot{V} = \frac{\mathrm{d}V}{\mathrm{d}t} = \frac{\mathrm{d}E_k}{\mathrm{d}t} + \frac{\mathrm{d}E_p}{\mathrm{d}t} \tag{6.23}$$

The derivative of the kinetic energy can be calculated from equation 6.18 as:

$$\frac{\mathrm{d}E_k}{\mathrm{d}t} = \frac{\partial E_k}{\partial \Delta\omega}\frac{\mathrm{d}\Delta\omega}{\mathrm{d}t} = M\Delta\omega\frac{\mathrm{d}\Delta\omega}{\mathrm{d}t} = \left[M\frac{\mathrm{d}\Delta\omega}{\mathrm{d}t}\right]\Delta\omega \tag{6.24}$$

The factor in the square brackets corresponds to left-hand-side of equation 6.14. Replacing it by the right-hand-side gives

$$\frac{\mathrm{d}E_k}{\mathrm{d}t} = \frac{\partial E_k}{\partial \Delta\omega}\frac{\mathrm{d}\Delta\omega}{\mathrm{d}t} = +[P_m - b\sin\delta']\Delta\omega - D\Delta\omega^2 \tag{6.25}$$

Differentiating the potential energy defined by equation 6.18 gives

$$\frac{\mathrm{d}E_p}{\mathrm{d}t} = \frac{\partial E_p}{\partial \delta'}\frac{\mathrm{d}\delta'}{\mathrm{d}t} = -[P_m - b\sin\delta']\Delta\omega \tag{6.26}$$

Substituting equations 6.25 and 6.26 into equation 6.23 yields

$$\dot{V} = \frac{dV}{dt} = -D\Delta\omega^2 \tag{6.27}$$

which illustrates how the total system energy decays at a rate proportional to the damping coefficient ($D > 0$) and the square of the speed deviation ($\Delta\omega^2$). As $\dot{V} < 0$, the function $V(\delta', \Delta\omega)$ is a Lyapunov function and the first equilibrium point $\hat{\delta}'_s$ is asymptotically stable.

The second equilibrium point $\hat{\delta}'_u = \pi - \hat{\delta}'_s$ is unstable as the matrix $\mathbf{H}$, equation 6.22, calculated at this point is not positive definite.

6.2.3 Transient stability area

The critical value of the Lyapunov function V introduced in equation 6.13 corresponds to the value of V at the nearest stationary point which, for the system considered here, is equal to the second equilibrium point ($\pi - \hat{\delta}'_s$, $\Delta\hat{\omega} = 0$). Substituting these values into equation 6.18 gives

$$V_{cr} = 2b\cos\hat{\delta}'_s - P_m(\pi - 2\hat{\delta}'_s) \tag{6.28}$$

According to equation 6.13 the generator-infinite busbar system is stable for all initial conditions (δ'_0, $\Delta\omega_0$) satisfying the condition

$$V(\delta'_0, \Delta\omega_0) < V_{cr} \tag{6.29}$$

In this context the initial conditions are the value of the transient rotor angle and the rotor speed deviation at the instant of fault clearance when the generator starts to swing freely.

Careful examination of equation 6.28 shows that the critical value of the Lyapunov function (total system energy) depends on the stable equilibrium point $\hat{\delta}'_s$ which, in turn, depends on the generator load P_m. At no-load, when $\hat{\delta}'_s = 0$, the value of V_{cr} is greatest. Increasing the load on the generator, and hence $\hat{\delta}'_s$, reduces V_{cr} until, for $\hat{\delta}'_s = \pi/2$, its value is $V_{cr} = 0$. Figure 6.19 shows equiscalar contours $V(\delta', \Delta\omega) = \text{constant}$ on the phase-plane $\Delta\omega$ versus δ'.

Figure 6.19a shows the case $\hat{\delta}'_s = 0$. For small values of $V(\delta', \Delta\omega) < V_{cr}$ the equiscalar contours $V(\delta', \Delta\omega) = \text{constant}$ are closed around the equilibrium point $\hat{\delta}'_s = 0$. The critical value $V(\delta', \Delta\omega) = V_{cr}$ corresponds to the closed contour shown in bold crossing the stationary points $\pm\pi$ equal to the unstable equilibrium points. As the equiscalar energy increases the contours start to open up. The equilibrium point $\hat{\delta}'_s = 0$ is the point of minimum total energy while the stationary points $\pm\pi$ are the saddle points. Now assume that at the instant of fault clearance (i.e. at the initial point δ'_0, $\Delta\omega_0$) the system trajectory is inside the contour $V(\delta', \Delta\omega) < V_{cr}$. If damping is neglected then, according to equation 6.27, the value of $V(\delta', \Delta\omega)$ remains constant and the trajectory will be a closed curve with $V(\delta', \Delta\omega) = V(\delta'_0, \Delta\omega_0) = \text{constant}$. If damping is included $D > 0$ then $V(\delta', \Delta\omega)$ will decrease with time and the trajectory will tend to spiral in towards the equilibrium point $\hat{\delta}'_s = 0$.

Figure 6.19b shows the case of $\hat{\delta}'_s > 0$ but less than $\pi/2$. In this case the curve $V(\delta', \Delta\omega) = V_{cr} = \text{constant}$ corresponds to a contour crossing the stationary point

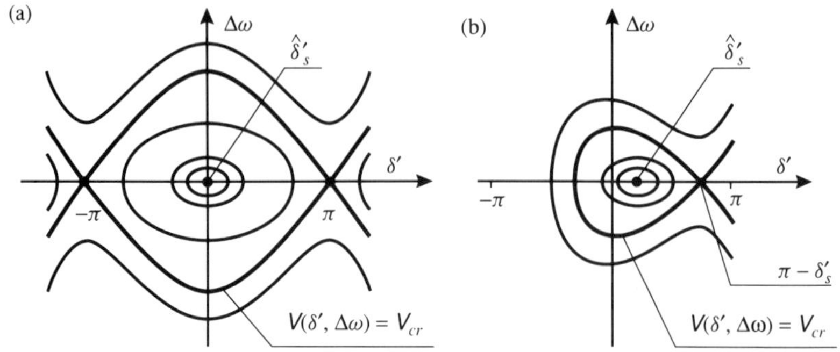

Figure 6.19 Equiscalar contours of the total system energy for the two equilibrium points: (a) $\hat{\delta}'_s = 0$; (b) $\hat{\delta}'_s < \pi/2$

$(\pi - \hat{\delta}'_s)$ equal to the unstable equilibrium point. The area enclosed by this boundary contour is now much smaller than before.

If, at the instant of fault clearance, the transient stability condition equation 6.29 is satisfied then the calculated value of the total energy $V_0 = V(\delta'_0, \Delta\omega_0)$ can be used to calculate a *transient stability margin*

$$K_{energy} = \frac{V_{cr} - V_0}{V_{cr}} \tag{6.30}$$

that describes the relative difference between the critical value of the total energy (transient stability boundary) and the energy released by the disturbance. K_{energy} determines the distance from the actual contour to the boundary contour of the transient stability area in terms of energy.

6.2.4 Equal area criterion

The equal area criterion was introduced in equation 6.6 as a means of assessing transient stability and, like the Lyapunov stability condition in equation 6.29, is also based on energy considerations. Both approaches can be shown to be equivalent by considering Figure 6.20 which shows the same three-phase fault as that illustrated in Figures 6.1 and 6.2a.

The first term of the energy based Lyapunov function defined in equation 6.18 corresponds to the kinetic energy

$$\tfrac{1}{2}M\Delta\omega^2 = \text{area } (1, 2, 3, 4) = \text{area } A \tag{6.31}$$

The second component, equal to potential energy, can be expressed as:

$$-\int_{\hat{\delta}'_s}^{\delta'} (P_m - b\sin\delta')\mathrm{d}\delta' = -[\text{area}(1, 7, 8, 4) - \text{area}(1, 7, 8, 5)] = \text{area } B \tag{6.32}$$

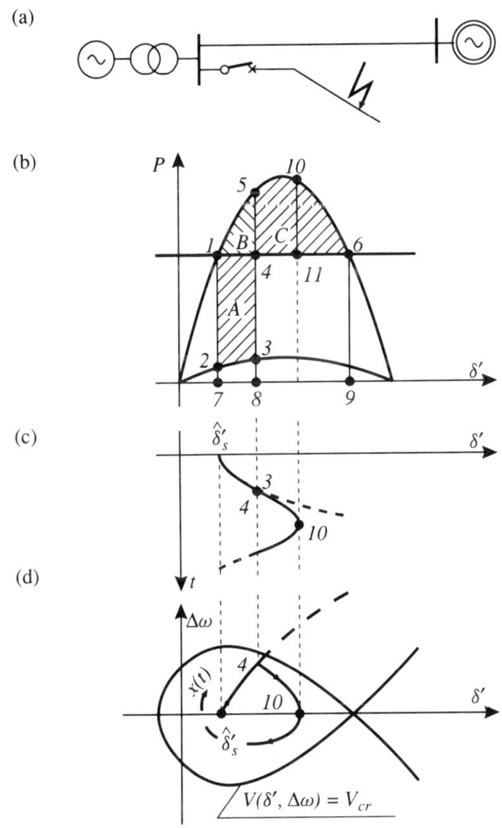

Figure 6.20 Equivalence between Lyapunov's direct method and the equal area criterion: (a) schematic diagram of the system; (b) transient power-angle characteristics with acceleration and deceleration areas; (c) angle variation; (d) stability area and trajectory of rotor motion

The Lyapunov function is equal to the sum of the two components:

$$V(\delta', \Delta\omega) = \text{area } A + \text{area } B \tag{6.33}$$

The critical value of the Lyapunov function is equal to the value of the potential energy at the unstable equilibrium point, that is

$$V_{cr} = -\int_{\hat{\delta}'_s}^{\pi-\hat{\delta}'_s} (P_m - b\sin\delta')\mathrm{d}\delta' = \text{area}(1, 7, 9, 6, 5)$$
$$- \text{area}(1, 7, 9, 6) = \text{area } B + \text{area } C \tag{6.34}$$

Thus, the stability condition $V(\delta'_0, \Delta\omega_0) < V_{cr}$ can be expressed as

$$\text{area } A + \text{area } B < \text{area } B + \text{area C} \quad \text{or} \quad \text{area } A < \text{area } C \tag{6.35}$$

which is equivalent to the equal area criterion.

The equiscalar contour $V(\delta', \Delta\omega) = V_{cr}$ in the $(\delta', \Delta\omega)$ plane determines the stability area, Figure 6.20c. If, during the short-circuit, the trajectory $\mathbf{x}(t)$ is contained within the area then, after clearing the fault, the trajectory will remain within the area and the system is stable. When $D > 0$ the trajectory tends towards the equilibrium point and the system is asymptotically stable. As the clearing time increases the initial point approaches the critical contour and reaches it when the fault time is equal to the critical clearing time. If the clearing time is longer than this then area $A >$ area C, the trajectory leaves the stability area, and the system is unstable.

For the case considered here the transient stability margin defined in equation 6.30 is equal to

$$K_{energy} = \frac{V_{cr} - V_0}{V_{cr}} = \frac{\text{area } B + \text{area } C - \text{area } A}{\text{area } B + \text{area } C} = \frac{\text{area } (10, 6, 11)}{\text{area } (1, 5, 6)} = K_{area} \quad (6.36)$$

which is exactly the same as the transient stability margin defined by the equal area criterion in equation 6.7. Consequently, for the generator-infinite busbar system, the equal area criterion is equivalent to Lyapunov's direct method based on an energy-type Lyapunov function.

6.3 Synchronisation

Section 4.4 described the electromagnetic dynamics occurring when a generator is synchronised to the system. This section expands this discussion to include the effect of the electromechanical dynamics that accompany synchronisation and in determining those conditions necessary for re-synchronisation.

Consider the circuit shown in Figure 4.30. When the synchronising switch is closed the resulting electromagnetic torque will attempt to pull the rotor into synchronisation with the power system by either slowing down or accelerating the rotor until eventually the generator reaches its final equilibrium position defined by the steady-state rotor angle $\hat{\delta}_s = 0$ and the rotor speed $\omega = \omega_s$. To analyse these rotor dynamics only the transient aperiodic component of the torque will be considered as the subtransient interval is normally too short to have any significant effect on the rotor swings. However the subtransient torques are vital when determining shaft torque rating and fatigue strength as described in sub-Section 4.2.7.

The generator is represented by the classical model, equations 5.15 and 5.40. As the generator is on no-load prior to synchronisation, it holds that $E' = E_f$ and $\delta = \delta'$. Now consider the case illustrated in Figure 6.21 when all the synchronisation conditions are satisfied apart from the phasors $\overline{V}_s$ and $\overline{E}_f$ being slightly out of phase, that is $P_m \approx 0$; $\omega_0 \approx \omega_s$; $\delta'(t = 0^+) = \delta'_0 > 0$. When the synchronising switch is closed the initial operating point lies at point *1* in Figure 6.21a. The equilibrium point is defined by $P_m = P_e$ and $\hat{\delta}'_s = 0$, that is point *3*. At the instant of synchronisation $P_e(\delta') > 0$, the right hand side of the swing equation is negative, the rotor decelerates and the rotor angle δ' decreases. This causes the average value of the electrical torque and power to decrease, with the energy used to decelerate the rotor being proportional to the shaded area *1-2-3*. At the equilibrium point *3* rotor inertia effects ensure that the rotor angle continues to decrease towards negative δ' until it reaches point *4*, where area *3-4-5* equals area *1-3-2*. At point *4* the accelerating power is positive and the rotor swings back towards δ'_0. The oscillations continue until the damping torques cause the rotor to settle at its equilibrium angle $\hat{\delta}'_s = 0$.

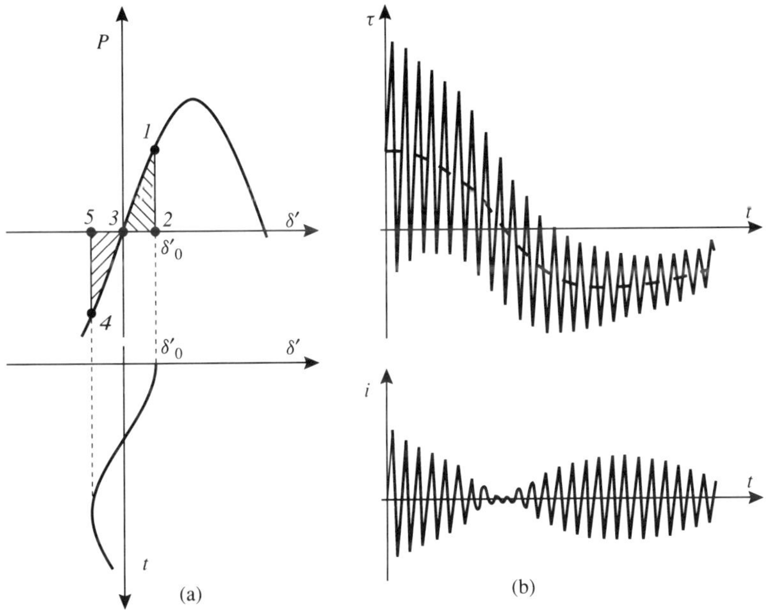

Figure 6.21 Synchronisation when $\delta_0' > 0$, $\omega_0 \approx \omega_s$: (a) power-angle characteristic and the swing curve; (b) time variation of the torque and current

Changes in δ' due to the swinging rotor produce corresponding oscillations in the torque and current as shown in Figure 6.21b. Initially, as δ' is large, the average value of the torque, shown by the dashed line, and the amplitude of the current are both large. As energy is dissipated in the windings the generator changes from the subtransient state to the transient state to the steady-state, and both the periodic component of the torque, and the amplitude of the current, vanish with time.

If at the instant of synchronisation $\delta_0' \neq 0$ and $\omega_0 \neq \omega_s$, then the changes in δ' may be much greater than those discussed above. Figure 6.22 shows a case when $\omega_0 > \omega_s$ and the rotor has an excess of kinetic energy. On closing the synchronising switch this excess kinetic energy will drive the rotor towards an increasing δ' until the shaded area is equal to the excess kinetic energy. The rotor then swings back towards the equilibrium point $\hat{\delta}_s'$. The conditions necessary to achieve successful synchronisation can be determined using the equal area criterion. At the instant of synchronisation the rotor has an excess kinetic energy relative to the rotating reference frame (infinite busbar). This energy is equal to

$$E_k = \tfrac{1}{2} M \Delta\omega_0^2 \tag{6.37}$$

where M is the inertia coefficient and $\Delta\omega_0 = \dot{\delta}'(t = 0^+) = \omega_0 - \omega_s$.

The maximum deceleration work that can be done by the generator $W_{\max}$ is proportional to the area *4-1-2-3* in Figure 6.22:

$$W_{\max} = \int_{\delta_0'}^{\pi - \hat{\delta}_s'} \left[\frac{E'V_s}{x_d'} \sin\delta' - P_m \right] \mathrm{d}\delta' = \frac{E'V_s}{x_d'} \left(\cos\hat{\delta}_s' + \cos\delta_0' \right) - P_m \left[(\pi - \hat{\delta}_s') - \delta_0' \right] \tag{6.38}$$

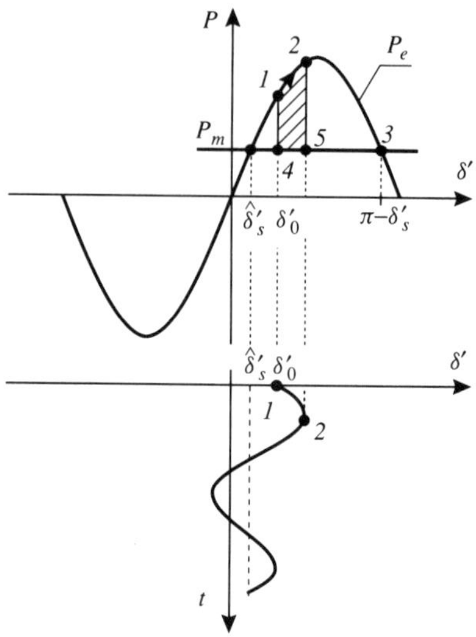

Line 1–2 corresponds to the increase in δ' caused by the excess of kinetic energy, whilst the shaded area corresponds to the work performed by the decelerating torque

Figure 6.22 Equal-area criterion applied to synchronisation with $\omega_0 \neq \omega_s$ and $\delta'_0 > \hat{\delta}'_s$

The equal area criterion stipulates that the generator can be synchronised only if $E_k < W_{\max}$. This, when using equations 6.37 and 6.38, gives

$$\Delta\omega_0^2 < \frac{2}{M}\left[\frac{E'V_s}{x'_d}\left(\cos\hat{\delta}'_s + \cos\delta'_0\right) - P_m\left(\pi - \hat{\delta}'_s - \delta'_0\right)\right] \tag{6.39}$$

The maximum permissible value of $\Delta\omega_0$ depends on the synchronisation angle δ'_0, the turbine power P_m and the parameters of the generator and transmission system as expressed by x'_d. Figure 6.23 shows two example of areas on the phase plane diagram ($\Delta\omega$,

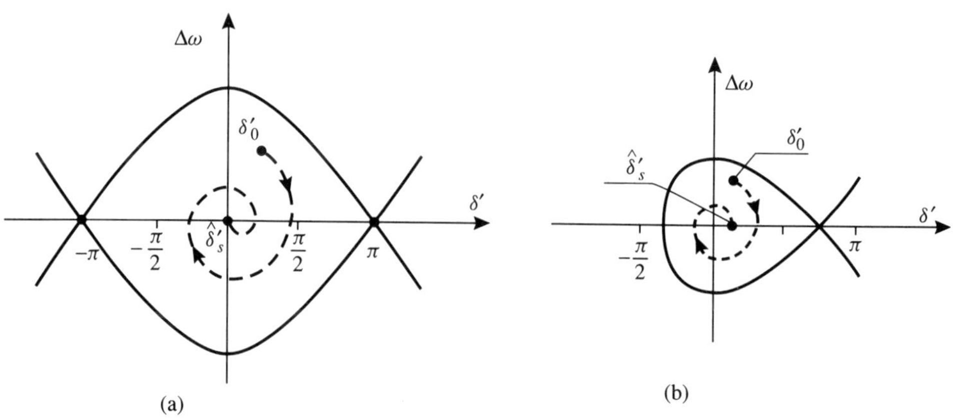

Figure 6.23 The trajectory of the operating point and the synchronisation areas on the phase plane for: (a) $P_m = 0$; (b) $P_m = 0.5E'V_s/x'_d$

δ') that satisfy the condition for successful synchronisation as defined by equation 6.39: one when $P_m = 0$ and the other when $P_m = 0.5E'V_s/x'_d$. The dashed lines show the trajectory of the operating point whilst the solid lines show the critical energy contours obtained from equation 6.39 for the limiting conditions on δ'_0 and ω_0. Figure 6.23 shows that an increase in P_m has a detrimental effect on synchronisation as the available decelerating area, defined by the area between P_m and P_e for $\delta'_0 < \delta' < \pi - \hat{\delta}'_s$, reduces as P_m increases. For $P_m > E'V_s/x'_d$ the system has no equilibrium point. Although it is not usual to synchronise a generator with the turbine developing large values of torque, similar conditions must be considered when synchronising certain types of wind turbine (Westlake, Bumby & Spooner, 1996).

The condition defined by equation 6.39 contains useful information regarding possible re-synchronisation after a fault is cleared by automatic reclose action, Figures 6.7 and 6.8. In this case the value of the rotor angle and the speed deviation at the moment the last switching occurs can be treated as the initial values δ_0 and $\Delta\omega_0$ in the synchronisation process.

Section 6.2 showed that the equal area criterion is equivalent to the direct Lyapunov method when an energy-type Lyapunov function is used. The synchronisation condition in equation 6.39 can therefore be derived from the stability condition in equation 6.29. The stability area shown in Figure 6.19 is the same as the synchronisation area shown in Figure 6.23. When the synchronisation condition in equation 6.39 is not satisfied, and the point (δ_0, $\Delta\omega_0$) lies outside the synchronisation area, the generator will start to operate *asynchronously*. This type of operation will be considered in the next section.

6.4 Asynchronous Operation and Resynchronisation

The previous section explained how a generator may lose synchronism with the rest of the system when it makes asynchronous rotations at a slip frequency a few Hz above synchronous speed. With such high speed changes the field winding time constant ensures that the net field winding flux linkage remains approximately constant so that the classical, constant flux linkage, generator model is valid with the generator synchronous power being given by $P_{E'}$, equation 5.40. However as $P_{E'}$ changes as $\sin\delta'$ and, as over one asynchronous rotation δ' goes through 360°, the average value of $P_{E'}$ over one rotation is zero. Furthermore as the speed increases above synchronous the action of the turbine-governing system comes into play while the asynchronous damping torques, proportional to $\Delta\omega$ are also significant. Indeed it is the interaction between the turbine-governor characteristic and the asynchronous damping torques that will determine the generator operating point.

As explained in sub-Section 2.2.3, the turbine static characteristic $P_m(\Delta\omega)$ is a straight line of droop ρ which crosses the vertical power axis in Figure 6.24 at point P_{m0} and the horizontal speed deviation axis at point $\rho\omega_s$. The form of the average asynchronous, damping, power characteristic $P_{Dav}(\Delta\omega)$ was discussed in Section 5.2 and is also shown on Figure 6.24. As the average value of synchronous power $P_{E'}$ over each asynchronous rotation is zero, the intersection between the static turbine characteristic and the damping power characteristic will define the generator operating point during asynchronous operation.

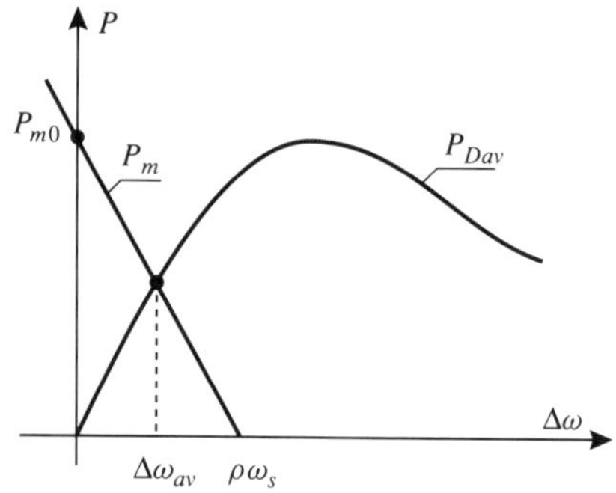

Figure 6.24 Static characteristics of the turbine power and the damping power

6.4.1 Transition to asynchronous operation

In order to explain what happens when a generator loses synchronism with the system consider the system shown in Figure 6.6. Now assume that a three-phase fault occurs on line $L2$ and is then cleared by tripping the faulted line (without auto-reclosing) after a time such that the acceleration area *1-2-3-4* shown on Figure 6.25a is larger than the available deceleration area *4-5-6*. The dashed $P_{E'}(\delta')$ curve is valid for the pre-fault state while the solid $P(\delta')$ curve is valid for the post-fault state. During the fault the rotor speed deviation increases but, when the fault is cleared, it starts to fall. Synchronism is lost when the rotor passes point *6* and follows the lower part of the power angle-characteristic. The speed deviation quickly increases because of the large difference between the turbine power and the synchronous power, shown by the shaded area in Figure 6.25a. As the speed deviation increases the turbine governing system starts to close the valves and the turbine mechanical power starts to decrease. Due to the long time constant of the turbine governing system the dynamic turbine power $P_m(t)$ does not follow the static characteristic in Figure 6.25b but moves above it. With increased speed deviation the average damping power P_{Dav} increases, according to the characteristic shown in Figure 6.24. As P_{Dav} increases and P_m decreases these two powers become equal at point B. Because of the delay introduced by the turbine time constants the mechanical power continues to decrease after passing point B and it is now less than P_{Dav}. Consequently the speed deviation starts to decrease and the turbine governor opens the valves to increase the turbine power until, at point C, the mechanical power and the asynchronous power are again equal. Again, because of the time delay in the turbine, the mechanical power continues to increase and is now greater than the asynchronous power. The speed deviation begins to increase, the governor valves close, and the cycle repeats until eventually the system operates at point D corresponding to the intersection of the static turbine characteristic and the asynchronous power characteristic. This point defines the average turbine speed deviation $\Delta\omega_{av}$ and power during asynchronous operation and is shown in Figure 6.24.

The effect of the synchronous power $P_{E'}$ is to produce speed deviation oscillations around the mean value as shown in the lower part of Figure 6.25b. The dashed lines in Figure 6.25b depict the envelopes of the power changes and the speed deviation changes due to the periodic acceleration and deceleration of the rotor caused by the synchronous power.

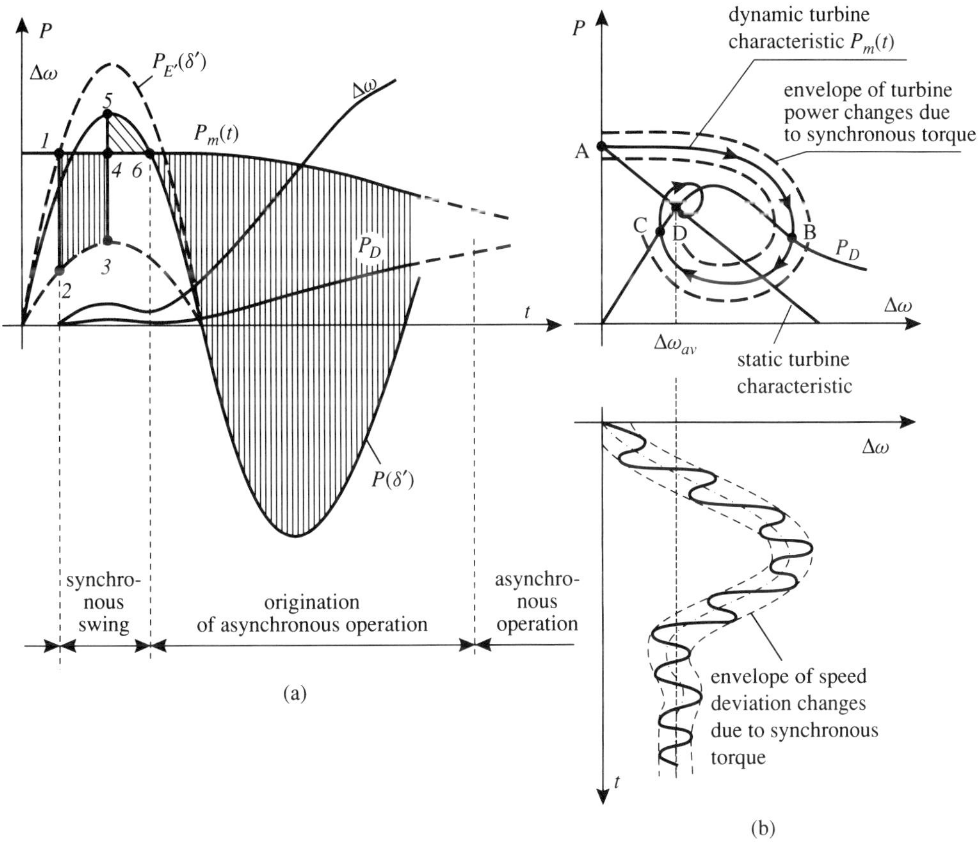

Figure 6.25 Transition to asynchronous operation: (a) development of high speed deviation and high asynchronous power; (b) settling down of the asynchronous operating point (Based on Venikov, 1978)

6.4.2 Asynchronous operation

Assuming that the generator is allowed to settle at the asynchronous operating point, the typical variation in some of the electric quantities are as shown in Figure 6.26. The voltage drop ΔV across the equivalent reactance x'_d can be calculated from the phasor diagram shown in Figure 5.8b. The cosine theorem gives

$$\Delta V^2 = (Ix'_d)^2 = (E')^2 + (V_s)^2 - 2E'V_s \cos\delta' \tag{6.40}$$

Accordingly, the stator current is seen to change as the rotor angle changes, with its maximum value being when $\delta' = \pi$ and its minimum when $\delta' = 0$ whilst the generator terminal voltage, equation (5.86), is obtained by adding the voltage drop across the system equivalent reactance to the infinite busbar voltage. Accounting for the variations in angle δ' shown in Figure 6.11 allows typical time variations of the current and voltage to be obtained as shown in Figure 6.26a.

As the armature current changes so does the armature reaction flux linking the closed rotor circuits and, according to the law of constant of flux linkage, currents must be

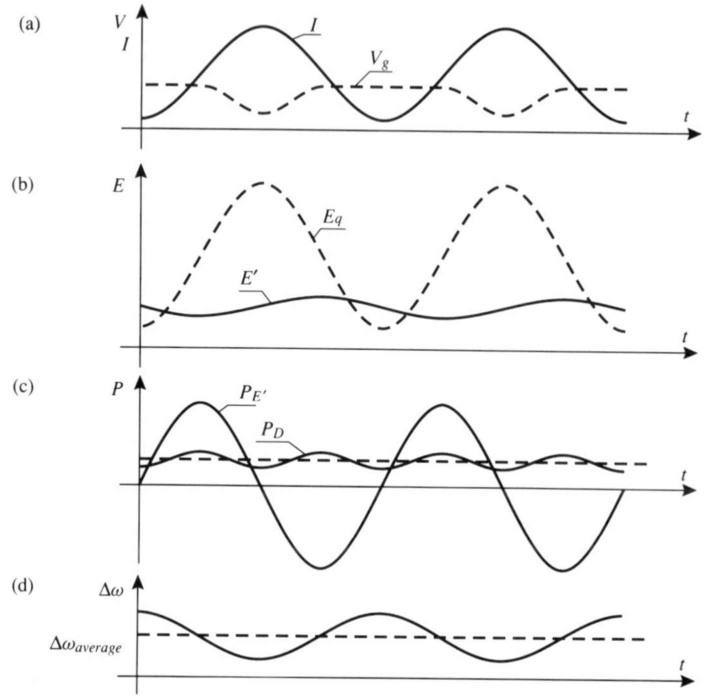

Figure 6.26 Changes in the electrical quantities of a generator operating at the asynchronous operating point: (a) current and terminal voltage; (b) internal emf and transient emf; (c) synchronous and asynchronous power; (d) speed deviation

induced in the rotor windings to maintain this flux linkage constant. Consequently the field current will follow the armature current fluctuations thereby inducing large fluctuations in the internal emf E_q, and smaller variations in E', as shown in Figure 6.26b. As the transient emf E' is almost constant the synchronous power $P_{E'}$ is given by equation 5.40 and varies as $\sin\delta'$, Figure 6.26c. The average value of the asynchronous damping power P_{Dav} is shown in Figure 6.26c and can be added to the synchronous power to obtain the resulting air-gap electrical power $P_e = P_{E'} + P_D$. This has the shape of a distorted sinusoid.

Figure 6.24 showed that the average speed deviation $\Delta\omega_{av}$ is determined by the intersection between the turbine characteristic $P_m(\Delta\omega)$ and the average asynchronous damping power characteristic $P_{Dav}(\Delta\omega)$. However due to the sinusoidal changes in the synchronous power $P_{E'}$ the actual speed deviation will oscillate around $\Delta\omega_{av}$ in the way shown in Figure 6.26d.

6.4.3 Possibility of resynchronisation

In order to resynchronise with the system the generator must fulfill the normal requirements for synchronisation, that is its speed deviation and power angle must be small, equation 6.39. This sub-section will investigate under what conditions these two requirements may be met.

Figure 6.25 was constructed assuming that the changes in the speed deviation produced by the synchronous power are small when compared to the average value $\Delta\omega_{av}$, whilst $\Delta\omega_{av}$ itself depends on the asynchronous power characteristic $P_{Dav}(\Delta\omega)$ shown in Figure 6.24. As the maximum value of $P_{Dav}(\Delta\omega)$ is inversely proportional to the square of the reactances $x'_d = X'_d + X$ and $x'_q = X'_q + X$, equation 5.24, then if the equivalent reactance X is small, the intersection between the $P_{Dav}(\Delta\omega)$ characteristic and the $P_m(\Delta\omega)$ characteristic may occur at a small value of speed deviation. Moreover a small value of X gives a large amplitude to the synchronous power $P_{E'}$ characteristic, equation 5.40, so that large variations of $P_{E'}$ during asynchronous operation produce large variations of speed deviation around the average value $\Delta\omega_{av}$. Large speed deviation variations, combined with a small average value, may at some point in time result in the speed deviation approaching zero as shown in Figure 6.27a. As the speed deviation approaches zero, the rotor loses its excess kinetic energy, the asynchronous power is zero, and the rotor behaviour is determined by the acceleration and deceleration work performed by the synchronous power $P_{E'}$. If the speed deviation reaches zero when $P_{E'}$ is sufficiently small then the generator may re-synchronise with the system.

Figure 6.27b shows two cases. In the first case the speed deviation becomes zero at point *1* when the synchronous power is large and negative. The synchronous power performs work proportional to area *1′-2-4* which is larger than the available deceleration area *4-5-6*. Re-synchronisation is not possible and the rotor makes another asynchronous

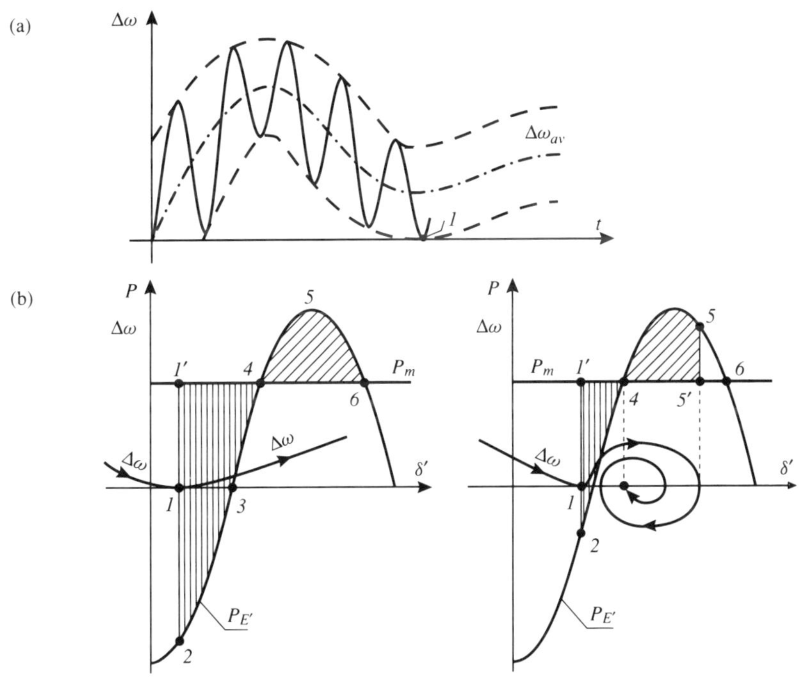

Figure 6.27 Re-synchronisation of a generator before the speed deviation reaches the average value $\Delta\omega_{av}$: (a) instantaneous value of the speed deviation approaches zero; (b) equal area criterion applied to check the condition for re-synchronisation. (Based on Venikov, 1978)

rotation. In the second case the speed deviation reaches zero when the synchronous power is much smaller. The acceleration area *1′-2-4* is now much smaller than the available deceleration area *4-5-6* and, after reaching point *5*, the decelerating power prevails and the rotor starts to swing back towards smaller values of the rotor angle. After a number of oscillations the rotor will settle at the synchronous equilibrium point *4*. Re-synchronisation has been achieved but whether or not the generator remains stable depends on the subsequent action of the turbine governing system and the voltage regulator.

Influence of the voltage regulator

Figure 6.26 showed that during each asynchronous rotation the voltage at the generator terminals drops quite significantly before recovering to a higher value. This means that the regulator error will oscillate at the frequency of the speed deviation oscillations. Normally this frequency is too high for even a fast AVR to follow. Consequently the AVR reacts, more or less, to the average value of the error signal so maintaining a high value of excitation voltage as shown in Figure 6.28a. As the average voltage value is lower than the pre-fault value, the AVR increases the excitation voltage V_f almost to its ceiling value so that the amplitude of the $P_{E'}(\delta')$ characteristic increases.

It has already been explained that the larger the magnitude of the speed deviation oscillations around the average value then the greater is the possibility that the speed deviation will reach zero value. By increasing the excitation voltage, the AVR increases the amplitude of the $P_{E'}(\delta')$ characteristic and so produces larger changes in the speed deviation. This increases the chance of the speed deviation reaching a zero value as shown in Figure 6.28b. Moreover, as the amplitude of the synchronous power increases,

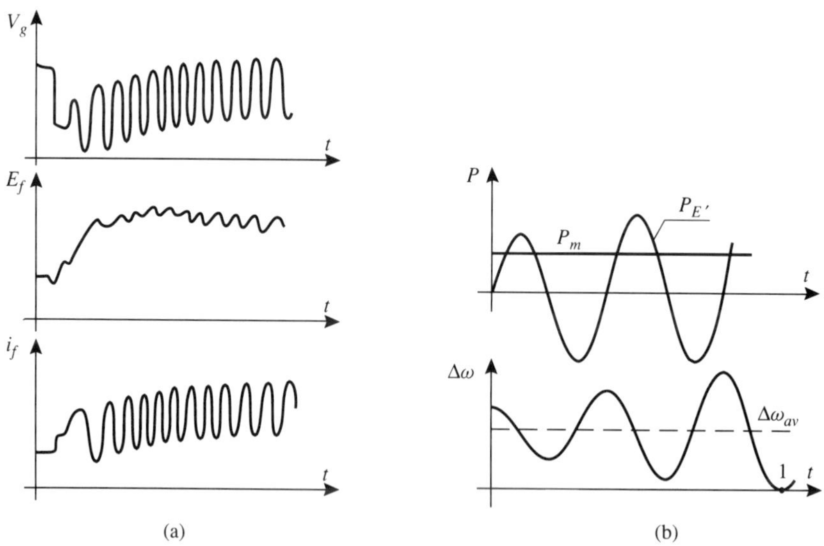

Figure 6.28 Effect of the AVR on re-synchronisation: (a) examples of the changes in the generator terminal voltage, excitation voltage and field current during asynchronous operation following a fault; (b) increase in the amplitude of the synchronous power oscillations and the speed deviation oscillations as a result of voltage regulation

the available deceleration area in Figure 6.27b increases and the range of rotor angles which are acceptable for re-synchronisation also increases.

Obviously re-synchronisation will be followed by a period of large rotor swings which may be damped or aggravated by the AVR action for the reasons described in Section 6.1.5.

Other possibilities of re-synchronisation

Successful re-synchronisation depends on the action of the AVR and the turbine governing system. Turbine governors can be equipped with supplementary control loops which allow the turbine power to be quickly reduced after a fault in order to protect the generator from loss of synchronism and/or assist in re-synchronisation. This is referred to as *fast valving* and is discussed in detail in Section 9.2. If a generator is not equipped with such a governor the operating staff may try to re-synchronise the asynchronously rotating generator by manually reducing the turbine power to reduce the average speed deviation value. However, in order to avoid large power swings in the system, and possible damage to the generator, generators are usually tripped after a few asynchronous rotations.

6.4.4 Out-of-step relaying

Usually the basic protection systems provide insufficient protection during power swings and must be augmented by additional *out-of-step relays*. According to their function such relays can be classified as either *out-of-step blocking relays or out-of-step tripping relays*.

Out-of-step blocking relays are installed in the transmission network with the aim of blocking the distance protection during power swings. For asynchronous, or deep synchronous, power swings the apparent impedance measured by the distance relay can move away from the normal load area into the distance protection zone causing unnecessary tripping.

Out-of-step tripping relays may be installed either as part of the generator protection system or as part of the transmission network protection. When installed in the generator protection system they isolate the generator in the event of asynchronous operation. The relay operates after a set number of asynchronous cycles and is also called *pole-slipping protection*. Out-of-step tripping relays can also be used inside the transmission network in order to split the system at predetermined points should asynchronous power swings occur.

Loci of relay impedance during power swings

The effect of power swings on the impedance loci can be studied using the simple equivalent circuit shown in Figure 6.29 in which two equivalent synchronous generators are linked via impedances $\overline{Z'}_a$ and $\overline{Z'}_b$. The voltage and current is measured at the relay point between these impedances to obtain the apparent impedance $\overline{Z}(t)$. The current and voltage at the relay point are given by

$$\overline{I} = \frac{\overline{E'}_a - \overline{E'}_b}{\overline{Z'}_a + \overline{Z'}_b} \qquad \overline{V} = \overline{E'}_a - \overline{I}\overline{Z'}_a \tag{6.41}$$

where $\overline{E'}_a$, $\overline{E'}_b$ are the equivalent generator transient emfs and $\overline{Z'}_a$, $\overline{Z'}_b$ are the equivalent impedances that include the impedance of the transmission network and the generator

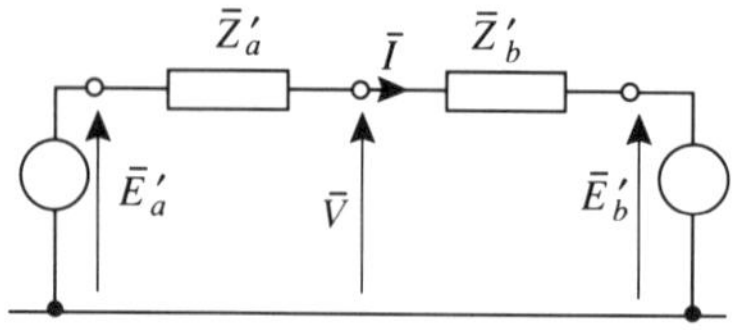

Figure 6.29 Equivalent circuit of the system with a relay point

transient reactances. If the magnitudes of the equivalent transient emfs are assumed to be constant then

$$\frac{\overline{E'}_a}{\overline{E'}_b} = \frac{|E'_a|}{|E'_b|}e^{j\delta'} = ke^{j\delta'} \tag{6.42}$$

where $k = |E'_a|/|E'_b| =$ constant and the transient power angle δ' is the difference between the arguments of the emfs.

The apparent impedance measured by the relay is $\overline{Z}(t) = \overline{V}/\overline{I}$ which, when substituting for $\overline{V}$ and $\overline{I}$ from the equations 6.41 and including equation 6.42, gives

$$\overline{Z}(t) = \frac{\overline{V}}{\overline{I}} = \frac{\overline{Z'}_a + \overline{Z'}_b ke^{j\delta'(t)}}{ke^{j\delta'(t)} - 1} \tag{6.43}$$

Note that during power swings the power angle and the apparent impedance are both time dependent. Solving equation 6.43 with respect to the power angle gives

$$ke^{j\delta'(t)} = \frac{\overline{Z}(t) - (-\overline{Z'}_a)}{\overline{Z}(t) - \overline{Z}'_b} \tag{6.44}$$

With the exception of the coefficient k, all the variables in this equation are complex. Taking the absolute value of both sides of equation 6.44 and recognising that $\left|e^{j\delta'}\right| = 1$ gives

$$\left|\frac{\overline{Z}(t) - (-\overline{Z'}_a)}{\overline{Z}(t) - \overline{Z'}_b}\right| = k = \text{ constant } = \frac{|\overline{E'}_a|}{|\overline{E'}_b|} \tag{6.45}$$

Equation 6.45 determines the locus on the complex plane of all points with the same value of k. This is illustrated in Figure 6.30. The impedances $(-\overline{Z'}_a)$ and $\overline{Z'}_b$ determine the points A and B respectively whilst the apparent impedance $\overline{Z}(t)$ determines point C, Figure 6.30a. As $|\overline{Z}(t) - (-\overline{Z'}_a)| = AC$ and $|\overline{Z}(t) - \overline{Z'}_b| = BC$, the ratio $k = AC/BC =$ constant. As $\overline{Z}(t)$ varies, a locus of constant value k will be traced out by point C. The locus for a given value of k is a circle that surrounds either point A or point B. The diameter of this circle, and the location of its centre, depends on the value of k and the impedances $\overline{Z'}_a$, $\overline{Z'}_b$. The centre of each circle lies on the straight line that passes through points A, B and is shown dotted in Figure 6.30b. For $k = 1$ the diameter of the circle tends to infinity and the circle becomes a straight line that bisects AB. For $k < 1$, $|\overline{E'}_a| < |\overline{E'}_b|$, the circle lies in the lower part of the complex plane and encircles point A. For $k > 1$, $|\overline{E'}_a| > |\overline{E'}_b|$, the circle lies in the upper part of the complex plane and encircles point B.

Now consider the effect of a power swing on the trajectory of $\overline{Z}(t)$. For the pre-fault operating conditions the voltage V is close to the rated voltage and the current I is small when compared to the fault current. Consequently Z is high and predominantly

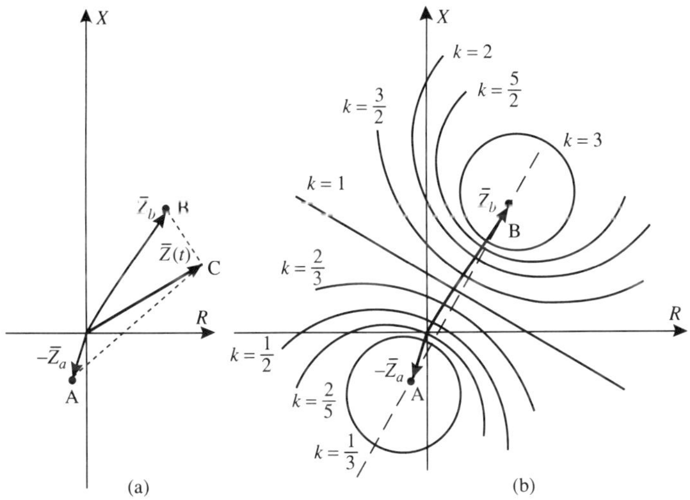

Figure 6.30 Graphical interpretation of equation 6.45: (a) impedances on the complex plane; (b) circles determining the impedance loci

resistive (as normally the power factor is close to unity). A fault results in a significant drop in voltage and an increase in current so that Z drops in value. Assuming that $k = |E'_a|/|E'_b| =$ constant then, when the fault is cleared, the value of Z rises again. Further changes in Z are solely due to the variation of the power angle δ' and must follow one of the circles shown in Figure 6.30.

Figure 6.31a illustrates the trajectory of $\overline{Z}(t)$ in the case of an asynchronous power swing, and Figure 6.31b in the case of a synchronous power swing, following a fault outside the relay protection zone. In both cases the trajectory starts from point O in the normal load area and, when the fault occurs, it jumps to point F lying outside protection zones *1* and *2*. During the short-circuit the impedance trajectory moves to point C where the fault is cleared. On fault clearance the trajectory jumps to point P which does not quite lie on the same circle as point O as the emfs change their values slightly during the fault. Further changes in δ' will cause $\overline{Z}(t)$ to follow a circle of a constant ratio k or, more precisely, a family of closed circles corresponding to slightly varying emfs and hence ratios k.

When $\delta'(t)$ increases, $Z(t)$ decreases and may encroach into the distance relay tripping zone. For the asynchronous power swing, shown in Figure 6.31a, the trajectory will trace a full circle as $\delta'(t)$ completes a full 360° cycle. For a synchronous power swing, shown in Figure 6.31b, the trajectory will reach point $B1$ on an impedance circle and then move back towards point $B2$ as $\delta'(t)$ decreases on the back swing. For constant $|\overline{E'}_a|/|\overline{E'}_b|$ the trajectory always lies on one circle. A change in the transient emfs due to AVR control will move the trajectory from circle to circle so that when power swings are damped the trajectory tends towards a post-fault equilibrium point in the load area.

Power swing blocking relays

A power swing can be detected by monitoring the speed at which the impedance locus approaches the outermost characteristic of the distance relay, usually the fault detection

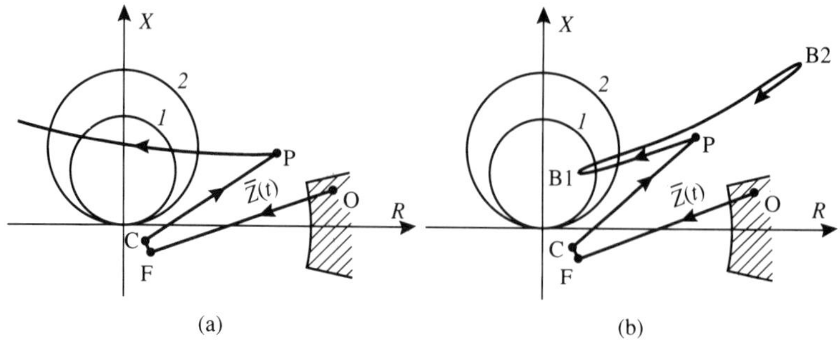

Figure 6.31 Example of the impedance trajectory for: (a) asynchronous power swing; (b) synchronous power swing (1, 2 — tripping zones of the distance relay, load area is shaded)

zone. If the fault is applied so that point F lies outside the relay protection zone (as in Figure 6.31) then the protection zones are approached by $\overline{Z}(t)$ with a finite speed. On the other hand if the fault is applied so that point F lies inside one of the protection circles shown in Figure 6.31 then the impedance locus jumps from point O to a point inside the protection zone almost instantaneously.

Figure 6.32 illustrates the operation of a distance relay with an offset Mho type characteristic. The power swing blocking relay has the same type of characteristic as the distance relay and surrounds the distance relay characteristic. A timer measures the time Δt taken for the trajectory to pass between the two characteristics. If the fault is inside the protection zone then the power swing relay and the fault detection relay will operate practically simultaneously and a blocking command is not generated. If the fault is outside the protection zone then, during a resulting power swing, the trajectory will take a finite time to move between the two relay characteristics and a blocking command is generated to hold up operation of the circuit breaker over a time when the impedance locus stays inside the fault detection zone.

The above blocking principle is also used in distance relays with other types of characteristic such as quadrilateral, rhombus and oval. Obviously the power swing relay must have the same type of characteristic as the distance relay.

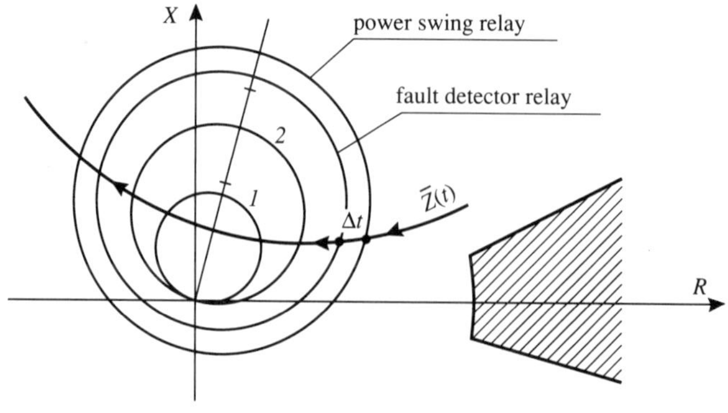

Figure 6.32 Power swing blocking and distance relay characteristics of the offset Mho type

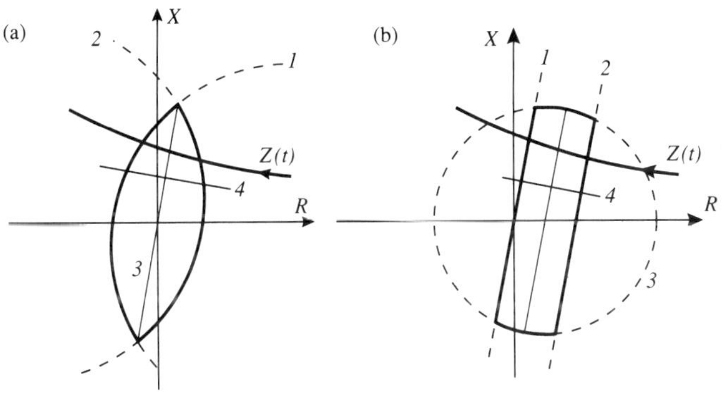

Figure 6.33 Two types of pole-slip protection characteristics: (a) lenticular type; (b) offset Mho type

Pole-slip protection

The purpose of this protection is to detect a loss of synchronism. Equation 6.43 shows that for $\delta' = 180°$ the apparent impedance seen by the relay is proportional to the difference of the equivalent impedances, that is

$$\underline{Z}(\delta' = 180°) = \frac{\underline{Z}'_a - k\underline{Z}'_b}{-k-1} = \frac{k\underline{Z}'_b - \underline{Z}'_a}{k+1} \tag{6.46}$$

Because the resistance of the equivalent impedances $\underline{Z}'_a$ and $\underline{Z}'_b$ is small, the locus of the impedance seen by the relay for $\delta' = 180°$ lies near the imaginary axis. For example if $\underline{Z}'_a = jX_a$ and $\underline{Z}'_b = jX_b$ then the impedances are purely reactive and the impedance trajectory $Z(t)$ crosses the imaginary axis when δ' reaches 180°. This means that an asynchronous cycle can be identified by a relay with an impedance characteristic that surrounds the imaginary axis. Two such characteristics are shown in Figure 6.33.

The lenticular characteristic shown in Figure 6.33a is obtained by using two impedance relays with offset Mho characteristics shown by the dashed circles *1* and *2*. The second type, Figure 6.33b, has an offset Mho characteristic, *3*, with left and right parts cut off by two directional relays *1* and *2*.

For both types of pole-slip protection an asynchronous cycle is identified if the impedance locus passes completely through the relay characteristic in either direction. Pole-slip protection usually has an additional impedance relay that divides $Z(t)$ into two zones as shown by line *4* in Figure 6.33. In this case the number of asynchronous cycles identified by the relay are counted with the output tripping signal depending on the number of cycles and the zone. Details can be found in Imhof (1977) and Stalewski, Goody & Downes (1980).

6.5 Swings in Multi-machine Systems

Although the simplification of a multimachine power system to a single generator-infinite busbar model enables many important conclusions to be made concerning the electromechanical dynamics and stability of the system, this simplification is only possible when the fault affects one generator and has little effect on other generators in the system.

Contemporary power systems have a well developed transmission network with power stations located relatively close to each other so that these conditions will not always be met. In these circumstances a fault near one of the power stations will also distort the power balance at neighbouring stations. The resulting electromechanical swings in the power system can be compared to the way the masses swing in the mechanical system shown in Figure 6.34.

Sub-section 5.4.7 showed that a single swinging rotor can be compared to a mass-spring-damper system. Hence a multimachine system can be compared to a number of masses (representing the generators) suspended from a "network" consisting of elastic strings (representing the transmission lines). In the steady-state each of the strings is loaded below its breaking point (steady-state stability limit). If one of the string is suddenly cut (representing a line tripping) the masses will experience coupled transient motion (swinging of the rotors) with fluctuation in the forces in the strings (line powers).

Such a sudden disturbance may result in the system reaching a new equilibrium state characterised by a new set of string forces (line powers) and string extensions (rotor angles) or, due to the transient forces involved, one string may break so weakening the network and producing a chain reaction of broken strings and eventual total system collapse.

Obviously this mechanical analogue of the power system has a number of limitations. Firstly, the string stiffnesses should be non-linear so as to model correctly the non-linear synchronising power coefficients. Secondly the string stiffness should be different in the steady-state from the transient state to model correctly the different steady-state and transient state models of the generators.

In practice a disturbance may affect the stability of a power system in one of four ways:

(i) The generator (or generators) nearest to the fault may lose synchronism without exhibiting any synchronous swings; other generators affected by the fault undergo a period of synchronous oscillations until they eventually return to synchronous operation

(ii) The generator (or generators) nearest to the fault lose synchronism after exhibiting synchronous oscillations

(iii) The generator (or generators) nearest to the fault is the first to lose synchronism and is then followed by other generators in the system

(iv) The generator (or generators) nearest to the fault exhibit synchronous swings without losing stability but one, or more, of the other generators remote from the fault loses synchronism with the system

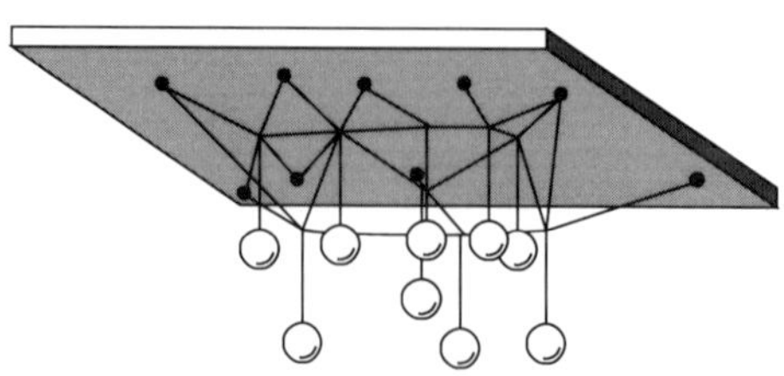

Figure 6.34 Mechanical analogue of swings in a multi-machine system (Based on Elgerd, 1982)

So far only the first case has been described as it may be represented by the generator-infinite busbar system. In the other three cases instability arises due to interaction with other generators located further from the fault. In the second case the generator initially has a chance to retain stability but, as the rotors of the other generators start to oscillate, conditions deteriorate and the generator loses synchronism. In the third case the generator nearest the fault is the first to lose synchronism; this has a major affect on the other generators in the system so that they may also lose synchronism. The fourth case is typical of the situation where some generators, located far from the point of fault, are weakly connected to the system. As the oscillations spread, the operating conditions of the weakly-connected generators deteriorate so that they may lose stability. Another example of this fourth instability case is when the network configuration is changed following the fault clearance. Although the tripped line may be just one of the lines that connects the nearest generator to the system it could also be the main connection for a neighbouring power station. The following example illustrates all the above cases (Omahen, 1994):

Example 6.1

Figure 6.35 shows a schematic diagram of a power system that comprises three subsystems connected together at node *2*. The capacity of generator *5* is large and can be considered as an infinite busbar. Figure 6.36a,b,c,d shows the power angles of all the generators, measured with respect to generator *5*, for faults located at points *a*, *b*, *c*, and *d* respectively. Each of these points corresponds to the four cases of instability. The faults are assumed to be cleared without auto-reclosing.

When line *1–2* is faulted, case *a*, generator *1* quickly loses synchronism and the rotors of the other generators undergo a period of synchronous oscillation. When the fault is on line *3–13*, case *b*, then, for a given clearing time, generator *2* could remain in synchronism just as if it were operating on the infinite busbars. However in this particular case the oscillations of the other generator rotors increase the relative difference in the power angles, and the situation with regards generator *2* deteriorates and it loses

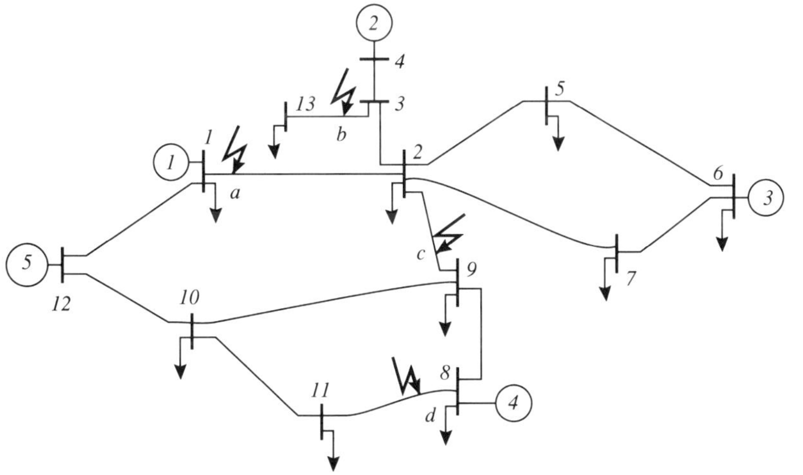

Figure 6.35 Schematic diagram of the test system

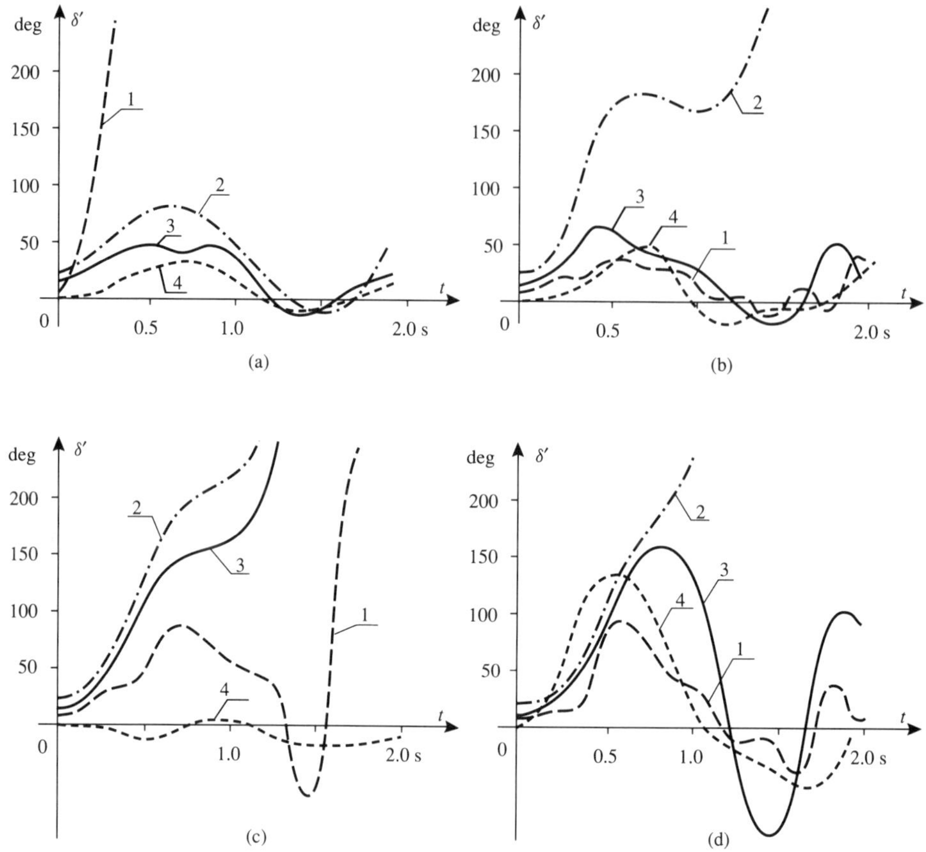

Figure 6.36 Rotor angle variations, measured with respect to generator rotor *5*, for different fault locations (Based on Omahen, 1994)

stability on its second swing. When line *9–2* is faulted, case c, the weakly-connected generators, *2* and *3*, immediately lose stability while generator *1*, which initially remains in synchronism with generator *5*, loses synchronism some time later. When line *8–11* is faulted, case *d*, the power angle of all the generators considerably increase. After the fault is cleared all the system loads take power from their neighbouring generators *1*, *3*, *4* and *5* but as generator *2* is far from these loads it loses stability.

6.6 Torsional Oscillations in the Drive Shaft

In this section the effect system faults can have on the torques in the shafts and couplings connecting each turbine stage, the turbine to the generator and the generator to the exciter are discussed. In the previous sections of this chapter these shafts and couplings, shown in Figure 5.1, were assumed to be very stiff when the total drive system moves as a rigid body and can be represented by a single inertia. In practice the drive shafts and couplings have a finite stiffness so that each turbine mass will be slightly displaced relative to each other. Consequently any change in the external torque on any of the rotor components initiates a movement of the equivalent mass. The shaft connecting this element to its neighbours twists slightly and the resulting torsion transmits torque to the neighbouring element(s). This process is repeated down the shaft creating torsional oscillations in the shaft. These torsional oscillations are superimposed on the external torque which may either increase or reduce the torque in the shaft sections, depending on the direction of twist and the phase and frequency of the changes in the external torque, so that under certain conditions these shaft torques can become very large. Such large shaft torques can result in a reduction in the shaft fatigue life and possibly shaft failure. It is therefore important to understand the mechanisms leading to these high shaft torques and in what situations they might arise.

6.6.1 The torsional natural frequencies of the turbine-generator rotor

As the turbine-generator rotor can be considered as a number of discrete masses connected together by springs, a *lumped-mass* model of the rotor can be developed in order to calculate the *torsional natural frequencies* of the rotor system. At each of these natural frequencies the rotor system will vibrate in a particular way defined by the *mode shape.*

Figure 6.37 shows the torques acting on each rotor mass, when applying Newton's Second Law gives the equation of motion for the general mass l as

$$J_l \frac{\mathrm{d}^2\delta_{ml}}{\mathrm{d}t^2} = \tau_l + \tau_{l,l+1} - \tau_{l,l-1} \tag{6.47}$$

where J_l is its moment of inertia, τ_l is the external applied torque to mass l and $\tau_{l,l+1}$ and $\tau_{l,l-1}$ are the torques in the two shafts $(l, l+1)$ and $(l, l-1)$. These shaft torques are respectively

$$\tau_{l,l+1}(t) = k_{l,l+1}(\delta_{ml+1} - \delta_{ml}) + D_{l,l+1}\left(\frac{\mathrm{d}\delta_{ml+1}}{\mathrm{d}t} - \frac{\mathrm{d}\delta_{ml}}{\mathrm{d}t}\right) \tag{6.48}$$

and

$$\tau_{l,l-1}(t) = k_{l,l-1}(\delta_{ml} - \delta_{ml-1}) + D_{l,l-1}\left(\frac{\mathrm{d}\delta_{ml}}{\mathrm{d}t} - \frac{\mathrm{d}\delta_{ml-1}}{\mathrm{d}t}\right) \tag{6.49}$$

where $k_{l,l-1}$ is the stiffness of the shaft section between masses l and $l-1$, δ_{ml} is the angle of the l-th mass in mechanical radians and $D_{l,l-1}$ is the shaft damping coefficient. The shaft damping is due to the energy lost in the shaft as it oscillates and goes through

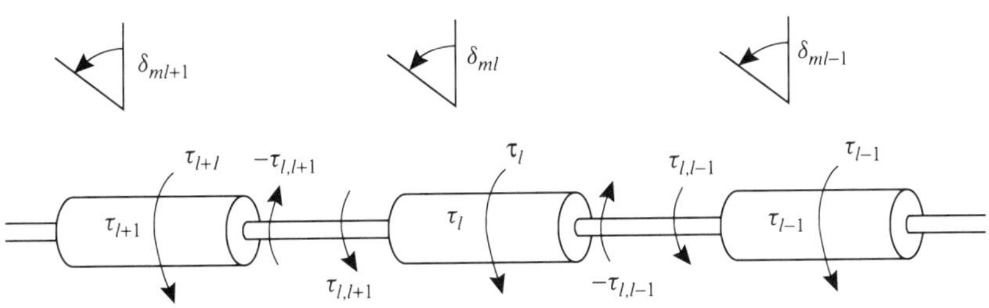

Figure 6.37 Torques acting on mass l: τ_l is the applied torque; $\tau_{l,l-1}$ and $\tau_{l,l+1}$ are the shaft torques; δ_{ml} the displacement of the mass; and J_l the inertia

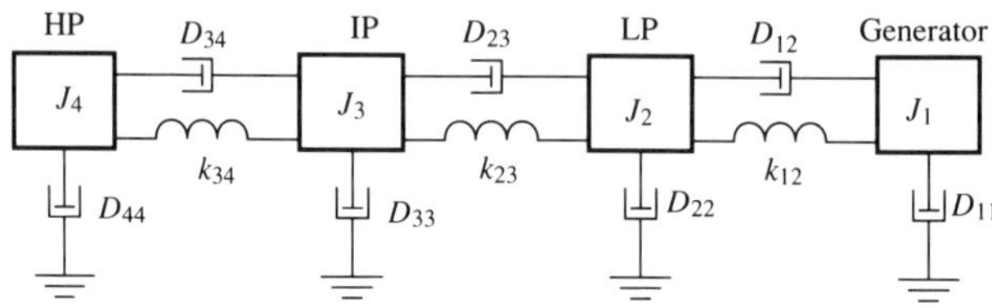

Figure 6.38 Four mass turbine-generator rotor model

a stress/strain hysteresis cycle. The above two equations can now be substituted into equation 6.47 to give the equation of motion

$$J_l \frac{d^2\delta_{ml}}{dt^2} = \tau_1(t) - k_{l,l-1}(\delta_{ml} - \delta_{ml-1}) - k_{l,l+1}(\delta_{ml} - \delta_{ml+1})$$
$$- D_{l,l+1}\left(\frac{d\delta_{ml}}{dt} - \frac{d\delta_{ml+1}}{dt}\right) - D_{l,l-1}\left(\frac{d\delta_{ml}}{dt} - \frac{d\delta_{ml-1}}{dt}\right) - D_{l,l}\frac{d\delta_{ml}}{dt} \tag{6.50}$$

In this equation $D_{l,l}$ is an additional damping term used to represent the damping effect arising at each turbine stage due to the flow of steam through the turbine. Generally both the shaft damping and the steam damping mechanisms are small. Equation 6.50 is in MKSA units but a per unit version can be obtained by following a similar procedure to that in Section 5.1[†]

Now consider a simple turbine-generator rotor consisting of three turbine stages and a generator as shown in Figure 6.38. A static exciter is assumed to act with this system so that no rotational mass is required to represent this component. Applying equation 6.50 to each rotor mass in turn gives a set of four second order differential equations

$$J_1 \frac{d\delta_{m1}^2}{dt^2} = \tau_1 - k_{12}(\delta_{m1} - \delta_{m2}) - D_{12}\left(\frac{d\delta_{m1}}{dt} - \frac{d\delta_{m2}}{dt}\right) - D_{11}\frac{d\delta_{m1}}{dt}$$

$$J_2 \frac{d\delta_{m2}^2}{dt^2} = \tau_2 - k_{12}(\delta_{m2} - \delta_{m1}) - k_{23}(\delta_{m2} - \delta_{m3}) - D_{12}\left(\frac{d\delta_{m2}}{dt} - \frac{d\delta_{m1}}{dt}\right)$$
$$- D_{23}\left(\frac{d\delta_{m2}}{dt} - \frac{d\delta_{m3}}{dt}\right) - D_{22}\frac{d\delta_{m2}}{dt}$$

$$J_3 \frac{d\delta_{m3}^2}{dt^2} = \tau_3 - k_{23}(\delta_{m3} - \delta_{m2}) - k_{34}(\delta_{m3} - \delta_{m4}) - D_{23}\left(\frac{d\delta_{m3}}{dt} - \frac{d\delta_{m2}}{dt}\right)$$
$$- D_{34}\left(\frac{d\delta_{m3}}{dt} - \frac{d\delta_{m4}}{dt}\right) - D_{33}\frac{d\delta_{m3}}{dt}$$

$$J_4 \frac{d\delta_{m4}^2}{dt^2} = \tau_4 - k_{34}(\delta_{m4} - \delta_{m3}) - D_{34}\left(\frac{d\delta_{m4}}{dt} - \frac{d\delta_{m3}}{dt}\right) - D_{44}\frac{d\delta_{m4}}{dt} \tag{6.51}$$

[†] In per unit form J becomes $2H/\omega_s$ whilst τ and k are expressed in per unit obtained by dividing the MKSA values by $T_{base} = \frac{S_{base}}{\omega_{sm}} = \frac{S_{base}}{\omega_s}\frac{p}{2}$ and $k_{base} = T_{base}\frac{p}{2} = \frac{S_{base}}{\omega_s}\frac{p^2}{4}$ respectively. δ_l is now expressed in electrical radians.

This set of second order equations is then written as a series of eight first order equations by noting that for mass i

$$\frac{d\delta_{mi}}{dt} = \omega_{mi} - \omega_{sm} = \Delta\omega_{mi}, \qquad \frac{d^2\delta_{mi}}{dt^2} = \frac{d\Delta\omega_{mi}}{dt} \tag{6.52}$$

Using this substitution, and taking mass 1 as an example, gives

$$\begin{aligned} J_1\frac{d\Delta\omega_{m1}}{dt} &= \tau_1 - k_{12}(\delta_{m1} - \delta_{m2}) - D_{11}\Delta\omega_{m1} - D_{12}(\Delta\omega_{m1} - \Delta\omega_{m2}) \\ \frac{d\delta_{m1}}{dt} &= \Delta\omega_{m1} \end{aligned} \tag{6.53}$$

By considering small perturbations and defining a set of state variables $\boldsymbol{x}$ so that

$$x_1 = \Delta\delta_{m1},\ x_2 = \Delta\delta_{m2},\ x_3 = \Delta\delta_{m3},\ x_4 = \Delta\delta_{m4}, \text{ and}$$

$$x_5 = \Delta\omega_{m1},\ x_6 = \Delta\omega_{m2},\ x_7 = \Delta\omega_{m3},\ x_8 = \Delta\omega_{m4}$$

and substituting into the equations 6.53 gives a linearised matrix equation of the form

$$\dot{\mathbf{x}} = \mathbf{A}\mathbf{x} + \mathbf{B}\mathbf{u} \tag{6.54}$$

where $\boldsymbol{A}$ is the *plant* or *state matrix*, $\boldsymbol{B}$ the *driving matrix*, $\boldsymbol{x}$ a *vector of state variables* and $\boldsymbol{u}$ a *vector of inputs*, in this case the change in the external torque $\Delta\tau$ applied to each mass. The plant matrix is given by

$$\mathbf{A} = \begin{bmatrix} \mathbf{0} & \mathbf{I} \\ \mathbf{K} & \mathbf{D} \end{bmatrix} \tag{6.55}$$

where $\mathbf{0}$ is a null matrix, $\boldsymbol{I}$ the identity matrix, $\boldsymbol{K}$ the stiffness matrix and $\boldsymbol{D}$ the damping matrix. For the four mass problem being considered these matrices are

$$\mathbf{K} = \begin{bmatrix} \frac{-k_{12}}{J_1} & \frac{k_{12}}{J_1} & & \\ \frac{k_{12}}{J_2} & \frac{-k_{12}-k_{23}}{J_2} & \frac{k_{23}}{J_2} & \\ & \frac{k_{23}}{J_3} & \frac{-k_{23}-k_{34}}{J_3} & \frac{k_{34}}{J_3} \\ & & \frac{k_{34}}{J_4} & \frac{-k_{34}}{J_4} \end{bmatrix} \tag{6.56}$$

$$\mathbf{D} = \begin{bmatrix} \frac{-D_{12}-D_{11}}{J_1} & \frac{D_{12}}{J_1} & & \\ \frac{D_{12}}{J_2} & \frac{-D_{12}-D_{23}-D_{22}}{J_2} & \frac{D_{23}}{J_2} & \\ & \frac{D_{23}}{J_3} & \frac{-D_{23}-D_{34}-D_{33}}{J_3} & \frac{D_{34}}{J_3} \\ & & \frac{D_{34}}{J_4} & \frac{-D_{34}-D_{44}}{J_4} \end{bmatrix} \tag{6.57}$$

whilst the driving matrix $\mathbf{B} = [\mathbf{0}\ \mathbf{J}^{-1}]^{T}$, where $\mathbf{J}^{-1}$ is simply a (4×4) diagonal matrix with entries $1/J_1$, $1/J_2$, $1/J_3$ and $1/J_4$. The structure of all these matrices is very clear allowing more inertia elements to be easily added if required.

The rotor torsional natural frequencies are found from equation 6.54 by assuming that there is no change in the applied torque when the input vector $\mathbf{u} = \mathbf{0}$ to give

$$\dot{\mathbf{x}} = \mathbf{A}\mathbf{x} \tag{6.58}$$

Assuming a solution of the form $\mathbf{x}(t) = \mathbf{x}_0 e^{\lambda_i t}$ gives

$$(\mathbf{A} - \lambda_i \mathbf{I}) = 0 \tag{6.59}$$

This is an eigenvalue problem with the eigenvalues λ_i giving the rotor natural frequencies. As the damping is small the system will be underdamped and complex conjugate pairs of eigenvalues will be returned of the form

$$\lambda_{i,i+1} = -\zeta_i \omega_{ni} \pm j\omega_{di} \tag{6.60}$$

where ζ_i is the damping ratio associated with the particular oscillation mode, ω_{di} is its damped natural frequency and ω_{ni} its undamped natural frequency. The damped, and the undamped, natural frequencies are related by the standard expression $\omega_d = \omega_n \sqrt{1 - \zeta^2}$.

To obtain the mode shapes all the damping coefficients are set to zero when the eigenvalues are purely imaginary and give the undamped natural frequencies. The eigenvector corresponding to each of these undamped natural frequencies is now real and defines how the different rotor masses will be displaced relative to each other if excited at this particular frequency. This is the mode shape and either the elements of the eigenvector associated with the speed deviation or the angle deviation can be used. It is also normal practice to normalise the eigenvectors so that the maximum displacement is set to unity.

The sensitivity of any particular shaft to a harmonic forcing torque on the generator can also be found via equation 6.54 by defining an output equation

$$\mathbf{y} = \mathbf{C}\mathbf{x} \tag{6.61}$$

where $\boldsymbol{y}$ is the output required, in this case the torque in a particular shaft or shafts, with the elements of the output matrix $\boldsymbol{C}$ following directly from equations 6.48 or 6.49. If all the forcing torques are set to zero, except the generator torque which is assumed to be sinusoidal, the transfer function relating the shaft torque, $\boldsymbol{y}$, to the input $\boldsymbol{u}$ can be found by substituting equation 6.61 into equation 6.54 to give the standard transfer function matrix equation

$$\mathbf{G}(s) = \mathbf{C}(s\mathbf{I} - \mathbf{A})^{-1}\mathbf{B} \tag{6.62}$$

where $\boldsymbol{B}$ now simply becomes $[0\ 0\ 0\ 0\ 0\ 1/J_1\ 0\ 0\ 0]$ and $G_{11}(s) = y_1/u$, $G_{12}(s) = y_2/u$ etc. The inertia entry in $\boldsymbol{B}$ is that of the generator so that its position in the matrix will depend on which mass represents the generator.

The evaluation of the frequency response associated with equation 6.62 is best achieved using a standard software package such as MATLAB (Hicklin & Grace, 1992) when the matrices $\boldsymbol{A}$, $\boldsymbol{B}$, $\boldsymbol{C}$ and $\boldsymbol{D}$ need only be defined and standard subroutines used to do the rest of the work.

Example 6.2

The shaft configuration for a 577 MVA, 3000 rpm thermal power plant is shown in Figure 6.39 with the parameters detailed in Table 6.2.

The plant matrix, equation 6.55, is formed using the sub-matrices defined in equations 6.56 and 6.57 and the eigenvectors and eigenvalues found using standard MATLAB routines. Figure 6.39 shows the undamped natural frequencies and mode shapes. As is standard practice, all the mode shapes have been normalised with the maximum deflection set to 1.0. Mode *0* represents free body rotation and, when the generator is connected to the system, the frequency of this mode would increase and become the rotor frequency of oscillation. This can easily be shown by connecting the generator to the infinite system through an "electromagnetic spring" of stiffness $K_{E'}$ where $K_{E'}$ is the transient synchronising power coefficient. This concept was introduced in section 5.4. As the value of $K_{E'}$ depends on the system loading condition, a figure of $K_{E'} = 2.0$ pu has been assumed. This "electromagnetic" spring will replace the change in generator torque. As $\Delta\tau_2 = -K_{E'}\Delta\delta_2$, $K_{E'}$ must now be included in the stiffness matrix so that element $\mathbf{K}[2,2]$ becomes $\mathbf{K}[2,2] = -(k_{12} + k_{23} + K_{E'})/J_2$. The eigenvalues now give the rotor frequency of oscillations as 1.14 Hz but the rotor system continues to rotate as a rigid body as shown by the mode shape at the bottom of Figure 6.39.

Following the mode shapes through for each individual shaft indicates some degree of movement in the Generator-Exciter shaft at all the natural frequencies. In Mode *1*,

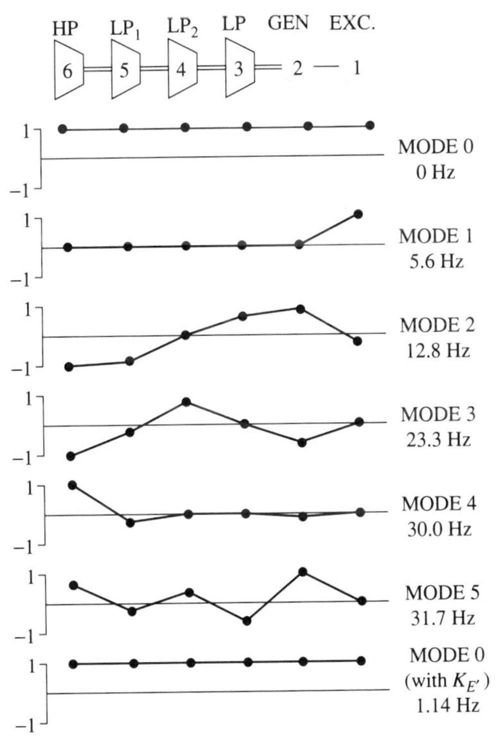

Figure 6.39 577 MVA, 3000 rpm, turbine-generator torsional natural frequencies and associated mode shapes

Table 6.2 577 MVA, 3000 rpm, turbine generator parameters (Ahlgren, Johnson & Gadhammar, 1978)

node	H_i	k_{ij}	D_{ii}	D_{ij}
	s	pu torque/rad	pu torque/rad/s	pu torque/rad/s
1	0.09	—	0.0	—
1-2	—	0.7	—	0.00007
2	0.74	—	0.0	—
2-3	—	110	—	0.001
3	1.63	—	0.001	—
3-4	—	95	—	0.001
4	1.63	—	0.001	—
4-5	—	87	—	0.001
5	1.63	—	0.001	—
5-6	—	40	—	0.001
6	0.208	—	0.001	—

at 5.6 Hz, the turbine and the generator components remain stationary and the exciter moves relative to them. Mode 4 shows a large amount of movement in the HP-LP_1 shaft while the LP_3-Generator shaft displays movement in all Modes except Mode 1.

To further study the frequency sensitivity of the Generator-Exciter shaft and the LP_3-Generator shaft the frequency response of these two shafts to a sinusoidal generator torque is shown in Figure 6.40. To obtain this frequency response the ***B*** and ***C*** matrices are defined as

$$\mathbf{B} = [0, 0, 0, 0, 0, 0, 0, \frac{1}{J_2}, 0, 0, 0, 0]$$

$$\mathbf{C} = \begin{bmatrix} -k_{12} & k_{12} & 0 & 0 & 0 & 0 & -D_{12} & D_{12} & 0 & 0 & 0 & 0 \\ 0 & -k_{23} & k_{23} & 0 & 0 & 0 & 0 & -D_{23} & D_{23} & 0 & 0 & 0 \end{bmatrix}$$

As expected the Generator-Exciter shaft is sensitive to all five Modes but the torque is somewhat less than that in the LP_3-Generator shaft which is particularly sensitive to frequencies around 31.7 Hz but is insensitive to Mode 1 type oscillation. The distinct nature of these curves is due to the low degree of damping present in the system. Additional damping would reduce the peak values of these curves but unfortunately is not usually available although a small amount of additional damping may be introduced by the electrical system.

Note that the natural frequencies are substantially removed from 50 and 100 Hz so that the periodic changes in external torque at fundamental or double frequency which occurs under fault conditions (Chapter 4) induce a relatively low value of shaft torque.

Such eigenvalue techniques can readily be extended to include details of the power system and are invaluable when conducting full system stability studies or when assessing the feasibility of new generator designs (Bumby, 1982, Bumby & Wilson, 1983, Westlake, Bumby & Spooner, 1996).

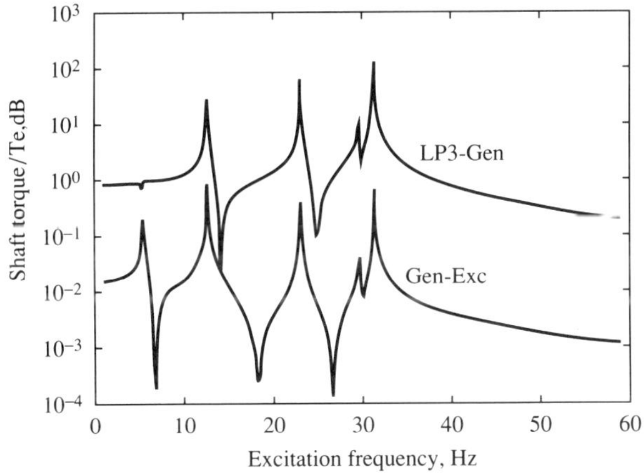

Figure 6.40 Frequency sensitivity of the LP_3-Generator and the Generator-Exciter drive shafts for the 577 MVA, 3000 rpm generator

6.6.2 Effect of system faults

When a system fault occurs the highest shaft torque is usually in the main shaft connecting the generator and the turbine as the applied torque on these two rotor components are in opposing directions. The shaft between the last two turbine stages, the low-pressure and the intermediate-pressure sections shown in Figure 5.1, can also experience high torques. Figure 6.41 shows typical changes in the shaft torque which may occur in the LP-Generator shaft (denoted as *I* in Figure 5.1) when a generator is subjected to three different electromagnetic torques: a step change, a 50 Hz variation and a 100 Hz variation. The highest momentary values of shaft torque occur when the rotor is subject to an aperiodic external torque and is due to the oscillatory shaft torque being reinforced by the external torque on each back swing of the rotor masses. As a result the shaft torque oscillates around the average value of the external torque at a frequency determined by the different torsional natural frequencies. In this example a frequency response similar to that in Figure 6.40 showed the LP-Generator shaft to be particularly sensitive to 12 and 24 Hz oscillations and these two frequencies can be observed to dominate the time response.

The above considerations lead to the conclusion that the highest shaft torque corresponds to an electromagnetic torque with a high aperiodic component which, as was shown in the previous sections, appears either after synchronisation of a generator when the synchronising angle is high or after clearing a fault in the network. Figure 6.42 shows examples of the torque in the main coupler *I* during a phase-to-phase short-circuit on the generator terminals, after synchronisation at a large angle and after clearing a three phase short-circuit in the network.

It is interesting to compare the last two cases. The average value of the shaft torque shown by the dashed line is the same in both cases, but the maximum value is much higher in case (c). This is a direct result of the shaft already being twisted when the fault is cleared and this twist being such that the aperiodic electromagnetic torque appearing on fault clearance reinforces the twist. The influence of the degree of shaft twist on the shaft torque, after the fault is cleared, is shown in Figure 6.43 where it can seen that, when the

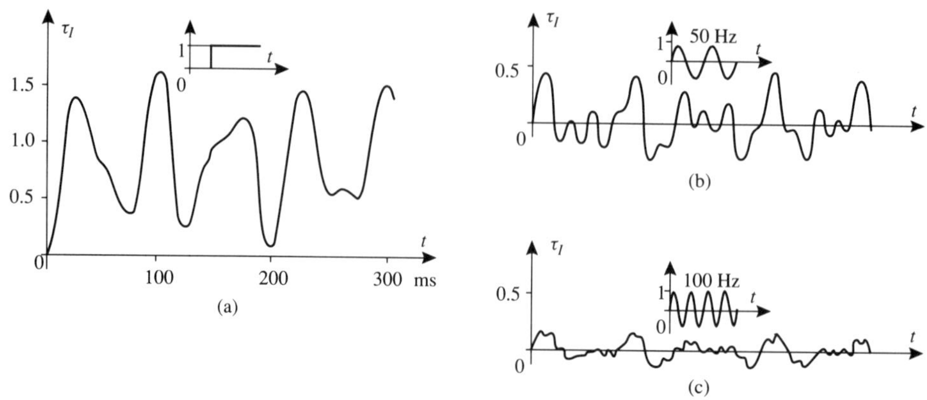

Figure 6.41 Examples of the changes in the torque in the main coupler due to: (a) aperiodic component of the electromagnetic torque; (b) fundamental-frequency component; (c) double-frequency component (Based on Bölder, Kulig & Lambrecht, 1975)

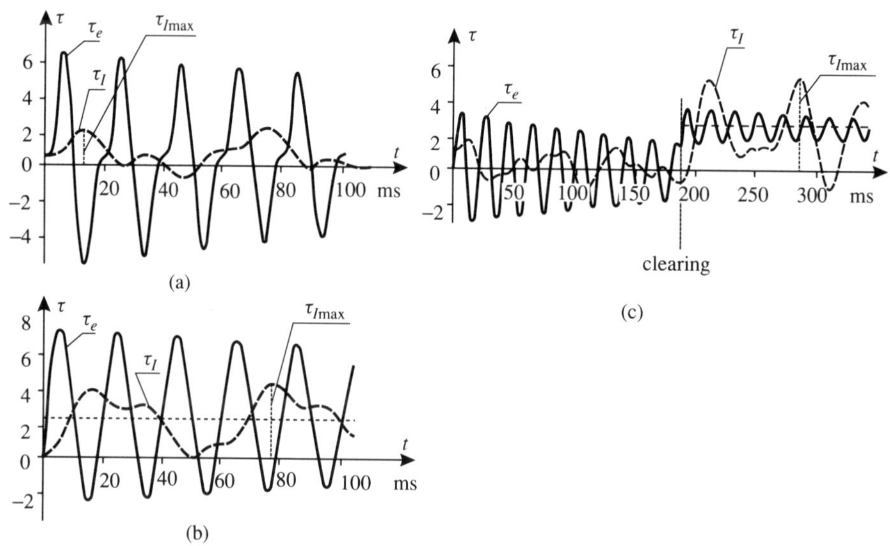

Figure 6.42 Torque acting on the main coupler for the case of: (a) phase-to-phase short-circuit on the generator terminals; (b) synchronisation when $\delta = 2\pi/3$; (c) fault in the network cleared after 0.187 s (Based on Läge & Lambrecht, 1974)

shaft torque has a negative value at the instant of clearing, the resulting shaft torque is substantially increased. This effect may be compounded by unsuccessful auto-reclosing.

The example shown in Figure 6.42a corresponded to a phase-to-phase short-circuit on the generator terminals applied when the generator torque loading was high. However the corresponding torque in the main coupler is much smaller than that occurring after clearing a three phase system fault, Figure 6.42c. This is characteristic of big machines whereas in medium and low power machines the highest shaft torques generally accompany faults on the generator terminals. An approximate relationship between the maximum torque in the main coupler and the clearing time is shown in Figure 6.44. The greater susceptibility

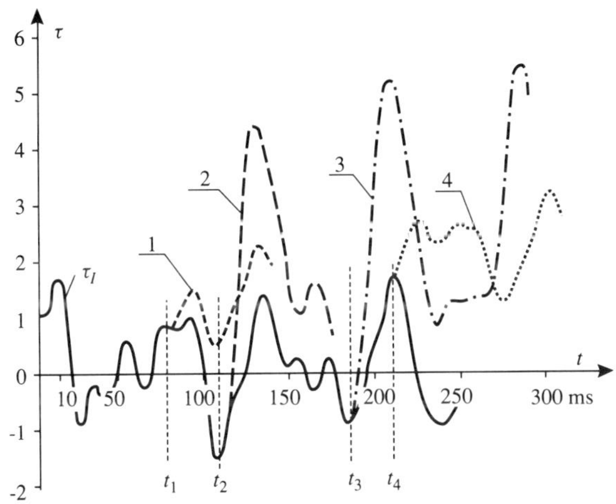

Figure 6.43 Influence of the initial shaft twist on the shaft torque when the fault is cleared after: 1: $t_2 = 80$ms; 2: $t_2 = 110$ms; 3: $t_2 = 187$ms; 4: $t_2 = 213$ms (Based on Kulicke & Webs, 1975)

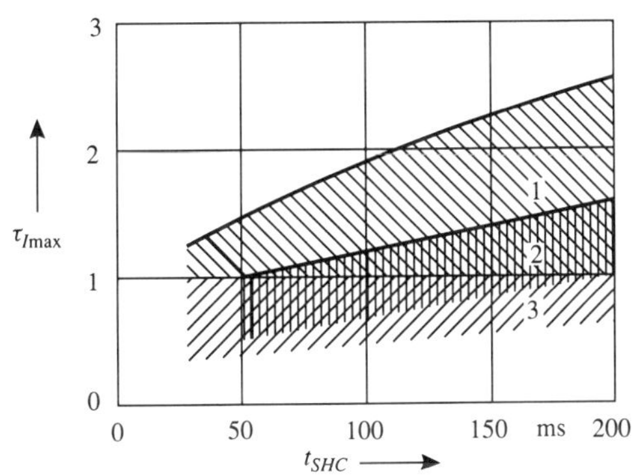

Figure 6.44 Approximate relationship between the maximum torque in the main coupler and the clearing time when the rating of the generator is: 1: $S >> 500$ MVA; 2: $S < 500$ MVA; 3: $S < 300$ MVA. The value of τ_l is relative with respect to the value of torque accompanying a phase-to-phase fault on the generator terminals (Based on Bölder Kulig & Lambrecht, 1975)

of large machines to short-circuits in the network is primarily due to the lower per unit impedance of the generator transformer associated with these large machines so that its influence in the transfer reactance is less than for smaller rated generators. Also the per-unit rotor inertia decreases as unit rating increases so that with large units even short clearing times may result in a relatively large angle increase and a high value of aperiodic torque when the fault is cleared.

6.6.3 Subsynchronous resonance

Section 4.2 explained how, when a fault occurs on a power system, large currents are produced comprising a dc offset and an ac term. The ac term depends on the type of fault and, in the case of a symmetrical fault, will consist of a positive sequence component whilst with unsymmetrical faults both positive sequence and negative sequence components will be present. Due to the relative speed of the rotor with respect to the stator, and the space distribution of the stator armature windings, these different current components produce rotor torques at f_s , 0 and $2f_s$ Hz respectively. Such observations assume that the transmission system is purely inductive and no capacitance compensation is present in the system. Such a system is often referred to as uncompensated.

If now the transmission line reactance is compensated by a series capacitor SC as shown in Figure 6.45a then, depending on the relative magnitude of the total circuit resistance R, inductance L and capacitance C, the transient component of any current may now be an oscillation at a frequency f_d decaying slowly with time. This oscillation frequency is closely related to the undamped natural frequency of the line, $f_n = 1/2\pi\sqrt{LC}$ Hz. Such oscillations produce rotor currents and torques at a frequency $(f_s - f_n)$ which, if close to one of the rotor torsional natural frequencies, can produce very high shaft torques and ultimate shaft failure. The frequency $(f_s - f_n)$ is known as the *frequency complement* of the line natural frequency f_n as both these frequencies sum to f_s the system frequency.

The way such a damped oscillatory current is produced can be understood by considering the equivalent circuit in Figure 6.45b and calculating the current when the switch S is suddenly closed. The differential equation governing the flow of current is

$$L\frac{di}{dt} + Ri + \frac{l}{C}\int i dt = E_m \sin(\omega_s t + \theta_0) \tag{6.63}$$

Differentiating gives

$$\frac{d^2 i}{dt^2} + \frac{R}{L}\frac{di}{dt} + \frac{1}{LC}i = \omega_s E_m \cos(\omega_s t + \theta_0) \tag{6.64}$$

the solution of which is of the form

$$i(t) = i_{trans}(t) + \frac{E_m}{Z}\sin(\omega_s t + \theta_0 + \psi) \tag{6.65}$$

where $Z = \sqrt{R^2 + (X_L - X_C)^2}$, $X_L = \omega_s L$, $X_C = 1/(\omega_s C)$, and the phase angle $\psi =$ arctan$[(X_C - X_L)/R]$. The left hand side of equation 6.64 is of the standard second order form

$$\frac{d^2 x}{dt^2} + 2\zeta\omega_n\frac{dx}{dt} + \omega_n^2 = 0 \tag{6.66}$$

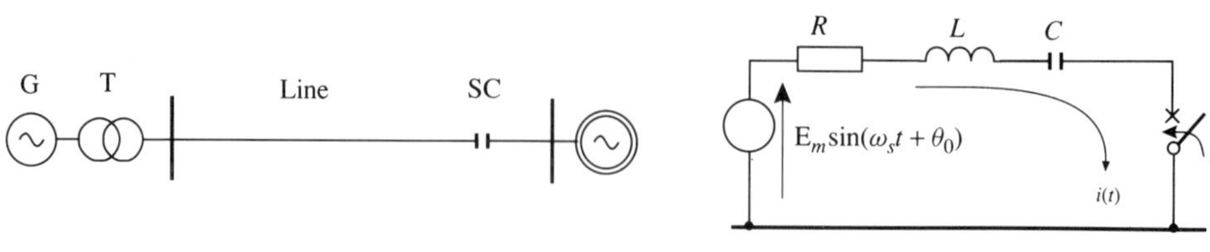

Figure 6.45 Series compensated transmission line: (a) schematic diagram; (b) simple equivalent circuit to calculate the form of the current when shorting the line

the solution of which determines the form of transient current. Equation 6.66 will have a different type of solution depending on the value of the damping ratio ζ. Of particular interest is the underdamped solution obtained when $0 < \zeta < 1$ which is of the form

$$i_{trans}(t) = Ae^{-\zeta\omega_n t}\sin(\omega_d t + \psi_2) \tag{6.67}$$

where A and ψ_2 are constants and the damped natural frequency $\omega_d = \omega_n\sqrt{1-\zeta^2}$. The undamped natural frequency ω_n and the damping ratio ζ can be found in terms of the circuit parameters by comparing equation 6.66 with the left-hand side of equation 6.64 to give

$$\omega_n = \sqrt{\frac{1}{LC}} = \omega_s\sqrt{\frac{X_c}{X_L}} \text{ and } \zeta = \frac{R}{2}\sqrt{\frac{C}{L}} \tag{6.68}$$

Substituting equation 6.67 into equation 6.65 gives the total current for the underdamped condition as

$$i(t) = Ae^{-\zeta\omega_n t}\sin(\omega_d t + \psi_2) + \frac{E_m}{Z}\sin(\omega_s t + \theta_0 + \psi) \tag{6.69}$$

With currents flowing in each armature phase at frequencies ω_d and ω_s, torques are induced in the rotor at the sum and difference of these frequencies to produce rotor currents and torques at frequencies below (subsynchronous) and above (supersynchronous) system frequency given by $(\omega_s - \omega_d)$ and $(\omega_s + \omega_d)$ respectively. Of particular importance are the subsynchronous torques and the way in which they interact with the turbine and generator rotor. If any subsynchronous torques appear on the rotor at, or near to, any of the rotor torsional natural frequencies, additional energy can be fed into the mechanical vibrations producing a resonance effect in the mechanical system and large shaft torques. Such large shaft torques produce high shaft stresses, a reduction in shaft fatigue life and, possibly, shaft failure. To avoid such *subsynchronous resonance* effects it is important to ensure that no subsynchronous torques are present at, or near to, any of the shaft torsional natural frequencies. Typically these torsional natural frequencies will be in the range 10–40 Hz for a 3000 rpm, 50 Hz, generator.

For the simple system shown in Figure 6.45a the subsynchronous frequencies can be estimated using $\omega_n = \omega_s\sqrt{X_C/X_L}$, equation 6.68, assuming that the degree of compensation employed is known. The results of such a calculation are shown in Table 6.3 for a 50 Hz system. Assuming the generator used in example 6.2 to be operating onto a simple system such as that shown in Figure 6.45 with ~30% compensation, the results

Table 6.3 Line natural frequency as a function of the degree of compensation

Degree of compensation X_C/X_L	Line natural frequency f_n Hz	Frequency complement $(50-f_n)$ Hz
0.1	15.8	34.2
0.2	22.4	27.6
0.3	27.4	22.6
0.4	31.6	18.4
0.5	35.4	14.6

of Table 6.3 would suggest that subsynchronous resonance with the 23.3 Hz, Mode *3*, torsional natural frequency might be a problem.

A fuller analysis of subsynchronous resonance problems that includes all the system damping mechanisms would require the use of a full system eigenvalue study such as that conducted by Ahlgren, Johansson & Gadhammar, (1978). This and other techniques can be used to study more practical, interconnected, power systems where a number of subsynchronous resonant frequencies may be present (Anderson, Agrawal & Van Ness, 1990).

7
Voltage Stability

Chapter 2 explained how an electric power network is made up from many different elements, is divided into transmission, subtransmission and distribution levels, and is often organised in such a way that each level belongs to a different company. Because of its immense size, analysing a complete power system is not possible, even using supercomputers, and normally the system is divided into sensible parts some of which are modelled in detail and others more superficially. Obviously the system model used must represent the problem being studied so that when analysing distribution networks the transmission or subtransmission networks are treated as *sources* operating at a given voltage. On the other hand, when analysing the transmission or subtransmission networks, the distribution networks are treated as *sinks* of real and reactive power and referred to as *composite loads* or simply *loads*.

A distribution network is connected to the transmission network at the *grid supply point* where a change in voltage may cause complicated dynamic interactions inside the distribution network itself due to:

- voltage control action due to transformer tap-changing
- control action associated with reactive power compensation and/or small embedded generators
- a low supply voltage causing changes in the power demand due to induction motors stalling and/or the extinguishing of discharge lighting
- operation of protection equipment by overcurrent or undervoltage relays, electromechanically held contactors, etc.
- re-ignition of discharge lighting and self-start induction motors when the supply voltage recovers

In this chapter the effect actions such as these can have on the voltage stability is examined using the static characteristics of the composite loads introduced in Section 3.4. This simplified analysis will help to give an understanding of both the different mechanisms that may ultimately lead to voltage collapse and the techniques that may be used to assess the voltage stability of a particular system. These techniques can then be extended to the analysis of large systems using computer simulation methods (Taylor, 1994 and Kundur, 1994).

7.1 Network Feasibility

The system shown in Figure 7.1 is representative of the general power supply problem in that it shows a generator supplying some composite load. Generally there will be some limits on the power that can be supplied to the load and this will determine how stable the supply is. To analyse this stability problem the composite load will be represented by its static voltage characteristics when the problem becomes that of determining the solutions to the power network equations, if they exist, and, if they do, determining what limits are placed on the solutions. This process is often referred to as determining the *network feasibility* or determining the *network loadability*.

The network feasibility problem can be explained by using the simple system shown in Figure 7.1. In this system the network is represented by an equivalent generator that can be modelled in the steady-state by an equivalent voltage source E behind an equivalent reactance X_g. Under normal operating conditions the generator AVR will keep the terminal voltage constant when the equivalent voltage source has a value equal to the terminal voltage V_g and the equivalent reactance is zero. However if the AVR is not operative, or the equivalent generator is operating near its excitation limit, then the field voltage will remain constant and the equivalent generator must be modelled by its synchronous emf E_f acting behind its synchronous reactance X_d. In general the resistance of the generator and transmission link is small and can be neglected whilst the equivalent reactance X must combine the source reactance with that of the transformer and the transmission line. The real and reactive power absorbed by the load, $P_L(V)$ and $Q_L(V)$, can be calculated from the phasor diagram in Figure 7.1 by noting that $IX\cos\varphi = E\sin\delta$ and $IX\sin\varphi = E\cos\delta - V$. This gives

$$\begin{aligned} P_L(V) &= VI\cos\varphi = V\frac{IX\cos\varphi}{X} = \frac{EV}{X}\sin\delta \\ Q_L(V) &= VI\sin\varphi = V\frac{IX\sin\varphi}{X} = \frac{EV}{X}\cos\delta - \frac{V^2}{X} \end{aligned} \tag{7.1}$$

The angle δ between the $\overline{E}$ and $\overline{V}$ phasors can be eliminated using the identity $\sin^2\delta + \cos^2\delta = 1$ to give

$$\left(\frac{EV}{X}\right)^2 = [P_L(V)]^2 + \left[Q_L(V) + \frac{V^2}{X}\right]^2 \tag{7.2}$$

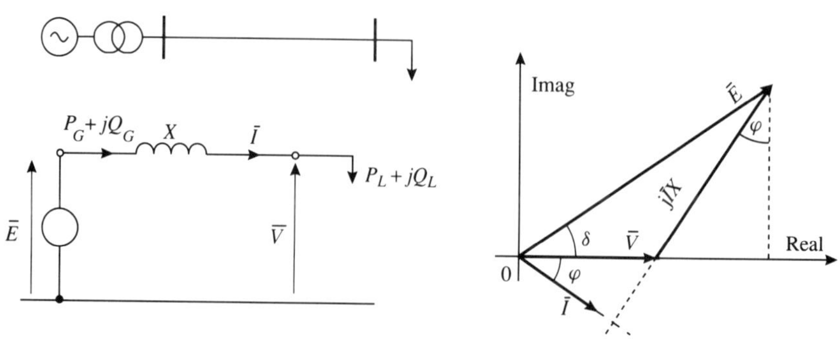

Figure 7.1 Equivalent circuit of the transmission link and its phasor diagram

This static power-voltage equation determines all the possible network solutions when the voltage characteristics $P_L(V)$ and $Q_L(V)$ are taken into account.

Stiff load

If the power demand of the load is independent of voltage and is constant then

$$P_L(V) = P_n \qquad \text{and} \qquad Q_L(V) = Q_n \tag{7.3}$$

where P_n and Q_n are the real and reactive power demand of the load at the rated voltage V_n. Such a load is said to be *ideally stiff*. Equation 7.2 can now be re-written as

$$\left(\frac{EV}{X}\right)^2 = P_n^2 + \left[Q_n + \frac{V^2}{X}\right]^2 \tag{7.4}$$

For a given value of V this equation describes a circle in the (P_n, Q_n) plane as shown in Figure 7.2a. The centre of the circle lies on the Q_n axis and is shifted vertically down from the origin by V^2/X. Increasing the voltage V produces a family of circles of increasing radius and downward shift, bounded by an envelope as shown in Figure 7.2b.

For each point inside the envelope, for example point A, there are two possible solutions to equation 7.4 at voltage values V_1 and V_2, as defined by the two circles whereas for any point B on the envelope there is only one value of V for which equation 7.4 is satisfied. An equation for this envelope can therefore be obtained by determining those values of P_n and Q_n for which there is only one solution of equation 7.4 with respect to V. Re-arranging equation 7.4 gives

$$\left(\frac{V^2}{X}\right)^2 + \left(2Q_n - \frac{E^2}{X}\right)\left(\frac{V^2}{X}\right) + (P_n^2 + Q_n^2) = 0 \tag{7.5}$$

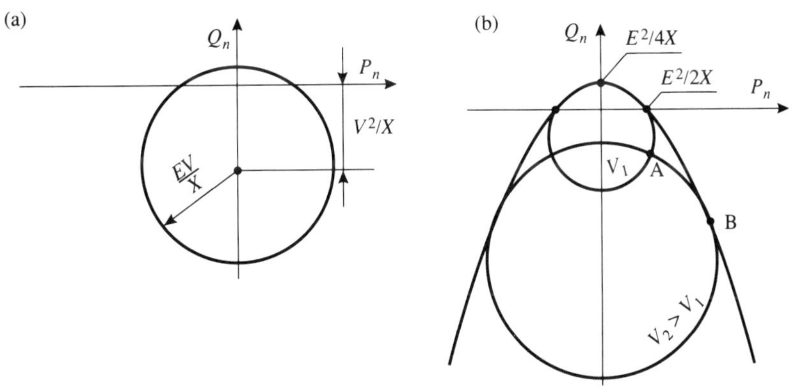

Figure 7.2 Circles determining the power that can be delivered to a stiff load: (a) one circle for a given voltage V; (b) a family of circles and their envelope

This is a quadratic equation in (V^2/X) and has only one solution when

$$\left(2Q_n - \frac{E^2}{X}\right)^2 - 4(P_n^2 + Q_n^2) = 0 \tag{7.6}$$

Solving for Q_n gives

$$Q_n = \frac{E^2}{4X} - \frac{P_n^2}{\dfrac{E^2}{X}} \tag{7.7}$$

which is the equation of an inverted parabola that crosses the P_n axis at $P_n = E^2/2X$ and has its maximum at

$$P_n = 0 \qquad \text{and} \qquad Q_{n\,MAX} = \frac{E^2}{4X} \tag{7.8}$$

The parabola described by equation 7.7 is important as it defines the shape of the envelope in Figure 7.2b that encloses all the possible solutions to the network equation 7.4. Each point (P_n, Q_n) inside the parabola satisfies two network solutions corresponding to two different values of the load voltage V while each point on the parabola satisfies one network solution corresponding to only one value of voltage. There are no network solutions outside the parabola. In other words, it is not possible to deliver power equal to P_n and Q_n corresponding to any point outside the parabola.

Influence of the load characteristics

For the more general case of a non-stiff load the power demand will depend on the voltage as described by the voltage characteristics $P_L(V)$ and $Q_L(V)$. The possible solutions to equation 7.2 will not now be bounded by a simple parabola, as for $P_L(V) = P_n$, $Q_L(V) = Q_n$, but the shape of the solution area will vary depending on the actual voltage characteristics as shown in Figure 7.3. In general the less stiff the load the more open the solution area. For the constant load discussed above, the solution area corresponds to a parabola, Figure 7.3a. If the reactive power characteristic is a square function of the voltage, $Q_L(V) = \left(\frac{V}{V_n}\right)^2 Q_n$, then the solution area opens up from the top, Figure 7.3b, so that for $P_n = 0$ there is no limit on Q_n. If the real power characteristic is linear $P_L(V) = \left(\frac{V}{V_n}\right) P_n$ as in Figure 7.3c, then the solution area is bounded by two parallel, vertical lines. Assuming both real and reactive power characteristics to be square functions of voltage then $P_L(V) = \left(\frac{V}{V_n}\right)^2 P_n$, $Q_L(V) = \left(\frac{V}{V_n}\right)^2 Q_n$ and there are no limits on the values of P_n and Q_n as shown in Figure 7.3d.

Consider again the characteristics of Figure 7.3d where there are no limits on P_n and Q_n. This can be proved by expressing

$$P_L(V) = P_n\left(\frac{V}{V_n}\right)^2 = \frac{P_n}{V_n^2}V^2 = G_n V^2 \quad ; \quad Q_L(V) = Q_n\left(\frac{V}{V_n}\right)^2 = \frac{Q_n}{V_n^2}V^2 = B_n V^2 \tag{7.9}$$

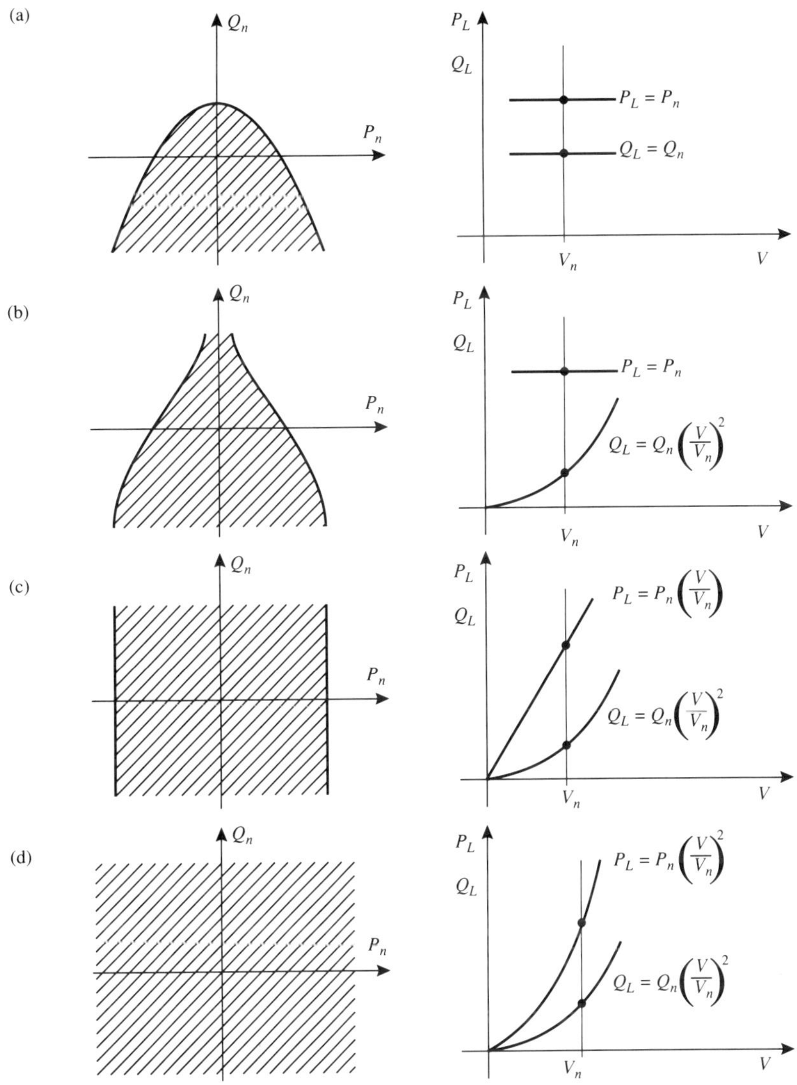

Figure 7.3 Dependence of the network solution area on the shape of the load characteristics

which allows the load to be represented by an equivalent load admittance $\overline{Y}_n = G_n + jB_n$ shown in Figure 7.4. Varying the value of P_n and Q_n from zero to infinity corresponds to changing the equivalent admittance from zero (open-circuit) to infinity (short-circuit). As current will flow in the circuit of Figure 7.4 for any value of $\overline{Y}_n$ so a solution of equation 7.2 exists for any P_n and Q_n. This can be proved mathematically by substituting equation 7.9 into equation 7.2 when the following formula is obtained

$$V = \frac{E}{\sqrt{(G_n X)^2 + (B_n X + 1)^2}} \tag{7.10}$$

confirming that for any G_n, B_n a network solution for V always exists.

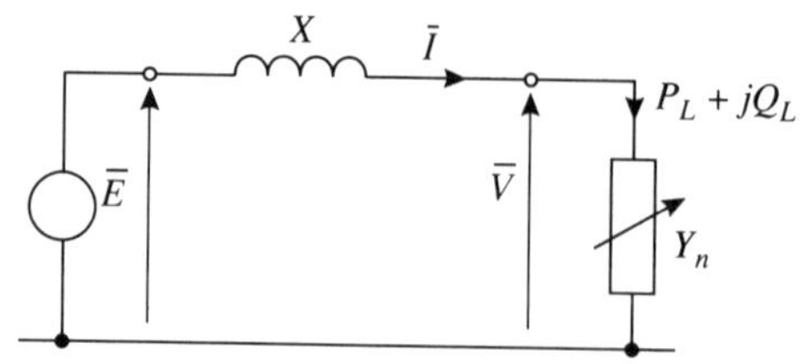

Figure 7.4 Transmission link loaded with an equivalent variable admittance

7.2 Stability Criteria

For each point inside the envelope of network solutions, for example point A in Figure 7.2, there are two solutions with respect to the voltage V: one with a higher and one with a lower value of voltage. It is now necessary to examine which of these solutions corresponds to a stable equilibrium point. This problem of *voltage stability* is considered in this section when different, but equivalent, voltage stability criteria are derived.

7.2.1 dΔQ/dV Criterion

This classic voltage stability criterion (Venikov, 1968, Weedy, 1968) is based on the capability of the system to supply the load with reactive power for a given real power demand. To explain this criterion it is convenient to notionally separate the reactive power demand from the real power demand as shown in Figure 7.5. To distinguish between the power supplied by the source at the load node and the load demand itself, let $P_L(V)$ and $Q_L(V)$ be the load demand and $P_S(V)$ and $Q_S(V)$ be the powers supplied by the source to the load.

As the real power is always connected to the transmission link then it holds that $P_L(V) = P_S(V)$. Similarly, during normal operation $Q_L(V) = Q_S(V)$ but, for the purposes of stability analysis, the link between $Q_L(V)$ and $Q_S(V)$ is notionally separated. $Q_S(V)$ is treated as the reactive power supplied by the source and is assumed not to be determined by the reactive power demand of the load. The real and reactive load powers are given by expressions similar to those in equation 7.1

$$P_L(V) = P_S(V) = \frac{EV}{X}\sin\delta \quad \text{and} \quad Q_S(V) = \frac{EV}{X}\cos\delta - \frac{V^2}{X} \tag{7.11}$$

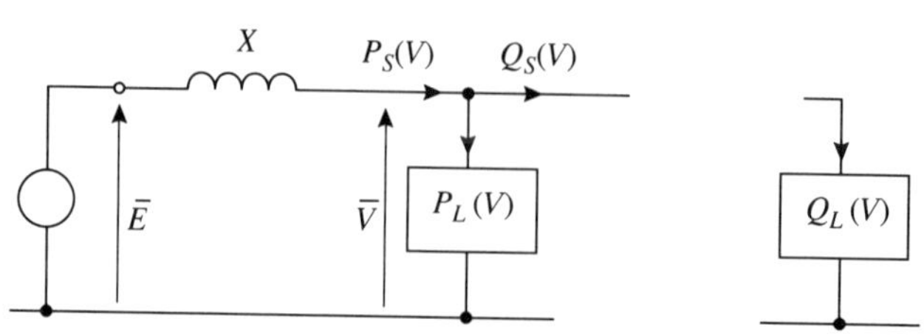

Figure 7.5 Equivalent circuit for determining the reactive power characteristic of the system

Eliminating the trigonometric functions using the identity $\sin^2\delta + \cos^2\delta = 1$ gives

$$\left(\frac{EV}{X}\right)^2 = P_L^2(V) + \left[Q_S(V) + \frac{V^2}{X}\right]^2 \tag{7.12}$$

and solving for $Q_S(V)$ gives

$$Q_S(V) = \sqrt{\left[\frac{EV}{X}\right]^2 - [P_L(V)]^2} - \frac{V^2}{X} \tag{7.13}$$

This equation determines the reactive power-voltage characteristic and shows how much reactive power will be supplied by the source if the system is loaded only with the real power $P_L(V)$ and the load voltage is treated as a variable. For a stiff real power load $P_L(V) = P_L = constant$ and equation 7.13 takes the form of an inverted parabola as shown in Figure 7.6. The first term in equation 7.13 depends on the equivalent system reactance X and the load real power P_L and has the effect of shifting the parabola downwards and towards the right as illustrated in Figure 7.6. For $P_L = 0$ the parabola crosses the horizontal axis at $V = E$ and $V = 0$ and has a maximum at $Q_{MAX} = E^2/4X$ and $V = E/2$. For $P_L > 0$ the maximum value of Q_S occurs at a voltage $V = \sqrt{[E/2]^2 + [P_L(V)X/E]^2}$ which is greater than $E/2$.

If the reactive power of the notionally separated load is now re-connected to the system then both the $Q_S(V)$ and $Q_L(V)$ characteristics can be drawn on the same diagram as in Figure 7.7a. At equilibrium the supply must be equal to the demand, that is $Q_S(V) = Q_L(V)$, and is satisfied by the two equilibrium points V^s and V^u. This corresponds to the situation shown in Figure 7.2b where for one value of power demand, point A, there are two possible, but different, values of voltage $V_1 \neq V_2$.

The stability of both equilibrium points can be tested using the small disturbance method. Recall from sub-Section 3.1.4, Figure 3.4, that an excess of reactive power was shown to produce an increase in voltage whilst a deficit of reactive power resulted in a decrease in voltage. Now consider equilibrium point s in Figure 7.7a and assume that there is a small negative voltage disturbance ΔV. This will result in the supplied reactive power $Q_S(V)$ being greater than the reactive power demand $Q_L(V)$. This excess of reactive power will tend to increase the voltage and therefore force the voltage to return to point s.

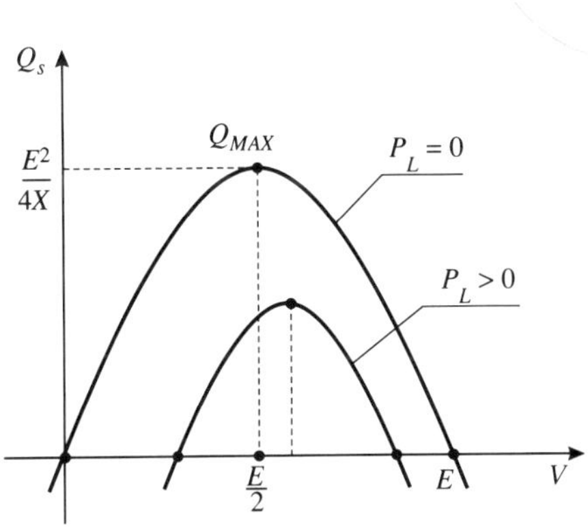

Figure 7.6 $Q_s(V)$ characteristic for $P_L = 0$ and $P_L > 0$

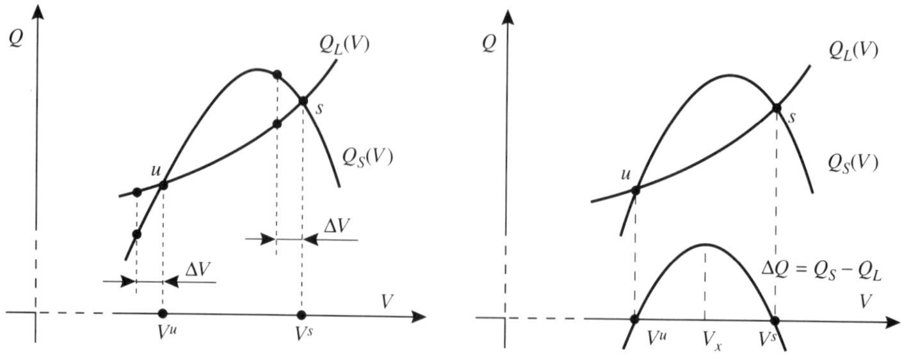

(a) (b)

Figure 7.7 $Q_S(V)$ and $Q_L(V)$ characteristics: (a) two equilibrium points; (b) classic stability criterion

If the disturbance produces an increase in voltage, the resulting deficit in reactive power will force the voltage to decrease and again return to point s. The conclusion is that equilibrium point s is stable.

On the other hand a disturbance in the vicinity of the second equilibrium point u which decreases the voltage will produce a deficit of reactive power with $Q_S(V) < Q_L(V)$ which will force a further decrease in voltage. As the disturbed system does not return to the equilibrium point, the equilibrium point u is unstable.

The classic voltage stability criterion is obtained by noting from Figure 7.7b that the derivative of the surplus of reactive power $\mathrm{d}(Q_S - Q_L)/\mathrm{d}V$ is of opposite sign at the two equilibrium points: it is negative at the stable point s and positive at the unstable point u. This is the essence of the classic $\mathrm{d}\Delta Q/\mathrm{d}V$ stability criterion:

$$\frac{\mathrm{d}(Q_S - Q_L)}{\mathrm{d}V} < 0 \quad \text{or} \quad \frac{\mathrm{d}Q_S}{\mathrm{d}V} < \frac{\mathrm{d}Q_L}{\mathrm{d}V} \tag{7.14}$$

In the simple system shown in Figure 7.5 the supplied real and reactive powers, expressed by equations 7.11, are functions of the two variables V and δ with increments given by

$$\begin{aligned} \Delta Q_S &= \frac{\partial Q_S}{\partial V}\Delta V + \frac{\partial Q_S}{\partial \delta}\Delta\delta \\ \Delta P_L = \Delta P_S &= \frac{\partial P_S}{\partial V}\Delta V + \frac{\partial P_S}{\partial \delta}\Delta\delta \end{aligned} \tag{7.15}$$

Eliminating $\Delta\delta$ from these two equations, and dividing the result by ΔV, gives

$$\frac{\Delta Q_S}{\Delta V} = \frac{\partial Q_S}{\partial V} + \frac{\partial Q_S}{\partial \delta}\left(\frac{\partial P_S}{\partial \delta}\right)^{-1}\left[\frac{\Delta P_L}{\Delta V} - \frac{\partial P_S}{\partial V}\right] \tag{7.16}$$

or

$$\frac{\mathrm{d}Q_S}{\mathrm{d}V} \approx \frac{\partial Q_S}{\partial V} + \frac{\partial Q_S}{\partial \delta}\left(\frac{\partial P_S}{\partial \delta}\right)^{-1}\left[\frac{\mathrm{d}P_L}{\mathrm{d}V} - \frac{\partial P_S}{\partial V}\right] \tag{7.17}$$

where the partial derivatives are obtained from the equations in (7.11) as

$$\frac{\partial P_S}{\partial \delta} = \frac{EV}{X}\cos\delta; \quad \frac{\partial P_S}{\partial V} = \frac{E}{X}\sin\delta; \quad \frac{\partial Q_S}{\partial \delta} = -\frac{EV}{X}\sin\delta; \quad \frac{\partial Q_S}{\partial V} = \frac{E}{X}\cos\delta - 2\frac{V}{X} \tag{7.18}$$

Substituting these partial derivatives into equation 7.17 gives

$$\begin{aligned}\frac{dQ_S}{dV} &\approx \frac{E}{X}\cos\delta - \frac{2V}{X} - \frac{EV}{X}\sin\delta\frac{X}{EV\cos\delta}\left[\frac{dP_L}{dV} - \frac{E}{X}\sin\delta\right] \\ &= \frac{E}{X\cos\delta} - \left(\frac{2V}{X} + \frac{dP_L}{dV}\tan\delta\right)\end{aligned} \tag{7.19}$$

This allows the stability condition defined in equation 7.14 to be expressed as

$$\frac{dQ_L}{dV} > \frac{E}{X\cos\delta} - \left(\frac{2V}{X} + \frac{dP_L}{dV}\tan\delta\right) \tag{7.20}$$

where the derivatives dQ_L/dV and dP_L/dV are calculated from the functions used to approximate the load characteristics.

Generally, for a multi-machine system, it is not possible to derive an analytical formula (similar to equation 7.20) for the stability criterion. However, by using a load flow program it is possible to obtain the system supply characteristic $Q_S(V)$ by defining the load node under investigation as a PV node and executing the program several times for different values of the node voltage V. The resulting $Q_S(V)$ characteristic can then be compared with the load characteristic $Q_L(V)$ to check the stability condition.

7.2.2 dE/dV Criterion

The system equivalent emf E can be expressed as a function of the load voltage by solving equation 7.2 for E to give:

$$E(V) = \sqrt{\left(V + \frac{Q_L(V)X}{V}\right)^2 + \left(\frac{P_L(V)X}{V}\right)^2} \tag{7.21}$$

where $Q_L(V)X/V$ is the in-phase and $P_L(V)X/V$ is the quadrature component of the voltage drop $\bar{I}X$ shown in Figure 7.1.

An example of an $E(V)$ characteristic is shown in Figure 7.8 with the load normally operating at a high voltage corresponding to the right-hand-side of the characteristic. As V is large, and much greater than both the in-phase and quadrature components of the voltage drop, then, according to equation 7.21, a decrease in voltage will cause the emf $E(V)$ to fall. As V continues to decrease, the $Q_L(V)X/V$ and $P_L(V)X/V$ components become more important and below a certain value of V, they will cause $E(V)$ to rise. Consequently, each value of $E(V)$ may correspond to two possible solutions of the network equations with respect to V. As before the stability of these solutions can be examined using the small disturbance method.

First consider the system behaviour at point s lying on the right-hand-side of the characteristic $E(V)$ in Figure 7.8 and assume that the source emf E is maintained at a constant value. A decrease in the load voltage by ΔV will produce a reduction in the emf

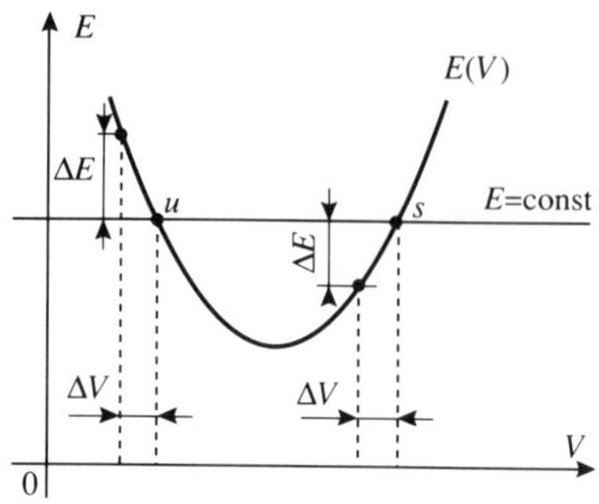

Figure 7.8 Illustration to the stability criterion $dE/dV > 0$

$E(V)$ that is less than the source emf E. As E is too large to maintain the reduced load, voltage V is forced to return to the initial equilibrium value. Similarly a voltage increase by ΔV results in the source emf E being smaller than the emf required to maintain the increased load voltage V so that the source emf E again forces the voltage to return to its initial value.

Now consider a disturbance that produces a voltage decrease by ΔV from the equilibrium point u on the left-hand-side of the characteristic. This disturbance results in the source emf E being less than the emf $E(V)$ required to maintain the lowered voltage. As E is too small, the load voltage further declines and does not return to its initial value. The conclusion is that point u is unstable.

From these discussions it is apparent that the system is stable if the equilibrium point lies on the right-hand-side of the characteristic, that is when

$$\frac{dE}{dV} > 0 \tag{7.22}$$

Venikov (1968) and Abe *et al* (1975) have both shown this condition to be equivalent to the classic stability condition defined in equation 7.14.

Using the stability condition defined in equation 7.22 to study the stability of a multi-node system is rather inconvenient when a load flow program is used. The derivative dE/dV is based on the equivalent circuit of Figure 7.1. In the load flow program, the load node should be of the PQ type (specified P_L and Q_L, unknown V and δ) which does not allow the voltage to be set. Consequently the voltage increase necessary to obtain the derivative dE/dV has to be enforced from the generator nodes (PV type) for which the voltage can be set. As there is normally a large number of generator nodes in the system, a large number of load flows would have to be run to account for all the possible combinations of changes in the voltage settings.

7.2.3 dQ_G/dQ_L Criterion

To understand this criterion it is necessary to analyse the behaviour of the reactive power generation $Q_G(V)$ as the load reactive demand $Q_L(V)$ varies. This approach is somewhat different to that taken in the two previous sub-sections in that $Q_G(V)$ now includes the reactive power demand of both the load, $Q_L(V)$, and the network, I^2X, whereas previously only the reactive power $Q_S(V)$ supplied by the source at the load node was considered.

The equation determining $Q_G(V)$ is similar to equation 7.1 but with E and V interchanged, that is

$$Q_G(V) = \frac{E^2}{X} - \frac{EV}{X}\cos\delta \tag{7.23}$$

where both V and δ depend on the demand $P_L(V)$ and $Q_L(V)$. Equation 7.1 allows the second component in this equation to be substituted to give

$$Q_G(V) = \frac{E^2}{X} - \frac{V^2}{X} - Q_L(V) \text{ or } \frac{V^2}{X} = \frac{E^2}{X} - Q_L(V) - Q_G(V) \tag{7.24}$$

Substituting this expression into equation 7.2 and performing some simple algebra gives

$$Q_G^2(V) - \frac{E^2}{X}Q_G(V) + P_L^2(V) + \frac{E^2}{X}Q_L(V) = 0 \tag{7.25}$$

or

$$Q_L(V) = -\frac{Q_G^2(V)}{\frac{E^2}{X}} + Q_G(V) - \frac{P_L^2(V)}{\frac{E^2}{X}} \tag{7.26}$$

For the case of the stiff real power load with $P_L(V) = P_L = constant$ this equation describes a horizontal parabola in the (Q_G, Q_L) plane as shown in Figure 7.9a. The vertex of the parabola is at a constant Q_G value equal to $E^2/2X$ whilst the maximum value of Q_L depends on P_L and for $P_L = 0$ the maximum is at $E^2/4X$. Increasing P_L shifts the parabola to the left along the Q_L axis but without any corresponding shift with respect to the Q_G axis.

It is worth noting that the vertex of the parabola (the maximum value of Q_L for a given P_L) corresponds to a point on the envelope of the $Q_L(P_L)$ characteristics shown in Figure 7.2b and to the vertex of the $Q_S(V)$ characteristic in Figure 7.6. Obviously for $P_L = 0$ all three characteristics give the same maximum value $Q_{n\,\mathrm{MAX}} = E^2/4X$ corresponding to equation 7.8.

Figure 7.9b shows how the $Q_G(Q_L)$ characteristic can be used to analyse the system stability. Assuming that the reactive load demand Q_L is lower than its maximum value

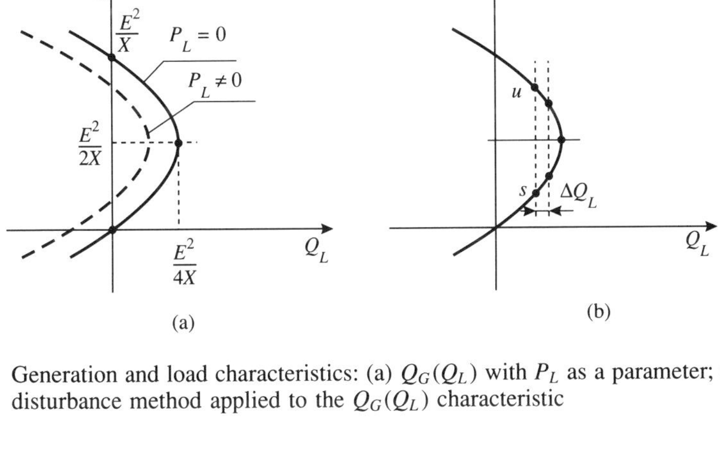

Figure 7.9 Generation and load characteristics: (a) $Q_G(Q_L)$ with P_L as a parameter; (b) the small disturbance method applied to the $Q_G(Q_L)$ characteristic

there are always two equilibrium points, i.e. two values of reactive generation corresponding to a given demand. At the lower point s, a momentary disturbance ΔQ_L that increases the load reactive power demand results in an increase in the generated reactive power whilst a disturbance that reduces the demand results in a corresponding drop in the reactive power generation. As the generation follows the demand the lower equilibrium point s is stable. The situation is reversed at the upper equilibrium point u. Here an increase in Q_L produces a reduction in Q_G whilst a reduction in Q_L produces an increase in Q_G. As the changes in reactive generation are now in the opposite direction to the changes in demand the upper equilibrium point u is unstable.

Consequently the system is stable if a small change in reactive load demand produces a change in the generation which has the same sign or, in other words, the derivative $\mathrm{d}Q_G/\mathrm{d}Q_L$ is positive:

$$\frac{\mathrm{d}Q_G}{\mathrm{d}Q_L} > 0 \tag{7.27}$$

It is worth noting that at the maximum loading point at the nose of the $Q_G(Q_L)$ characteristic the derivative $\mathrm{d}Q_G/\mathrm{d}Q_L$ tends to infinity.

The characteristic defined by equation 7.26, and shown in Figure 7.9, is only a parabola for the case of an ideal stiff real power load $P_L(V) = P_L = constant$. For a voltage-dependent load characteristic $P_L(V)$ will vary with voltage when it is not possible to obtain an explicit expression for $Q_L(Q_G)$ with $P_L(V)$ as a parameter. However the $Q_L(Q_G)$ characteristic can be obtained iteratively by solving the network equations for given values of load demand, P_L and Q_L, and emf E at the generator node.

The main advantage of the $\mathrm{d}Q_G/\mathrm{d}Q_L$ criterion is the ease with which it can be used with load-flow programs to analyse a multi-node system (Carpentier, 1984, Taylor, 1994). The generated reactive power Q_G is replaced by the sum of all the generated reactive powers at all the generator nodes, while the derivative is replaced by the quotient of the sum of the all the generated reactive power increments over the increment of the reactive load demand at the examined load node, that is $\Sigma\Delta Q_{Gi}/\Delta Q_L$.

7.3 Critical Load Demand and Voltage Collapse

Figure 7.3 showed how the network solution area depends on the shape of the load characteristic while Figure 7.7 illustrated the classical $\mathrm{d}\Delta Q/\mathrm{d}V$ stability criterion. Figure 7.10 extends this discussion and shows how the network equations may have two solutions, one solution or no solution at all depending on the relative position and shape of the $Q_L(V)$ and $Q_S(V)$ characteristics.

In Figure 7.10a there are two equilibrium points, corresponding to the intersection of the $Q_L(V)$ and $Q_S(V)$ characteristics, of which only point s is stable. For the special case of $Q_L(V) =$ constant and $P_L(V) =$ constant these two voltage values correspond to the voltages V_1 and V_2 designated by point A in Figure 7.2b, only one of which will be stable. For the more general case of $Q_L(V) \neq$ constant and $P_L(V) \neq$ constant shown in Figure 7.3b, c and d then a point equivalent to A in Figure 7.2b will exist somewhere inside the envelope with two voltage solutions depicted by the general points u and s in Figure 7.10a. If the point lies on the envelope then there is only one equilibrium point and this corresponds to the one intersection point shown in Figure 7.10b. This point is referred to as the *critical point* and its co-ordinates are referred to as the *critical power*

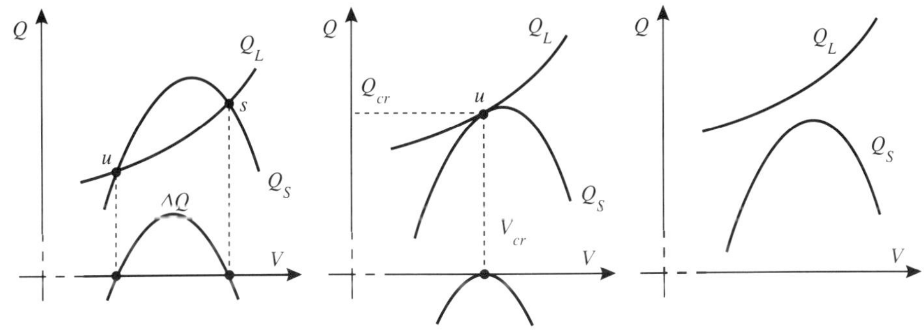

Figure 7.10 Relative position of the generation and load characteristics: (a) two equilibrium points; (b) a single, critical, equilibrium point; (c) no equilibrium points

and the *critical voltage*. Outside the network solution area in Figure 7.3 there are no equilibrium points and this corresponds to the $Q_L(V)$ and $Q_S(V)$ characteristics having no point of intersection as in Figure 7.10c where the $Q_L(V)$ characteristic lies above the $Q_S(V)$ characteristic. In general the area of network solution shown in Figure 7.3 is known as the *steady-state voltage stability area*.

It is worth remembering at this point that Figure 7.10 has been drawn for the special case of a real power stiff load $P_L(V) = P_L = constant$ so that the critical power Q_{cr}, P_{cr} is defined by the co-ordinates (Q_{cr}, P_L) and only Q_{cr} and V_{cr} are marked on Figure 7.10b.

7.3.1 Effects of increasing demand

A slow increase in the system demand, such as that due to the normal daily load variations, can have two detrimental effects on the voltage stability. According to equation 7.13 an increase in the real power lowers the $Q_S(V)$ characteristic as shown in Figure 7.6, whilst an increase in the reactive power raises the $Q_L(V)$ characteristic. As a consequence the stable equilibrium point s moves towards smaller values of voltage, whilst the unstable equilibrium point u moves towards larger values of voltage. As the demand further increases the equilibrium points move closer together until they finally merge at the critical equilibrium point shown in Figure 7.10b.

When a load operates at this critical point then any small increase in the reactive power demand will produce a deficit in the reactive power, the reactive power demand will be greater than supply, and the voltage will reduce. As the voltage reduces the deficit in reactive power increases and the voltage falls even further until it eventually falls to a very small value. This phenomenon is generally known as *voltage collapse* although in some countries the more graphic term of *voltage avalanche* is used. Two forms of voltage collapse are identified in the literature. When the voltage collapse is permanent some authors refer to it as a *total voltage collapse* (Taylor, 1994). On the other hand the term *partial voltage collapse* is used when a large increase in demand causes the voltage to fall below some technically acceptable limit. As a partial voltage collapse does not correspond to system instability it is perhaps better to consider the system as being in an emergency state as the system still operates although at a reduced voltage.

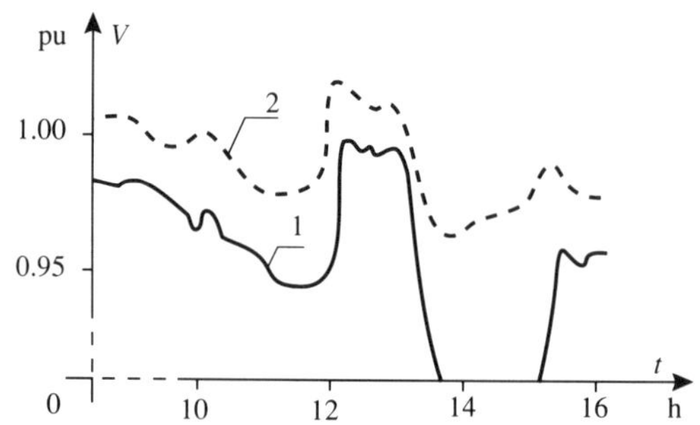

Figure 7.11 An example of the voltage collapse (Nago, 1975). *1* — voltage variations during the day of the voltage collapse, *2* — voltage variations during the previous day

An example of an actual voltage collapse is shown in Figure 7.11 (Nago 1975). In this figure curve *2* represents a typical morning period when the voltage drops slightly as the power demand increases but then recovers by a few percent during the lunch break as the power demand reduces. At about 13.00 hrs, after the lunch break, the power demand again builds up and the voltage drops. However on one particular day, curve *1*, the overall system load is greater, and the local voltage smaller, than normal so that as the load starts to increase after the lunch break, the voltage reaches its critical value and then collapses. The system operators then intervene and, after a long interruption, manually restore the normal operating conditions.

From the power system security point of view a knowledge of the critical power and voltage is very important as the operating voltage, and power, at the system nodes should be kept as far away as possible from their critical values. Unfortunately the non-linearity of the voltage characteristics make it impossible to derive a general formula for the critical voltage that is valid for any type of active and reactive power variations even assuming a simple power system model. However some simple iterative formula can be developed if the following three assumptions are made:

1. That when the load demand increases, the power factor is maintained constant by reactive power compensation in the consumer load so that

$$\frac{P_n(t)}{P_0} = \frac{Q_n(t)}{Q_0} = \xi \tag{7.28}$$

   where $P_n(t)$ and $Q_n(t)$ are the nominal values of the load demand at the time instant t, P_0 and Q_0 are the initial values of the demand and ξ is the coefficient of demand increase.

2. That the power characteristic of an industrial composite load, with a large number of induction motors, can be approximated by the polynomials (see sub-Section 3.4.4)

$$\frac{Q_L}{Q_n} = a_2\left(\frac{V}{V_n}\right)^2 - a_1\left(\frac{V}{V_n}\right) + a_0, \qquad \frac{P_L}{P_n} = b_1\left(\frac{V}{V_n}\right) \tag{7.29}$$

   that is by a parabola and a straight line respectively.

3. That the load composition is constant thereby making the coefficients a_0, a_1, a_2 and b_1 constant.

Equation 7.28 allows the characteristics defined in equation 7.29 to be converted to the following form

$$Q_L = \xi[\alpha_2 V^2 - \alpha_1 V + \alpha_0], \qquad P_L = \xi\beta_1 V \tag{7.30}$$

where

$$\alpha_2 = \frac{Q_0}{V_n^2}a_2, \qquad \alpha_1 = \frac{Q_0}{V_n}a_1, \qquad \alpha_0 = Q_0 a_0, \qquad \beta_1 = \frac{P_0}{V_n}b_1$$

Substituting the second of the equations in 7.30 into equation 7.13 gives

$$Q_S(V) = V\sqrt{\left(\frac{E}{X}\right)^2 - \xi^2\beta_1^2} - \frac{V^2}{X} \tag{7.31}$$

Figure 7.10b shows how, at the critical point, the supply characteristic and the demand power characteristic are tangent to each other which allows the value of the critical voltage V_{cr} to be found by solving the following two equations

$$Q_S(V_{cr}) = Q_L(V_{cr}) \tag{7.32}$$

$$\left.\frac{dQ_S}{dV}\right|_{V=V_{cr}} = \left.\frac{dQ_L}{dV}\right|_{V=V_{cr}} \tag{7.33}$$

Substituting for Q_S and Q_L from equations (7.30) and (7.31) gives

$$V_{cr}\sqrt{\left(\frac{E}{X}\right)^2 - \xi_{cr}^2\beta_1^2} - \frac{V_{cr}^2}{X} = \xi_{cr}(\alpha_2 V_{cr}^2 - \alpha_1 V_{cr} + \alpha_0) \tag{7.34}$$

$$\sqrt{\left(\frac{E}{X}\right)^2 - \xi_{cr}^2\beta_1^2} - 2\frac{V_{cr}}{X} = \xi_{cr}\ (2\alpha_2 V_{cr} - \alpha_1) \tag{7.35}$$

Equation 7.35 determines the critical voltage as a function of the system parameters and an unknown coefficient ξ_{cr} of the demand increase as

$$V_{cr} = \frac{\sqrt{\left(\frac{E}{\beta_1 X}\right)^2 - \xi_{cr}^2} + \frac{\alpha_1}{\beta_1}\xi_{cr}}{2\frac{\alpha_2}{\beta_1}\xi_{cr} + \frac{2}{\beta_1 X}} \tag{7.36}$$

Multiplying equation 7.35 by $(-V_{cr})$ and adding the result to equation 7.34 gives

$$\frac{V_{cr}^2}{X} = \xi_{cr}[\alpha_0 - \alpha_2 V_{cr}^2] \tag{7.37}$$

from which finally

$$\xi_{cr} = \frac{1}{\frac{\alpha_0 X}{V_{cr}^2} - \alpha_2 X} \tag{7.38}$$

Equations 7.36 and 7.38 can now be used in an iterative calculation to find the critical voltage V_{cr} and the critical demand increase ξ_{cr}.

Example 7.1

A composite load of rating $S_n = 150 + j100$ MVA and $V_n = 110$ kV is supplied from the network by a 130 km, 220 kV, transmission line with a reactance of 0.4 Ω/km via two parallel 160 MVA transformers, each with a reactance of 0.132 p.u. Assume that the system can be replaced by a 251 kV voltage source with a reactance of 13Ω and that the load characteristics in kV and MVA are: $P_L = 0.682\xi\ V$ and $Q_L = \xi\ (0.122V^2 - 4.318V + 460)$. Find the critical voltage and the critical coefficient of the load demand increase.

Simple calculation leads to the parameter values shown in Figure 7.12. Equations 7.36 and 7.38 then give

$$V_{cr} = \frac{\sqrt{18.747 - \xi_{cr}^2} + 6.331\xi_{cr}}{0.03578\xi_{cr} + 0.0345}; \quad \xi_{cr} = \frac{1}{\dfrac{39100}{V_{cr}^2} - 1.037} \tag{7.39}$$

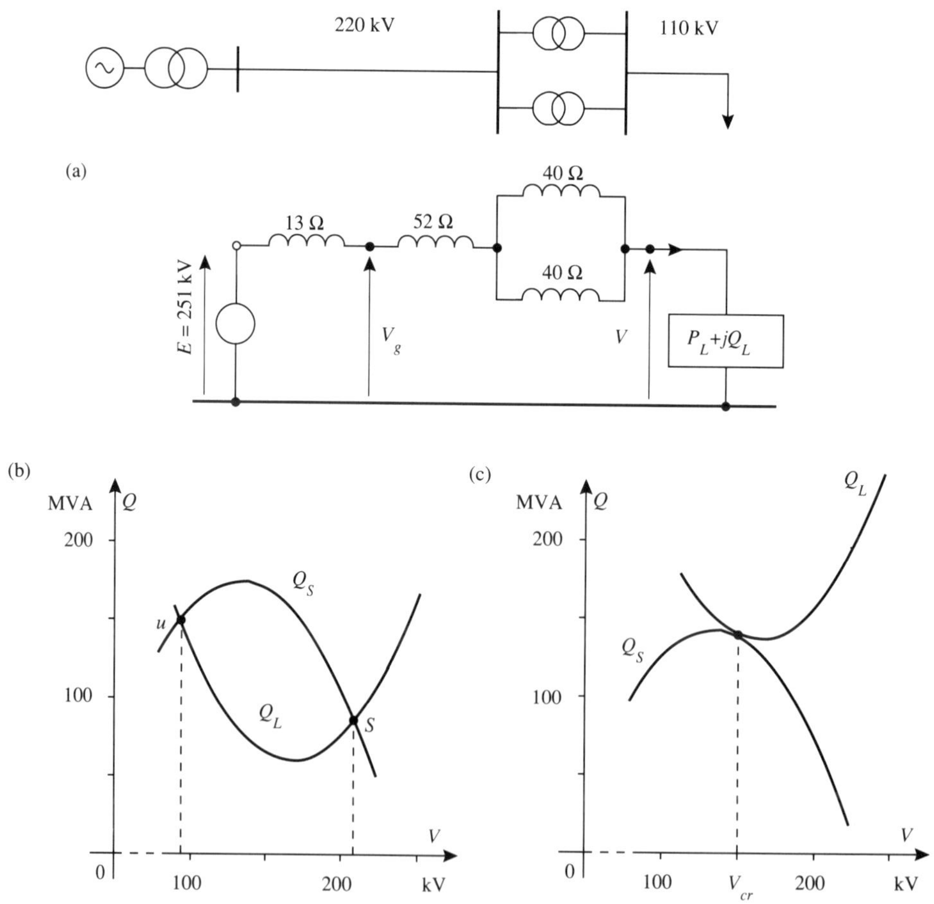

Figure 7.12 Illustration to Example 7.1: (a) equivalent circuit with parameters calculated at 220 kV; (b) initial characteristics; (c) characteristics at the critical state

The initial substitution of $\xi_{cr} = 1$ into the first of these equations gives $V_{cr} = 151.69$. Substituting this value to the second equation gives $\xi_{cr} = 1.51$. After 5 such iterations, the final solution of $\xi_{cr} = 1.66$ and $V_{cr} = 154.5$ kV is obtained indicating that the system would lose stability if the demand increased 1.66 times. The primary voltage $V_{cr} = 154.5$ kV referred to the transformer secondary winding gives 77.25 kV, that is 70% of the rated voltage of 110 kV. Figure 7.13b, c shows the system and the load characteristic in the initial and the critical states. Initially the system operated at point s at a voltage of 208 kV, or 104 kV at the transformer secondary, that is at 94.5% of the rated voltage.

7.3.2 Effect of network outages

The relative position of the supply characteristic defined in equation 7.13 depends on the equivalent system reactance. Large changes in this reactance, such as those caused by network outages, may lower the generation characteristic and cause voltage stability problems.

Example 7.2

A composite load of $S_n = (240 + j160)$ MVA and $V_n = 110$ kV is supplied from the network by a double circuit transmission line of length 130 km and reactance 0.4 Ω/km via two parallel 160 MVA transformers each with a reactance of 0.132 p.u. The power system may be represented by an emf of 251 kV acting behind an equivalent reactance of 13Ω. The load characteristics are: $P_L = 1.09V$ and $Q_L = 0.0195V^2 - 6.9V + 736$. Check the system stability after one of the lines is tripped.

Figure 7.13 shows the load characteristic and the generation characteristic before, Q'_S, and after, Q''_S, the line is tripped. Tripping the line causes a reduction in the load voltage to about 170 kV, or 85 kV on the secondaries of the transformers, that is to about 77% of the rated voltage. The system is on the verge of losing stability. Calculations similar

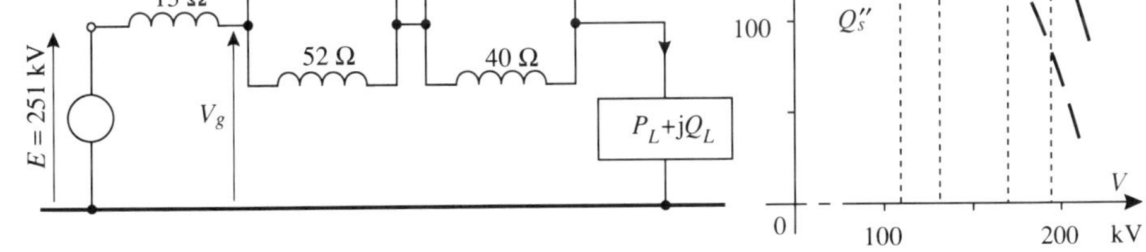

(a) (b)

Figure 7.13 Illustration to Example 7.2: (a) equivalent circuit with parameters calculated at 220 kV; (b) voltage characteristics

to those in Example 7.1 show that a demand increase by 3.8% will result in the system losing stability. Due to the unacceptably low voltage the system is now in an emergency state with a high risk of instability.

7.3.3 Influence of the shape of the load characteristics

Figure 7.6 and equation 7.13 demonstrate how an increase in the real power demand shifts the vertex of the reactive power supply characteristic downwards. This observation has important implications with regards the voltage stability of systems where the real power varies with voltage. For example if the real power demand reduces with reducing voltage then the stability of the system will be improved compared to when the real power characteristics is independent of voltage. This situation is examined further in Example 7.3.

Example 7.3

For the system in Example 7.2 determine the supply characteristics after one of the transmission lines is tripped assuming that the real power characteristics of the load (in MW) are as follows:

(1)$P_L = 240 =$ constant, (2)$P_L = 16.18\sqrt{V}$, (3)$P_L = 1.09\ V$, (4)$P_L = 0.004859\ V^2$.

The four real power characteristics are shown graphically in Figure 7.14a with curve *1* representing an ideal stiff load. Curves *2*, *3* and *4* represent loads of decreasing stiffness. The corresponding reactive power supply characteristics can be easily determined using equation 7.13 and are shown in Figure 7.14b. For the ideal stiff load, curve *1*, the system has no equilibrium points and line tripping results in immediate voltage collapse. The slightly less stiff load *2* is in the critical state and will also lose stability should the line trip. Load *3* is close to the critical state and will lose stability when the demand increases by 3.8% (see Example 7.2) whilst load *4* is quite flexible and operates with a greater safety margin. In this case stability will be lost if the demand increases by about 15%.

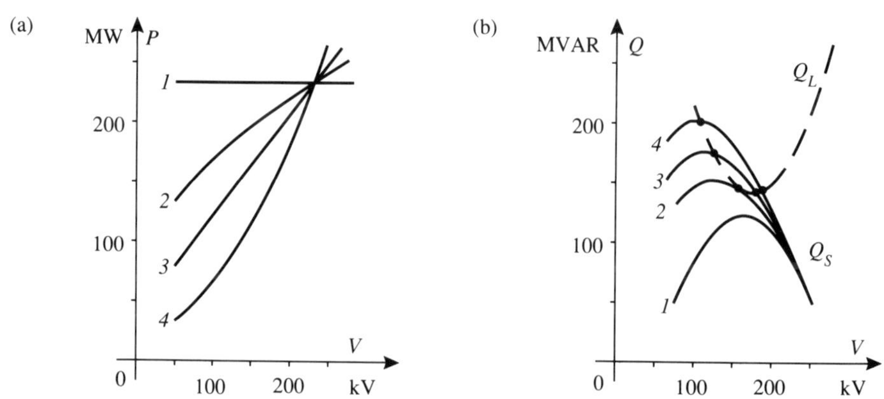

Figure 7.14 Real and reactive power characteristics of Example 7.3

Absolutely stable loads

Figure 7.3 showed how the network solution area depends on the shape of the load characteristic with Figure 7.3d corresponding to a constant-admittance load. The equivalent circuit for such a load is shown in Figure 7.4 and Section 7.1 explained how for a load with such a characteristic every equilibrium point for $V > 0$ is stable and the system cannot suffer voltage collapse. This can also be proved using the classical dQ/dV criterion.

For a constant admittance load the $P_L(V)$ characteristic is given by equation 7.9 which, when substituted into equation 7.13, gives the reactive power supply characteristic as

$$Q_S = V\left[\sqrt{\left(\frac{E}{X}\right)^2 - (G_n V)^2} - \frac{V}{X}\right] \tag{7.40}$$

The resulting characteristic is drawn in Figure 7.15 and shows for any value of the load conductance G_n the supply characteristic always crosses the origin at $V = 0$. Again from equation 7.9 the reactive power demand is given as $Q_L = B_n V^2$ so that the origin is always one of the equilibrium points for any value of the load susceptance B_n. As the load susceptance increases, the parabola defined by $Q_L = B_n V^2$, becomes steeper but the stability condition 7.14 is satisfied for all equilibrium points when $V > 0$. This is true even for the equilibrium point $S_{(4)}$ on the left-hand-side of the supply parabola. The constant-admittance load is therefore absolutely stable as no value of Y_L can cause the voltage to collapse although a large value of the load admittance Y_L will result in an inadmissible low operating voltage.

7.3.4 Influence of the voltage control

Section 7.1 showed that the reactance of the equivalent system generator, and therefore the point of constant voltage in the network model, depends on the voltage control capability

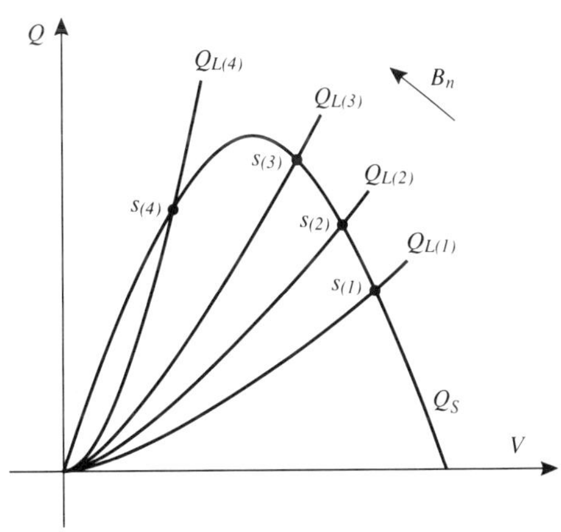

Figure 7.15 A family of reactive demand characteristics for a fixed load conductance G_n and increased load susceptance B_n

of the generator. Provided that neither the excitation limits or the stator current limit are exceeded then the AVR will keep the generator terminal voltage constant and there is no need to add the generator synchronous reactance X_d to the network equivalent reactance X in Figure 7.1. However if any of the generator current limits are met then the field voltage will be maintained constant and the generator must be modelled by and emf E_f behind X_d. The modelling reactance must now be increased to include the effect of X_d. Consequently as the equivalent system reactance in the generator-line-load model depends significantly on the voltage control capability of the generator so to will the voltage stability.

Both the area of network feasibility, Section 7.1, and the characteristics used to analyse the voltage stability, Section 7.2, depend on the equivalent reactance of the system. This reactance occurs in two terms in equation 7.13, in the $-V^2/X$ term and in the EV/X term both of which determine the steepness of the parabola and the position of its vertex. If X is large then the supply characteristic parabola is shallow and its vertex is lowered as shown in Figure 7.6. This results in a reduction in the maximum reactive power and hence the critical power.

Example 7.4

For the system in Example 7.2 determine the supply characteristic when the system is operating with only one transmission line. Assume that the equivalent system generator is capable of keeping the voltage at the sending-end of the line constant.

Before the line is tripped the load operated at 208 kV (referred to the primaries of the transformers). Figure 7.13 shows the voltage drop across the source reactance gives the terminal voltage $V_g = 245$ kV. Assuming that the AVR keeps this voltage constant, the system can now be represented by a voltage $E = V_g = 245 \text{ kV} = \text{constant}$ with zero internal reactance. The reactance of one line, and two parallel transformers, give a total $X = 52 + 40/2 = 72\ \Omega$. Figure 7.16 shows that the supply characteristic is now much higher compared to when voltage control was neglected. Calculations similar to those in Example 7.1 show that the system would lose stability when the demand increases by about 12%. Neglecting voltage control gave the critical demand increase as 3.8%.

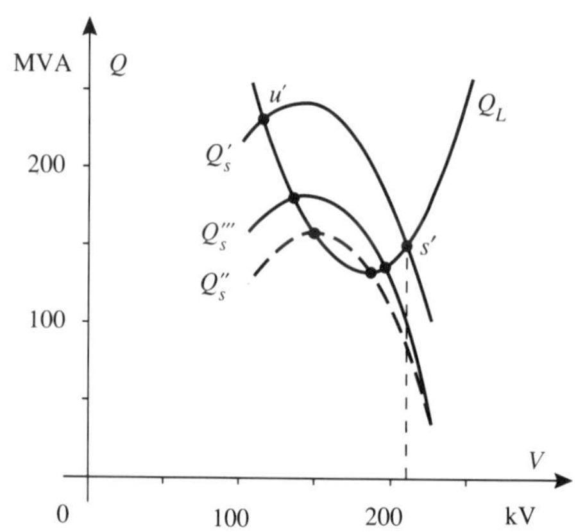

Figure 7.16 Generation and load characteristics analysed in Example 7.4: Q'_S — before tripping the line; Q''_S — after tripping with $E =$ constant; Q'''_S — after tripping with $V_g =$ constant

7.4 Voltage Proximity Indices

For network planning purposes the critical load power can be calculated off-line using a modified load-flow computer program and the $\mathrm{d}Q/\mathrm{d}V$ or $\mathrm{d}Q_G/\mathrm{d}Q_L$ stability criterion (Van Cutsem, 1991, Ajjarapu *et al* 1992, Taylor 1994). The procedure is similar to that described in Section 7.3 for the simple generator-line-load system and the idea is to bring the system to a critical state by increasing the power demand of a chosen load or a group of loads. The use of a load-flow program allows a variety of effects such as generator voltage control, reactive power compensation, the distribution of the real power among the generation units with active governors, the voltage characteristics of the loads etc. to be included Unfortunately the method is computationally time consuming and cannot be used for on-line applications, such as steady-state security assessment, which require a fast assessment of the voltage stability conditions and an estimation of how far a given operating point is from the critical state. The distance from the critical state is usually quantified by one of the so-called *voltage stability indices* (Kessel *et al*, 1986, Tiranuchit *et al*, 1987, Löf *et al*, 1992 and many others). A number of different indices are discussed by Taylor (1994).

A voltage stability index based on the classical $\mathrm{d}Q/\mathrm{d}V$ criterion can be constructed by observing that as the load demand gets closer to the critical value, both the equilibrium points shown in Figure 7.10 move towards each other until they become one unstable point. As shown in Figure 7.7b there is always a point between the equilibrium points, of voltage V_x, such that

$$\left.\frac{d(Q_S - Q_L)}{\mathrm{d}V}\right|_{V=V_x} = 0 \tag{7.41}$$

As the power demand of the composite load increases, the voltage V_x tends towards the critical voltage V_{cr}. A voltage proximity index can be therefore defined as (Venikov, 1968)

$$k_v = \frac{V_s - V_x}{V_s} \tag{7.42}$$

where V_x must satisfy equation 7.41. In practice calculating the proximity index defined by equation 7.42 is quite cumbersome as it requires a fragment of the generation characteristic $Q_S(V)$ to be determined by means of a load flow program. An alternative proximity index that is easier to determine is the value of the derivative

$$k_{\Delta Q} = \frac{\mathrm{d}(Q_S - Q_L)}{\mathrm{d}V}\frac{V_s}{Q_s} \tag{7.43}$$

calculated near a given equilibrium point. As the load demand tends towards the critical value, the index in equation 7.43 tends towards zero.

Yet another voltage proximity index can be derived directly from the $\mathrm{d}Q_G/\mathrm{d}Q_L$ criterion as

$$k_Q = \frac{\mathrm{d}Q_G}{\mathrm{d}Q_L} \tag{7.44}$$

When the network is lightly loaded the reactive power absorbed by the network is small so that the increment in the generation caused by the increment in the load demand is almost equal to the increment in the load demand itself and the index defined in equation 7.44

is near unity. When the system approaches the critical state, the index tends to infinity, Figure 7.9. The value of this index can be calculated for a multi-node system using a load-flow program.

Currently the problem of voltage stability is a rapidly developing research area with many new papers appearing either proposing new proximity indices, improved methods of calculation, or suggesting how proximity indices can be identified on-line using local measurements.

7.5 The Dynamics of Voltage Collapse

The voltage stability analysis presented so far in this chapter has assumed that the loads can be represented by their static voltage characteristics. Although this is useful to help understand the principles of voltage stability, and its relation to the network feasibility problem, the static load voltage characteristics can only approximate the real behaviour of the composite load for slow voltage variations. In practice the actual behaviour of both the composite load and the system is a tightly coupled dynamic process that is influenced by the load dynamics, especially induction motor dynamics, the automatic voltage and frequency control equipment, and by the operation of the protection systems. Any, or all, of these factors may speed-up, slow-down, or even prevent voltage collapse. To complement the static considerations of the previous sections of this Chapter a few examples of some typical voltage collapse scenarios are now briefly discussed (Taylor 1994, Bourgin *et al* 1993).

Scenario 1: Load build up

Section 7.3 explained how voltage collapse can result from a very large build up of load, particularly during periods of heavy demand, so that the power demand exceeds the critical value as determined by the network parameters. In this scenario the main factors contributing to voltage collapse are:

- the stiffness of the load characteristics continuing to demand high values of active and reactive power despite voltage dips in the load area. As Chapter 3 explained induction motors are mainly responsible for producing a stiff load characteristic.
- the control of tap-changing transformers in distribution and subtransmission networks maintaining constant voltage, and therefore high active and reactive power demand, when the supply voltage dips. High demand is undesirable in the emergency state.
- the limited ability for reactive power control by the generators. Due to field and armature current limits a high reactive power demand by the system loads may cause the generators to lose their ability to act as a constant voltage source. The generator then behaves like a voltage source behind the synchronous reactance and its terminal voltage reduces.

A voltage collapse due to load build up may be caused by some, or all of, the above factors. The dynamics of the various voltage control devices (generators, compensaters, transformers) may interact in such a way that the actual voltage collapse is different to that predicted by static considerations.

Scenario 2: Network Outages

As seen in previous sections of this chapter the network parameters play a crucial role in determining the maximum power that can be delivered to the load areas. Tripping one of the lines in the power grid increases the equivalent reactance between the equivalent voltage source and the load, reduces the critical power and increases the probability of voltage collapse. Generator tripping has a similar effect in that it not only increases the equivalent system reactance but also reduces the systems capability to generate real and reactive power. The dynamics of the various voltage control devices in the system can again influence the actual scenario.

Scenario 3: Phenomena inside the composite load

As already mentioned the stiffness of the load characteristics is one of the dominant factors contributing to voltage collapse. However the dynamic response of the composite load may result in the dynamic and static load characteristics being different. This difference is mainly attributed to induction motors and may result in a reduction in the system stability ultimately leading to voltage collapse. For example, a rapid, severe, voltage dip, such as that which occurs during a slowly-cleared short-circuit, can cause a reduction in the motor torque and consequent motor stalling. As shown in Section 3.4 a stalled induction motor demands reactive power further reducing the voltage stability conditions. This may lead to other, nearby, motors stalling. In this scenario the voltage continues to fall until the protection equipment, or the electromechanically-held contactors, trip the motors from the system thereby reducing the reactive power demand. The voltage will then start to recover but an uncontrolled restoration of the composite load by, for example, heavy induction motor self starts, can again reduce the voltage and lead to a total voltage collapse.

Scenario 4: Voltage collapse and asynchronous operation

Voltage collapse at one, or a few, of the network nodes may cause the voltage to dip at neighbouring nodes leading to voltage collapse at these nodes. The voltage then dips

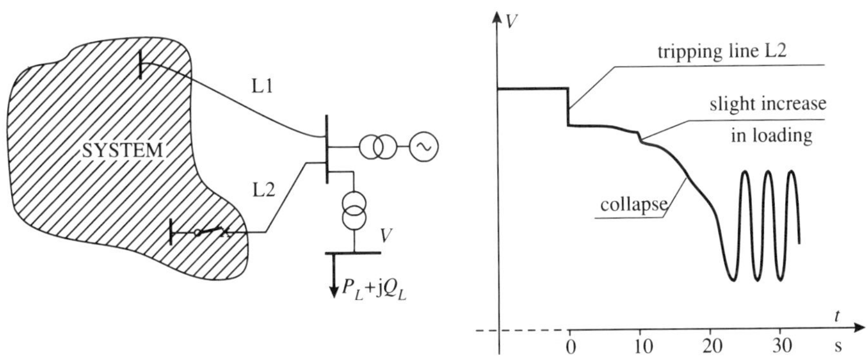

Figure 7.17 Loss of synchronism caused by the voltage collapse: (a) schematic of the system; (b) voltage variations. Based on (Venikov, 1978)

at other nodes so propagating throughout the network and affecting the synchronous generators. If the affected generators are weakly connected with the system they may lose synchronism.

A voltage dip, accompanied by a reduction in the real power demand and an increase in the reactive power demand, has a similar effect on the synchronous generator as a short-circuit in the network. An example of just such behaviour is shown in Figure 7.17 where the tripping of line *L*2 results in the generator and the load operating through one, quite long, line. The load is in the critical state and a small increase in the load demand results in voltage collapse and the generator losing synchronism with the rest of the system. The load voltage then undergoes periodic voltage variations characteristic of asynchronous operation. The asynchronously operating generator must now be tripped from the system which further deteriorates the situation at the load node leading to an eventual total voltage collapse.

8
Frequency Variations

Previous chapters have described how the generators in a power system respond when there is a momentary disturbance in the power balance between the electrical power consumed in the system and the mechanical power delivered by the turbines. Such disturbances are caused by short-circuits in the transmission network and are normally cleared without the need to reduce the generated or consumed power. However, if a large load is suddenly connected to (or disconnected from) the system, or if a generating unit is suddenly disconnected by the protection equipment, there will be a long-term distortion in the power balance between that delivered by the turbines and that consumed by the loads and, as a result, the frequency in the system will change. This frequency change can be conveniently divided into a number of stages allowing the dynamics associated with each of these stages to be described separately. This helps illustrate how the different dynamics develop in the system. However it is first necessary to describe the operation of the automatic generation control (AGC) as this is fundamental in determining the way in which the frequency will change in response to a change in load.

It is worth adding that Section 8.1 refers to traditional vertically-integrated utilities which control generation, transmission and often distribution in their own service area and where power interchanges between utilities are presumed to be prescheduled and well defined. Over the last decade a new trend has emerged where utilities make their transmission networks available for *third party access* to allow power interchange between third parties (so-called *power wheeling*). This coincides with the development of *spot markets* for electricity in many countries where utilities have no direct control over power stations in their service area and have to purchase energy and control functions (often referred to as *ancillary services*) from individual power stations. These new developments, known under the generic name of *deregulation,* will require new control paradigms which are yet to emerge (Wood and Wollenberg, 1996, Christie and Bose, 1996, Bialek, 1996).

8.1 Automatic Generation Control

Section 2.1 explained how an electrical power system consists of many generating units and many loads while its total power demand varies continuously throughout the day in a more or less anticipated manner. The large, slow, changes in demand are met centrally by deciding at regular intervals which generating units will be operating, shut-down, or in an intermediate hot reserve state. This process of *unit commitment* may be conducted once per day to give the daily operating schedule whilst at shorter intervals, typically every 30 minutes, *economic dispatch* determines the actual power output required from each of the committed generators. Smaller, but faster, load changes are dealt with by automatic generation control so as to

- maintain frequency at the scheduled value (frequency control)
- maintain the net power interchanges with neighbouring control areas at their scheduled values (tie-line control)
- maintain power allocation among the units in accordance with area dispatching needs (economic, security or emergency)

In some systems the role of automatic generation control (AGC) may be restricted to one or two of the above objectives. For example, tie-line power control is only used where a number of separate power systems are interconnected and operate under mutually beneficial contractual agreements.

8.1.1 Generation characteristic

Section 2.2.3 discussed operation of turbines and their governing systems. In the steady-state the idealised power-speed characteristic of a single generating unit is given by equation 2.3. As the rotational speed is proportional to frequency, equation 2.3 may be re-written for the i-th generating unit as

$$\frac{\Delta f}{f_n} = -\rho_i \frac{\Delta P_{mi}}{P_{ni}}, \qquad \frac{\Delta P_{mi}}{P_{ni}} = -K_i \frac{\Delta f}{f_n} \tag{8.1}$$

In the steady-state all the generating units operate synchronously at the same frequency when the overall change in the total power ΔP_T generated in the system can be calculated as the sum of changes at all generators:

$$\Delta P_T = \sum_{i=1}^{N_G} \Delta P_{mi} = -\frac{\Delta f}{f_n} \sum_{i=1}^{N_G} K_i P_{ni} = -\Delta f \sum_{i=1}^{N_G} \frac{K_i P_{ni}}{f_n} \tag{8.2}$$

where N_G is the number of generating units in the system. The subscript 'T' indicates that ΔP_T is the change in generated power as supplied by the turbines. Figure 8.1 illustrates how the characteristics of individual generating units can be added according to equation 8.2 to obtain the equivalent *generation characteristic*. This characteristic defines the ability of the system to compensate for a power imbalance at the cost of a deviation in frequency. For a power system with a large number of generating units, the generation characteristic is almost horizontal such that even a relatively large power change only results in a very small frequency deviation. This is one of the benefits accruing from combining generating units into one large system.

To obtain the equivalent generation characteristic of Figure 8.1 it has been assumed that the speed-droop characteristic of the individual turbine generator units is linear over the full range of power and frequency variations. In practice the output power of each

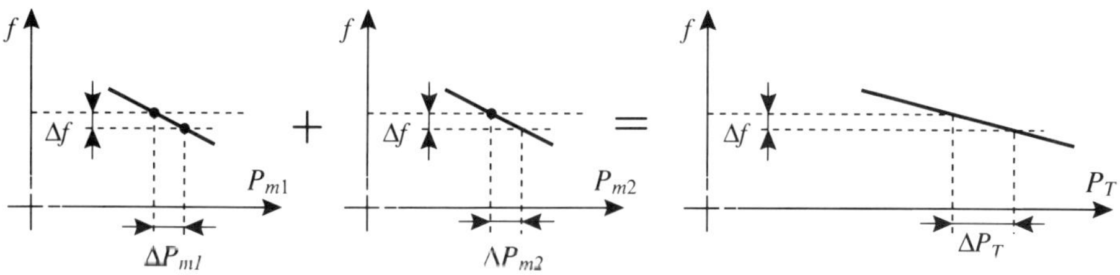

Figure 8.1 Generation characteristic as the sum of the speed-droop characteristics of all the generation units

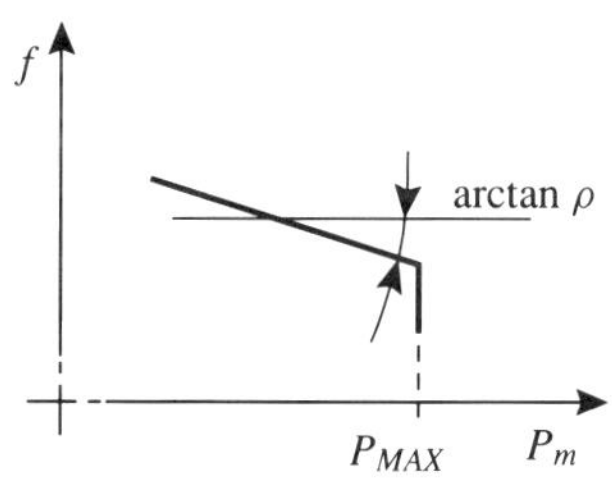

Figure 8.2 Speed-droop characteristic of a turbine with an upper limit

turbine is limited by its technical parameters. For example, coal-burn steam turbines have a lower power limit due to the need to maintain operational stability of the burners and an upper power limit that is set by thermal and mechanical considerations. In the remainder of this section only the upper limit will be considered when the turbine characteristic will be as shown in Figure 8.2.

If a turbine is operating at its upper power limit then a decrease in the system frequency will not produce a corresponding increase in its power output. At the limit $\rho = \infty$ or $K = 0$ and the turbine does not contribute to the equivalent system characteristic. Consequently the generation characteristic of the system will be dependent on the number of units operating away from their limit at part load, that is it will depend on the *spinning reserve*, where spinning reserve is the difference between the sum of the power rating of all the operating units and their actual load. The allocation of spinning reserve is an important factor in power system operation as it determines the shape of the generation characteristic. This is demonstrated in Figure 8.3 which shows a simple case of two generating units. In Figure 8.3a the spinning reserve is allocated proportionally to both units so that they both reach their upper power limit at the same frequency f_1. In this case the equivalent characteristic, obtained from equation 8.2, is linear until the upper power limit for the whole system is reached. In Figure 8.3b the same amount of spinning reserve is available but is allocated solely to the second generator with the first unit operating at full load. The generation characteristic is now non-linear and consists of two sections of different slope.

Similarly, the generation characteristic of an actual power system is non-linear and consists of many short sections of increasing slope as more and more generating units reach their operating limit as the total load increases until, at the maximum system load, there is no more spinning reserve available. The generation characteristic then becomes a vertical line. For small power and frequency disturbances it is convenient to approximate

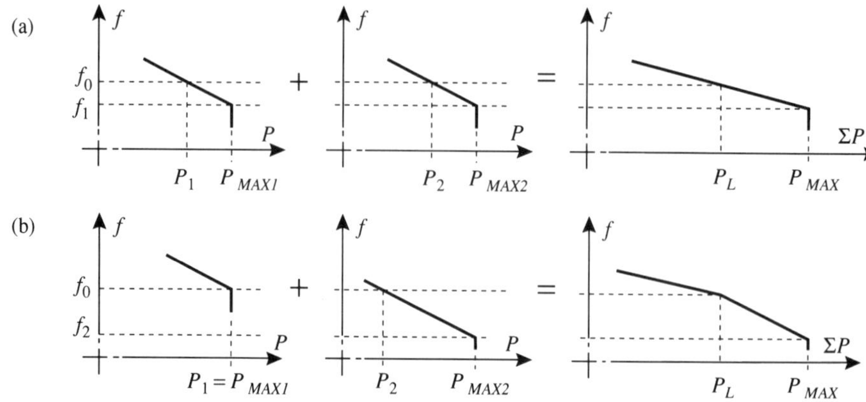

Figure 8.3 Influence of the turbine upper power limit and the spinning reserve allocation on the generation characteristic (a) proportional allocation of spinning reserve; (b) allocation of spinning reserve to the second unit only

this non-linear generation characteristic in the vicinity of the operation point by a linear characteristic with a local droop value. The total system generation is equal to the total system load P_L (including transmission losses):

$$\sum_{i=1}^{N_G} P_{mi} = P_L \tag{8.3}$$

where N_G is the number of the generating units. Dividing equation 8.2 by P_L gives:

$$\frac{\Delta P_T}{P_L} = -K_T \frac{\Delta f}{f_n} \quad \text{or} \quad \frac{\Delta f}{f_n} = -\rho_T \frac{\Delta P_T}{P_L} \tag{8.4}$$

where

$$K_T = \frac{\sum_{i=1}^{N_G} K_i P_{ni}}{P_L}, \qquad \rho_T = \frac{1}{K_T} \tag{8.5}$$

Equation 8.4 describes the linear approximation of the generation characteristic calculated for a given total system demand. Consequently the coefficients in equation 8.5 are calculated with respect to the total demand, not the sum of the power ratings, so that ρ_T is the local speed droop of the generation characteristic and depends on the spinning reserve and its allocation in the system as shown in Figure 8.3.

8.1.2 Frequency control

When the total generation is equal to the total system demand (including losses) then the frequency is constant, the system is in equilibrium, and the generation characteristic is approximated by equation 8.4. However, as discussed in Section 3.4, system loads are also frequency-dependent and an expression similar to equation 8.4 can be used to obtain a linear approximation of the frequency response characteristic of the total system load as

$$\frac{\Delta P_L}{P_L} = K_L \frac{\Delta f}{f_n} \tag{8.6}$$

where K_L is the *frequency sensitivity coefficient of the power demand.* A similar coefficient k_{Pf} was defined in equation 3.90 and referred to as the frequency sensitivity of a single composite load and should not be confused with K_L which relates to the total system demand. Tests conducted on actual systems indicate that the generation response characteristic is much more frequency dependent than the demand response characteristic. Typically K_L is between 0.5 and 3 (see Table 3.3) whilst $K_T \approx 20(\rho = 0.05)$. In equations 8.4 and 8.6 the coefficients K_T and K_L have opposite sign so that an increase in frequency corresponds to a drop in generation and an increase in electrical load.

On the (P, f) plane the intersection between the generation and the load characteristic equations 8.4 and 8.6 defines the system equilibrium point. A change in the total power demand ΔP_{demand} corresponds to a shift of the load characteristic in the way shown in Figure 8.4 so that the equilibrium point is moved from point *1* to point *2*. The increase in the system load is compensated for in two ways. Firstly by the turbines increasing the generation by ΔP_T and secondly by the system loads reducing the demand by ΔP_L from that required at point *3* to that required at point *2*. Figure 8.4 shows that taking both increments into account gives

$$\Delta P_{demand} = \Delta P_T - \Delta P_L = -(K_T + K_L) P_L \frac{\Delta f}{f_n} = -K_f P_L \frac{\Delta f}{f_n} \tag{8.7}$$

The new value of the system frequency f_2 is lower than the old one f_1. Equation 8.7 represents the frequency response of the system whilst the coefficient $K_f = K_T + K_L$ is referred to as the *stiffness*. It should be emphasised that $\Delta P_T \gg \Delta P_L$.

If the turbine-generators are equipped with governing systems, such as those described in sub-Section 2.2.3, then, following a change in the total power demand, the system will not be able to return to the initial frequency unaided. According to Figure 8.4 in order to return to the initial frequency the generation characteristic must be shifted to the position shown by the dashed line. Such a shift can be enforced by changing the P_{ref} setting in the turbine governing system (the load reference set point in Figure 2.13). As shown

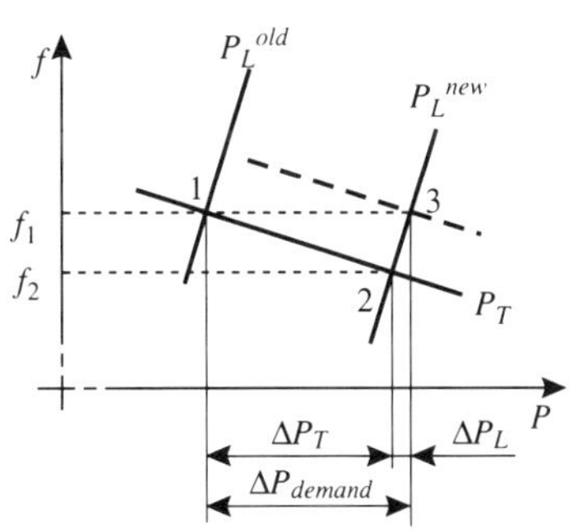

Figure 8.4 Equilibrium points for an increase in the power demand

in Figure 8.5 changes in the settings $P_{ref(1)}$, $P_{ref(2)}$ and $P_{ref(3)}$ enforce a corresponding shift of the characteristic to the positions $P_{m(1)}$, $P_{m(2)}$ and $P_{m(3)}$. Obviously no change of settings can force a turbine to exceed its maximum power rating P_{MAX}. The more turbines that have their P_{ref} setting changed, the more the generation characteristic of the system will be shifted. Eventually this may lead to a restoration of the frequency but now at the required increased value of power demand. Such control action active on the governing systems of many turbines is referred to as *frequency control.*

In an isolated power system automatic frequency control may be implemented as a *decentralised control* function by adding a supplementary control loop to the turbine-governor system. This modifies the block diagram of the turbine governor, Figure 2.13, to that shown in Figure 8.6 where P_{ref} and P_m are expressed as a fraction of the nominal or rated power P_n. The supplementary control loop, shown by the dashed line, consists of an integrating element which adds a control signal ΔP_ω that is proportional to the integral of the speed (or frequency) error to the load reference point. This signal modifies the value of the setting in the P_{ref} circuit thereby shifting the speed-droop characteristic in the way shown in Figure 8.5.

Not all the generating units in a system that implements decentralised control need be equipped with supplementary loops and participate in frequency control. Usually medium-sized units are used for frequency regulation while large, base load, units are independent and are set to operate at a prescribed generation level. In combined-cycle gas and steam turbine power plants the supplementary control may affect only the gas turbine or both the steam and the gas turbine.

In an interconnected power system comprising a number of different control areas, frequency control cannot be decentralised because the supplementary control loops have

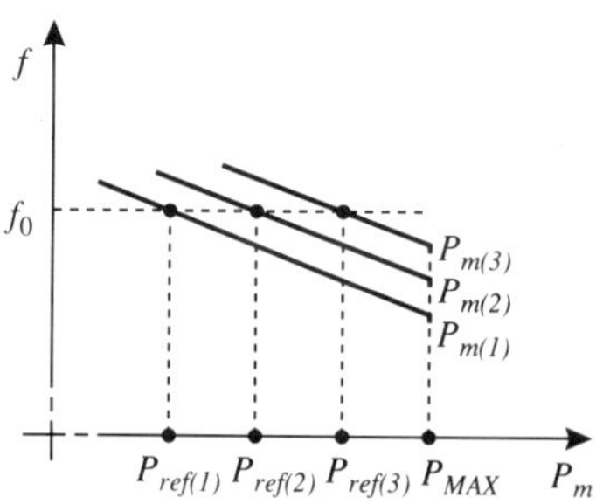

Figure 8.5 Turbine speed-droop characteristics for various settings of P_{ref}

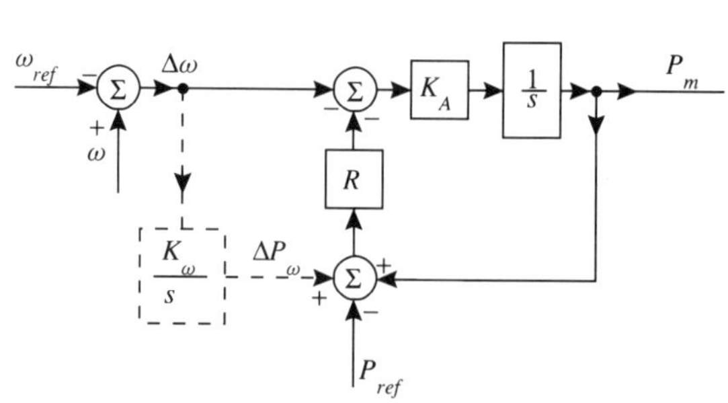

Figure 8.6 Supplementary control added to the turbine governing system

no information as to where the power imbalance occurs so that a change in the power demand in one area would result in regulator action in all the other areas. Such decentralised control action would cause undesirable changes in the power flows in the tie-lines linking the systems and the consequent violation of the contracts between the co-operating systems. To avoid this a centralised frequency control system is used as discussed in the next sub-section.

8.1.3 Tie-line control

In interconnected power systems automatic generation control is implemented in such a way that each area, or subsystem, has its own central regulator. As shown in Figure 8.7, the power system is in equilibrium if, for each area, the total power generation P_T, the total power demand P_L and the net tie-line interchange power P_{tie} satisfy the condition

$$P_T - (P_L + P_{tie}) = 0 \tag{8.8}$$

The objective of each area regulator is to maintain frequency at the scheduled level (frequency control) and to maintain net tie-line interchanges from the given area at the scheduled values (tie-line control). The regulation is executed by changing the power output of the turbines in the area through varying P_{ref} in their governing systems. Figure 8.8 shows a functional diagram of the central regulator. Frequency is measured in the local low voltage network and compared with the reference frequency to produce a signal that is proportional to the frequency deviation Δf. The information on power flows in the tie-lines is sent via telecommunication lines to the central controller which compares it with the reference value in order to produce a signal proportional to the tie-line interchange error ΔP_{tie}. Before adding these two signals together the frequency

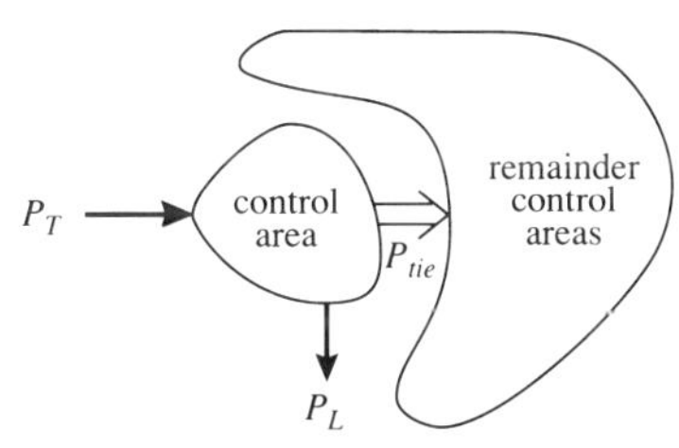

Figure 8.7 Power balance of a control area

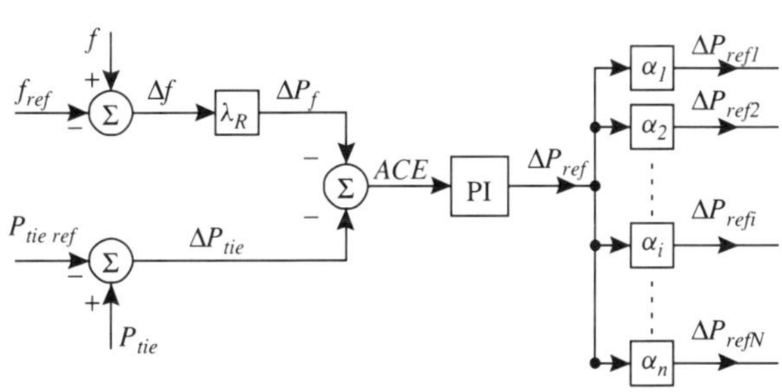

Figure 8.8 Functional diagram of a central regulator

deviation is amplified by the factor λ_R, called the *frequency bias factor*, to obtain

$$\Delta P_f = \lambda_R \Delta f \tag{8.9}$$

This represents the change in the generation power which must be forced upon the controlled area in order to compensate for the frequency deviation resulting from the power imbalance in this area. Assuming that the power imbalance is caused by a change in the power demand it is possible to assess the bias λ_R by substituting for ΔP_f in equation 8.9 by the value of ΔP_{demand} from equation 8.7. This will lead to the following formula

$$\lambda_R = K_f \frac{P_L}{f_n} \tag{8.10}$$

indicating that the value of the bias can be assessed provided that the stiffness of a given area and its total demand are known. In practice it is difficult to evaluate the exact value of the stiffness so that imprecise setting of the bias λ_R in the central regulator may have some undesirable effects on the regulation process. This will be discussed later in this chapter.

The signal ΔP_f is now added to the net tie-line interchange error ΔP_{tie} so that the *area control error* (ACE) is created

$$\text{ACE} = -\Delta P_{tie} - \lambda_R \Delta f \tag{8.11}$$

As in the decentralised regulator shown in Figure 8.6, the central regulator must have an integrating element in order to remove the error, and this is usually supplemented by a proportional element to give a better regulation process. For such a PI regulator the output signal is

$$\Delta P_{ref} = \beta_R\,(\text{ACE}) + \frac{1}{T_R}\int_0^t (\text{ACE})\mathrm{d}t \tag{8.12}$$

where β_R and T_R are the regulator parameters. The area control error ACE corresponds to the power by which the total area power generation must be changed in order to maintain both the frequency and the tie-line flows at their scheduled values. The regulator output signal ΔP_{ref} is then multiplied by the *participation factors* $\alpha_1, \alpha_2, \ldots, \alpha_n$ which define the contribution of the individual generating units to the total generation control, Figure 8.8. The control signals $\Delta P_{ref1}, \Delta P_{ref2}, \ldots, \Delta P_{refn}$ obtained in this way are then transmitted to the power plants and delivered to the reference set points of the turbine governing systems (Figure 2.11). As with decentralised control not all generating units need participate in generation control.

In the past the central control of frequency and power interchanges was implemented using large analogue controllers but this has now been replaced by a central computer. In normal operating conditions this computer also performs *security assessment* analysis and *economic dispatch*, i.e. allocation of the generated power among the power plants in a way that minimises the production cost. Should an emergency occur the computer may change the allocation of power in order to prevent system blackout. A detailed description of these facilities is beyond the scope of this book but interested readers are referred to Sterling (1978), Wood & Wollenberg (1996) and Debs (1988). The ΔP_{ref} signals, resulting from the frequency and tie-line control, and the P_{ref} signals, resulting from the economic dispatch or emergency control, are transmitted to the power plants using the same telecommunication links.

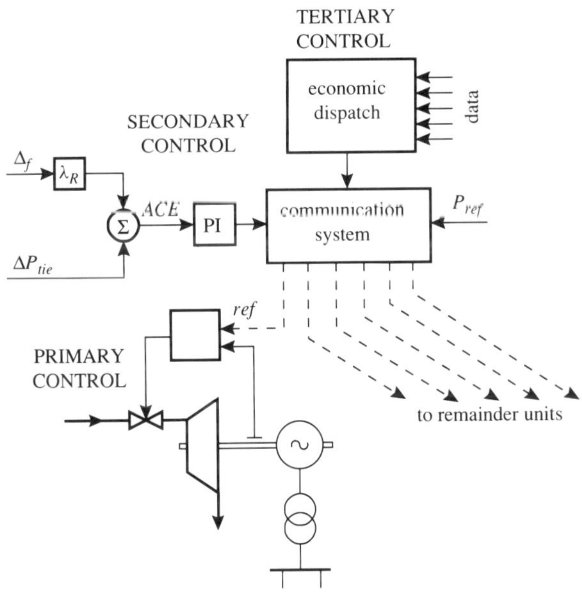

Figure 8.9 Levels of automatic generation control

8.1.4 AGC as a multi-level control

Automatic generation control is an excellent example of a multi-level control system whose overall structure is shown in Figure 8.9. The turbine governing system with its load reference point is at the lowest *primary control* level and all the commands from the upper levels are executed at this level. Primary control is decentralised because it is installed in power plants situated at different geographical sites. Frequency control and tie-line control constitutes *secondary control* and "forces" the primary control to eliminate the frequency and net tie-line interchange deviations. In isolated systems secondary control is limited to frequency control and could be implemented locally without the need to go through the central controller. In interconnected systems secondary control of frequency and tie-line flows is implemented using a central computer. Economic dispatch constitutes a third level of control referred to as *tertiary control* with the main difference between secondary and tertiary control being the response time. Values of ΔP_{ref}, set by the frequency and tie-line control, are updated relatively frequently (every few seconds) whilst the values of P_{ref}, set by the economic dispatch, are updated once every few minutes.

8.2 Stage I — Rotor Swings in the Generators

Having described the AGC, it is now possible to analyse the response of a power system to a power imbalance. This response can be divided into four stages depending on the duration of the dynamics involved:

Stage I rotor swings in the generators (first few seconds)

Stage II frequency drop (a few seconds to several seconds)

Stage III primary control by the turbine governing systems (several seconds)

Stage IV secondary control by the central regulators (several seconds to a minute).

The dynamics associated with each of these four stages will be described separately in order to illustrate how they develop in the system. To begin the discussion the power system shown schematically in Figure 8.10a will be considered where a power station transmits its power to the system via two parallel transmission lines. The power station itself is assumed to consist of two identical generating units operating onto the same busbar. The disturbance considered will be the disconnection of one of the generating units. In Stage I of the disturbance the way in which the remaining generating unit contributes to the production of the lost power will be given special attention.

The sudden disconnection of one of the generators will initially produce large rotor swings in the remaining generating unit and much smaller rotor swings in the other generators within the system. For simplicity these small oscillations will be neglected to allow the rest of the system to be replaced by an infinite busbar. The time scale of the rotor swings are such that the generator transient model applies and the mechanical power supplied by the turbine remains constant. To simplify considerations, the classical model representation of the generator will be used, equations 5.16 and 5.40. Figure 8.10b shows the equivalent pre-disturbance circuit diagram for the system. As both the generators are identical they can be represented by the same transient emf $\overline{E}'$ behind an equivalent reactance that combines the generator transient reactance X'_d, the transformer reactance X_T, and the system reactance X_s.

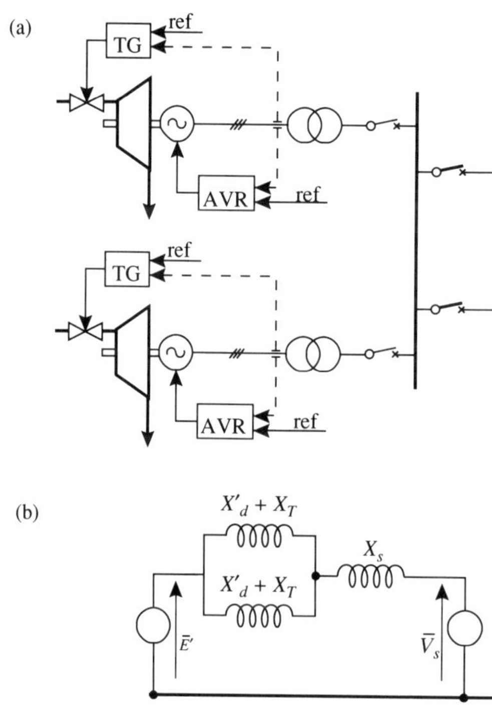

Figure 8.10 Parallel generating units operating onto an infinite busbar: (a) schematic diagram; (b) equivalent circuit

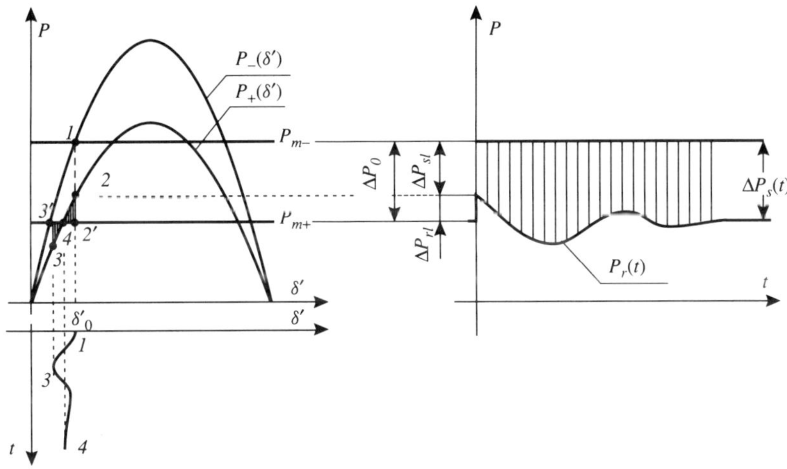

Figure 8.11 Application of the equal area criterion to determine the first stage of the dynamics (The duration of the phenomena is the first few seconds)

Figure 8.11 shows how the equal area criterion can be applied to this problem. In this diagram $P_-(\delta')$ and $P_+(\delta')$ are the power-angle characteristics, and P_{m-} and P_{m+} the mechanical powers, before and after the disturbance occurs. Initially the plant operates at point *1* and the equivalent power angle with respect to the infinite busbar is δ'_0. Disconnection of one of the generators has two effects. Firstly, the equivalent reactance of the system increases so that the amplitude of the power-angle characteristic decreases. Consequently the pre- and post-disturbance power-angle characteristics are

$$P_-(\delta'_0) = \frac{E'V_s}{\dfrac{X'_d + X_T}{2} + X_s} \sin\delta'_0, \qquad P_+(\delta'_0) = \frac{E'V_s}{X'_d + X_T + X_s} \sin\delta'_0 \tag{8.13}$$

Secondly, the mechanical power delivered to the system drops by an amount equal to the power of the lost unit, that is $P_{m+} = 0.5P_{m-}$.

As the rotor angle of the remaining generator cannot change immediately after the disturbance occurs, the generator electrical power is greater than the mechanical power delivered by its prime mover, point *2*. The rotor is decelerated and loses kinetic energy corresponding to the area *2-2′-4*. Due to its momentum the rotor continues to decrease its angle past point *4* until it stops at point *3* when the area *4-3-3′* equals the area *2-2′-4*. The damping torques then damp out subsequent oscillations and the rotor tends towards its equilibrium point *4*.

The amplitude of the rotor oscillations of any given generator will depend on the amount of the lost generation it picks up immediately after the disturbance occurs. Using the notation of Figure 8.11 gives

$$\Delta P_0 = P_-(\delta'_0) - P_{m+} = P_{m-} - P_{m+}$$

$$\Delta P_{rI} = P_+(\delta'_0) - P_{m+}, \qquad \Delta P_{sI} = \Delta P_0 - \Delta P_{rI} \tag{8.14}$$

where ΔP_0 is the lost generating power and ΔP_{rI} and ΔP_{sI} are the contributions of the generating units remaining in operation and of the system, respectively, in meeting the power imbalance ΔP_0 at the very beginning of the disturbance. The subscript I indicates that these equations apply to Stage I of the disturbance. Using the first of equations 8.14 a formula for ΔP_{rI} can be re-written as

$$\Delta P_{rI} = P_+(\delta_0') - P_{m+} = [P_+(\delta_0') - P_{m+}]\frac{1}{P_-(\delta_0') - P_{m+}}\Delta P_0 \tag{8.15}$$

Substituting equation 8.13 into equation 8.15 and noting that $P_{m+} = 0.5P_{m-}$ gives

$$\Delta P_{rI} = \frac{1}{1+\beta}\Delta P_0 \tag{8.16}$$

where $\beta = (X_d' + X_T)/X_s$. The amount that the system contributes in order to meet the lost generation can now be calculated as

$$\Delta P_{sI} = \Delta P_0 - \Delta P_{rI} = \frac{\beta}{\beta+1}\Delta P_0 \tag{8.17}$$

with the ratio of the contributions in equations 8.16 and 8.17 being

$$\frac{\Delta P_{rI}}{\Delta P_{sI}} = \frac{1}{\beta} = \frac{X_s}{X_d' + X_T} \tag{8.18}$$

This equation shows that ΔP_r, the contribution of the unit remaining in operation in meeting the lost power, is proportional to the system equivalent reactance X_s. Both contributions ΔP_r and ΔP_s are depicted in Figure 8.11. Due to the fact that the inertia of the unit is much smaller than that of the power system, the generator quickly decelerates, loses kinetic energy, and both its rotor angle and generated power decrease. As a consequence the power imbalance starts to increase and is met by the system which starts to increase its contribution ΔP_s. This power delivered by the system has been shaded in Figure 8.11 and, when added to the generator's share $P_r(t)$, must equal the load existing before the disturbance. As can be seen from this figure the proportion of the lost generation picked up by the system changes with time so that the system contributes more and more as time progresses.

Although the formulae in equation (8.18) has been derived for two parallel generating units operating onto the system a similar expression can be derived for the general multi-machine case. This expression would lead to a conclusion that is similar to the two machine case in that at the beginning of Stage I of the dynamics the share of any given generator in meeting the lost load will depend on its electrical distance from the disturbance (coefficient β). In the case of Figure 8.10 the imbalance appeared at the power plant busbar. Thus X_s is a measure of the electrical distance of the system from the disturbance whilst $(X_d' + X_T)$ is a measure of the electrical distance of the unit remaining in operation.

8.3 Stage II — Frequency Drop

The situation shown in Figure 8.11 can only last for a few seconds before the power imbalance causes all the generators in the system to slow down and the system frequency to drop. So begins Stage II of the dynamics. During this stage the share of any one generator in meeting the power imbalance depends solely on its inertia and not on its electrical distance from the disturbance. Assuming that all the generators remain in synchronism, they will slow down at approximately the same rate after a few rotor swings in Stage I of the dynamics. This may be written as

$$\frac{d\Delta\omega_1}{dt} \approx \frac{d\Delta\omega_2}{dt} \approx \ldots \approx \frac{d\Delta\omega_{N_G}}{dt} = \varepsilon \tag{8.19}$$

where $\Delta\omega_i$ is the speed deviation of i-th generator, ε is the average acceleration and N_G is the number of generators.

According to the swing equation 5.15 the derivative of the speed deviation can be replaced by the ratio of the accelerating power ΔP_i of the i-th unit over the inertia coefficient M_i. This modifies equation 8.19 to

$$\frac{\Delta P_1}{M_1} \approx \frac{\Delta P_2}{M_2} \approx \ldots \approx \frac{\Delta P_n}{M_n} \approx \varepsilon \tag{8.20}$$

If the change in the system load due to the frequency change is neglected, then the sum of the extra load taken by each generator must be equal to the power lost ΔP_0

$$\Delta P_0 = \sum_{i=1}^{N_G} \Delta P_i \tag{8.21}$$

Substituting for the increment ΔP_i in this equation by $\Delta P_i = M_i\varepsilon$ gives

$$\Delta P_0 = \varepsilon \sum_{i=1}^{N_G} M_i \text{ or } \varepsilon = \frac{\Delta P_0}{\sum_{i=1}^{N_G} M_i} \text{ so that } \Delta P_i = M_i\varepsilon = \frac{M_i}{\sum_{k=1}^{N_G} M_k}\Delta P_0 \tag{8.22}$$

This equation determines the contribution of the i-th generator in meeting the lost power in Stage II of the dynamics when each generator contributes an amount of power proportional to its inertia. In practice the inertia constant H_i is similar for all the generators so that substituting from equation 5.13 for $M_i = 2H_iS_{ni}/\omega_s$ equation 8.22 can be written as

$$\Delta P_i \approx \frac{S_{ni}}{\sum_{i=1}^{N_G} S_{ni}}\Delta P_0 \tag{8.23}$$

During Stage II the contribution of the generator remaining in operation and the rest of the power system to meeting the lost power can be expressed, using equation 8.22, as

$$\Delta P_{rII} = \frac{M_r}{M_r + M_s}\Delta P_0 \quad \text{and} \quad \Delta P_{sII} = \frac{M_s}{M_r + M_s}\Delta P_0 \tag{8.24}$$

Subscript II has been added here to emphasise that these equations are valid during Stage II. In a similar way as in equation 8.18 the ratio of the contributions is

$$\frac{\Delta P_{rII}}{\Delta P_{sII}} = \frac{M_r}{M_s} \approx \frac{S_{nr}}{S_{ns}} \tag{8.25}$$

and is equal to the ratio of the inertia coefficients or, approximately, to the ratio of the rated powers. As $S_{ns} \gg S_{nr}$ (infinite busbar), the contribution by the generator remaining in

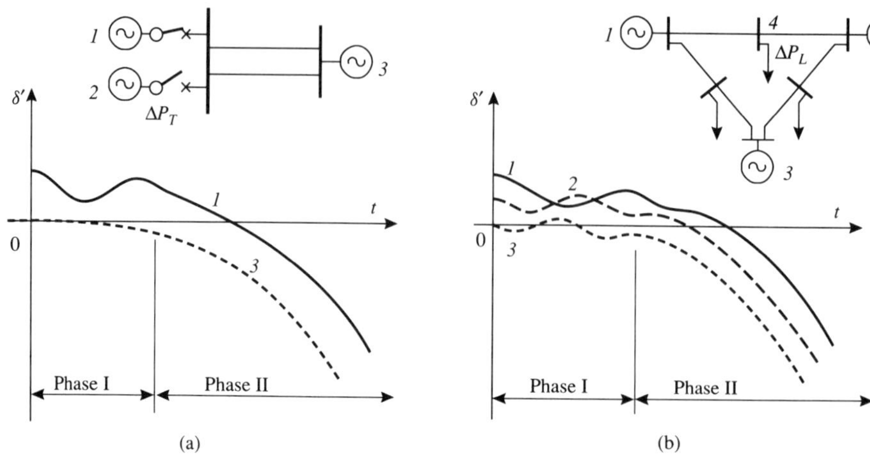

Figure 8.12 Examples of changes in the rotor angles in the case of a real power disturbance: (a) disconnection of generator *2*; (b) sudden increase of the load at node *4* (The duration of Stage II is a few to several seconds)

operation in covering the lost power is very small. The assumption made when preparing Figure 8.11 was that $\Delta P_{rII} \approx \Delta P_0 S_{nr}/S_{ns} \approx 0$.

Figure 8.12a illustrates the case when the system has a finite equivalent inertia and the rotor angle of both the remaining generator and the system equivalent generator decrease as the frequency drops. The initial rotor angle oscillations are the same as those shown for the first stage of the dynamics in Figure 8.11. The angles then decrease together as the generators operate synchronously.

Figure 8.12b shows the case of a three-generator system where the load at node *4* suddenly increases. Generators *1* and *2* are close to the disturbance and so they participate more strongly in the first stage of the oscillations than does generator *3*. During the second stage the power angles synchronously decrease and the system frequency drops.

8.4 Stage III — Primary Control

Stage III of the dynamics depends on how the generating units and the loads react to the drop in frequency. Sub-Section 2.2.3 explained how as frequency (speed) drops the turbine governor opens the main control valves to increase the flow of working fluid through the turbine and so increase the turbine mechanical power output. In the steady-state, and during very slow changes of frequency, the increase in mechanical power for each generating unit is inversely proportional to the droop of the static turbine characteristic. Equations 8.4 and 8.6 can be re-written as

$$
\begin{aligned}
P_T &= P_{T0} + \Delta P_T = P_{T0} - K_T \Delta f \frac{P_{L0}}{f_n} \\
P_L &= P_{L0} + \Delta P_L = P_{L0} + K_L \Delta f \frac{P_{L0}}{f_n}
\end{aligned}
\tag{8.26}
$$

The operating frequency of the system is determined by the point of intersection between these two characteristics.

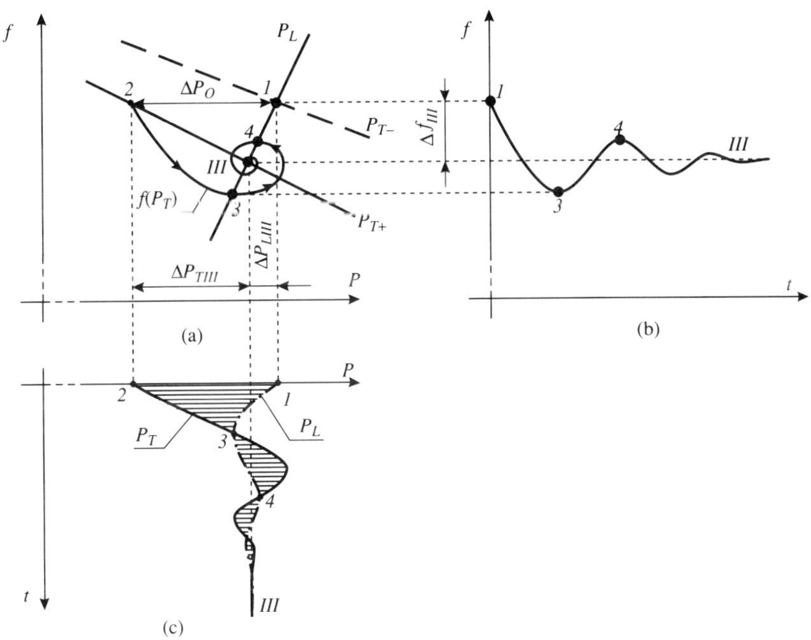

Figure 8.13 Stage III of the dynamics caused by an imbalance in real power: (a) the generation characteristic and the frequency response characteristic; (b) frequency changes; (c) power changes (The duration of Stage III is several seconds)

Typical generation and load characteristics are shown in Figure 8.13 where the generation characteristic before the disturbance is denoted P_{T-} and after by P_{T+}. If a generating unit is lost the system generation characteristic P_T moves to the left by the value of the lost power ΔP_0 and, according to equation 8.4, the slope of the characteristic increases slightly as the value of K_T decreases.

Before the disturbance occurs the system operates at point *1* corresponding to the intersection of the P_L and P_{T-} characteristics. After the generator is disconnected the frequency initially remains unchanged and the generation operating point is shifted to point *2*. Then the generation tends to move towards point *III* which corresponds to the intersection of the P_L and P_{T+} characteristics but, because of the delay introduced by the time constants of the turbines and their governors, this point cannot be reached immediately. Initially the difference between the generated power, point *2*, and the load power, point *1*, is large and the frequency starts to drop as described in Stage I and Stage II of the dynamics. In Stage III the turbine reacts to the drop in frequency by increasing its power output but, because of the aforementioned time delays in the turbine regulator system, the trajectory of the turbine power $f(P_T)$ lies below the static generation characteristic P_{T+}. As the frequency drops the generated power increases while the power taken by the load decreases. The difference between the load and the generation is zero at point *3*. According to the swing equation a zero value of deceleration power means

$$\frac{\mathrm{d}\Delta\omega}{\mathrm{d}t} = 2\pi\frac{\mathrm{d}\Delta f}{\mathrm{d}t} = 0 \tag{8.27}$$

and the frequency $f(t)$ reaches a local minimum, Figure 8.13b.

Because of the inherent inertia of the turbine regulation process the mechanical power continues to increases after point *3* so that the generated power exceeds the load power and the frequency starts to rise. At point *4* the balance of power is again zero and corresponds to the local maximum in the frequency shown in Figure 8.13b. The oscillations continue until the steady-state value of the frequency f_{III} is reached at point *III* corresponding to the intersection of the P_L and P_{T+} static characteristics. The value of f_{III} can be determined from Figure 8.13a

$$\begin{aligned}\Delta P_0 = \Delta P_{TIII} - \Delta P_{LIII} &= -K_T \frac{P_L}{f_n} \Delta f_{III} - K_L \frac{P_L}{f_n} \Delta f_{III} \\ &= -P_L(K_T + K_L)\frac{\Delta f_{III}}{f_n}\end{aligned} \tag{8.28}$$

with the frequency error f_{III} being

$$\frac{\Delta f_{III}}{f_n} = \frac{-1}{K_T + K_L}\frac{\Delta P_0}{P_L} \quad \text{or} \quad \frac{\Delta f_{III}}{f_n} = \frac{-1}{K_f}\frac{\Delta P_0}{P_L} \tag{8.29}$$

where the coefficient $K_f = K_T + K_L$ is the system stiffness introduced in Equation 8.7. Figure 8.13c shows how the generated power and the load power change during this period. The difference between the two is shaded and corresponds to the power that decelerates ($P_T < P_L$) or accelerates ($P_T > P_L$) the rotor.

For the system shown in Figure 8.10 the contributions of the generator remaining in operation and of the system in covering the lost generation in Stage III of the dynamics can be obtained from equation 8.4 as

$$\Delta P_{rIII} = -K_{Tr}\frac{\Delta f_{III}}{f_n} P_{nr}, \qquad \Delta P_{sIII} = -K_{Ts}\frac{\Delta f_{III}}{f_n} P_{ns} \tag{8.30}$$

where the subscript *III* is added to emphasise that these equations are valid during Stage III of the dynamics. The contributions ratio resulting from these equations are

$$\frac{\Delta P_{rIII}}{\Delta P_{sIII}} = \frac{K_{Tr}}{K_{Ts}}\frac{P_{nr}}{P_{ns}} \approx \frac{P_{nr}}{P_{ns}} \tag{8.31}$$

The approximate equality in this equation is valid when the droop of all the turbine characteristics is approximately the same. In practice the ratios in equations 8.31 and 8.25 are very similar. This physically corresponds to the fact that, during the transition period between the second and the third stage, the generator is in synchronism with the system and there are no mutual oscillations between them.

8.4.1 The importance of the spinning reserve

The discussion has so far assumed that, within the range of the frequency variations, each of the generating units has a linear turbine characteristic and that the overall generation characteristic is also linear. In practice each generating unit must operate within the limits that are placed on its thermal and mechanical performance. To ensure that these limits are adhered to its governing system is equipped with the necessary facilities to make certain that the unit does not exceed its maximum power limit or the limit placed on the speed at which it can take up power. These limits can have a substantial impact on how a unit behaves when the system frequency changes.

Figure 8.3 showed how the spinning reserve, and its allocation in the system, influences the shape of the generation characteristic. In order to help quantify the influence of spinning reserve the following coefficients are defined

$$r = \frac{\sum_{i=1}^{N_G} P_{ni} - P_L}{P_L}, \qquad p = \frac{\sum_{i=1}^{R} P_{ni}}{\sum_{i=1}^{N_G} P_{ni}} \tag{8.32}$$

where $\Sigma_{i=1}^{N_G} P_{ni}$ is the sum of the power ratings of all the generating units connected to the system and $\Sigma_{i=1}^{R} P_{ni}$ is the sum of the power ratings of all the units operating on the linear part of their characteristics, that is loaded below their power limit. The coefficient r is the *spinning reserve coefficient* and defines the relative difference between the maximum power capacity of the system and the actual load.

A simple expression for the local droop of the generation characteristic can be obtained assuming that the droop of all the units which are not fully loaded are approximately identical, that is $\rho_i = \rho$ and $K_i = K = 1/\rho$. For the units operating at their limits $\rho_i = \infty$ and $K_i = 0$. Under these conditions

$$\begin{aligned} \Delta P_T &= -\sum_{i=1}^{N_G} K_i P_{ni} \frac{\Delta f}{f_n} = -\sum_{i=1}^{R} K_i P_{ni} \frac{\Delta f}{f_n} \cong -K \sum_{i=1}^{R} P_{ni} \frac{\Delta f}{f_n} \\ &= -K p \sum_{i=1}^{N_G} P_{ni} \frac{\Delta f}{f_n} = -K p(r+1) P_L \frac{\Delta f}{f_n} \end{aligned} \tag{8.33}$$

Dividing by P_L gives

$$\frac{\Delta P_T}{P_L} = -K_T \frac{\Delta f}{f_n} \tag{8.34}$$

where

$$K_T = p(r+1)K \quad \text{and} \quad \rho_T = \frac{\rho}{p(r+1)} \tag{8.35}$$

Equation 8.34 is similar to equation 8.4 and describes the linear approximation of the non-linear generation characteristic for a given load. The local droop ρ_T increases as the spinning reserve decreases. At the limit, when the load P_L is equal to the system generating capacity, both the coefficients r and p are zero and $\rho_T = \infty$. This corresponds to all the generating units being fully loaded.

The influence of the spinning reserve on the frequency drop in Stage III of the dynamics can now be determined. For the linear approximation to the generator characteristic given in equation 8.34 the drop in frequency can be determined from equation 8.29 as

$$\frac{\Delta f_{III}}{f_n} = \frac{-1}{p(r+1)K + K_L} \frac{\Delta P_0}{P_L} \tag{8.36}$$

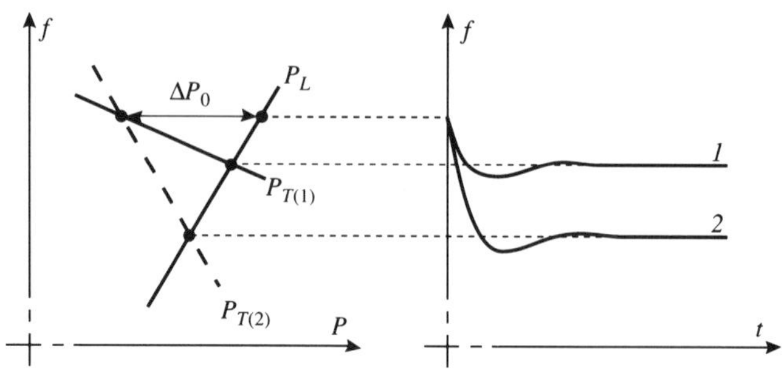

Figure 8.14 The influence of spinning reserve on the frequency drop due to a power imbalance (*1* — large spinning reserve, *2* — small spinning reserve)

indicating that the smaller the spinning reserve coefficient r the bigger will be the drop in frequency due to the loss of power ΔP_0. With a large spinning reserve the static characteristic $P_{T(1)}$ shown in Figure 8.14 has a shallow slope and the drop in frequency is small. On the other hand when the spinning reserve is small, the $P_{T(2)}$ characteristic is steep, and the frequency drop is increased for the same power disturbance ΔP_0. In the extreme case when no spinning reserve is available, the generating units would be unable to increase their generation and the whole power imbalance ΔP_0 could only be covered by the frequency effect of the loads. As the frequency sensitivity K_L of the loads is generally small, the frequency drop would be very high.

Equation 8.33 defines the way in which each of the generating units contribute to the power imbalance at the end of Stage III, with equation 8.34 quantifying the net effect of the primary turbine control in meeting the lost load.

Example 8.1

Consider a 50 Hz system with a total load $P_L = 10\,000$ MW in which $p = 60\%$ of the units give $r = 15\%$ of the spinning reserve. The remaining 40% of the units are fully loaded. The average droop of the units with spinning reserve is $\rho = 7\%$ and the frequency sensitivity coefficients of the loads is $K_L = 1$. If the system suddenly loses a large generating unit of $\Delta P_0 = 500$ MW calculate the frequency drop and the amount of power contributed by the primary control.

From equation 8.35 $K_T = 0.6(1 + 0.15)1/0.07 = 9.687$ and, according to equation 8.29, the primary control will give the frequency drop in Stage III as

$$\Delta f_{III} = \frac{-1}{9.867 + 1} \times \left[\frac{500}{10\,000}\right] \times 50 \approx 0.23 \text{ Hz}$$

where the turbine governor primary control contributes

$$\Delta P_{TIII} = 9.867 \times \frac{0.23}{50} \times 10\,000 = 454 \text{ MW}$$

with the remaining deficit of

$$\Delta P_{LIII} = 1 \times \frac{0.23}{50} \times 10\,000 = 46 \text{ MW}$$

being covered by the frequency effect of the loads. If there were no spinning reserve, the deficit would be covered entirely by the frequency effect of the loads giving the frequency drop of

$$\Delta f_{III} = \frac{-1}{0+1} \times \left[\frac{500}{10\,000}\right] \times 50 = 2.5 \text{ Hz}$$

which is about ten times greater than in the case with (only) 15% spinning reserve.

8.4.2 Frequency collapse

Spinning reserve is much more important than is suggested by the approximate formula in equation 8.36. In practice the power output of the turbine is frequency dependent so that if the frequency is much lower than the nominal frequency the turbine power will be less than that given by equation 8.34 and the frequency may drop further and further until the system suffers a *frequency collapse*.

The hidden assumption made in the generation characteristics of Figure 8.14 is that when a unit is fully loaded the mechanical driving power delivered by the turbine does not depend on the frequency deviation. This assumption is only true for small frequency deviations whilst a larger frequency drop reduces mechanical power due to deterioration in performance of boiler feed pumps, see sub-Section 2.2.3. In this case the static generation characteristic takes the form shown in Figure 2.12. Adding the power-frequency characteristics of individual generating units now gives the system generation characteristic P_T shown in Figure 8.15.

On the upper part of the generation characteristic the local droop at each point is positive and depends on the system load according to the formulae given in equation 8.35. The lower part of the characteristic corresponds to the decrease in the power output due to the deterioration of the performance of the boiler feed pumps. For a given load frequency characteristic P_L the inflexion in the generation characteristic produces two equilibrium points. At the upper point s any small disturbance that produces an increase

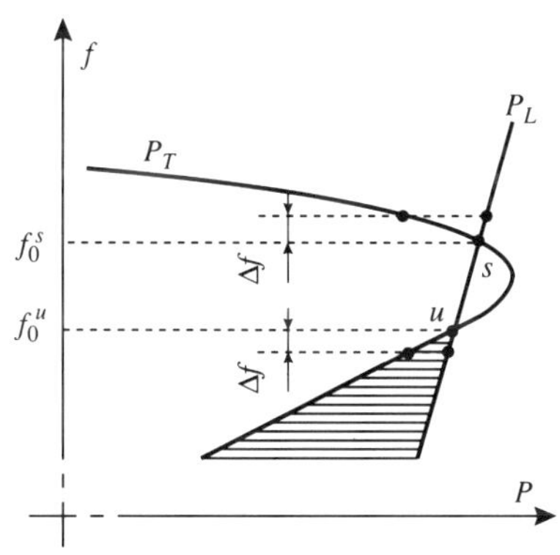

Figure 8.15 The static generation and load characteristics of the power system and the equilibrium points

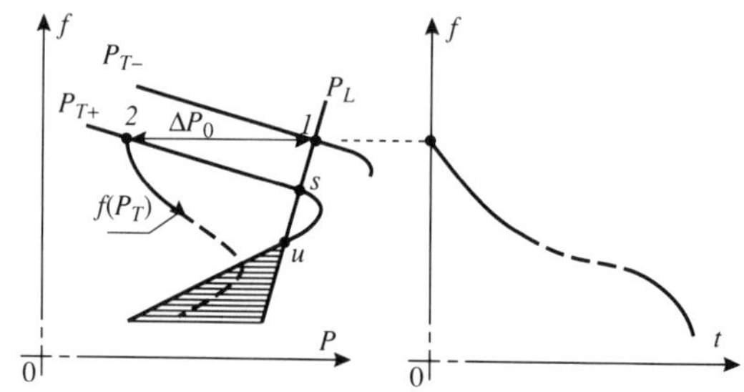

Figure 8.16 An example of frequency collapse

in the frequency $\Delta f > 0$ will result in the load power exceeding the generation power; the generators are decelerated and the system returns to the equilibrium point s. Similarly, any small disturbance that causes a decrease in frequency will result in the generation power exceeding the load power, the rotors are accelerated, the frequency increases and the system again returns to the equilibrium point. Point s is therefore *locally stable*, as for any small disturbance within the vicinity of this point the system returns to point s. The region in which this condition holds is referred to as the *area of attraction*.

In contrast the lower point u is *locally unstable* as any disturbance within the vicinity of this point will result in the system moving away from the equilibrium point. The region in which this happens is referred to as the *area of repulsion*. For example if a disturbance reduces the frequency $\Delta f < 0$ then the system moves into the shaded area below point u, the generated power is reduced so that the system load is greater than the system generation, the rotors decelerate, and the frequency drops further.

With these qualifications on turbine performance in mind assume that the system operates with a low spinning reserve at point *1* on Figure 8.16. A loss ΔP_0 in generation now occurs and the operating point moves from point *1* to point 2. The excess load over generation is large and produces an initial rapid drop in frequency. As the difference between the load and generation reduces, the rate at which the frequency drops slows down and the system generation trajectory $f(P_T)$ approaches the equilibrium point u. When the trajectory $f(P_T)$ enters the area of repulsion of point u it is forced away and the system suffers a frequency collapse.

The frequency collapse illustrated in Figure 8.16 occurred following the *sudden* appearance of the power imbalance ΔP_0. Had the system generation characteristic changed slowly from P_{T+} to P_{T-} by gradually reducing the generation then it would have met the locally stable point s and stayed there. Obviously if the power imbalance ΔP_0 is greater than the spinning reserve then the post-disturbance demand characteristic P_L lies to the right of the nose of the generation characteristic P_{T+} and the frequency would collapse regardless of how sudden the power change.

8.4.3 Under frequency load shedding

Many systems can be protected from frequency collapse by importing large blocks of power from neighbouring systems to make up for the lost generation. However in an islanded system, or in an interconnected system with a shortage of tie line capacity, this will not be possible and the only way to prevent a frequency collapse following a large disturbance is to employ *automatic load shedding*.

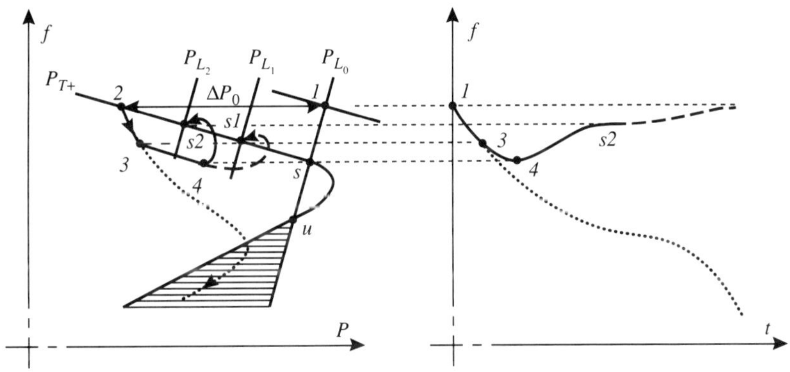

Figure 8.17 Two-stage load shedding to protect against frequency collapse

Automatic load shedding is implemented using under-frequency relays. These relays detect the onset of decay in the system frequency and shed appropriate amounts of system load until the generation and load are once again in balance and the power system can return to its normal operating frequency. Load shedding relays are normally installed in distribution and subtransmission substations as it is from here that the feeder loads can be controlled.

As load shedding is a somewhat drastic control measure it is usually implemented in stages with each stage triggered at a different frequency level to allow the least important loads to be shed first. Figure 8.17 shows the effect of load shedding when a power imbalance ΔP_0 appears on a system with a low spinning reserve. Without load shedding the system would suffer a frequency collapse as shown by the dotted line in Figure 8.17. The first stage of load shedding is activated at point *3* and limits the load to a value corresponding to characteristic P_{L_1}. The rate of frequency drop is now much slower as the difference between the load and generation is smaller. At point *4* the second stage of load shedding is triggered further reducing the load to a value corresponding to characteristic P_{L_2}. Generation is now higher than the load, the frequency increases, and the system trajectory tends towards point *s2* where the P_{L_2} and P_{T+} characteristics intersect.

Besides protecting against frequency collapse, load shedding may also be used to prevent deep drops in system frequency.

8.5 Stage IV — Secondary Control

In Stage IV of the dynamics the drop in the system frequency, and the deviation in the tie-line power flows, activate the central automatic generation control, the basic operation of which has been described in Section 8.1.

8.5.1 Islanded systems

In an islanded system there are no tie-line connections to neighbouring systems so that the central regulator only controls the system frequency. As the frequency drops the central regulator transmits control signals to the participating generating units to force

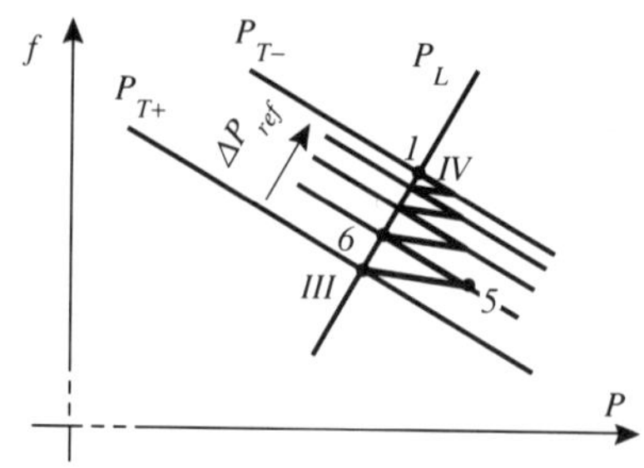

Figure 8.18 Stepped shifting of the generation characteristic by a slow acting central regulator

them to increase their power output. This lifts the system generation characteristic on the (f, P) plane.

Figure 8.18 illustrates the action of a very slow acting central regulator which transmits its first control signal at the end of Stage III corresponding to point *III*. This first control signal shifts the generation characteristic upwards a small amount so that at a given frequency, point *5*, there is an excess of generation over load. The generators start to accelerate and the frequency increases until point *6* is reached. The central regulator now sends a further signal to increase power output and the generation characteristic is shifted further up. After a few such steps point *IV* is reached at which the frequency returns to its reference value and the central regulator ceases operating.

Although the zigzag line in Figure 8.18 is only a rough approximation to the actual trajectory it illustrates the interaction between the secondary control action of the central regulator, which shifts the generation characteristic upwards, and the primary control of the turbine governing systems, which moves the operating point along the static generation characteristic. In a real power system the inertia within the power regulation process ensures smooth changes in the power around the zigzag line to produce the type of response shown in Figure 8.19. Stage I, II and III of the frequency change are as described in the previous sections and, at the end of Stage III, the trajectory tends to wrap itself around the temporary equilibrium point *III*, Figure 8.19a, but does not settle at this point. The automatic generation control of Stage IV now comes into operation and the trajectory tends towards the new equilibrium point *IV*. Although the difference between the frequency at points *III* and *IV* is small, the central regulator acts slowly so that correction of the frequency drop during Stage IV of the dynamics may take a long time as shown by the broken curves in Figure 8.19b and c.

If the central frequency control acts faster than as shown in Figure 8.19, it will come into operation before the end of Stage III. The trajectory $f(P_T)$ does not now wrap around point *III* and the frequency starts to increase earlier as shown by curve *2* in Figure 8.20.

The way in which the frequency is returned to its nominal value depends on the dynamics of the central regulator shown in Figure 8.8 and defined by equation 8.12. The regulator dynamics comprise of proportional and integral action, both of which increase the regulator output signal ΔP_{ref} as the frequency drops. The amount of integral action is determined by the integral time constant T_R while the proportional action is dependent on the coefficient β_R. Careful selection of these two coefficients ensures that the frequency returns smoothly to its reference value as shown in Figure 8.20. Particularly problematic is a small integral time constant as the smaller this time constant the faster the regulation

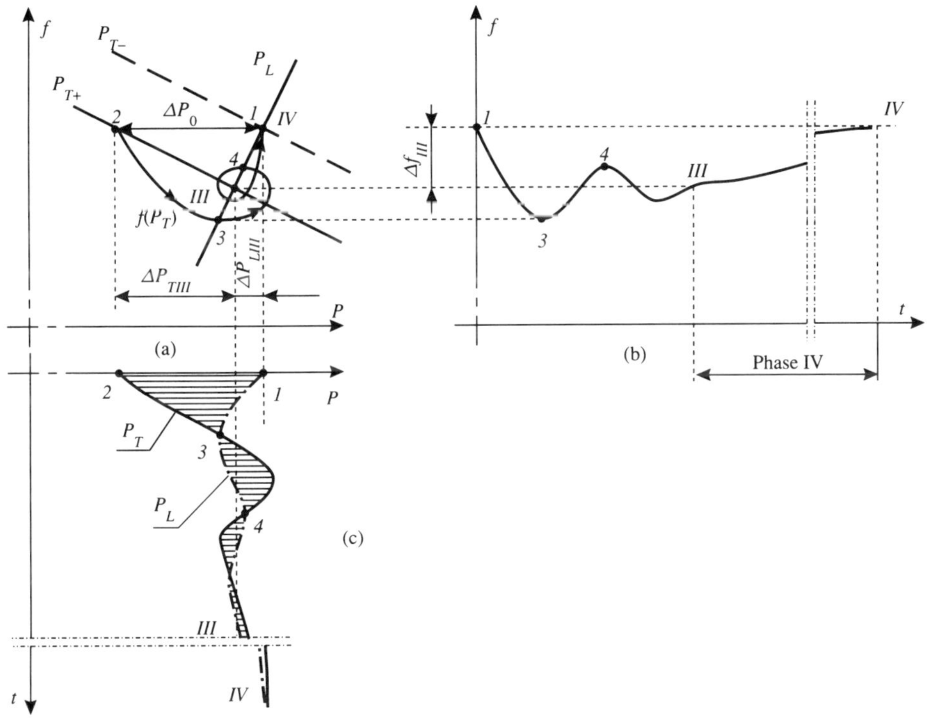

Figure 8.19 Stage III and IV of the frequency variations produced by a disturbance in the balance of real power: (a) generation and load characteristics and the system trajectory; (b) changes in frequency; (c) changes in power (The time duration of Stage IV is several seconds to a minute)

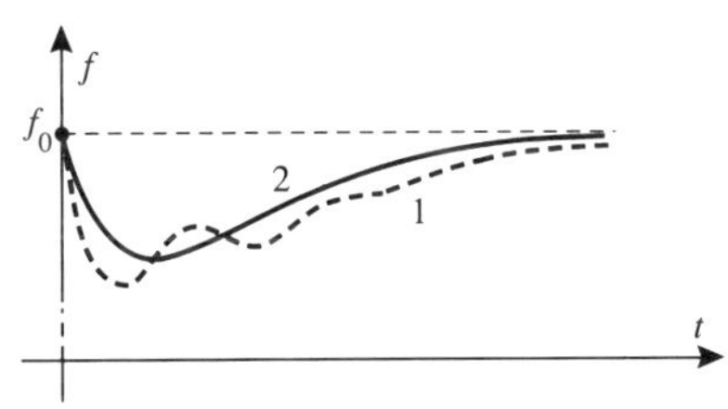

Figure 8.20 Examples of changes in frequency: *1* — with a slow central regulator, *2* — with a fast central regulator

signal increases. This can be compensated to some extent by the coefficient β_R which will start to decrease the signal ΔP_{ref} as soon as the frequency starts to rise. In the extreme case of small values of β_R and T_R the response is underdamped and the reference frequency value will be reached in an oscillatory way.

Section 8.1 explained how not all the generating units necessarily participate in automatic generation control. This means that only a part of the spinning reserve can be activated by the central regulator during secondary control. That part of the spinning reserve that belongs to the generators participating in the secondary control is referred to as the *available regulation power*.

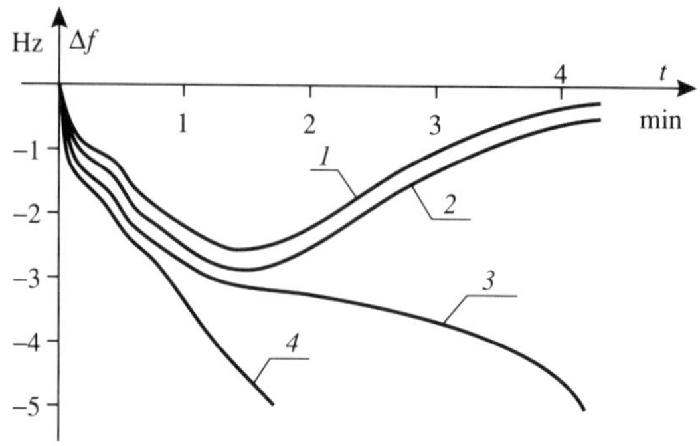

curve 1 corresponds to $r = 16\%$; *2* corresponds to $r = 14\%$; *3* corresponds to $r = 12\%$; and *4* corresponds to $r = 8\%$

Figure 8.21 An example of the frequency variations when $\Delta P_0 = 10\%\ P_L$

If the available regulation power is less than the lost power, then Stage IV of the dynamics will terminate when all the available regulation power has been used. This corresponds to the system trajectory settling at an equilibrium point somewhere between points *III* and *IV* on Figure 8.19 at a frequency that is lower than the reference value. The system operators may now verbally instruct other generating stations not on central control to increase their power output to help remove the frequency offset. In the case of a large power deficit further action would consist of connecting new generating units from the cold reserve and, when connected, will release the capacity controlled by the central regulator. The frequency changes associated with this process are very slow and are not considered here.

Equation 8.31 defines the contribution of each of the generating units in covering the power imbalance at the end of Stage III. In Stage IV the increase in generation is enforced by the central regulator and the contribution of each unit, and the ratio $\Delta P_{rIV}/\Delta P_{sIV}$, will depend on whether or not the particular generating unit participates in the central control.

Provided that the spinning reserve and the available regulation power are large enough then in many cases they can prevent the system suffering a frequency collapse. Figure 8.21 shows an example of how the frequency variations depend on the value of the spinning reserve coefficient. The disturbance consists of losing generation ΔP_0 equivalent to 10% of the total load power P_L. In the first two cases $r \geq 14\%$ and the frequency returns to its reference value thanks to the operation of the primary and secondary control. The third case corresponds to the frequency collapse shown in Figure 8.16. In the fourth case there is no intersection point between the generation and the load characteristic and the frequency quickly collapses.

The energy balance over the four stages

When a power system loses a generating unit it loses a source of both electrical and mechanical energy. By the end of the power system dynamic response, this lost energy has been recovered as illustrated in Figure 8.22. The upper bold curve shows the variations of the mechanical power provided by the system while the lower bold curve shows the

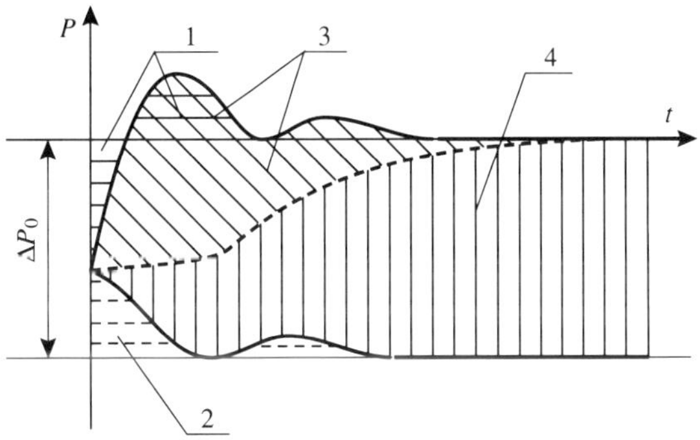

Figure 8.22 Share of the individual components in covering the power imbalance: *1* — rotating masses of the generating units; *2* — rotating masses of the loads; *3* — primary control; *4* — secondary control (Based on Welfonder 1980)

variations of electrical power of the loads due to frequency variations. All these variations are similar to those shown in Figure 8.19c. Initially the energy shortfall is produced by converting the kinetic energy of the rotating masses of the generating units and the loads to electrical energy, Area 1 and Area 2. This reduction in kinetic energy causes a drop in frequency which activates the turbine governor primary control so that the mechanical energy supplied to the system is increased but at a lower frequency, Area 3. This energy is used to partially generate the required energy shortfall, and partially to return the kinetic energy borrowed from the rotating masses. Secondary control then further increases the mechanical energy, Area 4, which is used to generate the required additional electrical energy and to increase the kinetic energy of the rotating masses so restoring the system frequency.

8.5.2 Interconnected systems and tie-line oscillations

This discussion will be limited to the interconnected system shown in Figure 8.23 consisting of two subsystems of disproportionate size: system A and system B, referred to as the big system and the small system respectively. The tie-line interchange power P_{tie} will be assumed to flow from the big system to the small system and an imbalance of power ΔP_0 assumed to arise in the small system. During the first three stages of the dynamics the influence of the central regulators in both system A and system B may be neglected.

Both systems are replaced by an equivalent generator, as in Figure 8.10b, to obtain the equivalent circuit of Figure 8.24a. In this circuit the reactance X combines the reactances of the tie-line connecting the two systems, the equivalent network reactance of both systems and the transient reactances of all the generators. As one of the subsystems is large compared to the other the generator-infinite busbar model and the equal area criterion can be used to analyse Stage I of the dynamics.

The power-angle characteristic $P_B(\delta')$ of the small system is a sinusoid, corresponding to a power transfer between the systems, shifted by a constant value corresponding to the power demand in the small system P_{LB}. It is assumed that system B imports power from A so that $P_{TB} < P_{LB}$ and its power angle is negative with respect to system A. Before

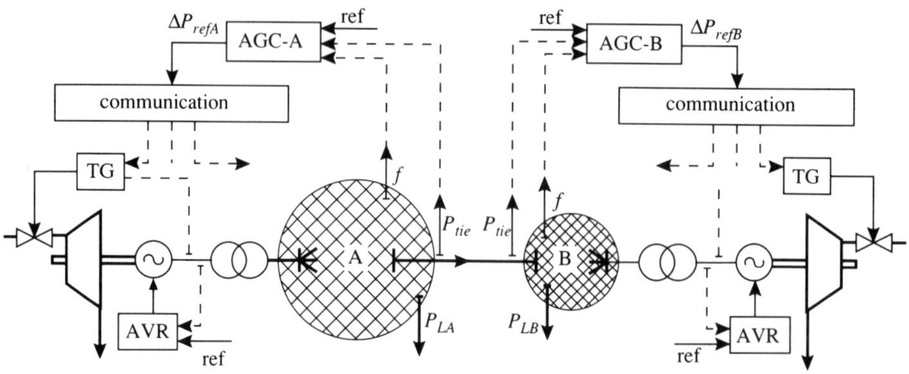

Figure 8.23 The functional diagram of an interconnected system consisting of two subsystems: a big system A and a small system B

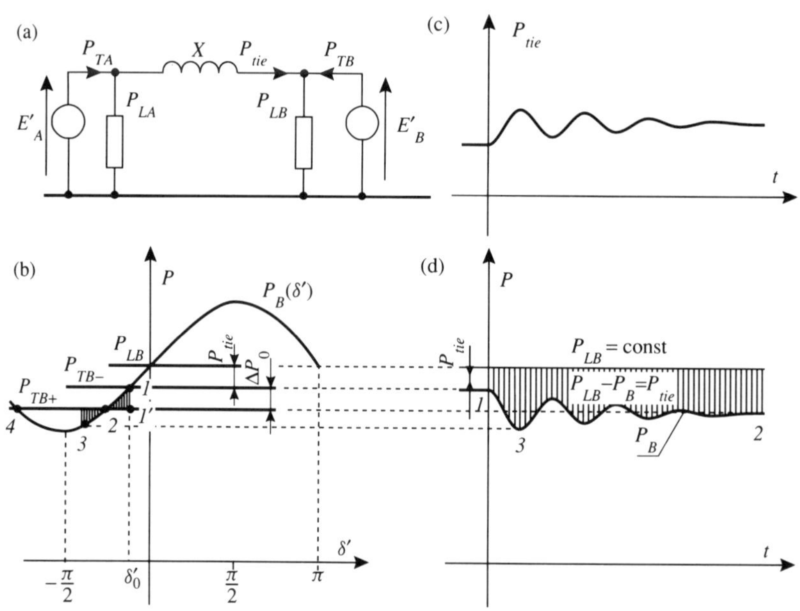

Figure 8.24 Application of the equal area criterion to determine the tie-line power during the first stage of the dynamics: (a) equivalent circuit; (b) the power-angle characteristic taking into account equivalent loads; (c) changes in the tie-line power; (d) changes in the generated power (The time duration of the phenomena is a few seconds)

the disturbance occurs the system operates at point *1* corresponding to the intersection of the $P_B(\delta')$ characteristic with the horizontal line representing the mechanical power P_{TB-}. The small system now loses generation equal to ΔP_0 and the power generated in this system drops to P_{TB+}. The electrical power exceeds the mechanical power and the rotors of the generators in system B slow down. The equivalent system rotor moves from point *1* to point *2* and then to point *3* in Figure 8.24b. Along this motion the electrical power generated in system B decreases and, as a result, additional power starts to flow from system A, Figure 8.24d. The maximum change in the value of the instantaneous power in the tie-lines increases to almost double the value of the lost power, the difference between points *1* and *3*. Oscillations in the power begin, during which kinetic energy in both systems is used to cover the lost generation. The angular velocities of the generator rotors drop and the system enters Stage II of the dynamics.

Stage I of the dynamics is dangerous from a stability point of view in that if the area *1-1′-2* is greater than the area *2-3-4* than the system loses stability and the small system will operate asynchronously with respect to the big system with $\omega_B < \omega_A$. Such a situation may occur when the tie-line capacity is small and the disturbance in power balance large. During asynchronous operation large oscillations in the power tie-line power will occur with the difference between the maximum and the minimum value being twice the amplitude of the sinusoid shown in Figure 8.24b. As the possibility of re-synchronising the systems is small they would usually be disconnected to avoid damaging equipment in either system.

In Stage II the power imbalance is covered in proportion to the inertia coefficients of the equivalent systems as determined by equations 8.24, i.e.

$$\Delta P_{AII} = \frac{M_A}{M_A + M_B}\Delta P_0 \quad \text{and} \quad \Delta P_{BII} = \frac{M_B}{M_A + M_B}\Delta P_0 \tag{8.37}$$

where $M_A = \Sigma_{i=1}^{N_{GA}} M_i$ and $M_B = \Sigma_{i=1}^{N_{GB}} M_i$ are the sums of inertia coefficients of the rotating masses in each systems. Equations 8.37 suggest that the big system A will almost entirely cover the power imbalance as $M_A \gg M_B$, so that in Stage II of the frequency variations $\Delta P_{tieII} \approx \Delta P_0$ and the tie-lines will be additionally loaded with a value of the power lost in system B.

At the end of Stage III the frequency drops by a value Δf_{III} that can be calculated from the system stiffness K_f defined in equation 8.29. ΔP_0 is given by

$$\Delta P_0 = (\Delta P_{TAIII} - \Delta P_{LAIII}) + (\Delta P_{TBIII} - \Delta P_{LBIII}) \tag{8.38}$$

and, when substituting the values for K_{TA}, K_{LA}, K_{TB} and K_{LB} obtained from equations 8.35 and 8.6, a similar formulae to that in equation 8.29 for the frequency drop in Stage III is obtained

$$\frac{\Delta f_{III}}{f_n} = \frac{-1}{K_{fA}P_{LA} + K_{fB}P_{LB}}\Delta P_0 \tag{8.39}$$

where $K_{fA} = K_{TA} + K_{LA}$ and $K_{fB} = K_{TB} + K_{LB}$ are the stiffnesses of the big and the small system respectively. The increase in the tie-line interchange can be determined from a power balance on one of the systems, in this case the big system A, when

$$\begin{aligned}\Delta P_{tieIII} &= \Delta P_{TAIII} - \Delta P_{LAIII} = -(K_{TA} + K_{LA})P_{LA}\frac{\Delta f_{III}}{f_n} \\ &= -K_{fA}P_{LA}\frac{\Delta f_{III}}{f_n} = \frac{K_{fA}P_{LA}}{K_{fA}P_{LA} + K_{fB}P_{LB}}\Delta P_0\end{aligned} \tag{8.40}$$

Under the assumption $P_{LA} \gg P_{LB}$ this formula simplifies to $\Delta P_{tieIII} \approx \Delta P_0$ when, during Stage III, the tie-line power interchange is increased by the value of the lost power, as during Stage II. Such an increase in the power flow may result in system instability when the interconnected system will split into two asynchronously operating subsystems. Also with such a large power transfer the thermal limit on the line may be exceeded when the overcurrent relays will trip the line again leading to asynchronous operation of both systems.

Assuming that the tie-line remains intact the increase in the tie-line interchange power, combined with the drop of frequency, will force the central regulators in both systems to intervene. The area control error at the end of Stage III can be calculated from equation 8.11 using the formulae given in equations 8.39 and 8.40 as

$$ACE_A = -\Delta P_{tieIII} - \lambda_{RA}\Delta f_{III} \quad \text{and} \quad ACE_B = +\Delta P_{tieIII} - \lambda_{RB}\Delta f_{III} \tag{8.41}$$

As explained in Section 8.1 (equation 8.10) the ideal regulator bias settings are

$$\lambda_{RA} = K_{fA}\frac{P_{LA}}{f_n} \quad \text{and} \quad \lambda_{RB} = K_{fB}\frac{P_{LB}}{f_n} \tag{8.42}$$

but this is difficult to achieve in practice as the value of the stiffnesses K_{fA} and K_{fB} can only be estimated. Consequently assuming that K_{RA} and K_{RB} are estimates of K_{fA} and K_{fB} respectively then the bias settings are

$$\lambda_{RA} = K_{RA}\frac{P_{LA}}{f_n} \quad \text{and} \quad \lambda_{RB} = K_{RB}\frac{P_{LB}}{f_n} \tag{8.43}$$

and equations 8.41 become

$$ACE_A = -\Delta P_{tieIII} - \lambda_{RA}\Delta f_{III} = \frac{-K_{fA}P_{LA} + K_{RA}P_{LA}}{K_{fA}P_{LA} + K_{fB}P_{LB}}\Delta P_0 \tag{8.44}$$

$$ACE_B = +\Delta P_{tieIII} - \lambda_{RB}\Delta f_{III} = \frac{K_{fA}P_{LA} + K_{RB}P_{LB}}{K_{fA}P_{LA} + K_{fB}P_{LB}}\Delta P_0 \tag{8.45}$$

Ideal settings of the regulators

Assume that the stiffnesses K_{fA} and K_{fB} of both systems are known and that the central regulator bias settings λ_{RA} and λ_{RB} are selected such that

$$K_{RA} = K_{fA}, \qquad K_{RB} = K_{fB} \tag{8.46}$$

In this case, equations 8.44 and 8.45 give

$$ACE_A = 0 \quad \text{and} \quad ACE_B = \Delta P_0 \tag{8.47}$$

and the central regulator of the big system, system A, will not demand an increase in the power generation in its area. Only the small system will increase its generation in order to cover its own power imbalance. If the available regulation power in the small system is high enough to cover the lost generation then the big system will not intervene at all and, as the generation in the small system is increased, the tie-line power interchange will drop to its scheduled value.

The value of the system stiffness $K_f = K_T + K_L$ is never constant as it depends on the composition of the system load and generation and on the spinning reserve. Consequently the condition defined by equation 8.46 is almost never satisfied and the central regulator of the bigger system will take part in the secondary control to an amount defined by equation 8.44.

Example 8.2

An interconnected system consists of two subsystems of different size. The data of the subsystems are: $f_n = 50$ Hz, $P_{LA} = 37\,500$ MW, $K_{TA} = 8$ ($\rho_{TA} = 0.125$), $K_{LA} \approx 0$, $K_{RA} = K_{TA}$, $P_{LB} = 4000$ MW, $K_{TB} = 10$ ($\rho_{TB} = 0.1$), $K_{LB} \approx 0$, $K_{RB} = K_{TB}$.
Two large generating units are suddenly lost in the smaller system producing a power deficit of $\Delta P_0 = 1300$ MW, that is 32.5% of the total generation in this subsystem. The resulting frequency and power variations for both subsystems, assuming no limits on the power generation (linear generation characteristics), are shown in Figure 8.25.
The characteristic P_{TA} has a small slope corresponding to $K_{TA}P_{LA}/f_n = 6000$ MW/Hz. The slope of the characteristic of the smaller system is much higher and corresponds to $K_{TB}P_{LB}/f_n = 800$ MW/Hz. The characteristic P_{TB+} is inverted and shifted by $\Delta P_0 = 1300$ MW, so that intersection of the characteristics P_{TB+} and P_{TA} determines the frequency deviation Δf_{III} at point *III*. The central regulator of the small system acts fast enough for the P_{TB} characteristic to move, so that point *III* is not reached. Further action due to the integral term in the regulator shifts the characteristic P_{TB} slowly to the position $P_{TB\infty}$ corresponding to nominal frequency.
It is worth noting the small frequency oscillations characteristic of the third stage of the dynamics. Because of the fast-acting central regulator these oscillations take place above the value Δf_{III}, Figure 8.25b. The tie-line power flow initially increases very quickly in Stage I of the dynamics, and then oscillates around the scheduled value increased by the value of the lost generation. The period of oscillations is around 3 s. Then, as the central regulator in the small system enforces an increase in the generation, the tie-line flow slowly decreases to its scheduled value.

The central PI regulator quite quickly creates the output signal ΔP_{refB} as shown in Figure 8.25c. The generating units respond to this increased power demand by increasing the generation by ΔP_{TB}, Figure 8.25d. Because of the primary turbine control the mechanical power of the big system P_{TA} initially increases to cover the power imbalance and then quickly drops to the instantaneous equilibrium point corresponding to the end of Stage III. After that P_{TA} slowly decreases as the generation P_{TB}, forced by the secondary control, increases. At the same time the frequency increases so that the frequency error of the central regulator of the big system becomes negligibly small and the regulator becomes inactive.

Non-ideal settings of the central regulator

If $K_{RA} > K_{fA}$, then the signal $\lambda_{RA}\Delta f$ is initially larger than the signal ΔP_{tie} and the regulator tries to increase the generation in system A. Although the increased generation speeds up the rate at which the frequency increases it slows down the rate at which the tie-line error ΔP_{tie} decreases and in some cases can lead to a temporary increase in this error. Consequently the signal ΔP_{tie} becomes greater than the signal $\lambda_{RA}\Delta f$ and the central regulator starts to reduce the generation in system A, thus essentially withdrawing this subsystem from the secondary control.

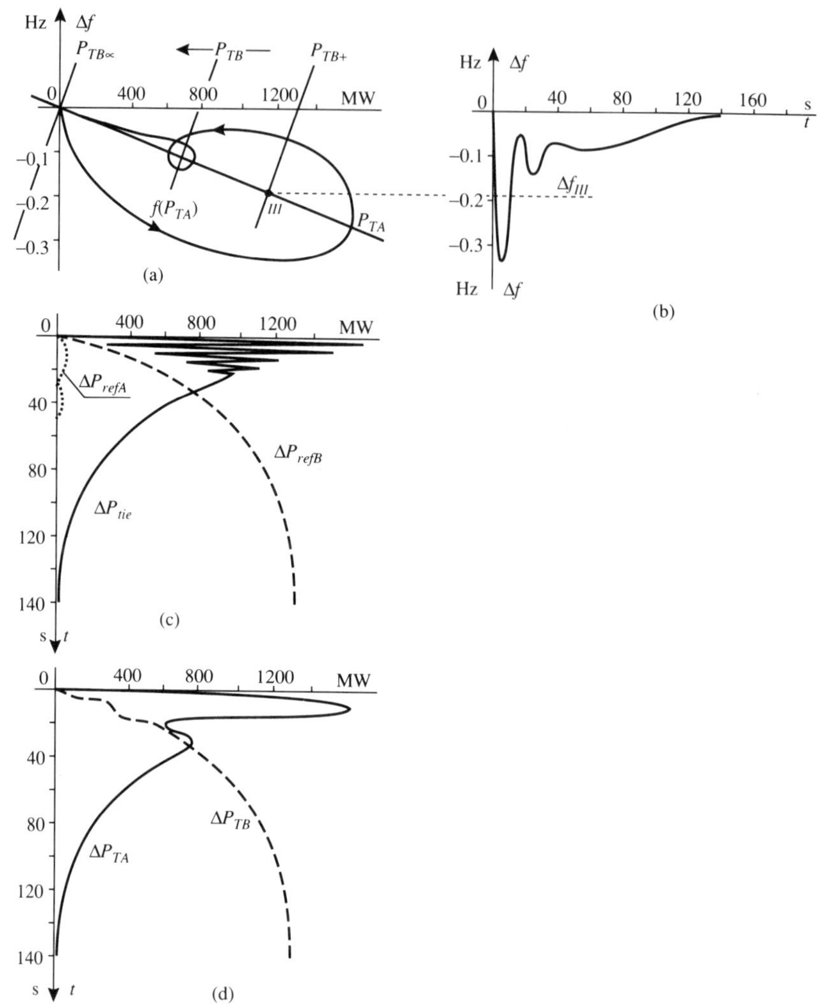

Figure 8.25 Illustration to example 8.2: (a) the static characteristics of both systems and the trajectory $f(P_{TA})$; (b) frequency variations; (c) tie-line power interchange oscillations and the change of power demanded by the central regulators of both subsystems; (d) change in the mechanical power supplied by the turbines

If $K_{RA} < K_{fA}$, then the signal ΔP_{tie} is initially higher than the signal $\lambda_{RA}\Delta f$ and the regulator of the big system tries to reduce its generation despite the fact that frequency is smaller than the nominal. This drop in generation is not desirable from the frequency regulation point of view but does reduce the tie-line flows. When $\lambda_{RA}\Delta f$ becomes larger than ΔP_{tie} the regulator will start to increase the recently reduced generation so as to re-establish the required power in system A.

In both the above cases the inaccuracy of the frequency bias setting causes unnecessary intervention of the bigger system in covering the generation loss in the smaller system. In the above example a large amount of regulation power was available so that the use of non-ideal regulator settings was not dangerous. However, should the amount of regulation power available be insufficient to cover the lost power then the consequence of non-ideal regulator settings becomes more significant as described below.

Insufficient available regulation power

When the available regulation power ΔP_{regB} in the small system is less than the generation loss ΔP_0 then system B is unable to cover the power loss on its own and the big system A must intervene to cover part of the lost power. Initially the dynamics are identical to the case in which the regulation power in the small system was unlimited. The difference between the two cases appears when the small system has used up all its available regulation power. Further changes can only take place via regulation of the big system. Its central regulator is now subject to two error signals of opposite sign (Figure 8.8). The signal $\lambda_{RA}\Delta f$ generated by the frequency error demands an increase in the generation whilst the signal ΔP_{tie} generated by the tie-line power deviation demands a decrease in the generation. The regulation process terminates when the signals balance each other and the total error signal is zero. Denoting the final steady-state values of the error signals as Δf_∞ and $\Delta P_{tie\infty}$, the regulation equation is

$$ACE_A = -\Delta P_{tie\infty} - K_{RA}P_{LA}\frac{\Delta f_\infty}{f_n} = 0 \tag{8.48}$$

On the other hand the tie-line power interchange must satisfy the overall power balance of the small system

$$\Delta P_0 - \Delta P_{regB} = \Delta P_{tie\infty} - (K_{TB} + K_{LB})P_{LB}\frac{\Delta f_\infty}{f_n} \tag{8.49}$$

Physically this means that the power imbalance in the small system B may be covered partly by an increase in the power import from the big system A, partly by a change in internal generation and partly by a decreased demand resulting from the frequency drop in the whole interconnected system. Solving equations 8.48 and 8.49 gives

$$\Delta P_{tie\infty} = \frac{K_{RA}P_{LA}}{K_{RA}P_{LA} - K_{fB}P_{LB}}(\Delta P_0 - \Delta P_{regB}) \tag{8.50}$$

$$\frac{\Delta f_\infty}{f_n} = -\frac{1}{K_{RA}P_{LA} + K_{fB}P_{LB}}(\Delta P_0 - \Delta P_{regB}) \tag{8.51}$$

Under the assumption $P_{LA} \gg P_{LB}$ equations 8.50 and 8.51 may be simplified to

$$\Delta P_{tie\infty} \cong (\Delta P_0 - \Delta P_{regB}) \tag{8.52}$$

$$\frac{\Delta f_\infty}{f_n} \cong -\frac{1}{K_{RA}P_{LA}}(\Delta P_0 - \Delta P_{regB}) \tag{8.53}$$

Additional validity is given to this simplification if the small system is fully loaded ($\Delta P_{regB} = 0$ and $K_{TB} = 0$) when the stiffness $K_{fB} = K_{TB} + K_{LB}$ has a small value corresponding to the frequency sensitivity of the loads K_{LB}. In this situation the whole power imbalance is covered by the tie-line interchange.

The frequency deviation Δf_∞ is inversely proportional to the coefficient $\lambda_{RA} = K_{RA}P_{LA}/f_n$ set at the central regulator. If the available regulation power is not large

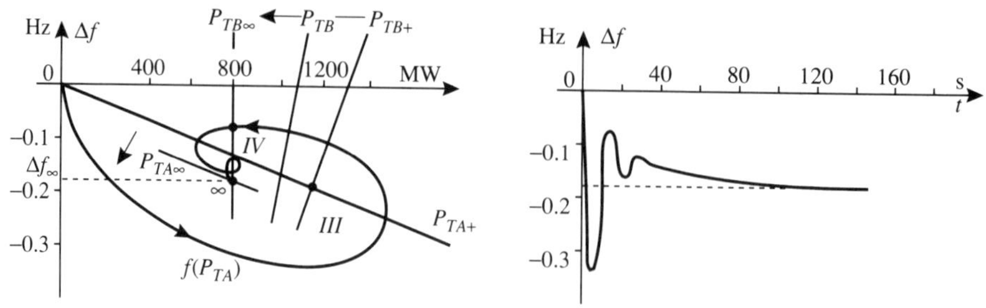

Figure 8.26 Illustration to Example 8.3

enough then too low a setting of the central regulator produces a steady-state frequency error. If $K_{RA} = K_{fA}$ then the final value of the frequency will correspond to the frequency level at which the available regulation power of the small system has run out. If $K_{RA} > K_{fA}$ then the regulator of the big system will increase its generation decreasing the frequency error and allowing the tie-line interchange error to increase. If $K_{RA} < K_{fA}$, then the regulator of the big system will decrease its generation increasing the frequency error and the tie-line interchange error will not be allowed to increase. This can be illustrated by the following example.

Example 8.3

The available regulating power of the small subsystem considered in Example 8.2 is $\Delta P_{regB} = 500$ MW. The settings of the central regulators are $K_{RA} = 5.55 < K_{TA}$ and $K_{RB} = 12.5 > K_{TB}$. Neglecting the frequency sensitivity of the load, equations 8.52 and 8.53 give: $\Delta P_{tie\infty} = 800$ MW and $\Delta f_{\infty} = -0.16$ Hz. The power and frequency variations are illustrated in Figure 8.26. During the first 20 s of the disturbance the trajectory is the same as the one shown in Figure 8.25, but now the characteristic P_{TB} eventually settles at $P_{TB\infty}$. Trajectory $f(P_{TA})$ begins to wrap around the instantaneous equilibrium point *IV*, but the big system regulator decreases its generation shifting the characteristic from position P_{TA+} to position $P_{TA\infty}$. The dynamics end up at point ∞ where the frequency of the interconnected system is lower, by about $\Delta f_{\infty} = -0.16$ Hz, than the required value. A small portion of the trajectory $f(P_{TA})$, between point *IV* and ∞, corresponds to a slow reduction of frequency over several tens of seconds. The variations in the tie-line power interchange are similar to those shown in Figure 8.25, but settle down at a level corresponding to point ∞, i.e. $\Delta P_{tie\infty} = 800$ MW (the generation lost was $\Delta P_0 = 1300$ MW).

If the tie-line power interchange determined by equation 8.50 is greater than the maximal thermal capacity of the tie-line, then the line will be tripped and the systems separated. The imbalance of power in the small system will cause a further drop in frequency which, in the absence of automatic load shedding, may lead to frequency collapse in that sub-system.

9
Stability Enhancement

The stability of a power system can be enhanced, and its dynamic response improved, by correct system design and operation. For example, the following features help to improve stability:

- the use of protection equipment and circuit breakers that ensure the fastest possible fault clearing
- the use of single-pole circuit breakers so that during single-phase faults only the faulted phase is cleared and the unfaulted phases remain intact
- the use of a system configuration that is suitable for the particular operating conditions (e.g. avoiding long, heavily loaded, transmission links)
- ensuring an appropriate reserve in transmission capability
- avoiding operating the system at low frequency and/or voltage
- avoiding weakening the network by the simultaneous outage of a large number of lines and transformers

In practice financial considerations determine the extent to which any of these features can be implemented and there must always be a compromise between operating a system near to its stability limit and operating a system with an excessive reserve of generation and transmission. The risk of losing stability can be reduced by using additional elements inserted into the system to help smooth the system dynamic response. This is commonly referred to as *stability enhancement* and is the subject of this chapter.

9.1 Power System Stabilisers

A power system stabiliser (PSS) is a device which provides additional supplementary control loops to the Automatic Voltage Regulator (AVR) system and/or the turbine-governing system of a generating unit. A PSS is also one of the most cost-effective methods of enhancing power system stability.

9.1.1 PSS applied to the excitation system

Adding supplementary control loops to the generator AVR is one of the most common ways of enhancing both small-signal (steady-state) stability and large-signal (transient) stability. Adding such additional control loops must be done with great care as Section 5.5 explained how an AVR (without supplementary control loops) can weaken the damping provided by the damper and field windings. This reduction in the damping torque is primarily due to the voltage regulation effects inducing additional currents in the rotor circuits that oppose the currents induced by the rotor speed deviation $\Delta\omega$. This phase relationship was illustrated in Figure 5.26 for the field winding (and in Figure 5.27 for the d-axis damper winding) and gives an immediate insight into what is required from the PSS.

The main idea of power system stabilisation is to recognise that in the steady-state, that is when the speed deviation is zero or nearly zero, the voltage controller should be driven by the voltage error ΔV only. However in the transient state the generator speed is not constant, the rotor swings, and ΔV undergoes oscillations caused by the change in rotor angle. The task of the PSS is to add an additional signal which compensates for the ΔV oscillations and provides a damping component that is in phase with $\Delta\omega$. This is illustrated in Figure 9.1a where the signal V_{PSS} is added to the main voltage error signal ΔV. In the steady-state V_{PSS} must be equal to zero so that it does not distort the voltage regulation process. Figure 9.1b shows the phasor diagram of the signals in the transient state. As in Section 5.5.3 it is assumed that each signal varies sinusoidally with the frequency of rotor swings and may therefore be represented by a phasor. The phasor $\underline{V}_{PSS}$ directly opposes $\Delta\underline{V}$ and is larger than it. The net voltage error phasor $\Delta\underline{V}_{\Sigma}$ now leads the speed deviation phasor $\Delta\underline{\omega}$ instead of lagging it as in Figure 5.26b. As explained in Section 5.5.3 the phasor of the incremental excitation emf $\Delta\underline{E}_f$ lags $\Delta\underline{V}_{\Sigma}$ by an angle introduced by the AVR and the exciter so that the quadrature component (with respect to $\Delta\underline{\delta}$) of the phasor $\Delta\underline{E}'_{q(\Delta E_f)}$, due to the excitation control is now in phase with $\Delta\underline{\omega}$. This, together with $\Delta\underline{E}'_{q(\Delta\delta)}$, introduces a large damping torque into the system. However if the magnitude of $\underline{V}_{PSS}$ is less than that of $\Delta\underline{V}$ then only partial compensation of the negative damping component introduced by the AVR is achieved.

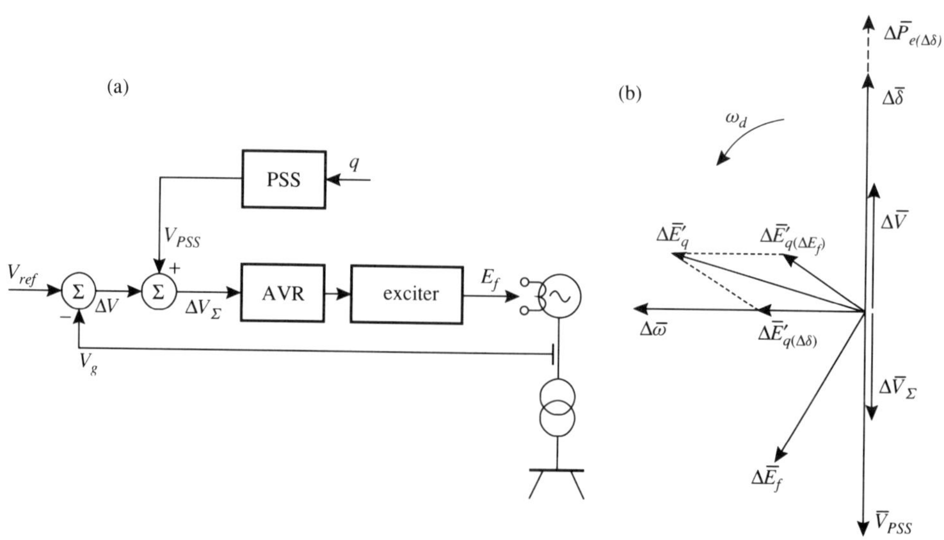

Figure 9.1 Supplementary control loop for the AVR system: (a) block diagram; (b) phasor diagram

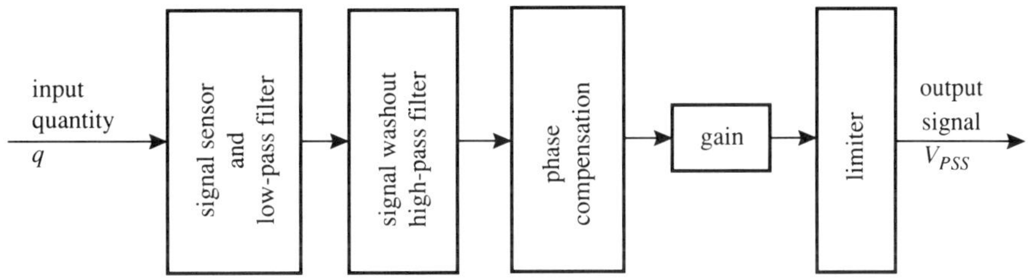

Figure 9.2 The major elements of a PSS

The general structure of the PSS is shown in Figure 9.2 where the PSS signal V_{PSS} can be provided from a number of different input signals measured at the generator terminals. The measured quantity (or quantities) is passed through low- and high-pass filters. The filtered signal is then passed through a lead and/or lag element in order to obtain the required phase shift and, finally, the signal is amplified and passed to a limiter. When designing the phase compensation it is necessary to take into account the phase shift of the input signal itself and that introduced by the low- and high-pass filters. Sometimes the filters are designed in such a way that they give a net zero-phase-shift for the frequency of rotor oscillations (Huwer, 1992). The mathematical model of the PSS will be described in more detail in Section 10.2.

Typically the measured quantities used as input signals to the PSS are the rotor speed deviation, the generator active power or the frequency of generator terminal voltage. There are a number of possible ways of constructing a PSS depending on the signal chosen.

PSS based on $\Delta\omega$

The oldest types of PSS use a measurement of the speed deviation of the generator shaft. Obviously this signal must be processed in order to filter out all the measurement noise. The main problem with this method, when applied to turbogenerators with long shafts prone to torsional oscillations, relates to selecting a measurement position on the shaft that properly represents the speed deviation of the rotor magnetic poles. For long shafts it is necessary to measure the speed deviation at a number of points along the shaft and use this information to calculate the average speed deviation. Moreover the stabiliser gain is constrained by the influence the PSS has on the torsional oscillations. These problems are described by Watson & Coultes (1973) and Kundur, Lee & Zein El-Din (1981).

PSS based on $\Delta\omega$ and P_e

The need to measure the speed deviation at a number of points along the shaft can be avoided by calculating the average speed deviation from measured electrical quantities. The method calculates the equivalent speed deviation $\Delta\omega_{eq}$ indirectly from the integral of the accelerating power:

$$\Delta\omega_{eq} = \frac{1}{M}\int(\Delta P_m - \Delta P_e)\mathrm{d}t \tag{9.1}$$

and ΔP_e is calculated from measurements of the generated real power P_e. The integral of the change in the mechanical power ΔP_m can be obtained from

$$\int \Delta P_m \mathrm{d}t = M \Delta\omega_{measured} + \int \Delta P_e \mathrm{d}t \tag{9.2}$$

where $\omega_{measured}$ is based on the end-of-shaft speed sensing system. Because the mechanical power changes are relatively slow the derived integral of the mechanical power can be passed through a low-pass filter to remove the torsional frequencies from the speed measurement. The resulting PSS contains two input signals, $\Delta\omega_{measured}$ and ΔP_e, which are used to calculate $\Delta\omega_{eq}$. The final V_{PSS} signal is designed to lead $\Delta\omega_{eq}$. The block diagram of the system is shown in Figure 9.3 (Kundur, 1994) where $G(s)$ is the transfer function of the torsional filter. This type of PSS with two input signals allows a large gain to be used so that good damping of power swings is obtained (Lee, *et al.*, 1981).

PSS based on P_e

A simplified form of the PSS shown in Figure 9.3 can be obtained by neglecting the shaft speed measurement and only measuring the generator real power P_e. With this arrangement only one input signal is required but it can only be used when the mechanical power can be assumed to be constant. If the mechanical power does change, for example due to secondary frequency control, this solution produces transient oscillations in the voltage and reactive power that are unnecessarily forced by the PSS as it sees the change in the mechanical power as a power swing.

PSS based on f_{Vg} and $f_{E'}$

The measurement of shaft speed can be replaced by a measurement of the generator terminal voltage frequency f_{Vg} (Larsen & Swan, 1981). A disadvantage of this solution is that the terminal voltage waveform can contain noise produced by large industrial loads such as arc furnaces. The accuracy of this measured speed signal can be improved by adding the voltage drop across the transient reactance to the generator voltage to obtain the transient emf E' and its frequency $f_{E'}$. The PSS now receives two signals, the generator current and voltage. As in the case of the PSS utilising the measured shaft speed deviation, the PSS gain is limited by the effect of shaft torsional oscillations. The advantage of this solution compared with other types of stabilisers is that it improves the damping of inter-area oscillations in interconnected power systems.

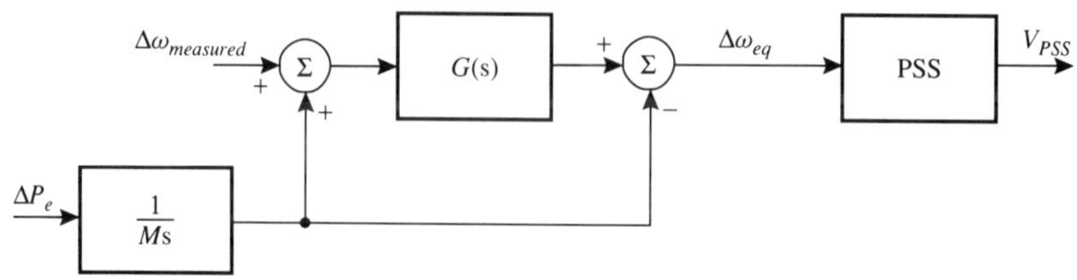

Figure 9.3 Block diagram of PSS using speed deviation and real power as input signals

PSS design

Designing and applying the PSS is not simple and requires a thorough analysis of the regulator structure and its parameters. A badly designed PSS can become a source of a variety of undesired oscillations. It should be remembered that the phasor diagrams in Figures 5.26, 5.27 and 9.1 are valid only for a simple generator-infinite busbar system with all the resistances and local loads neglected. A more detailed analysis shows that the phase shift between $\Delta\overline{E}_f$ and $\Delta\overline{E}'_{q(\Delta E_f)}$ is not exactly $\pi/2$ and depends on the loading and system parameters (De Mello & Concordia, 1969). This requires a more precise matching of the phase compensation with the actual loading and system parameters.

The parameters of a PSS are usually optimised with respect to the damping of small disturbance power swings. However a properly designed PSS also improves the damping under large-disturbance conditions. In order to enhance the first-swing transient stability an additional control loop can be added to the PSS that acts in a similar way to the *forced excitation* in old electromechanical AVR systems. Such forced excitation was executed by short-circuiting resistors in the excitation winding in order to increase E_f to its ceiling value for about 0.5 s. The resistors were then re-inserted and E_f decreased. A similar solution is used in so-called *discontinuous excitation control* systems. In the solution described by Lee & Kundur (1986) and Kundur (1994) an additional element is switched in by a relay which supplies, in parallel to the PSS, a signal to force the excitation to an increased value. This element is switched off when the sign of the speed deviation changes (that is the rotor decelerates).

In the former Soviet Union supplementary PSS control loops are not used in the AVR but multi-variable regulators with internal feedback loops are used instead (1970).

9.1.2 PSS applied to the turbine governor

As all the generators in the power system are linked by the transmission network, voltage control on one of the generators influences the dynamic response of all the other generators. Consequently, a PSS that improves the damping of one generator does not necessarily improve the damping of the other generators. Therefore a local design may not provide the global optimal solution and a co-ordinated synthesis procedure is desirable. This co-ordination increases the design computation and is usually valid only for typical network configurations and loading conditions. When a severe fault occurs the post-fault network configuration, and load, may be significantly different from the pre-fault conditions and poorly damped swings may result. Because of these factors interest has focused on utilising the turbine governor for the damping of local and inter-area oscillations.

Including a PSS signal in the turbine governing systems with the aim of improving damping is not new. Moussa & Yu (1972) describe some solutions regarding hydro-turbines. The principle of providing an additional damping torque from the turbine-governor is similar to that used when adding a PSS loop to the excitation system. The time constants in the turbine-governor introduce a phase shift between the oscillations in the speed deviation $\Delta\omega$ and the turbine mechanical power. As the input signal to the supplementary PSS control loop is proportional to $\Delta\omega$ the PSS transfer function must be chosen in such a way that at the frequency of rotor oscillations it compensates the phase shift introduced by the turbine-governor. Consequently the PSS will force changes in the mechanical power ΔP_m that are in phase with $\Delta\omega$ and, according to the swing equation 5.15, provide positive damping.

The main advantage of applying a PSS loop to the turbine governor lies in the fact that the turbine-governor dynamics are weakly coupled with those of the rest of the system. Consequently the parameters of the PSS do not depend on the network parameters. Wang *et al.* (1993) show interesting simulation results for systems equipped with PSS applied to the governors of steam turbines. Although this type of PSS is not currently used in practice, such solutions should not be ruled out in the future.

9.2 Fast Valving

Chapter 6 explained how a large disturbance near to a generator (e.g. a sudden short-circuit) will produce a sudden drop in the generator output power followed by rapid acceleration of the generator rotor. The natural action to counteract this drop in electrical power would be to rapidly reduce the mechanical input power thereby limiting the acceleration torque. The effect of such action can be explained by considering Figure 9.4 which shows the equal area criterion applied to the system shown in Figure 6.6 when a fault appears on line *L2* and is cleared without auto-reclosure. If it is assumed that the

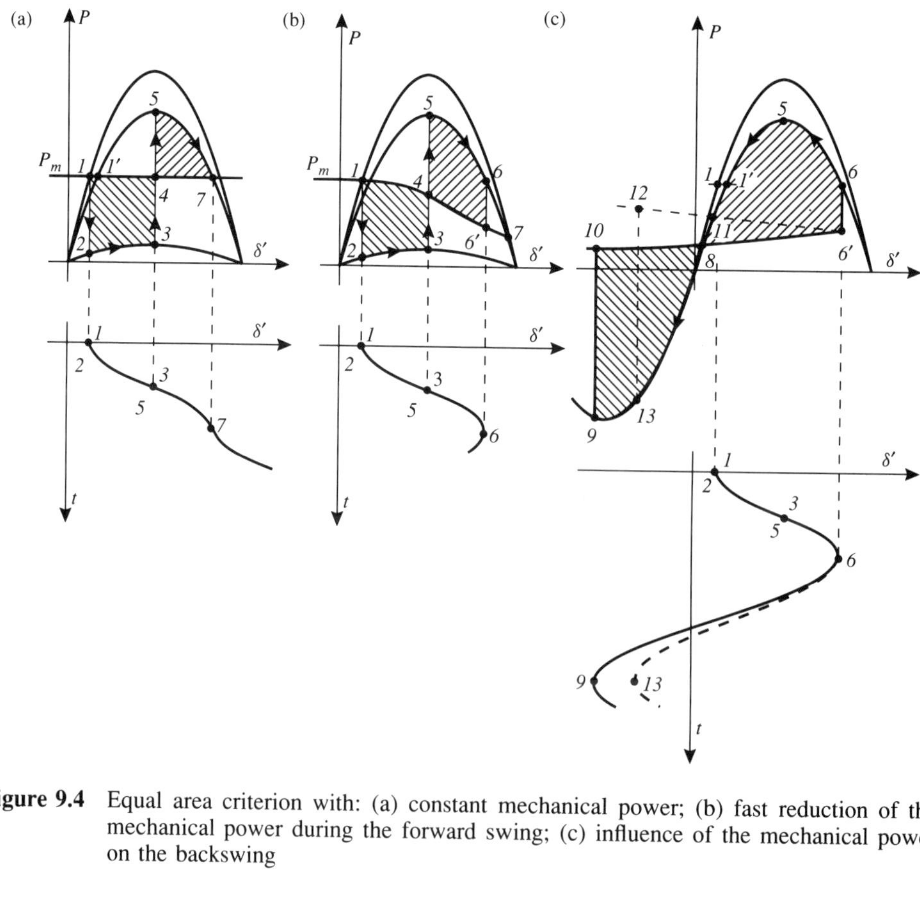

Figure 9.4 Equal area criterion with: (a) constant mechanical power; (b) fast reduction of the mechanical power during the forward swing; (c) influence of the mechanical power on the backswing

accelerating area *1-2-3-4* in Figure 9.4a is greater than the maximum possible decelerating area *4-5-7* then, when the line is tripped, the rotor will make an asynchronous rotation and the system will lose stability. Now assume that the mechanical power P_m is reduced immediately after the disturbance occurs. The technical possibilities of providing such a fast power reduction will be discussed later. Figure 9.4b shows that reduction of P_m during the forward swing reduces the accelerating area *1-2-3-4* and increases the decelerating area to area *4-5-6-6′*. The system remains stable with the stability margin being proportional to area *6-7-6′*. The maximum improvement in stability is obtained when the reduction in P_m takes place as early, and as fast, as possible.

Figure 9.4c shows the situation to be somewhat different during the backswing. With the mechanical input power reduced the system returns towards a rotor angle δ' that is smaller than the initial value and performs deceleration work equal to the integral of the difference between the electrical and mechanical power. During this part of the oscillation a further decrease in the mechanical power, shown by the solid line *6′-8-10*, has a detrimental effect on the system dynamics as the deceleration work *6′-6-5-8* performed during the backswing can only be balanced by a large accelerating area *8-9-10* giving a large rotor deflection towards negative δ'. The backswing is increased by the drop in P_m. The dashed line in Figure 9.4c shows how the amplitude of the backswing could be reduced by increasing P_m during the backswing. This would result in a smaller decelerating area *6′-6-5-11* which is then balanced by the smaller accelerating area *11-13-12*.

After a disturbance the system shown in Figure 9.4c recovers to the pre-fault value of the mechanical power. In systems operating near to their steady-state stability limit it may happen that the fault clearance increases the system equivalent reactance to a value that would correspond to steady-state instability, that is when the mechanical power is greater than the amplitude of the post-fault characteristic. In order to ensure steady-state stability in such situations it is necessary to reduce the final, post-fault, value of the mechanical power. This is illustrated in Figure 9.5a where, on fault clearance, the post-fault power-angle characteristic lies below the dashed line $P_{m0} =$ constant and the system would lose stability even if the fault were cleared instantaneously. To prevent such an instability the final, post-fault, value of mechanical power must be reduced to $P_{m\infty}$ as illustrated in Figure 9.5b.

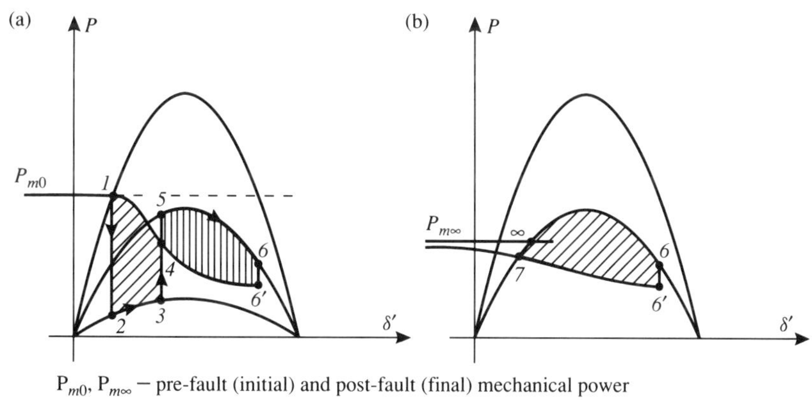

Figure 9.5 Equal area criterion with fast changes of mechanical power in a system operating near its steady-state stability limit: (a) forward swing; (b) backswing

Fast changes in the mechanical power as shown in Figure 9.4 and Figure 9.5 require a very fast response from the turbine. The examples shown above indicate that the decrease in the turbine power should take place within the first third of the swing period, that is during the first few tenths of a second following a disturbance. Restoring the power to the required post-fault value should take about half of the swing period, that is less than a second. Such fast control is not possible with hydro-turbines due to the large change in pressure, and huge torques, necessary to move the control gates. However a steam turbine can be used for stability enhancement as it can be made to respond almost as fast as required. The control action needed within the turbine to produce this fast response is referred to as *fast valving*. When the mechanical power is restored to a final value that is equal to the pre-fault value the fast valving is said to be *momentary*. When the final value of mechanical power is smaller than the pre-fault value, the fast valving is said to be *sustained*. Mechanical power variations corresponding to both these cases are shown in Figure 9.6.

Steam turbine fast valving cannot be achieved using the standard closed-loop turbine governor shown in Figure 2.11 as a change in the valve setting only occurs after a speed error appears at the controller input. Due to the large inertia of the turbine-generator rotor the speed can only change slowly so that the response of the closed-loop arrangement, with speed as the control variable, is too slow. As control action is required immediately the fault is detected, open-loop control systems are normally used.

In a modern steam turbine fast valving is executed using the existing control valves. Consider, for example, the single reheat unit shown in Figure 2.6. Here rapid closure of the main governor control valves will not produce a large reduction in the turbine power because the high-pressure stage produces only about 30% of the power. The reheater stores a large volume of steam and, even with the main governor valve closed, the turbine still supplies about 70% of power through the intermediate and low-pressure stages. A large, rapid, reduction of power can only be obtained by closing the intercept control valves as it is these which control the steam flow to the intermediate and low-pressure parts of the turbine.

Momentary fast valving, shown in Figure 9.7a, is achieved by rapidly closing the intercept control valves, holding them closed for a short time and then re-opening them in order to restore the power to its initial, pre-fault, value. During the short time the intercept valves are shut the steam flows from the boiler through the high-pressure turbine and is accumulated in the reheater.

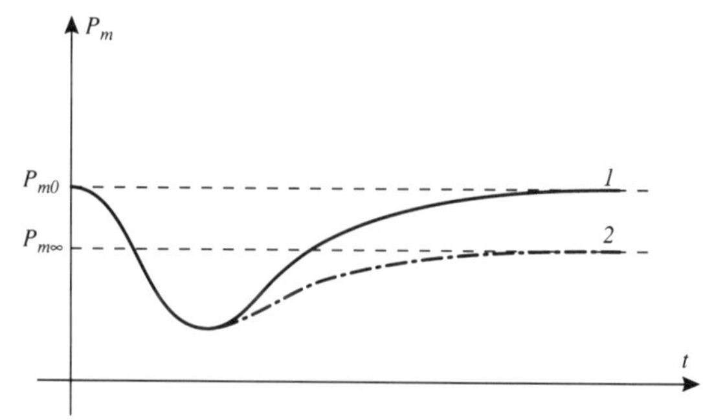

Figure 9.6 Mechanical power variations in the case of: *1* — momentary fast valving; *2* — sustained fast valving

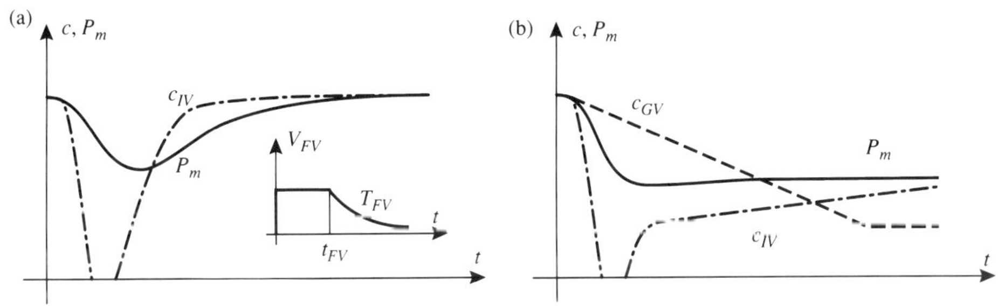

c_{GV} - position of the main governor control valve, c_{IV} - position of the intercept control valve

Figure 9.7 Variations in the valve position and output power of a single-reheat steam turbine due to : (a) momentary fast valving; (b) sustained fast valving

Sustained fast valving can be provided by a rapid closure of the intercept valves, Figure 9.7b, followed by slow partial closing of the main governor valve. After a short period the intercept valves are again re-opened. This guarantees a rapid initial reduction in the mechanical power followed by its restoration to a value smaller than the initial one.

Fast valving of modern turbines equipped with electro-hydraulic governing systems shown in Figure 2.11b can be achieved by feeding a fast valving signal $V_{FV}(t)$ directly to the coil of the electro-hydraulic converter. The signal is produced by an additional controller FV operating in open-loop mode. In the case of momentary fast valving the signal $V_{FV}(t)$ consists of two parts as shown in Figure 9.7a. The first part is a rectangular pulse necessary for rapid closure of the valves. The height of the rectangular signal determines the magnitude of the voltage applied to the converter coil and, if large enough, the servomotor will close the valve in about 0.1–0.4s. The width of the rectangular pulse t_{FV} determines the duration of the power reduction. The second part of the signal decays with a time constant T_{FV}, where T_{FV} is selected so that the speed at which the valves are re-opened does not exceed a pre-set maximum value. This limit on the speed of valve re-opening is mainly set by the strength of the rotor blades so that when re-opening the valves a 100% change in the valve position cannot normally be completed in less than about 1s.

Generally the fast valving controller FV operates with a pre-determined set of control parameters that are prepared off-line, by performing a large number of power system simulations, taking into account:

- the pre-fault network configuration
- the pre-fault loading condition
- the location of the fault, that is in which line the fault occurs
- the distance and the type of fault as measured by, for example, the accelerating power or voltage drop
- whether the fault clearance is with, or without, auto-reclosure

A control signal is then prepared off-line for a large number of possible fault scenarios. When a fault occurs the controller is fed real-time information on all the above factors so that it can pick a control strategy, from its predetermined set, that most closely matches the actual fault condition.

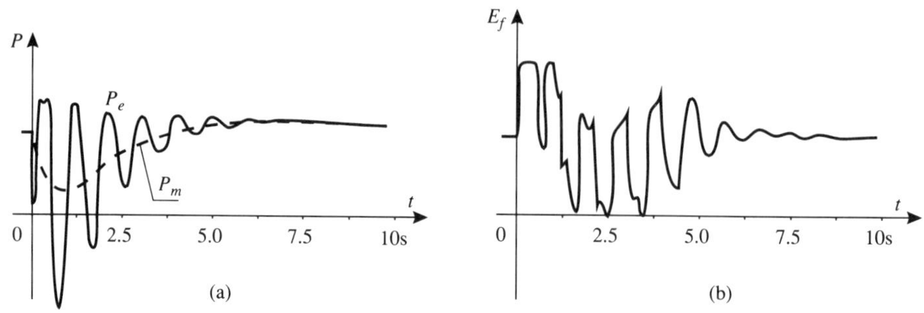

Figure 9.8 Influence of fast valving and PSS on the transient state: (a) mechanical and electrical power variations; (b) excitation voltage variations (Based on Brown *et al.*, 1970)

In practice the reduction, and restoration, of power shown in Figure 9.7 tends to be slower than that required from the stability enhancement point of view described in Figure 9.4. If the control valves are re-opened too late the backswing will increase and may lead to *second-swing instability*. When a generator is equipped with a fast AVR and a power system stabiliser (PSS) the situation can be improved as illustrated in Figure 9.8. Fast power reduction takes place within the first few tenths of a second after the fault so that the rotor does not lose stability but swings over the peak of the power-angle characteristic as indicated by the two humps in the first power swing (compare with Figure 6.9a). Restoration of the turbine power takes a few seconds which is too slow and so causes two deep rotor backswings that reach the motoring range of operation (negative electrical power). The depth of these backswings is reduced by fast control of the excitation voltage enforced by the AVR equipped with PSS. The average value of E_f is decreased and the average value of P_e follows the mechanical power P_m.

Although the cost of implementing fast valving is usually small, the adverse effect on the turbine and boiler may be serious. Generally fast valving is only used in difficult situations where the AVR and PSS cannot, on their own, prevent instability.

9.3 Generator Tripping

Tripping one or more generators from a group of generators that are operating in parallel on a common busbar is perhaps the simplest, and most effective, means of rapidly changing the torque balance on the generator rotors. Historically generator tripping was confined to hydro power stations where fast valving could not be used, but now many power companies have extended its use to both fossil-fuelled and nuclear-powered generating units in order to try to prevent system instability after severe disturbances.

When a generator is tripped it will normally go through a standard shut-down and start-up cycle which can take several hours. To avoid this long procedure the generator is usually disconnected from the network but still used to supply the station auxiliary demand. This allows the generator to be re-synchronised to the system and brought back to full or partial load in several minutes.

The main disadvantage of generator tripping is that it creates a long-term power imbalance characterised by variations in frequency and power interchange between interconnected systems as described in Chapter 8. Moreover generator tripping results in a sudden increase in the electromagnetic torque acting on the other generators which can lead to a reduction in rotor shaft fatigue life. Although unlikely, if two or more units are tripped at exactly the same time, then the shaft loading on the remaining generators can become very severe indeed.

Generator tripping falls into two different categories. It is said to be *preventive* when co-ordinated with fault clearing to ensure that the generators remaining in operation maintain synchronism. Tripping is said to be *restitutive* when one, or more, generators are tripped from a group of generators that have already lost synchronism. The objective here is to make re-synchronisation of the remaining generators easier.

Preventive tripping

Preventive generator tripping is illustrated in Figure 9.9. In this example both generators are assumed to be identical and, as they are connected in parallel, may be treated as one equivalent generator. This means that the equivalent generator will have a reactance half that of each individual generator but an input power twice that of an individual generator.

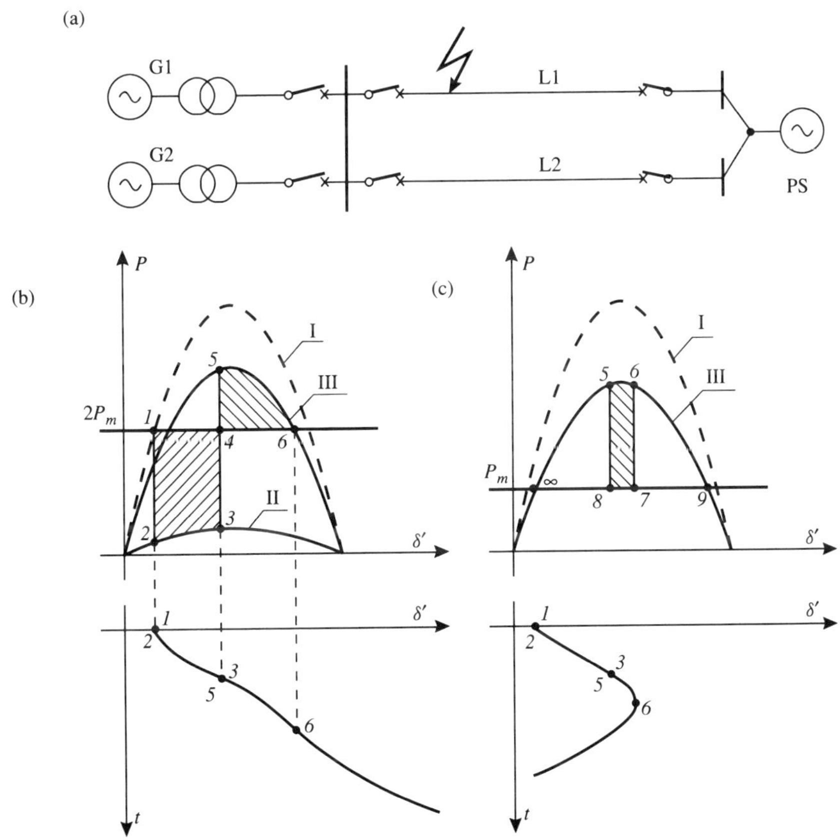

Figure 9.9 Illustration of generator tripping: (a) schemataic diagram of the system; (b) acceleration and deceleration area when no generator is tripped; (c) deceleration area when one of the generators is tripped

The pre-fault, fault and post-fault power-angle characteristics are denoted as *I*, *II* and *III*, respectively. In the case without tripping, Figure 9.9b, the acceleration area *1-2-3-4* is larger than the available deceleration area *4-5-6*. The system is unstable and both generators lose synchronism. In the second case, Figure 9.9c, one of the generators is assumed to be tripped at the same instant as the fault is cleared. The post-fault characteristic *III* now has a smaller amplitude than in Figure 9.9b as the tripping of one of the generators increases the system equivalent reactance. However the mechanical power of one generator is half that of the equivalent generator so that the acceleration area corresponding to one generator is half the acceleration area *1-2-3-4* shown in Figure 9.9b. The rotor now reaches synchronous speed (speed deviation equal to zero) at point *6* when area *8-5-6-7* equals half of area *1-2-3-4*. The system remains stable with a stability margin equal to the area *6-9-7*. The new stable equilibrium point ∞ corresponds to the intersection between the electrical characteristic *III* and the mechanical power P_m, and will be reached after a number of deep rotor swings. In this example the synchronism of one generator has been saved at the cost of tripping the other.

As the objective of generator tripping is to maintain the stability of a number of generators operating in parallel on the same busbar, tripping one generator may be insufficient and *multi-tripping* may be necessary. The number of generators that must be tripped depends on a number of factors including the pre-fault loading conditions, fault type and location and the fault clearing time. The control system that executes the tripping must be able to take into account all of these factors in order to prevent asynchronous operation whilst minimising the number of generators tripped.

Broadly, there are two types of control system that can be used to achieve the above objective. The first is similar to that used to select the fast valving control signal described in the previous section. Central to such a control scheme is the logic pre-determined off-line on the basis of a detailed stability analysis of the system. Then, just as in the fast valving control scheme, the control system obtains real-time information about the fault and on this information selects the number of generators to be tripped. As the pre-determined logic cannot always accurately assess the actual fault condition an over-pessimistic assessment may be made resulting in more generators than necessary being tripped. This is the main drawback of preventive generator tripping.

The second type of control system is more sophisticated and is based on real-time simulation by fast micro-computer systems (Kumano *et al.*, 1994). Briefly this system consists of a number of micro-computers connected by a fast telecommunications network. Each micro-computer obtains real-time measurements from the power plant and information about the disturbance. They then simulate the dynamic process faster than real-time in order to predict instability and compute the minimum number of generators that must be tripped to maintain stability.

Restitutive tripping

As explained above, preventive tripping systems can be too pessimistic and trip too many generators whilst the more sophisticated control schemes using faster than real-time simulation are very new and expensive. An alternative solution is to use restitutive tripping and utilise signals from the out-of-step relays described in sub-Section 6.4.4. When a group of generators operating in parallel on the same busbars lose synchronism, one of the generators is tripped to make re-synchronisation of the remaining generators easier. However if, after a set number of asynchronous rotations, re-synchronisation is unsuccessful another generator in the group is tripped and the process repeated until re-synchronisation is successful.

The main disadvantage of restitutive tripping is in allowing momentary asynchronous operation. The main advantage, when compared with preventive tripping, is that it never disconnects more generators than necessary.

Careful study of Figure 6.25a suggests that the optimum instant for restitutive tripping is when the rotor angle passes through the unstable equilibrium point when the large acceleration area (shaded) behind point 6 is reduced by a sudden change in P_m. The unstable equilibrium point is close to $\pi/2$ so that when the system trajectory passes through it, the apparent impedance measured by the out-of-step relay crosses the lenticular or offset Mho characteristic of the relay shown in Figure 6.33. The signal from the relay can therefore be used to trigger the generator tripping. The circuit breakers have a certain operating time so that the generator is tripped after a short delay. In order to avoid this delay, and provide for earlier tripping, the relay can be extended with the measurement of the derivative of the apparent resistance $\mathrm{d}R/\mathrm{d}t$ (Taylor *et al.*, 1983 and Taylor, Haner & Laughlin, 1986).

9.4 Shunt Elements

Power swings can also be damped by changing the parameters in the transmission network that links the generator with the system. Such a parameter change can be accomplished by using additional network elements, such as shunt capacitors and reactors switched on and off at appropriate moments. Optimal system performance can be achieved by correct control of the switching instants and has been the subject of a large number of publications. Rather than describing such optimal control schemes here the aim in this section is to explain how a proper switching strategy can damp power swings and what control signals can be used to execute the switching sequence. The following explanations are based on publications by Machowski & Nelles (1992–1994).

9.4.1 Power-angle characteristic

Figure 9.10 shows the general situation where a shunt element is connected at a point along the transmission link. The generator is represented by the classic model ($E_g = E' =$ constant behind the reactance X_g). The total reactance to the left, and to the right, of the shunt element is denoted respectively as X_g and X_s. Using the star-delta transformation a π-equivalent is obtained which contains a resistance in its equivalent series branch if the shunt element, $\overline{Y}_{sh} = G_{sh} + jB_{sh}$, contains a non-zero conductance. The resulting values of the π-equivalent circuit are

$$\overline{Z}_{gs} = \left(X_g + X_s\right)\left[-X_{SHC}G_{sh} + \mathrm{j}\left(1 - X_{SHC}B_{sh}\right)\right]$$

$$\tan\mu_{gs} = \frac{\mathrm{Re}\,\overline{Z}_{gs}}{\mathrm{Im}\,\overline{Z}_{gs}} = -\frac{X_{SHC}G_{sh}}{1 - X_{SHC}B_{sh}} \tag{9.3}$$

$$\overline{Y}_{g0} = \frac{1}{\overline{Z}_{g0}} = \frac{1}{X_g}\,\frac{X_{SHC}G_{sh} + \mathrm{j}X_{SHC}B_{sh}}{(1 - X_{SHC}B_{sh}) + \mathrm{j}X_{SHC}G_{sh}}$$

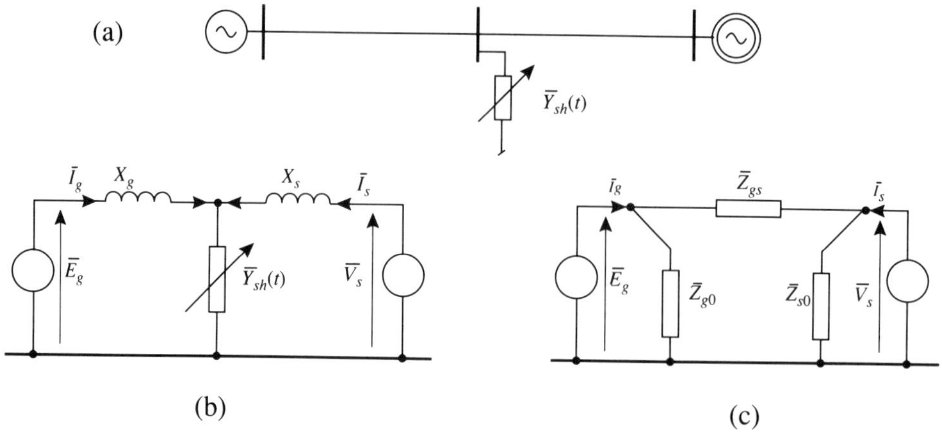

Figure 9.10 Generator-infinite busbar system with the shunt element and its equivalent circuits; (a) one-line diagram; (b) equivalent circuit; (c) π-equivalant circuit

where $X_{SHC} = X_g X_s / (X_g + X_s)$ is the short-circuit reactance of the system as seen from the node where the shunt element is installed. Normally the short-circuit power $S_{SHC} = V_n^2/X_{SHC}$ is between a few thousand and about twenty thousand MVA whilst the rated power of the shunt element $P_{n\,sh} = G_{sh}V_n^2$ and $Q_{n\,sh} = B_{sh}V_n^2$ is less than a hundred MVA. Therefore, as $P_{n\,sh}$ and $Q_{n\,sh}$ are at least ten times less than S_{SHC}, it can be safely assumed that

$$X_{SHC}G_{sh} \ll 1, \qquad X_{SHC}B_{sh} \ll 1 \tag{9.4}$$

Now consider the real number α such that for $\alpha \ll 1$ it can be assumed, with good accuracy, that

$$\frac{1}{1-\alpha} \cong 1+\alpha \qquad \text{and} \qquad \frac{1}{1+\alpha} \cong 1-\alpha \tag{9.5}$$

Equations 9.4, and the identity in equation 9.5, now allow equations 9.3 to be simplified to give

$$\overline{Y}_{gs} = G_{gs} + \mathrm{j}B_{gs} = \frac{1}{\overline{Z}_{gs}} \cong \frac{1}{X_g + X_s}\left[-X_{SHC}G_{sh} - \mathrm{j}(1 + X_{SHC}B_{sh})\right] \tag{9.6}$$

$$Y_{gs} \cong \frac{1}{X_g + X_s}(1 + X_{SHC}B_{sh}) \tag{9.7}$$

$$\tan \mu_{gs} \cong -X_{SHC}G_{sh}(1 + X_{SHC}B_{sh}) = -X_{SHC}G_{sh} \tag{9.8}$$

$$\overline{Y}_{g0} = G_{g0} + \mathrm{j}B_{g0} \cong \frac{1}{X_g}(X_{SHC}G_{sh} + \mathrm{j}X_{SHC}B_{sh}) \tag{9.9}$$

$$G_{gg} = G_{g0} + G_{gs} \cong \frac{1}{X_g + X_s}\frac{X_s}{X_g}X_{SHC}G_{sh} \tag{9.10}$$

Diligent readers may find the plus sign in equation 9.6 rather surprising but it is correct and arises due to the use of the small value approximations in equation 9.5. These derived parameters now allow a formula for the generator real power to be determined using the general formula given in equation 3.107† adapted to the π-equivalent circuit of Figure 9.10c to give

$$P(\delta') = G_{g0}E_g^2 + Y_{gs}E_g V_s \sin(\delta' - \mu_{gs}) \tag{9.11}$$

where $E_g = E'$ = constant and δ' is the transient rotor angle measured with respect to the infinite busbar. The sine function can be expressed as

$$\sin(\delta' - \mu_{gs}) = \sin\delta' \cos\mu_{gs} - \cos\delta' \sin\mu_{gs} = \cos\mu_{gs}\left(\sin\delta' - \tan\mu_{gs}\cos\delta'\right) \tag{9.12}$$

According to equations 9.4 and 9.8 the angle μ_{gs} is small and $\cos\mu_{gs} \approx 1$. Making this approximation, and substituting equation 9.8 into 9.12, gives

$$\sin(\delta' - \mu_{gs}) \cong \sin\delta' - X_{SHC}G_{sh}\cos\delta' \tag{9.13}$$

This equation, and the expressions in equations 9.7 and 9.9, can now be substituted into equation 9.11 to give

$$P(\delta') \cong b\sin\delta' + b(\xi + \cos\delta')X_{SHC}G_{sh} + (b\sin\delta')X_{SHC}B_{sh} \tag{9.14}$$

where $b = E_g V_s / \left(X_g + X_s\right)$ and $\xi = \left(E_g/V_s\right)\left(X_s/X_g\right)$. Careful examination of this equation shows that the coefficient b is the amplitude of the power-angle characteristic *without* the shunt element, while the coefficient ξ depends on the location of the shunt element along the transmission link. With the shunt element disconnected the characteristic defined by equation 9.14 is the transient power-angle characteristic defined in equation 5.40.

Figure 9.11 shows the influence of both G_{sh} and B_{sh} on the transient power-angle characteristic. Inserting a purely resistive shunt element, $G_{sh} \neq 0, B_{sh} = 0$, shifts the characteristic to the left, or the right, by the angle μ_{GS} depending on the sign of G_{sh},

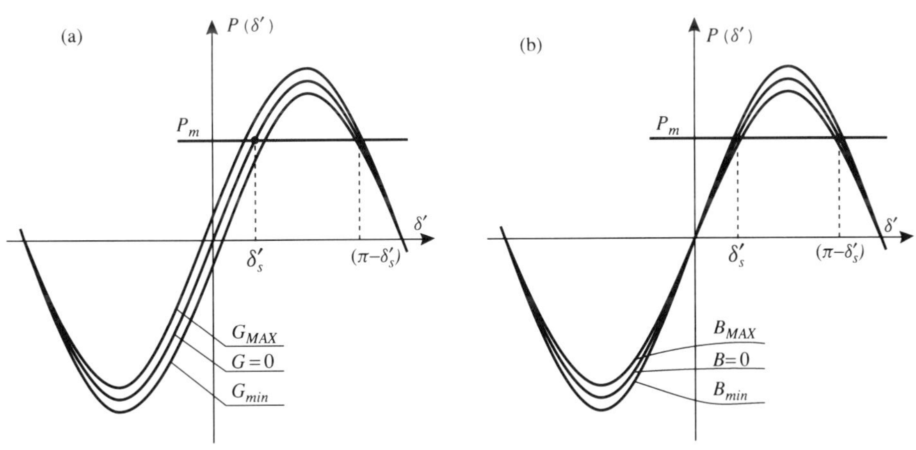

Figure 9.11 Influence of the shunt element on the power-angle characteristic: (a) pure conductance G_{sh}; (b) pure susceptance B_{sh}

† Equation 3.107 used the angle $\theta = \arg(\overline{Y})$ whilst it is more convenient here to use the angle $\mu = \pi/2 - \theta$.

Figure 9.11a. This shift by the angle μ_{GS}, and controlled by G_{sh}, is clearly demonstrated in equation 9.11. Figure 9.11b shows how inserting a pure reactive shunt element, $G_{sh} = 0, B_{sh} \neq 0$, increases the amplitude of the characteristic by $X_{SHC}B_{sh}$.

Further examination of Figure 9.11 shows that the main influence of G_{sh} is at small values of δ' before the peak of the characteristic is reached. For large values of δ', beyond the peak, the influence of the conductance is negligible. In contrast the influence of B_{sh} is mostly at large values of δ', near the peak of the characteristic.

9.4.2 State-variable control

As the shunt element influences the shape of the power-angle characteristic it can be used to damp rotor swings either by switching it on and off at appropriate moments (bang-bang control) or by suitable continuous control action (provided continuous control of the shunt element is possible). The method used to control the shunt element is called the *control strategy*. Control using the system state variables is referred to as *state-variable control*.

The system state equations can be obtained by substituting the generator real power given by equation 9.14 into the swing equation 5.15 to give:

$$\frac{\mathrm{d}\delta'}{\mathrm{d}t} = \Delta\omega$$

$$M\frac{\mathrm{d}\Delta\omega}{\mathrm{d}t} = (P_m - b\sin\delta') - D\frac{\mathrm{d}\delta'}{\mathrm{d}t} - b(\xi + \cos\delta')X_{SHC}G_{sh}(t) - (b\sin\delta')X_{SHC}B_{sh}(t) \tag{9.15}$$

The last two terms in the second equation depend on the conductance and susceptance of the controlled shunt element. Both these parameters are shown as functions of time to emphasise that they are the time-varying control variables.

Energy dissipation

In order to derive the required control strategy the energy approach described in Section 6.2 will be used. In this section it was shown that when a disturbance occurs in the network part of the kinetic energy stored in the rotating masses of the generators and loads is released, and this then undergoes oscillatory conversions between its kinetic and potential form during subsequent rotor swings. The oscillations continue until the damping torques dissipate all the released energy and the system trajectory returns to the equilibrium point. The goal is to control $G_{sh}(t)$ and $B_{sh}(t)$ in such a way as to maximise the speed of energy dissipation. This can be accomplished by maximising the value of the derivative $\mathrm{d}E/\mathrm{d}t$ of the total system energy with time along the trajectory of the differential equations 9.15.

The total system energy E for the generator-infinite busbar system is determined by equation 6.18. The speed of energy changes is

$$\frac{\mathrm{d}E}{\mathrm{d}t} = \frac{\partial E}{\partial \delta'}\frac{\mathrm{d}\delta'}{\mathrm{d}t} + \frac{\partial E}{\partial \Delta\omega}\frac{\mathrm{d}\Delta\omega}{\mathrm{d}t} \tag{9.16}$$

The partial derivatives $\partial E/\partial \delta'$ and $\partial E/\partial \Delta\omega$ can be calculated by differentiating the total energy given by equation 6.18 similarly as in equations 6.23 to 6.27. The ordinary time derivatives $d\delta'/dt$ and $d\Delta\omega/dt$ can be now substituted from equation 9.15 to give

$$\frac{dE}{dt} = -\left[D\Delta\omega^2 + \Delta\omega b(\xi + \cos\delta')X_{SHC}G_{sh}(t) + \Delta\omega(b\sin\delta')X_{SHC}B_{sh}(t)\right] \tag{9.17}$$

The first term in this equation corresponds to the energy dissipated by the natural damping torques (coefficient D) whilst the next two terms are contributed by the shunt element. Proper control of this element can contribute to a faster dissipation of energy. This will happen if $G_{sh}(t)$ and $B_{sh}(t)$ are varied so that the signs of the two last terms in 9.17 are always positive. This requires the sign of $G_{sh}(t)$ and $B_{sh}(t)$ to vary depending on the sign of the two state variables $\Delta\omega$ and δ'. This can be realised using either bang-bang control or continuous control.

Continuous control

The aim of the control is firstly to ensure that the last two terms in equation 9.17 are of positive sign and, secondly, to maximise their sum. This can be achieved by enforcing the following values on the control variables

$$\begin{aligned} G_{sh}(t) &= K\Delta\omega\left[b(\xi + \cos\delta')\right]X_{SHC} \\ B_{sh}(t) &= K\Delta\omega\left[b\sin\delta'\right]X_{SHC} \end{aligned} \tag{9.18}$$

where K is the gain of the controller. Substituting this equation into equation 9.17 gives

$$\frac{dE}{dt} = -D\Delta\omega^2 - D_{sh}\Delta\omega^2 \tag{9.19}$$

$$\text{where } D_{sh} = K\left\{\left[b(\xi + \cos\delta')\right]^2 + (b\sin\delta')^2\right\}X_{SHC}^2 \tag{9.20}$$

is the equivalent damping coefficient introduced by the shunt element control. For the control strategy in equation 9.18, the swing equation then becomes

$$\begin{aligned} \frac{d\delta'}{dt} &= \Delta\omega \\ M\frac{d\Delta\omega}{dt} &= (P_m - b\sin\delta') - D\frac{d\delta'}{dt} - D_{sh}\frac{d\delta'}{dt} \end{aligned} \tag{9.21}$$

In the first of the two equations 9.18 the expression $(\xi + \cos\delta')$ is positive over a large range of $\delta'(-\pi/2 > \delta' > \pi/2)$ so that the sign of $G_{sh}(t)$ will change at the same time as the rotor speed deviation $\Delta\omega$. On the other hand the sign of $B_{sh}(t)$ depends on both $\Delta\omega$ and $\sin\delta'$ and changes sign whenever the angle crosses through zero.

Figure 9.12 shows the changes in the sign of $G_{sh}(t)$ and $B_{sh}(t)$ that must occur along the system trajectory in order to produce positive damping. Whenever the trajectory crosses the horizontal axis, $\Delta\omega$ changes its sign and this will be accompanied by a change in sign of both $G_{sh}(t)$ and $B_{sh}(t)$. These points of sign changes have been denoted by small squares on the trajectory. Whenever the trajectory crosses the vertical axis, δ' changes

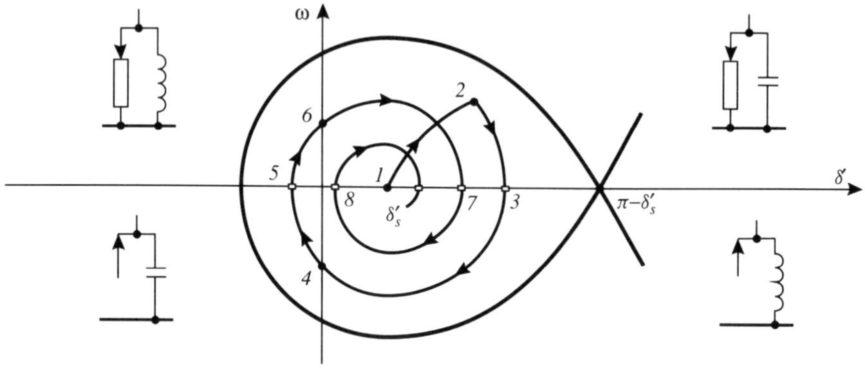

Figure 9.12 Stability area and the system trajectory

its sign and this is accompanied by a sign change of $B_{sh}(t)$ only. These points of sign changes have been denoted by the small, solid, circles on the trajectory.

In Figure 9.12 the characteristic states of the shunt element have been illustrated schematically in each quadrant on the phase plane. Negative susceptance is denoted by a coil symbol while positive susceptance is denoted by a capacitor symbol. Positive conductance is denoted by a resistance symbol while negative resistance (i.e. a source of real power) is denoted by an arrow. This obviously corresponds to the general case when the power of the shunt element can be adjusted in all four quadrants of the complex power plane.

Bang-bang control

The term bang-bang control refers to a control mode in which the element is switched on and off at appropriate moments. This type of control can be used with a shunt element whose parameters cannot be smoothly controlled. The switching then takes place at instants when the system trajectory crosses one of the phase axes shown in Figure 9.12 and passes into the next quadrant.

Equation 9.18 leads to the following bang-bang control strategies:

$$G_{sh}(t) = \begin{cases} G_{\mathrm{MAX}} & \text{for} \quad b(\xi + \cos\delta')\Delta\omega > +\varepsilon \\ G_{\min} & \text{for} \quad b(\xi + \cos\delta')\Delta\omega < -\varepsilon \end{cases} \tag{9.22}$$

$$B_{sh}(t) = \begin{cases} B_{\mathrm{MAX}} & \text{for} \quad \Delta\omega(b\sin\delta') > +\varepsilon \\ B_{\min} & \text{for} \quad \Delta\omega(b\sin\delta') < -\varepsilon \end{cases} \tag{9.23}$$

where G_{MAX} and B_{MAX} are the maximum, and $G_{\min}$ and $B_{\min}$ the minimum, values of the switching element. If the element cannot assume a negative value then the minimum value will be zero. The small positive number determines the dead-zone ($\pm\varepsilon$) where the control variables are set to zero. The dead-zone is necessary in order to avoid unstable operation of the controller for small signals.

As $(\xi + \cos\delta') > 0$ for a wide range of δ', the strategy for controlling the conductance $G_{sh}(t)$ can be simplified to

$$G_{sh}(t) = \begin{cases} G_{\text{MAX}} & \text{for} \quad \Delta\omega > +\varepsilon \\ G_{\min} & \text{for} \quad \Delta\omega < -\varepsilon \end{cases} \tag{9.24}$$

and the conductance switching is triggered by a change in sign of the speed deviation.

Interpretation of the shunt element control using the equal area criterion

As the energy approach is directly related to the equal area criterion (sub-Section 6.2.4) the control strategies derived above can be usefully interpreted using the equal area criterion.

Figure 9.13 shows how the bang-bang control strategy enlarges the available deceleration area and reduces the acceleration area during every rotor forward swing, whilst during the backswing the control reduces the deceleration area and enlarges the available acceleration area. The initial pre-fault state is point *1*. The fault reduces the electrical real power to a value corresponding to point *2* and the rotor accelerates. The rotor angle increases until the fault is cleared at point *3* giving the first acceleration area *1-2-3-4*. After clearing the fault, $\Delta\omega$ and δ' are positive so that G_{MAX} and B_{MAX} are switched in and the electrical power follows the higher $P(\delta')$ curve. As the maximum available decelerating area is *4-5-6-7* the generator remains stable with a stability margin proportional to area *6-7-8*. At point *6* the speed deviation changes sign, the shunt admittance values switch to $G_{\min}$ and $B_{\min}$, and the electrical power follows the lower $P(\delta')$ curve. This reduces the deceleration area during the backswing and enlarges the available acceleration area so that the amplitude of the backswing is reduced. The cycle then repeats and helps damp consecutive rotor swings.

9.4.3 Control based on local measurements

The control strategy given by equation 9.18 is based on the state variables δ' and $\Delta\omega$. As these quantities are not normally available at the shunt element busbar the practical implementation of the control must be based on other signals that can be measured locally.

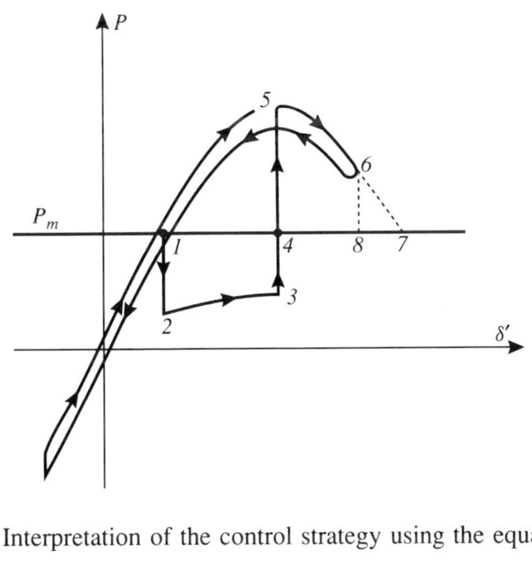

Figure 9.13 Interpretation of the control strategy using the equal area criterion

How exactly such a local control imitates the state-variable control depends on the choice of the measured quantities and the structure of the controller.

Dynamic properties of local measurements

Let q_G and q_B be some quantities used as input signals to the shunt element controller. In the control strategy given by equation 9.18 the shunt admittance depends on the rotor speed deviation $\Delta\omega$. If the magnitude of the transient emf is assumed constant (classical model), the derivative with respect to time of any electric quantity q_G can be expressed as

$$\frac{\mathrm{d}q_G}{\mathrm{d}t} = \frac{\partial q_G}{\partial \delta'}\Delta\omega + \alpha_{GG}\frac{\mathrm{d}G_{sh}}{\mathrm{d}t} + \alpha_{GB}\frac{\mathrm{d}B_{sh}}{\mathrm{d}t} \tag{9.25}$$

where the coefficients

$$\alpha_{GG} = \frac{\partial q_G}{\partial G_{sh}}, \qquad \alpha_{GB} = \frac{\partial q_G}{\partial B_{sh}} \tag{9.26}$$

determine the sensitivity of q_G to a change in the control variables $G_{sh}(t)$ and $B_{sh}(t)$. Equation 9.25 gives

$$\Delta\omega\frac{\partial q_G}{\partial \delta'} = \frac{\mathrm{d}q_G}{\mathrm{d}t} - \alpha_{GG}\frac{\mathrm{d}G_{sh}}{\mathrm{d}t} - \alpha_{GB}\frac{\mathrm{d}B_{sh}}{\mathrm{d}t} \tag{9.27}$$

If the sensitivity coefficients α_{GG} and α_{GB} are known, the right-hand-side of equation 9.27 can be computed in real-time and used to determine a signal proportional to the rotor speed deviation necessary for the control of $G_{sh}(t)$. Comparing the right-hand-side of the first of the equations 9.18 with the left-hand-side of equation 9.27 shows that the signal obtained from equation 9.27 is the same as the state-variable control signal if

$$\frac{\partial q_G}{\partial \delta'} = [b(\xi + \cos\delta')]X_{SHC} \tag{9.28}$$

Substitution of the right-hand-side of the first of equations 9.18 by equation 9.27 gives the following control principle

$$G_{sh}(t) = K\left[\frac{\mathrm{d}q_G}{\mathrm{d}t} - \alpha_{GG}\frac{\mathrm{d}G_{sh}}{\mathrm{d}t} - \alpha_{GB}\frac{\mathrm{d}B_{sh}}{\mathrm{d}t}\right] \tag{9.29}$$

This means that if a measured quantity q_G satisfies the condition in equation 9.28 then the modulation controller need simply differentiate q_G with respect to time and subtract from the result values proportional to the rate of change of the controlled variables $B_{sh}(t)$ and $G_{sh}(t)$.

The control principle for the shunt susceptance can be obtained in a similar way as

$$B_{sh}(t) = K\left[\frac{\mathrm{d}q_B}{\mathrm{d}t} - \alpha_{BG}\frac{\mathrm{d}G_{sh}}{\mathrm{d}t} - \alpha_{BB}\frac{\mathrm{d}B_{sh}}{\mathrm{d}t}\right] \tag{9.30}$$

where the coefficients

$$\alpha_{BG} = \frac{\partial q_B}{\partial G_{sh}}, \qquad \alpha_{BB} = \frac{\partial q_B}{\partial B_{sh}} \tag{9.31}$$

determine the sensitivity of q_B to changes in the control variables $B_{sh}(t)$ and $G_{sh}(t)$. Comparison with the second of equations 9.18 shows that the input quantity q_B should satisfy the following condition

$$\frac{\partial q_B}{\partial \delta'} = [b \sin \delta'] X_{SHC} \tag{9.32}$$

It now remains to determine what locally measurable quantities q_B and q_G will satisfy the conditions defined in equations 9.28 and 9.32.

Voltage based quantities

The current flowing from the network to the shunt element in Figure 9.10 is given by

$$\overline{I}_{sh} = \overline{V}_{sh}\overline{Y}_{sh} = \frac{\overline{E}_g - \overline{V}_{sh}}{\mathrm{j}X_g} + \frac{\overline{V}_s - \overline{V}_{sh}}{\mathrm{j}X_s} \tag{9.33}$$

where $\overline{Y}_{sh} = G_{sh}(t) + \mathrm{j}B_{sh}(t)$ and X_g and X_s are the equivalent reactances denoted in Figure 9.10. Multiplying the current by the short-circuit reactance X_{SHC} and moving $\overline{V}_{sh}$ to the left-hand-side gives

$$\overline{V}_{sh}\{[X_{SHC}G_{sh}(t)] - \mathrm{j}[1 - X_{SHC}B_{sh}(t)]\} = \frac{\overline{E}_g X_s + \overline{V}_s X_g}{\mathrm{j}(X_g + X_s)} \tag{9.34}$$

Substituting for the complex voltages

$$\overline{E}_g = E_g(\cos\delta' + \mathrm{j}\sin\delta') \text{ and } \overline{V}_s = V_s \tag{9.35}$$

and multiplying the resulting equation by its conjugate gives, after a little algebra,

$$V_{sh}^2 = \frac{bX_{SHC}\left(\xi + \dfrac{1}{\xi} + 2\cos\delta'\right)}{[X_{SHC}G_{sh}(t)]^2 + [1 - X_{SHC}B_{sh}(t)]^2} \tag{9.36}$$

where ξ is the coefficient defined in equation 9.14. When deriving equation 9.36 it is also possible to find the phase angle θ of the shunt element voltage measured with respect to the infinite bus:

$$\tan\theta = \frac{\left(\dfrac{1}{\xi} + \cos\delta'\right)X_{SHC}G_{sh}(t) + \sin\delta'[1 - X_{SHC}B_{sh}(t)]}{\sin\delta' X_{SHC}G_{sh}(t) + \left(\dfrac{1}{\xi} + \cos\delta'\right)[1 - X_{SHC}B_{sh}(t)]} \tag{9.37}$$

The inequalities in equation 9.4 allow equations 9.36 and 9.37 to be simplified to

$$V_{sh}^2 \cong b\left(\xi + \frac{1}{\xi} + 2\cos\delta'\right)X_{SHC}, \quad \tan\theta \cong \frac{\sin\delta'}{\dfrac{1}{\xi} + \cos\delta'} \tag{9.38}$$

These two simplified equations now allow the derivatives $\partial(V_{sh}^2)/\partial\delta'$ and $\partial\theta/\partial\delta'$ to be found as

$$\frac{\partial V_{sh}^2}{\partial\delta'} = -2[b\sin\delta']X_{SHC} \tag{9.39}$$

and

$$\frac{\partial\theta}{\partial\delta'} = \frac{\xi + \cos\delta'}{\xi + \dfrac{1}{\xi} + 2\cos\delta'} \tag{9.40}$$

The expression in the denominator of equation 9.40 is the same as the expression in brackets in the first of equations 9.38. Substitution gives

$$V_{sh}^2\frac{\partial\theta}{\partial\delta'} = [b(\xi + \cos\delta')]X_{SHC} \tag{9.41}$$

Equations 9.39 and 9.41 show that local measurements of the squared magnitude of the shunt element voltage and its phase angle can give good signals for controlling the shunt element. Comparing equations 9.39 and 9.32 shows that the signal $-\frac{1}{2}V_{sh}^2$ satisfies the condition in equation 9.32 for the control strategy of the shunt susceptance. Similarly comparing equations 9.41 and 9.28 shows that θ satisfies the condition in 9.28 for the required control of the shunt conductance provided that the derivative is multiplied by V_{sh}^2.

A sensitivity analysis of the effect of changes in V_{sh}^2 and θ on the changes in the controlled variables $B_{sh}(t)$ and $G_{sh}(t)$ can be conducted by evaluating the derivatives in equations 9.26 and 9.31 using equations 9.36 and 9.37. This involves a lot of simple, but arduous, algebraic and trigonometric transformations which finally leads to the following simplified formulas:

$$\begin{aligned} \alpha_{GG} &= \frac{\partial q_G}{\partial G_{sh}} \cong -X_{SHC}, &\qquad \alpha_{GB} &= \frac{\partial q_G}{\partial B_{sh}} \cong 0 \\ \alpha_{BG} &= \frac{\partial q_B}{\partial G_{sh}} \cong 0, &\qquad \alpha_{BB} &= \frac{\partial q_B}{\partial B_{sh}} \cong -V_{sh}^2X_{SHC} \end{aligned} \tag{9.42}$$

Zero values of α_{GB} and α_{BG} signify that changes in the shunt susceptance/conductance have a negligibly small effect on the given quantity.

Control schemes

Substituting equations 9.42 into equations 9.29 and 9.30 and taking the squared voltage magnitude and the voltage phase angle as control signals yields

$$\begin{aligned} G_{sh}(t) &= KV_{sh}^2\left[\frac{\mathrm{d}\theta}{\mathrm{d}t} + X_{SHC}\frac{\mathrm{d}G_{sh}}{\mathrm{d}t}\right] \\ B_{sh}(t) &= K\left[-\frac{1}{2}\frac{\mathrm{d}(V_{sh}^2)}{\mathrm{d}t} + X_{SHC}V_{sh}^2\frac{\mathrm{d}B_{sh}}{\mathrm{d}t}\right] \end{aligned} \tag{9.43}$$

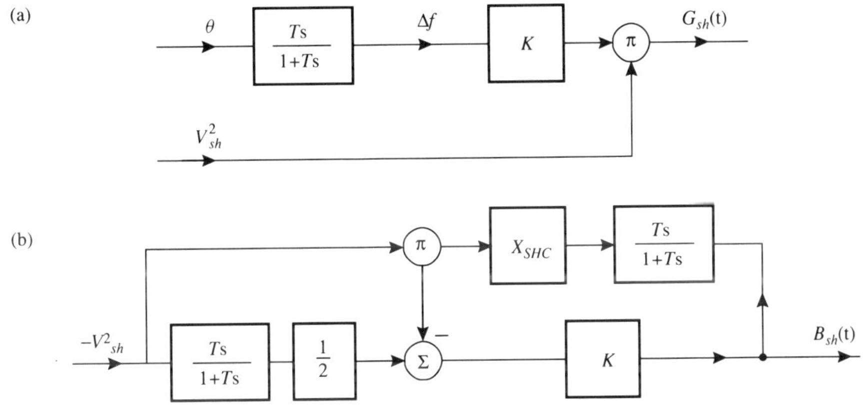

Figure 9.14 Modulation controller employing the frequency-voltage control scheme for: (a) $G_{sh}(t)$; (b) $B_{sh}(t)$

The rotor angle δ' and the voltage phase angle θ are measured with respect to the infinite busbar voltage or the synchronous reference frame. As the derivative of θ with time is equal to the deviation of the local frequency, that is $d\theta/dt = 2\pi\Delta f$, the control scheme given by equations 9.43 is referred to as *frequency and voltage based control.*

Figure 9.14 shows the block diagram of the appropriate control circuits. Differentiation has been replaced by a real differentiating element with a small time constant T. The shunt susceptance controller is non-linear as the output signal in the main feedback loop is multiplied by the main input signal. The shunt conductance controller is linear but its effective gain is modulated by the squared voltage magnitude which is also the main input signal for the shunt susceptance controller. The short-circuit reactance X_{SHC} plays only a corrective role and its value may be set with a large error. For practical applications its value can be assessed off-line and set as a constant parameter.

The input signals for the control system, the squared magnitude and the changes in the angle of the local voltage, may be measured using digital techniques described by Phadke, Thorap & Adamiak (1983) or Kamwa & Grondin (1992).

9.4.4 Examples of controllable shunt elements

Continuous state-variable control, and its practical implementation based on locally measurable quantities, is possible assuming that both $G_{sh}(t)$ and $B_{sh}(t)$ can be changed smoothly over a range of negative and positive values. Some of the different types of shunt elements that can be thyristor-controlled have been described in Section 2.4.4 where the ability to change $G_{sh}(t)$ and $B_{sh}(t)$ was seen to depend on the particular device in question. When using a particular shunt element any such limitation must be taken into account in the control structure by inserting appropriate limiters into the control circuits shown in Figure 9.14.

Modulation controller of SMES

The schematic diagram of a superconducting magnetic energy storage (SMES) system utilising two GTO thyristor converters was shown in Figure 2.23. The two GTO thyristor converters allow the SMES unit to generate power, for a short time, from any quadrant of the complex power plane assuming that the apparent power (or equivalent admittance) satisfies the inequality in equation 2.6. The frequency and voltage based controller shown in Figure 9.14 can be used as the modulation controller of a SMES system by forcing the required values on the real and reactive power in proportion to the signals $G_{sh}(t)$ and $B_{sh}(t)$. If the values of these signals are too large then they can be proportionally reduced to satisfy the limits.

Supplementary control of SVC

Figures 2.20 and 2.21 showed different types of static VAR compensaters (SVCs). In the steady-state, SVC normally controls the voltage at the bus where it has been installed. The conventional SVC regulator employs a proportional-plus-integral (PI) closed-loop controller to regulate the voltage magnitude. Such regulation is very effective in forcing the steady-state voltage error to zero. In the transient-state the regulators are incapable of providing enough damping as the voltage error carries no information about the system dynamic response. A more robust control can be obtained when the conventional regulator is equipped with the supplementary control loop shown in Figure 9.15. This additional supplementary loop can be used to force a control signal for $B_{sh}(t)$, equation 9.43, to help damp the power swings. This control loop acts much faster than the main PI voltage controller. The control circuit for $B_{sh}(t)$ was shown in Figure 9.14.

Control of braking resistors

Braking resistors are sometimes used to artificially increase the electrical load and thereby reduce the rotor acceleration when a fault occurs near the busbars of a power station. The resistors may be made of stainless steel wire strung on towers. A simple control solution is to switch on the brake immediately after the fault is cleared and switch it off after a short time, typically about 0.5 s (Kundur, 1994). This reduces the rotor acceleration during the first forward swing.

The damping provided by the braking resistors may be improved if their switching circuits are equipped with a bang-bang controller implementing the strategy defined in equation 9.24 with $G_{\min} = 0$. If the rotor speed deviation signal $\Delta\omega$ is not available it may

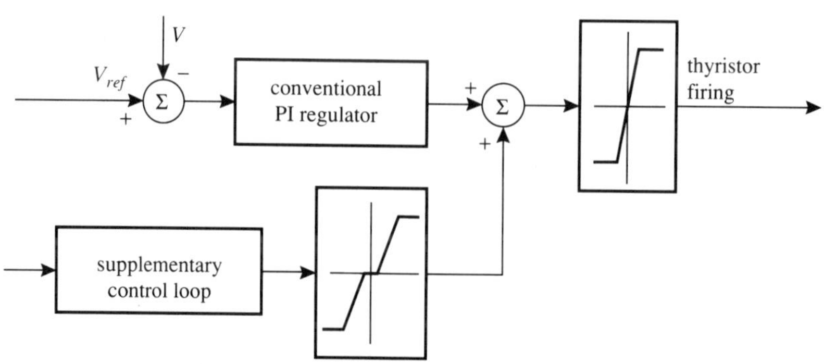

Figure 9.15 Conventional regulator of the SVC with supplementary control loop

be replaced by the local frequency deviation $2\pi\Delta f = d\theta/dt$, as in the shunt controller shown in Figure 9.14. Another, more expensive, possibility is to control the braking resistor using thyristors as in the control circuit of Figure 9.14a. The thyristor-controlled braking resistor was shown in Figure 2.22.

SVC plus braking resistors

Figure 9.11 showed that when δ' is small the influence of a shunt capacitive element on the real power is very small indeed. Consequently when the generator operates at a small power angle very little damping of the power swings using SVC alone can be obtained. This can be improved by adding a thyristor-controlled, or thyristor-switched, braking resistor to the static VAR compensater. SVC would then act alone during the steady-state operation providing a return on investment from the voltage control and reactive compensation. The additional braking resistor would support SVC in the transient state by providing additional damping of power swings.

9.4.5 Damping of power swings in multi-machine systems

The control strategy given by equation 9.43 was derived for the generator-infinite busbar system. Machowski & Nelles (1994) proved that this strategy is also valid for a multi-machine system and forces dissipation of the total system energy regardless of where the shunt element is located within the system. Figure 9.16 shows simulation results of such a multi-machine system with a SMES unit installed at bus *B4*. Generator *G3* is a large

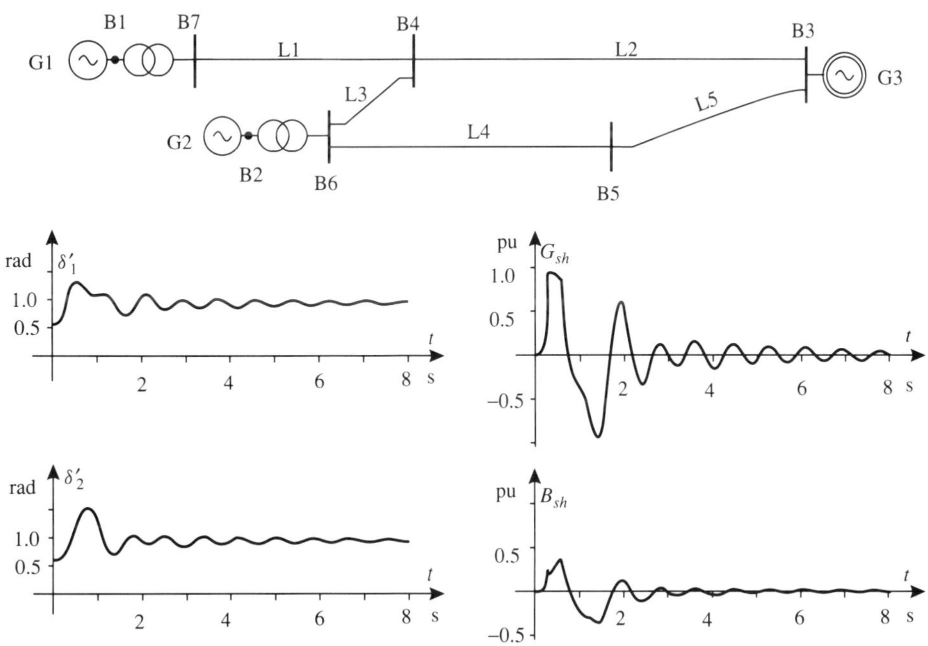

Figure 9.16 Three-machine test system and simulation results for SMES installed at bus *B4*

machine that effectively constitutes an infinite busbar. Generator *G2* has a much smaller inertia than generator *G1* so that its rotor has a higher natural frequency of oscillation. A shunt element installed at busbar *B4* will therefore have a difficult task in damping power swings as it will be exposed to superimposed signals from the two generators swinging at significantly different frequencies.

Figure 9.16 shows simulation results for a three-phase short-circuit in line *L4*, near generator *G2*, cleared without auto-reclosure. In the post-fault conditions, generator *G2* (with smaller inertia) is connected via line *L3* to the middle point of the link between generator *G1* (higher inertia) and the infinite busbar *G3*. The graphs of $\delta'_1(t)$ and $\delta'_2(t)$ show that within a few seconds the rotor swings have been quickly damped. In this example all the damping was provided by the SMES as for testing purposes all the inherent system damping was removed from the model. The graphs of $G_{sh}(t)$ and $B_{sh}(t)$ show that the controller acted mainly on the shunt conductance and, to a lesser extent, on the shunt susceptance.

Other tests have shown that incorporating additional shunt elements, equipped with the controllers described in the previous sub-sections and installed at various locations in the system, have a cumulative effect in damping rotor swings. Moreover, no negative interaction between the controllers was experienced, that is no shunt element was seen to excite power swings that had been suppressed by another.

9.5 Series Elements

Sub-Section 3.1.2 showed how the power-transfer capability of a long transmission line depends on its inductive reactance and how this reactance can be offset by inserting a series capacitor. Besides being useful in the steady-state such a reduction in the line reactance is also useful in the transient state as it increases the amplitude of the transient power-angle characteristic thereby increasing the available deceleration area. By proper control of a switched series capacitor this change in amplitude can be used to provide additional damping of power swings. In particular the conventional series capacitor, equipped with a zinc-oxide protective scheme, Figure 2.18, can be used in a bang-bang control mode as it can be almost instantaneously bypassed and re-inserted at appropriate moments. Advanced series capacitors (ASC), or thyristor-controlled series compensaters (TCSS), Figure 2.24, allow smooth control of the equivalent reactance and can be used in a continuous control mode.

9.5.1 State-variable control

Consider the generator-infinite busbar system shown in Figure 9.17a and assume that a three-phase short-circuit occurs on line *L2*. In this example line *L2* does not connect the generator to the infinite busbar. If the resistance of the network is neglected the generator real power is expressed by equation 5.40. For the system in Figure 9.17a this equation can be re-written as

$$P(\delta') = \frac{E_g V_s}{X_\Sigma} \sin \delta' \tag{9.44}$$

where $X_\Sigma = X'_d + X_T + X_{L1} + X_s - X_C(t)$ is the equivalent reactance and, as before, $E_g = E'$ is the constant transient emf. Assume that $X_C(t) = X_C - \Delta X_C(t)$ where X_C

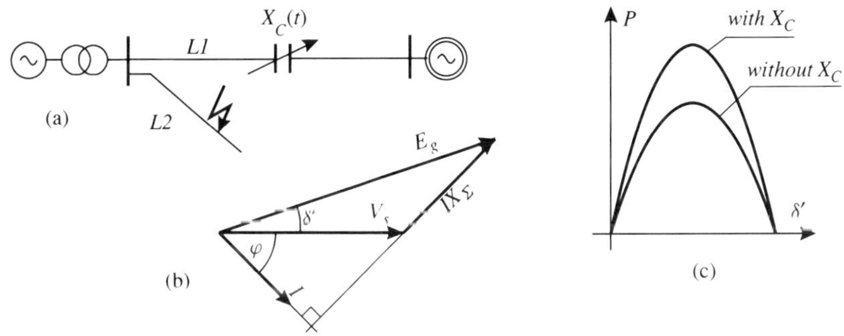

Figure 9.17 System with a shunt capacitor: (a) network diagram; (b) phasor diagram; (c) power-angle characteristic with and without series compensation

is the full value of the series reactance normally inserted in the steady-state while $\Delta X_C(t)$ is the controllable part. The goal is to determine the required changes in $\Delta X_C(t)$ in order to improve the damping of power swings. In the pre-fault state the capacitor is fully inserted and $\Delta X_C(t) = 0$.

Figure 9.17c shows the power-angle characteristic of the system with and without series compensation. With the capacitor switched in, the equivalent series reactance is reduced, and the amplitude of the characteristic increased. This effect is similar to that of the shunt capacitor except that series capacitors have a much larger effect on the amplitude of $P(\delta')$.

The equivalent reactance can be expressed as $X_\Sigma = X + \Delta X_C(t)$ where $X = X'_d + X_T + X_{L1} + X_s - X_C$ is the pre-fault reactance. Equation 9.44 can now be re-written as

$$P(\delta') = \frac{E_g V_s}{X} \sin\delta' - \frac{E_g V_s}{X + \Delta X_C(t)} \frac{\Delta X_C(t)}{X} \sin\delta' = b \sin\delta' - b_\Sigma \frac{\Delta X_C(t)}{X} \sin\delta' \quad (9.45)$$

where $b = E_g V_s / X$, $b_\Sigma = E_g V_s / (X + \Delta X_C(t))$ and $\Delta X_C(t)$ is the control variable. As series compensation is always less than 100%, b_Σ is always positive and, as $\Delta X_C(t) << X$, it can be regarded as constant. The first term in equation 9.45 corresponds to the pre-fault power, or to the transient power when $\Delta X_C(t) = 0$. The second term lowers the amplitude of the power angle characteristic and is a direct consequence of by-passing the regulated part of the capacitor.

Equation 9.45 now allows the swing equation 5.15 to be written as

$$\begin{aligned} \frac{d\delta'}{dt} &= \Delta\omega \\ M\frac{d\Delta\omega}{dt} &= P_m - b\sin\delta' - D\frac{d\delta'}{dt} + (b_\Sigma \sin\delta')\frac{\Delta X_C(t)}{X} \end{aligned} \quad (9.46)$$

The last term in this equation contains $\Delta X_C(t)$ and will provide additional damping if it can be made to be negative and proportional to the speed deviation $\Delta\omega = d\delta'/dt$. This condition can be met if $\Delta X_C(t)$ is made to satisfy

$$\Delta X_C(t) = -KX \sin\delta' \Delta\omega \quad (9.47)$$

where K is the gain of the controller. Substituting this value of $\Delta X_C(t)$ into the swing equation gives

$$\frac{d\delta'}{dt} = \Delta\omega$$
$$M\frac{d\Delta\omega}{dt} = P_m - b\sin\delta' - D\frac{d\delta'}{dt} - D_{ser}\frac{d\delta'}{dt} \tag{9.48}$$

where $D_{ser} = Kb_{\Sigma}(\sin\delta')^2 \geq 0$ is the additional damping coefficient introduced by controlling the series capacitor.

Bang-bang control

ΔX_C corresponds to the by-passed part of the series capacitor and its minimum value is zero (insertion of the whole capacitor). The bang-bang controller chooses either this minimum zero value or the maximum value $\Delta X_{C\ \mathrm{MAX}}$ according to the following strategy:

$$\Delta X_C(t) = \begin{cases} 0 & \text{for} \quad (\sin\delta')\Delta\omega \geq -\varepsilon \\ \Delta X_{C\ \mathrm{MAX}} & \text{for} \quad (\sin\delta')\Delta\omega < -\varepsilon \end{cases} \tag{9.49}$$

where ε determines the dead-zone.

Interpretation using equal area criterion

Consider again the example power system of Figure 9.17 with the fault occurring on line *L2*. Figure 9.18 shows the power-angle curves with $\Delta X_C = 0$, the upper characteristic, and with $\Delta X_C = \Delta X_{C\mathrm{MAX}}$, the lower characteristic. The fault causes the generator power to drop so that kinetic energy proportional to area *1-2-3-4* is released. At the instant of fault clearance the expression $(\Delta\omega\sin\delta')$ is positive so that the signal $\Delta X_C(t)$ is set to its minimal value $\Delta X_C = 0$ and the whole capacitor is inserted. The available deceleration area is *4-5-6-10*. The rotor reaches speed deviation $\Delta\omega = 0$ at point *6* when area *4-5-6-7* becomes equal to area *1-2-3-4* and then starts to swing back. The expression $(\Delta\omega\sin\delta')$ becomes negative and the control strategy in equation 9.49 causes part of the capacitor to be by-passed, $\Delta X_C(t)$ is increased and the rotor follows the lower $P(\delta')$ characteristic. The power jumps from point *6* to point *8* and the rotor swings back along path *8–9* performing deceleration work proportional to area *8-9-7*. The backswing deceleration area is now much smaller than area *6-7-1*, the deceleration area available when $X_C(t) = 0$. When the rotor starts to swing forward, the speed deviation changes sign, $X_C(t)$ is increased, and the acceleration area is reduced. The complete switching cycle is then repeated but with rotor swings of reduced amplitude. Thus the control principle is established. During the forward swing the acceleration area is minimised and the available deceleration area maximised whilst during the backswing the deceleration area is minimised and the available acceleration area maximised.

In some circumstances bang-bang control with a small dead-zone can lead to large power angle swings and instability. To explain this again consider Figure 9.18 and assume that the rotor reaches point *11* during its forward swing. Sudden bypassing of the capacitor causes the system trajectory to jump to point *12* from which the rotor would be further accelerated causing a subsequent asynchronous operation. This would not happen with continuous control as both the control signal and $X_C(t)$ change smoothly. Consequently at point *11* the signal $X_C(t)$ would be equal to zero and the $P(\delta')$ trajectory would start to

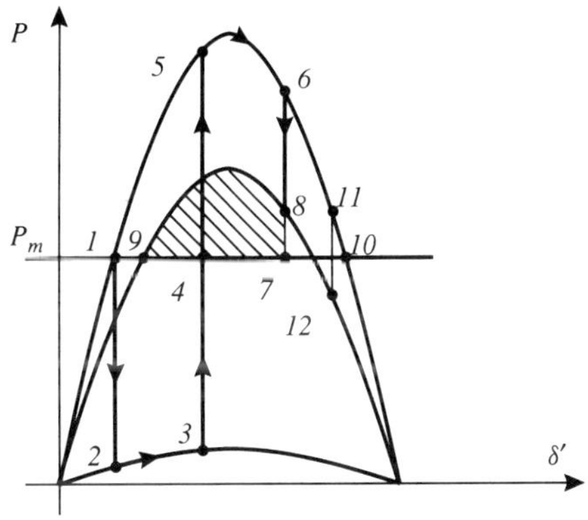

Figure 9.18 Interpretation of the series capacitor control using the equal area criterion

return towards the equilibrium point along a characteristic that is very close to the upper one. This would slowly change towards the lower characteristic as $\Delta\omega$ increases.

9.5.2 Control strategy based on local measurements

As the control strategy in equation 9.47 for the series capacitor is very similar to the control strategy defined in equation 9.18 for the shunt susceptance it can be implemented using similar local measurements. As with the shunt susceptance the squared voltage signal measured at the installation node can be used but, with the series capacitor, it is more convenient to use the square of the current flowing through the compensated line.

Applying the cosine theorem to the voltage triangle in the phasor diagram of Figure 9.17b gives

$$(IX_\Sigma)^2 = E_g^2 + V_s^2 - 2E_gV_s\cos\delta' \tag{9.50}$$

from which the squared current can be calculated as

$$I^2 = \frac{I}{X_\Sigma^2}(E_g^2 + V_s^2 - 2E_gV_s\cos\delta') \tag{9.51}$$

With both $E_g = E'$ and V_s constant this signal depends on X_Σ and δ'. The partial derivatives of the signal are:

$$\frac{\partial(I^2)}{\partial\delta'} = \frac{2}{X_\Sigma^2}E_gV_s\sin\delta', \quad \frac{\partial(I^2)}{\partial X_\Sigma} = -\frac{2}{X_\Sigma}I^2 \tag{9.52}$$

with the time derivative of the squared current signal being

$$\frac{\mathrm{d}(I^2)}{\mathrm{d}t} = \frac{\partial(I^2)}{\partial\delta'}\frac{\mathrm{d}\delta'}{\mathrm{d}t} + \frac{\partial(I^2)}{\partial X_\Sigma}\frac{\mathrm{d}X_\Sigma}{\mathrm{d}t} \tag{9.53}$$

As $X_\Sigma = X + \Delta X_C(t)$ with $X =$ constant, $\mathrm{d}X_\Sigma/\mathrm{d}t = \mathrm{d}\Delta X_C/\mathrm{d}t$. Substituting equations 9.52 into equation 9.53 gives

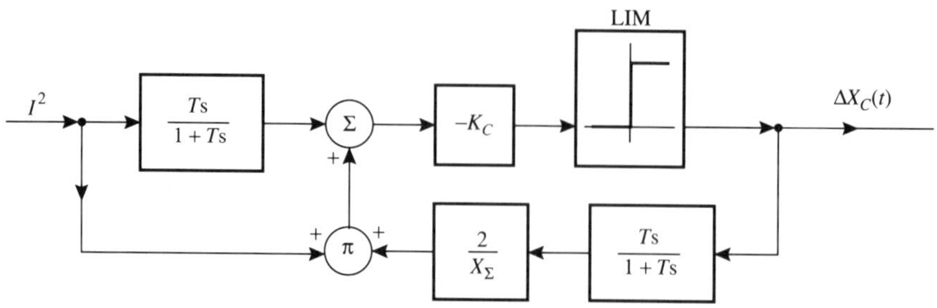

Figure 9.19 Modulation controller for $\Delta X_C(t)$

$$\frac{d(I^2)}{dt} = \frac{2}{X_\Sigma^2} E_g V_s (\sin\delta')\Delta\omega - \frac{2}{X_\Sigma} I^2 \frac{d\Delta X_C}{dt} \tag{9.54}$$

so that

$$E_g V_s (\sin\delta')\Delta\omega = \frac{X_\Sigma^2}{2}\frac{d(I^2)}{dt} + X_\Sigma I^2 \frac{d\Delta X_C}{dt} \tag{9.55}$$

which when substituted into equation 9.47 gives the following equivalent control strategy

$$\Delta X_C(t) = -K_C \left[\frac{d(I^2)}{dt} + \frac{2}{X_\Sigma} I^2 \frac{d\Delta X_C}{dt}\right] \tag{9.56}$$

where K_C is the equivalent gain and, as in equation 9.45 X_Σ can be assumed constant.

Figure 9.19 shows the block diagram of a modulation controller capable of producing the control strategy defined in equation 9.56. Again differentiation has been replaced by a real differentiating element with a small time constant T. As $\Delta X_C(t)$ is that part of the capacitor that is by-passed, it cannot be negative and $\Delta X_C(t) \geq 0$. This is realised in Figure 9.19 by a signal limiter. As for the controller shown in Figure 9.14 this controller is non-linear as the signal in the feedback loop is multiplied by the input signal. This feedback offsets the influence of changes in I^2 due to changes in $X_C(t)$. As the overall influence of the feedback signal $\Delta X_C(t)$ is of secondary importance compared with the influence of the main regulation path, the value of X_Σ can be replaced by an average value calculated off-line. The main advantage of using the squared current signal I^2 compared to the voltage squared signal V^2 is its low sensitivity to changes in the controlled element $\Delta X_C(t)$.

With thyristor-controlled capacitors it is also possible to use modulation controllers based on other input signals. For example an EPRI report published in 1991 (EPRI, 1991) describes a synthesised signal based on both voltage and current. The general trend is to use locally measurable quantities to avoid signal communication requirements.

9.6 Unified Power Flow Controller

Regulating transformers, particularly phase-shifting transformers with mechanical tap changers, have been used for many years to control steady-state power flows. Recent progress in power electronics has resulted in the development of several new types of thyristor-controlled FACTS devices including the unified power flow controller (UPFC) shown in Figure 2.25b.

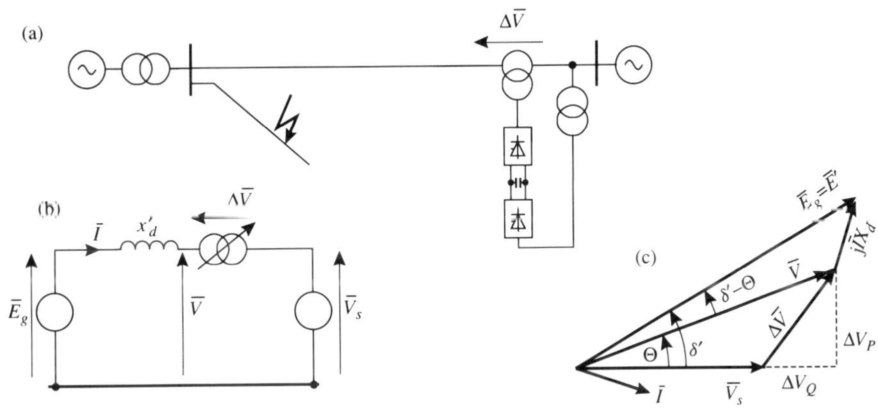

Figure 9.20 Generator-infinite busbar system with the UPFC; (a) one-line diagram; (b) equivalent circuit (c) phasor diagram

The way the UPFC can be used to damp power swings can be explained using the simple generator-infinite busbar system shown in Figure 9.20. To simplify considerations the UPFC is assumed to be installed near the end of the transmission link, close to the infinite busbar. The system equivalent reactance has been neglected in order to simplify the equation describing the equivalent circuit. The booster voltage ΔV has an in-phase component ΔV_Q and a quadrature component ΔV_P. Both components are proportional to the voltage at the point where the UPFC is installed and can be written as

$$\Delta V_Q = V_s\beta(t), \quad \Delta V_P = V_s\gamma(t) \tag{9.57}$$

where $\beta(t)$ and $\gamma(t)$ are the control variables. Figure 9.20c gives

$$\sin\Theta = \frac{\Delta V_P}{V} = \frac{V_s}{V}\gamma; \quad \cos\Theta = \frac{V_2 + \Delta V_Q}{V} = \frac{V_s}{V}(1+\beta) \tag{9.58}$$

Neglecting network losses, the electrical power can be expressed as

$$P(\delta') = \frac{E_g V}{x'_d}\sin(\delta' - \Theta) = \frac{E_g V}{x'_d}(\sin\delta'\cos\Theta - \cos\delta'\sin\Theta) \tag{9.59}$$

where x'_d is the equivalent transient reactance which also includes the reactance of the transformer and the transmission line. Substituting for $\sin\Theta$ and $\cos\Theta$ expressions obtained from equation 9.58 gives

$$\begin{aligned} P(\delta') &= \frac{E_g V_s}{x'_d}[(\sin\delta')(1+\beta) - (\cos\delta')\gamma] \\ &= b\sin\delta' - (b\cos\delta')\gamma(t) + (b\sin\delta')\beta(t) \end{aligned} \tag{9.60}$$

where $b = E_g V_s / x'_d$ is the amplitude of the power-angle characteristic with zero booster voltage. Equation 9.60 shows that control of the in-phase booster voltage (via the control variable $\beta(t)$) increases the amplitude of the power-angle characteristic while control of the quadrature booster voltage (via the control variable $\gamma(t)$) shifts the characteristic horizontally.

The generator swing equation is

$$\frac{d\delta'}{dt} = \Delta\omega$$
$$M\frac{d\Delta\omega}{dt} = P_m - (b\sin\delta') - D\frac{d\delta'}{dt} - P_{UPFC} \tag{9.61}$$

where $P_{UPFC} = -(b\cos\delta')\gamma(t) + (b\sin\delta')\beta(t)$ is the additional power component enforced by the controller. This component introduces additional damping to the system if it is made positive and proportional to the speed deviation $\Delta\omega = d\delta'/dt$. This can be achieved through the following control strategy:

$$\gamma(t) = -K(\cos\delta')\Delta\omega \tag{9.62}$$

$$\beta(t) = K(\sin\delta')\Delta\omega \tag{9.63}$$

Substituting equations 9.62 and 9.63 into the expression for P_{UPFC} gives $P_{UPFC} = Kb\Delta\omega = D_{UPFC}\Delta\omega$ where $D_{UPFC} = Kb$ is the damping coefficient due to the controller. This coefficient is constant for the control strategy defined in equations 9.62 and 9.63 and does not depend on δ'.

If the thyristor controlled phase angle regulator (TCPAR) shown in Figure 2.26a is used instead of a UPFC then only the quadrature booster voltage can be controlled and $\beta(t) = 0$. This modifies the damping power to

$$P_{UPFC}\big|_{\beta(t)=0} = P_{TCPAR} = Kb(\cos\delta')^2\Delta\omega = D_{TCPAR}\Delta\omega \tag{9.64}$$

where $D_{TCPAR} = Kb(\cos\delta')^2$ is the damping coefficient due to TCPAR. In this case the damping depends on $\cos\delta'$ and is weak for large δ' near the peak of the power-angle characteristic.

There is a close analogy between controlling the in-phase booster voltage, equation 9.63, controlling the susceptance of the shunt element, equation 9.18, and controlling the series capacitance, equation 9.47. A similar analogy exists between controlling the quadrature booster voltage, equation 9.62, and controlling the conductance of the shunt element, equation 9.18.

If the properties of the locally measurable signals are similarly analysed as for the shunt and series elements then it can be shown that the state-variable control defined by equations 9.62 and 9.63 can be approximately executed using the time derivatives of one of the following three pairs of signals:

- real and reactive power, P and Q, measured at the transmission link supplying the UPFC
- active and reactive components of the current, $I_P = VI\cos\varphi = P/V$ and $I_Q = VI\sin\varphi = Q/V$, flowing through the link
- the apparent conductance and susceptance as seen from the UPFC, $G = P/V^2$ and $B = Q/V^2$

All these signals pairs have similar properties when the UPFC is installed near the infinite busbar or near the generator terminals. Differences between them only appear if the UPFC is installed inside the transmission network itself.

10
Advanced Power System Modelling

In Chapter 4 the dynamic interactions taking place inside a generator following a disturbance were explained by considering the changes in the armature and rotor mmf and flux linkage. Although this type of approach allows the mechanisms by which the currents and torques are produced to be explained it is difficult to quantify the behaviour of the generator under all operating conditions. In this chapter a more general mathematical approach is adopted that can be used to quantify the changes in the currents and the torque but which is slightly more removed from the physics. To produce this *mathematical model* the generator will be represented by a number of electrical circuits, each with its own inductance and resistance and with mutual coupling between the circuits. By making some judicious assumptions with regard to the dominant changes taking place inside the generator during a particular type of disturbance, this detailed mathematical model can be simplified to produce a series of generator models. These simplified models can then be used in the appropriate situation.

To fully utilise these mathematical models of the generator it is also necessary to produce mathematical models of the turbine and its governor, as well as the AVR. These aspects are covered in the second part of the chapter. The chapter ends by considering suitable models of power system loads.

10.1 Synchronous Generator

In order to examine what happens inside a synchronous machine when it is subjected to an abrupt change in operating conditions Section 4.2 described the behaviour of the generator following a sudden short circuit on the machine terminals. The effect of the fault is to cause the current in, and the flux linking, the different windings to change in such a way that three characteristic states can be identified. These characteristic states are termed the subtransient-state, the transient-state and the steady-state. In each of these three characteristic states the generator may be represented by an emf behind a reactance, the value of which is linked to the reluctance of the armature reaction flux path as explained in Section 4.2.3. In reality the transition from one state to another takes place smoothly so that the values of the fictitious internal emfs also change smoothly with time. In previous chapters these smooth changes were neglected and the emfs assumed to be constant in each of the characteristic states. In this chapter the flux changes occurring within the synchronous machine will be analysed more rigorously with the resulting algebraic and differential equations constituting an advanced, dynamic, mathematical model of the synchronous generator.

Assumptions

A schematic cross section of a generator is shown in Figure 10.1. The generator is assumed to have a three-phase stator armature winding (A,B,C), a rotor field winding (f) and two rotor damper windings - one in the d-axis (D) and one in the q-axis (Q). Figure 10.1 also shows the relative position of the windings, and their axes, with the centre of phase A taken as the reference. The notation is the same as that used in Figure 4.3 and follows the normal IEEE convention (IEEE Committee Report, 1969). In developing the mathematical model the following assumptions are made:

1. The three-phase stator winding is symmetrical
2. The capacitance of all the windings can be neglected
3. Each of the distributed windings may be represented by a concentrated winding
4. The change in the inductance of the stator windings due to rotor position is sinusoidal and does not contain higher harmonics
5. Hysteresis loss is negligible but the influence of eddy-currents can be included in the model of the damper windings
6. In the transient and subtransient-state the rotor speed is near synchronous speed ($\omega \approx \omega_s$)
7. The magnetic circuits are linear (not saturated) and the inductance values do not depend on the current

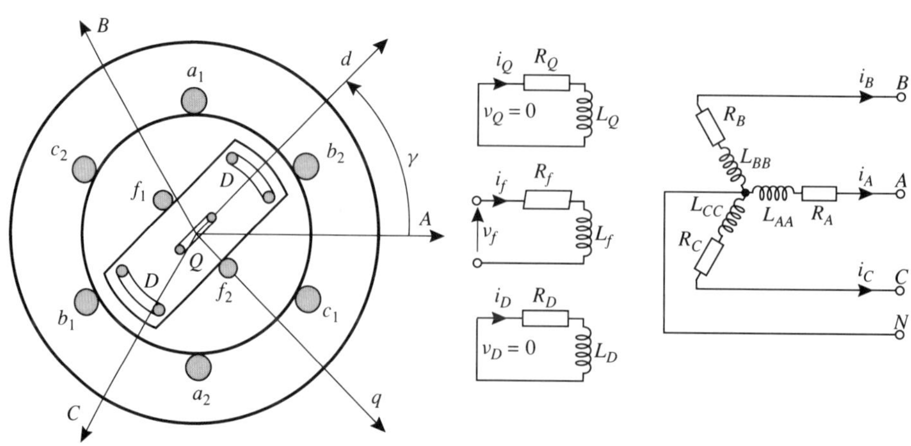

Figure 10.1 The windings in the synchronous generator and their axes

10.1.1 The flux-linkage equations in the stator reference frame

All the generator windings are magnetically coupled so that the flux in each winding depends on the currents in all the other windings. This is represented by the following matrix equation

$$\begin{bmatrix} \Psi_A \\ \Psi_B \\ \Psi_C \\ \hline \Psi_f \\ \Psi_D \\ \Psi_Q \end{bmatrix} = \left[\begin{array}{ccc|ccc} L_{AA} & L_{AB} & L_{AC} & L_{Af} & L_{AD} & L_{AQ} \\ L_{BA} & L_{BB} & L_{BC} & L_{Bf} & L_{BD} & L_{BQ} \\ L_{CA} & L_{CB} & L_{CC} & L_{Cf} & L_{CD} & L_{CQ} \\ \hline L_{fA} & L_{fB} & L_{fC} & L_{ff} & L_{fD} & L_{fQ} \\ L_{DA} & L_{DB} & L_{DC} & L_{Df} & L_{DD} & L_{DQ} \\ L_{QA} & L_{QB} & L_{QC} & L_{Qf} & L_{QD} & L_{QQ} \end{array}\right] \begin{bmatrix} i_A \\ i_B \\ i_C \\ \hline i_f \\ i_D \\ i_Q \end{bmatrix}$$

or

$$\begin{bmatrix} \mathbf{\Psi}_{ABC} \\ \hline \mathbf{\Psi}_{fDQ} \end{bmatrix} = \left[\begin{array}{c|c} \mathbf{L}_S & \mathbf{L}_{SR} \\ \hline \mathbf{L}_{SR}^T & \mathbf{L}_R \end{array}\right] \begin{bmatrix} \mathbf{i}_{ABC} \\ \hline \mathbf{i}_{fDQ} \end{bmatrix} \tag{10.1}$$

where $\mathbf{L}_S$ is a submatrix of the stator self and mutual inductances, $\mathbf{L}_R$ is a submatrix of the rotor self and mutual inductances and $\mathbf{L}_{SR}$ is a submatrix of the rotor to stator mutual inductances. Most of these inductances are subject to periodic changes due to both the saliency and rotation of the rotor. Consistent with the assumptions outlined above, the higher harmonics of these inductance changes will be neglected and the inductances represented by a constant component and a single periodic component.

In the case of the two-pole machine shown in Figure 10.1 the self inductance of each stator phase winding will reach a maximum value whenever the rotor d-axis aligns with the axis of the phase winding as, with the rotor in this position, the reluctance of the flux path is a minimum. This minimum reluctance condition occurs twice during each rotation of the rotor so that the stator self inductances are of the form

$$L_{AA} = L_S + \Delta L_S \cos 2\gamma,\ L_{BB} = L_S + \Delta L_S \cos\left(2\gamma - \frac{2}{3}\pi\right),$$

$$L_{CC} = L_S + \Delta L_S \cos\left(2\gamma + \frac{2}{3}\pi\right) \tag{10.2}$$

Both L_S and ΔL_S are constant and $L_S > \Delta L_S$.

As each of the stator windings are shifted in space relative to each other by 120 degrees the mutual inductance between each of the stator windings is negative. The magnitude of the inductance is a maximum when the rotor d-axis is midway between the axes of two of the windings. Referring to Figure 10.1 gives

$$\begin{aligned} L_{AB} = L_{BA} &= -M_S - \Delta L_S \cos 2\left(\gamma + \frac{1}{6}\pi\right) \\ L_{BC} = L_{CB} &= -M_S - \Delta L_S \cos 2\left(\gamma - \frac{1}{2}\pi\right) \\ L_{CA} = L_{AC} &= -M_S - \Delta L_S \cos 2\left(\gamma + \frac{5}{6}\pi\right) \end{aligned} \tag{10.3}$$

where $M_s > \Delta L_s$.

The mutual inductances between the stator and rotor windings change with rotor position and have a positive maximum value when the axes of a stator winding and the rotor winding align and have the same positive flux direction. When the flux directions are in opposition the value of the inductance is a negative minimum while when the axes are perpendicular the inductance is zero. Referring again to Figure 10.1 gives

$$L_{Af} = L_{fA} = M_f \cos\gamma$$
$$L_{AD} = L_{DA} = M_D \cos\gamma$$
$$L_{AQ} = L_{QA} = M_Q \sin\gamma$$

$$L_{Bf} = L_{fB} = M_f \cos\left(\gamma - \frac{2}{3}\pi\right), L_{Cf} = L_{fC} = M_f \cos\left(\gamma + \frac{2}{3}\pi\right)$$

$$L_{BD} = L_{DB} = M_D \cos\left(\gamma - \frac{2}{3}\pi\right), L_{CD} = L_{DC} = M_D \cos\left(\gamma + \frac{2}{3}\pi\right) \qquad (10.4)$$

$$L_{BQ} = L_{QB} = M_Q \sin\left(\gamma - \frac{2}{3}\pi\right), L_{CQ} = L_{QC} = M_Q \sin\left(\gamma + \frac{2}{3}\pi\right)$$

The self and mutual inductances of the rotor windings are constant and do not depend on rotor position. As the d- and q-axis windings are perpendicular to each other their mutual inductances are zero

$$L_{fQ} = L_{Qf} = 0 \text{ and } L_{DQ} = L_{QD} = 0 \qquad (10.5)$$

Most of the elements forming the inductance matrix **L** in the flux-linkage equation 10.1 depend on rotor position and are therefore functions of time.

10.1.2 The flux-linkage equations in the rotor reference frame

At any instant in time the position of the rotor relative to the stator reference axis is defined by the angle γ shown in Figure 10.1. Each phasor, whether voltage, current or flux linkage, in the stator reference frame (A,B,C) can be transformed into the (d,q) reference frame by projecting one reference frame onto the other using trigonometric functions of the angle γ. Using the notation of Figure 10.1 the current vectors are

$$i_d = \beta_d \left[i_A \cos\gamma + i_B \cos\left(\gamma - \frac{2}{3}\pi\right) + i_C \cos\left(\gamma + \frac{2}{3}\pi\right)\right]$$
$$i_q = \beta_q \left[i_A \sin\gamma + i_B \sin\left(\gamma - \frac{2}{3}\pi\right) + i_C \sin\left(\gamma + \frac{2}{3}\pi\right)\right] \qquad (10.6)$$

where β_d and β_q are arbitrary, non-zero, coefficients introduced due to the change of reference frame. Equation 10.6 describes a unique transformation from the stator (A,B,C) to the rotor (d,q) reference axis. The reverse transformation from (d,q) to (A,B,C) is not unique as the two equations 10.6 have three unknowns i_A, i_B, i_C. A unique transformation may be achieved by supplementing the (d,q) co-ordinates by an additional co-ordinate. It is convenient to assume this additional co-ordinate to be the zero-sequence co-ordinate defined in the same way as in the method of symmetrical components

$$i_0 = \beta_0(i_A + i_B + i_C) \qquad (10.7)$$

where β_0 is again an arbitrary coefficient introduced due to the change in reference frame. Combining equations 10.6 and 10.7 gives the following matrix equation

$$\begin{bmatrix} i_0 \\ i_d \\ i_q \end{bmatrix} = \begin{bmatrix} \beta_0 & \beta_0 & \beta_0 \\ \beta_d \cos\gamma & \beta_d \cos\left(\gamma - \frac{2}{3}\pi\right) & \beta_d \cos\left(\gamma + \frac{2}{3}\pi\right) \\ \beta_q \sin\gamma & \beta_q \sin\left(\gamma - \frac{2}{3}\pi\right) & \beta_q \sin\left(\gamma + \frac{2}{3}\pi\right) \end{bmatrix} \begin{bmatrix} i_A \\ i_B \\ i_C \end{bmatrix} \tag{10.8}$$

or $\qquad \mathbf{i}_{0dq} = \mathbf{W}\mathbf{i}_{ABC}$

The coefficients β_0, β_d and β_q are non-zero. Matrix $\mathbf{W}$ is non-singular and the inverse transformation is uniquely determined by

$$\mathbf{i}_{ABC} = \mathbf{W}^{-1}\mathbf{i}_{0dq} \tag{10.9}$$

Similar transformation can be defined for the phasors of stator voltage and flux linkage.

The rotor currents, voltages and flux linkages are already in the (d,q) reference frame and no transformation is necessary, allowing the transformation of all the winding currents to be written as

$$\begin{bmatrix} \mathbf{i}_{0dq} \\ \mathbf{i}_{fDQ} \end{bmatrix} = \begin{bmatrix} \mathbf{W} & \mathbf{0} \\ \mathbf{0} & \mathbf{1} \end{bmatrix} \begin{bmatrix} \mathbf{i}_{ABC} \\ \mathbf{i}_{fDQ} \end{bmatrix} \tag{10.10}$$

In this equation $\mathbf{i}_{fDQ}$ is a column vector of the currents i_f, i_D and i_Q, and $\mathbf{1}$ is a diagonal unity matrix. A similar transformation can be defined for the rotor voltages and flux linkages. The inverse transformation of equation 10.10 is

$$\begin{bmatrix} \mathbf{i}_{ABC} \\ \mathbf{i}_{fDQ} \end{bmatrix} = \begin{bmatrix} \mathbf{W}^{-1} & \mathbf{0} \\ \mathbf{0} & \mathbf{1} \end{bmatrix} \begin{bmatrix} \mathbf{i}_{0dq} \\ \mathbf{i}_{fDQ} \end{bmatrix} \tag{10.11}$$

Substituting this inverse transformation, along with a similar transformation for the fluxes, into the flux-linkage equation 10.1 gives

$$\begin{bmatrix} \mathbf{\Psi}_{0dq} \\ \mathbf{\Psi}_{fDQ} \end{bmatrix} = \begin{bmatrix} \mathbf{W} & \mathbf{0} \\ \mathbf{0} & \mathbf{1} \end{bmatrix} \begin{bmatrix} \mathbf{L}_S & \mathbf{L}_{SR} \\ \mathbf{L}_{SR}^T & \mathbf{L}_R \end{bmatrix} \begin{bmatrix} \mathbf{W}^{-1} & \mathbf{0} \\ \mathbf{0} & \mathbf{1} \end{bmatrix} \begin{bmatrix} \mathbf{i}_{0dq} \\ \mathbf{i}_{fDQ} \end{bmatrix} \tag{10.12}$$

which, after multiplying the three square matrices by each other, yields

$$\begin{bmatrix} \mathbf{\Psi}_{0dq} \\ \mathbf{\Psi}_{fDQ} \end{bmatrix} = \begin{bmatrix} \mathbf{W}\mathbf{L}_S\mathbf{W}^{-1} & \mathbf{W}\mathbf{L}_{SR} \\ \mathbf{L}_{SR}^T\mathbf{W}^{-1} & \mathbf{L}_R \end{bmatrix} \begin{bmatrix} \mathbf{i}_{0dq} \\ \mathbf{i}_{fDQ} \end{bmatrix} \tag{10.13}$$

The coefficients introduced due to the change in the reference frame are now chosen as $\beta_0 = 1/\sqrt{3}$ and $\beta_d = \beta_q = \sqrt{2/3}$ to give the following transformation matrix

$$\mathbf{W} = \sqrt{\frac{2}{3}} \begin{bmatrix} \frac{1}{\sqrt{2}} & \frac{1}{\sqrt{2}} & \frac{1}{\sqrt{2}} \\ \cos\gamma & \cos\left(\gamma - \frac{2}{3}\pi\right) & \cos\left(\gamma + \frac{2}{3}\pi\right) \\ \sin\gamma & \sin\left(\gamma - \frac{2}{3}\pi\right) & \sin\left(\gamma + \frac{2}{3}\pi\right) \end{bmatrix} \tag{10.14}$$

With this choice of transformation coefficients $\mathbf{W}^{-1} = \mathbf{W}^T$ where $\mathbf{W}^{-1}$ and $\mathbf{W}^T$ are respectively the inverse and the transpose of $\mathbf{W}$. For a matrix that satisfies the condition $\mathbf{W}^{-1} = \mathbf{W}^T$ then $\mathbf{W}\mathbf{W}^T = \mathbf{1}$ and the matrix is said to be *orthogonal*. As will be seen later such an orthogonal transformation is necessary to ensure that the power calculated in both the (A,B,C) and (d,q) reference frames is identical and the transformation is said to be *power invariant.* The transformation matrix $\mathbf{W}$ transforms the submatrix of self and mutual inductances of the stator windings $\mathbf{L}_S$ into

$$\mathbf{W}\mathbf{L}_S\mathbf{W}^{-1} = \frac{2}{3}\begin{bmatrix} \frac{1}{\sqrt{2}} & \frac{1}{\sqrt{2}} & \frac{1}{\sqrt{2}} \\ \cos\gamma & \cos\left(\gamma - \frac{2}{3}\pi\right) & \cos\left(\gamma + \frac{2}{3}\pi\right) \\ \sin\gamma & \sin\left(\gamma - \frac{2}{3}\pi\right) & \sin\left(\gamma + \frac{2}{3}\pi\right) \end{bmatrix} \begin{bmatrix} L_{AA} & L_{AB} & L_{AC} \\ L_{BA} & L_{BB} & L_{BC} \\ L_{CA} & L_{CB} & L_{CC} \end{bmatrix}$$

$$\begin{bmatrix} \frac{1}{\sqrt{2}} & \cos\gamma & \sin\gamma \\ \frac{1}{\sqrt{2}} & \cos\left(\gamma - \frac{2}{3}\pi\right) & \sin\left(\gamma - \frac{2}{3}\pi\right) \\ \frac{1}{\sqrt{2}} & \cos\left(\gamma + \frac{2}{3}\pi\right) & \sin\left(\gamma + \frac{2}{3}\pi\right) \end{bmatrix}$$

$$= \begin{bmatrix} L_0 & & \\ & L_d & \\ & & L_q \end{bmatrix} \quad (10.15)$$

This is a diagonal matrix in which $L_0 = L_S - 2M_S$, $L_d = L_S + M_S + 3\Delta L_S/2$ and $L_q = L_S + M_S - 3\Delta L_S/2$. Similarly the submatrix of the mutual inductances between the stator and the rotor windings is transformed into

$$\mathbf{W}\mathbf{L}_{SR} = \sqrt{\frac{2}{3}}\begin{bmatrix} \frac{1}{\sqrt{2}} & \frac{1}{\sqrt{2}} & \frac{1}{\sqrt{2}} \\ \cos\gamma & \cos\left(\gamma - \frac{2}{3}\pi\right) & \cos\left(\gamma + \frac{2}{3}\pi\right) \\ \sin\gamma & \sin\left(\gamma - \frac{2}{3}\pi\right) & \sin\left(\gamma + \frac{2}{3}\pi\right) \end{bmatrix} \begin{bmatrix} L_{Af} & L_{AB} & L_{AQ} \\ L_{Bf} & L_{BD} & L_{BQ} \\ L_{Cf} & L_{CD} & L_{CQ} \end{bmatrix}$$

$$= \begin{bmatrix} & & \\ kM_f & kM_D & \\ & & kM_Q \end{bmatrix}$$

where $k = \sqrt{3/2}$. The submatrix of the mutual inductances between the rotor and the stator windings is transformed into the same form, because $\mathbf{W}^{-1} = \mathbf{W}^T$ and

$$\mathbf{L}_{SR}^T\mathbf{W}^{-1} = \mathbf{L}_{SR}^T\mathbf{W}^T = (\mathbf{W}\mathbf{L}_{SR})^T$$

The matrix of the self and mutual inductances of the rotor windings is not changed. As a result of these transformations equation 10.13 becomes

$$\begin{bmatrix} \Psi_0 \\ \Psi_d \\ \Psi_q \\ \Psi_f \\ \Psi_D \\ \Psi_Q \end{bmatrix} = \begin{bmatrix} L_0 & & & & & \\ & L_d & & kM_f & kM_D & \\ & & L_q & & & kM_Q \\ & kM_f & & L_f & L_{fD} & \\ & kM_D & & L_{fD} & L_D & \\ & & kM_Q & & & L_Q \end{bmatrix} \begin{bmatrix} i_0 \\ i_d \\ i_q \\ i_f \\ i_D \\ i_Q \end{bmatrix} \quad (10.16)$$

An important feature of the above equation is that the matrix of inductances is symmetrical. This is due to the correct choice of the transformation coefficients β_0, β_d, and β_q ensuring orthogonality of the transformation matrix, **W**.

The transformation of all the generator windings into the rotor reference frame is referred to as *0dq transformation*, or *Park's transformation*. The original transformation matrix proposed by Park (Concordia, 1951) was not orthogonal and consequently the resulting matrix of equivalent inductances was not symmetrical. Concordia corrected this but customarily the transformation is still referred to as the *Park's* or the *modified Park's transformation*. All the elements of the inductance matrix in equation 10.16 are constant and independent of time. This is the main advantage of Park's transformation.

After reordering the variables equation 10.16 can be written as three independent sets of equations

$$\Psi_0 = L_0 i_0 \tag{10.17}$$

$$\begin{bmatrix} \Psi_d \\ \Psi_f \\ \Psi_D \end{bmatrix} = \begin{bmatrix} L_d & kM_f & kM_D \\ kM_f & L_f & L_{fD} \\ kM_D & L_{fD} & L_D \end{bmatrix} \begin{bmatrix} i_d \\ i_f \\ i_D \end{bmatrix} \tag{10.18}$$

$$\begin{bmatrix} \Psi_q \\ \Psi_Q \end{bmatrix} = \begin{bmatrix} L_q & kM_Q \\ kM_Q & L_Q \end{bmatrix} \begin{bmatrix} i_q \\ i_Q \end{bmatrix} \tag{10.19}$$

These equations describe three sets of magnetically coupled windings as shown in Figure 10.2. Each set of windings is independent of each other in that there is no magnetic coupling between the different winding sets. Figure 10.2 reflects this by showing the three winding sets perpendicular to each other. The first set of windings, represented by equation 10.18, consists of three windings in the d-axis. Two of these, f and D, correspond to the real field and damper windings of the rotor. The third winding, denoted as d, is fictitious and represents the effect of the three-phase stator winding in the d-axis of the rotor. This fictitious d-axis winding rotates with the rotor.

The second set of windings, represented by equation 10.19, consists of two windings. The first one, denoted by Q, corresponds to the real damper winding in the rotor q-axis, while the second, denoted by q, is a fictitious winding representing the effect of the three-phase stator winding in the q-axis. Obviously both windings rotate with the rotor.

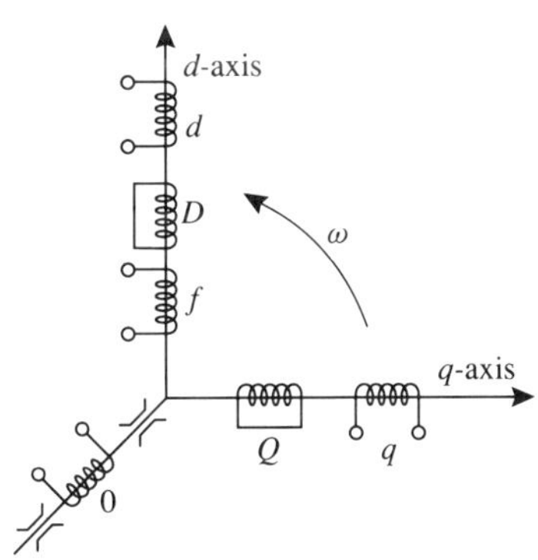

Figure 10.2 Three sets of fictitious perpendicular windings representing the synchronous generator

Equation 10.17 represents the third winding set consisting of a single winding which is magnetically separate from both of the other two sets. This winding is shown in Figure 10.2 to be perpendicular to both the d and q axes and to be along the axis of rotation of the equivalent rotor. This winding can be omitted if the three-phase stator winding is connected in star with the neutral point isolated, i.e. not earthed. With this winding connection the sum of the stator phase currents must be zero and, as $i_0 = (i_A + i_B + i_C)/\sqrt{3} = 0$, the current flowing in this third winding is also zero.

A physical interpretation of the (d,q) axes coils can be obtained by considering equation 10.16. This equation defines the flux linkages within the generator but with the actual three-phase stator armature winding replaced by one winding in the d-axis and another in the q-axis. As shown in Chapter 3 currents in the three-phase stator armature winding produce a rotating armature reaction flux which enters the rotor at an angle that depends on the armature loading condition. In the rotor (d,q) reference frame this rotating flux is simply represented by two dc flux components: one acting along the d axis and the other along the q-axis. These (d,q) component fluxes are produced by the currents flowing in the two fictitious (d,q) armature windings. In light of this, selecting the transformation coefficients $\beta_d = \beta_q = \sqrt{2/3}$ has important implications regarding the number of turns on the fictitious d and q armature windings. For balanced three-phase currents in the armature equation 3.42 showed that the value of the armature mmf rotating with the rotor was equal to 3/2 $N_a I_m$ where N_a is the effective number of turns in series per phase and $I_m = \sqrt{2}I_g$. However, as will be shown in equation 10.82, the same balanced three-phase current gives the maximum values of i_d and i_q as $\sqrt{3/2}I_m$. Consequently if i_d and i_q are to produce the same mmf as the actual three-phase armature winding the d and q armature windings must each have $\sqrt{3/2}$ more turns than an actual armature phase winding. This effect is reflected by the factor $k = \sqrt{3/2}$ which appears in the mutual inductance between the d axis armature winding and both d axis rotor windings. The same factor is present in the mutual inductance between the q axis armature winding and the q axis damper winding.

Other values of transformation coefficient can be used to produce an orthogonal transformation. In particular a transformation coefficient of 2/3 is favoured by a number of authors, for example Adkins (1957). The reason for this is discussed by Harris, Lawrenson & Stephenson (1970) where they compare different transformation systems. In their discussion these authors argue that a transformation coefficient of 2/3 is more closely related to flux conditions in the generator than a transformation coefficient of $\sqrt{2/3}$. With a current transformation coefficient of 2/3 both the d and q axis armature coils have the same number of turns as an individual phase winding. However in order to maintain a power invariant transformation the transformation coefficient used for voltages is different from that used for currents unless a per unit system is used. In such a per unit system it is necessary to have a different value of base current in the phase windings and the d,q windings; the (d,q) base current differing by a factor of 3/2 from the (A,B,C) base current.

Power in the (0dq) reference frame

The three-phase power output of the generator is equal to the scalar product of the stator voltages and currents

$$p_g = v_A i_A + v_B i_B + v_C i_C = \mathbf{v}_{ABC}^T \mathbf{i}_{ABC} \tag{10.20}$$

The orthogonal transformation from the (A,B,C) to the $(0,d,q)$ reference frame ensures that the transformation is power invariant so that the power is also given by

$$p_g = v_0 i_0 + v_d i_d + v_q i_q = \mathbf{v}_{0dq}^T \mathbf{i}_{0dq} \tag{10.21}$$

Equation 10.21 can be verified by substituting the voltage and current transformations $\mathbf{v}_{ABC} = \mathbf{W}^{-1}\mathbf{v}_{0dq}$ and $\mathbf{i}_{ABC} = \mathbf{W}^{-1}\mathbf{i}_{0dq}$ into equation 10.20 to give

$$\begin{aligned} p_g &= \mathbf{v}_{ABC}^T \mathbf{i}_{ABC} = (\mathbf{W}^{-1}\mathbf{v}_{0dq})^T \mathbf{W}^{-1}\mathbf{i}_{0dq} = \mathbf{v}_{0dq}^T (\mathbf{W}^{-1})\mathbf{W}^{-1}\mathbf{i}_{0dq} \\ &= \mathbf{v}_{0dq}^T (\mathbf{W}^T)^T \mathbf{W}^{-1}\mathbf{i}_{0dq} = \mathbf{v}_{0dq}^T \mathbf{W}\mathbf{W}^{-1}\mathbf{i}_{0dq} = \mathbf{v}_{0dq}^T \mathbf{i}_{0dq} \end{aligned}$$

noting that as the matrix $\mathbf{W}$ is orthogonal $\mathbf{W}^{-1} = \mathbf{W}^T$.

10.1.3 Voltage equations

The winding circuits shown in Figure 10.1 can be divided into two characteristic types. The first type, consisting of the stator windings (A, B, C) and the damper windings (D, Q), are circuits in which the emf induced in the winding drives the current in the winding. The application of the Kirchhoff's Voltage Law to such a circuit is shown in Figure 10.3a. The second type of circuit is represented by the rotor field winding f in which the current is supplied by an external voltage source. In this case an emf which opposes the current is induced in the winding. The equivalent circuit for this circuit is shown in Figure 10.3b. The convention for the direction of the voltages is the same as that used in Figure 10.1.

Using this convention the voltage equation in the (A,B,C) reference frame follows as

$$\begin{bmatrix} v_A \\ v_B \\ v_C \\ -v_f \\ 0 \\ 0 \end{bmatrix} = -\begin{bmatrix} R_A & & & & & \\ & R_B & & & & \\ & & R_C & & & \\ & & & R_f & & \\ & & & & R_D & \\ & & & & & R_Q \end{bmatrix} \begin{bmatrix} i_A \\ i_B \\ i_C \\ i_f \\ i_D \\ i_Q \end{bmatrix} - \frac{\mathrm{d}}{\mathrm{d}t} \begin{bmatrix} \Psi_A \\ \Psi_B \\ \Psi_C \\ \Psi_f \\ \Psi_D \\ \Psi_Q \end{bmatrix} \tag{10.22}$$

or, when written in compact matrix form,

$$\begin{bmatrix} \mathbf{v}_{ABC} \\ \mathbf{v}_{fDQ} \end{bmatrix} = -\begin{bmatrix} \mathbf{R}_{ABC} & \\ & \mathbf{R}_{fDQ} \end{bmatrix} \begin{bmatrix} \mathbf{i}_{ABC} \\ \mathbf{i}_{fDQ} \end{bmatrix} - \frac{\mathrm{d}}{\mathrm{d}t} \begin{bmatrix} \mathbf{\Psi}_{ABC} \\ \mathbf{\Psi}_{fDQ} \end{bmatrix} \tag{10.23}$$

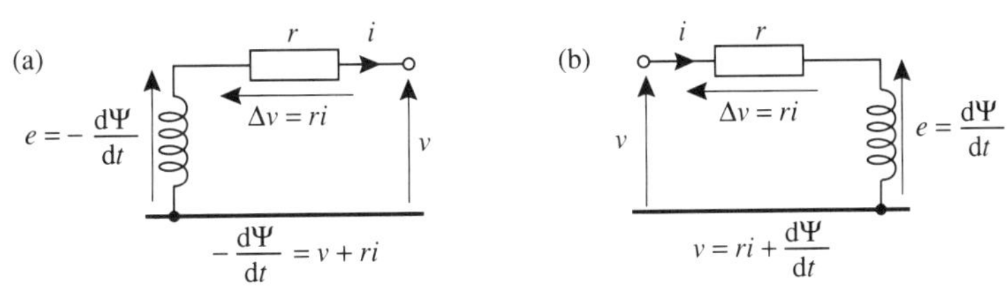

Figure 10.3 Kirchhoff's Voltage Law applied to the two types of circuits: (a) generator circuit; (b) motor circuit

where $\mathbf{R}_{ABC}$ and $\mathbf{R}_{fDQ}$ are diagonal resistance matrices. These equations can be transformed into the rotating reference frame using the transformation equation 10.11 for currents, voltages and flux linkages. After some algebra this gives

$$\begin{bmatrix} \mathbf{W}^{-1} & \\ & \mathbf{1} \end{bmatrix} \begin{bmatrix} \mathbf{v}_{0dq} \\ \mathbf{v}_{fDQ} \end{bmatrix} = - \begin{bmatrix} \mathbf{R}_{ABC} & \\ & \mathbf{R}_{fDQ} \end{bmatrix} \begin{bmatrix} \mathbf{W}^{-1} & \\ & \mathbf{1} \end{bmatrix} \begin{bmatrix} \mathbf{i}_{0dq} \\ \mathbf{i}_{fDQ} \end{bmatrix} - \frac{\mathrm{d}}{\mathrm{d}t} \begin{bmatrix} \mathbf{W}^{-1} & \\ & \mathbf{1} \end{bmatrix} \begin{bmatrix} \mathbf{\Psi}_{0dq} \\ \mathbf{\Psi}_{fDQ} \end{bmatrix}$$

which, when left multiplied by the transformation matrix $\mathbf{W}$, becomes

$$\begin{bmatrix} \mathbf{v}_{0dq} \\ \mathbf{v}_{fDQ} \end{bmatrix} = - \begin{bmatrix} \mathbf{W} & \\ & \mathbf{1} \end{bmatrix} \begin{bmatrix} \mathbf{R}_{ABC} & \\ & \mathbf{R}_{fDQ} \end{bmatrix} \begin{bmatrix} \mathbf{W}^{-1} & \\ & \mathbf{1} \end{bmatrix} \begin{bmatrix} \mathbf{i}_{0dq} \\ \mathbf{i}_{fDQ} \end{bmatrix} + \\ - \begin{bmatrix} \mathbf{W} & \\ & \mathbf{1} \end{bmatrix} \frac{\mathrm{d}}{\mathrm{d}t} \left\{ \begin{bmatrix} \mathbf{W}^{-1} & \\ & \mathbf{1} \end{bmatrix} \begin{bmatrix} \mathbf{\Psi}_{0dq} \\ \mathbf{\Psi}_{fDQ} \end{bmatrix} \right\} \tag{10.24}$$

If the resistance of each of the stator phases is identical $R_A = R_B = R_C = R$ and the product of the first three matrices on the right is a diagonal matrix, as

$$\mathbf{W}\mathbf{R}_{ABC}\mathbf{W}^{-1} = \mathbf{R}_{ABC} \tag{10.25}$$

According to equation 10.14 the transformation matrix $\mathbf{W}$ is a function of time and the derivative of the last term on the right in equation 10.24 must be calculated as the derivative of a product of two functions

$$\frac{\mathrm{d}}{\mathrm{d}t}(\mathbf{W}^{-1}\mathbf{\Psi}_{0dq}) = \dot{\mathbf{W}}^{-1}\mathbf{\Psi}_{0dq} + \mathbf{W}^{-1}\dot{\mathbf{\Psi}}_{0dq}$$

where the dot on the top of a symbol denotes a derivative with respect to time. Multiplication by the transformation matrix gives

$$\mathbf{W}\frac{\mathrm{d}}{\mathrm{d}t}(\mathbf{W}^{-1}\mathbf{\Psi}_{0dq}) = (\mathbf{W}\dot{\mathbf{W}}^{-1})\mathbf{\Psi}_{0dq} + \dot{\mathbf{\Psi}}_{0dq} = -(\dot{\mathbf{W}}\mathbf{W}^{-1})\mathbf{\Psi}_{0dq} + \dot{\mathbf{\Psi}}_{0dq} \tag{10.26}$$

as the derivative of the product $\mathbf{W}\mathbf{W}^{-1} = 1$ is $\dot{\mathbf{W}}\mathbf{W} + \mathbf{W}\dot{\mathbf{W}}^{-1} = \mathbf{0}$ and $\dot{\mathbf{W}}\mathbf{W}^{-1} = -\mathbf{W}\dot{\mathbf{W}}^{-1}$. Calculating $\dot{\mathbf{W}}$ as $\dot{\mathbf{W}} = \frac{\mathrm{d}}{\mathrm{d}t}\mathbf{W}$ and multiplying by $\mathbf{W}^{-1} = \mathbf{W}^T$ gives $\dot{\mathbf{W}}\mathbf{W}^{-1}$ as

$$\mathbf{\Omega} = \dot{\mathbf{W}}\mathbf{W}^{-1} = \omega \begin{bmatrix} 0 & 0 & 0 \\ 0 & 0 & -1 \\ 0 & 1 & 0 \end{bmatrix} \tag{10.27}$$

This matrix is referred to as the *rotation matrix* as it introduces terms into the voltage equations which are dependent on the speed of rotation.

The voltage equations in the (d,q) reference frame can be obtained after substituting the formulae in equations 10.25-10.27 into equation 10.24 to give

$$\begin{bmatrix} \mathbf{v}_{0dq} \\ \mathbf{v}_{fDQ} \end{bmatrix} = - \begin{bmatrix} \mathbf{R}_{ABC} & \\ & \mathbf{R}_{fDQ} \end{bmatrix} \begin{bmatrix} \mathbf{i}_{0dq} \\ \mathbf{i}_{fDQ} \end{bmatrix} - \begin{bmatrix} \dot{\mathbf{\Psi}}_{0dq} \\ \dot{\mathbf{\Psi}}_{fDQ} \end{bmatrix} + \begin{bmatrix} \mathbf{\Omega} & \\ & \mathbf{0} \end{bmatrix} \begin{bmatrix} \mathbf{\Psi}_{0dq} \\ \mathbf{\Psi}_{fDQ} \end{bmatrix} \tag{10.28}$$

This equation, without the term $\mathbf{\Omega}\mathbf{\Psi}_{0dq}$, describes Kirchhoff's Voltage Law for the equivalent generator circuits shown in Figure 10.2. The rotational term represents the emfs induced in the stator windings due to the rotation of the magnetic field. These *rotational emfs* are represented as

$$\mathbf{\Omega\Psi}_{0dq} = \omega \begin{bmatrix} 0 & 0 & 0 \\ 0 & 0 & -1 \\ 0 & 1 & 0 \end{bmatrix} \begin{bmatrix} \Psi_0 \\ \Psi_d \\ \Psi_q \end{bmatrix} = \begin{bmatrix} 0 \\ -\omega\Psi_q \\ +\omega\Psi_d \end{bmatrix} \tag{10.29}$$

Importantly this equation shows that the d-axis rotational emf is induced by the q-axis flux, while the q-axis rotational emf is induced by the d-axis flux. The plus and minus signs are a result of the assumed direction, and rotation, of the rotor axes and the fact that an induced emf must lag the flux which produces it by 90 degrees.

The armature emfs proportional to the rate of change of the flux, i.e. the $\dot{\Psi}$ terms, are referred to as the *transformer emfs* and are due to changing currents in coils on the same axis as the one being considered. They would be present even if the machine was stationary.

Equation 10.28 may be expanded to give

$$\left.\begin{aligned} v_0 &= -Ri_0 - \dot{\Psi}_0 \\ v_d &= -Ri_d - \dot{\Psi}_d - \omega\Psi_q \\ v_q &= -Ri_q - \dot{\Psi}_q + \omega\Psi_d \end{aligned}\right\} \text{ stator} \tag{10.30}$$

$$\left.\begin{aligned} v_f &= R_f i_f + \dot{\Psi}_f \\ 0 &= R_D i_D + \dot{\Psi}_D \\ 0 &= R_Q i_Q + \dot{\Psi}_Q \end{aligned}\right\} \text{ rotor} \tag{10.31}$$

If balanced operation only is considered then there are no zero sequence currents and the first of the stator equations, corresponding to the zero sequence, can be omitted*. Generally changes in the generator speed are small ($\omega \approx \omega_s$) whilst the transformer emfs ($\dot{\Psi}_d$ and $\dot{\Psi}_q$) are also small when compared with the rotation emfs ($-\omega\Psi_q$ and $+\omega\Psi_q$), whose values are close to the corresponding components of the generator voltage. Neglecting the transformer emfs allows the differential equations 10.30 describing the stator voltage to be replaced by the following two algebraic equations

$$\begin{bmatrix} v_d \\ v_q \end{bmatrix} \approx - \begin{bmatrix} R & 0 \\ 0 & R \end{bmatrix} \begin{bmatrix} i_d \\ i_q \end{bmatrix} + \omega \begin{bmatrix} -\Psi_q \\ +\Psi_d \end{bmatrix} \tag{10.32}$$

The differential voltage equations of the rotor windings (equations 10.31) remain unchanged and can be re-written as

$$\begin{bmatrix} \dot{\Psi}_f \\ \dot{\Psi}_D \\ \dot{\Psi}_Q \end{bmatrix} = - \begin{bmatrix} R_f & 0 & 0 \\ 0 & R_D & 0 \\ 0 & 0 & R_Q \end{bmatrix} \begin{bmatrix} i_f \\ i_D \\ i_Q \end{bmatrix} + \begin{bmatrix} v_f \\ 0 \\ 0 \end{bmatrix} \tag{10.33}$$

The equations 10.32 and 10.33 together with the algebraic flux linkage equations 10.18 and 10.19 can now be used to model the synchronous generator if the transformer emfs are neglected.

---

* As the 0-axis voltage and flux equations are decoupled from the other two axes these equations can be solved separately for unbalanced operation.

To be used for power system studies the generator equations must be interfaced to the equations describing the power system transmission network. If the armature transformer emfs are included in the model then the implication of having two differential equations to describe the armature voltage is that the power system transmission network equations themselves must be differential equations. For all but the simplest systems this introduces significant complexity into the system equations, requires a large computation time, implies a parameter accuracy that is often unrealistic and is usually not necessary when studying electromechanical dynamics, other than shaft torques. By neglecting the transformer emfs the armature differential equations are replaced by two algebraic equations which permits the power system to be modelled by a set of algebraic equations as described in Chapter 3, equation 3.103. This significantly simplifies the generator to power system interface.

For many power system studies it is possible, and highly desirable, to re-phrase and simplify the full set of generator equations, equations 10.32, 10.33, 10.18 and 10.19, so that they are in a more acceptable form and easier to interface to the power system network equations. Before examining how these changes can be made it is necessary to relate these circuit equations to the flux conditions inside the generator when it is in the steady-state, transient-state or the subtransient-state. These flux conditions and characteristic states were extensively described and discussed in Chapter 4.

10.1.4 Generator reactances in terms of circuit quantities

The d-axis consists of three R,L coupled circuits; one each for the d-axis armature coil, the field coil and the d-axis damper as shown in Figure 10.4a. Only two coils are on the q-axis; one each for the q-axis armature winding and the q-axis damper, Figure 10.4b. When viewed from the terminals of the armature the effective impedance of the armature coil to any current change will depend on the parameters of the different circuits, their mutual coupling and whether or not the circuits are open or closed.

Steady-state

When in the steady-state, the armature flux has penetrated through all the rotor circuits, the field and damper winding currents are constant and the armature current simply sees the synchronous inductance L_d in the direct axis and L_q in the quadrature axis.

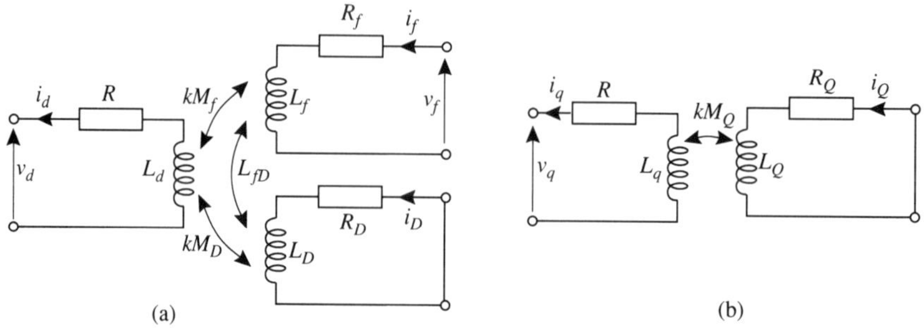

Figure 10.4 The d and q axis coupled circuits: (a) d-axis (b) q-axis

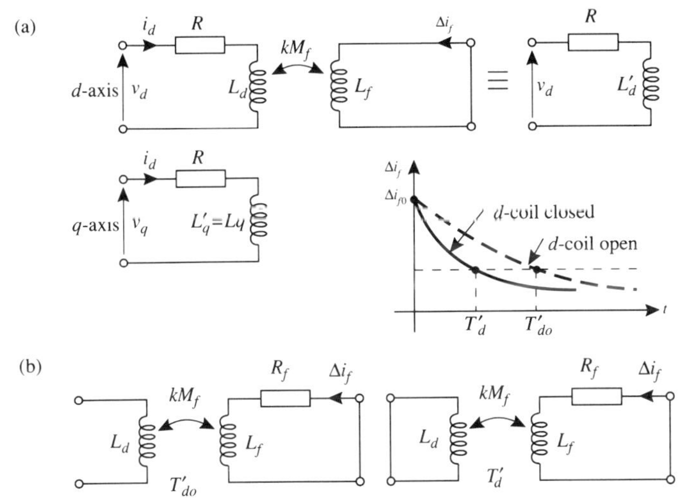

Figure 10.5 The d and q axis coupled circuits in the transient-state: (a) for determining transient inductance; (b) for determining the field winding time constants

Transient-state

In the transient-state, armature flux has penetrated the damper circuits and the field winding screens the rotor body from the armature flux. The damper circuits are no longer effective and can be removed from the model whilst the screening behaviour of the field winding is modelled by short circuiting the field winding and setting its resistance to zero, Figure 10.5a. This effectively represents the current changes that would occur in the field winding in order to maintain the flux linkage of this winding constant; the definition of the transient-state. The circuit equations for the d-axis can be written

$$\begin{aligned} v_d &= Ri_d + L_d\frac{\mathrm{d}i_d}{\mathrm{d}t} + kM_f\frac{\mathrm{d}\Delta i_f}{\mathrm{d}t} \\ \Delta v_f &= 0 = L_f\frac{\mathrm{d}\Delta i_f}{\mathrm{d}t} + kM_f\frac{\mathrm{d}i_d}{\mathrm{d}t} \end{aligned} \tag{10.34}$$

At this point it is convenient to use Laplace transform techniques to simplify these simultaneous differential equations. As the initial conditions are all zero then d/dt can be replaced by the Laplace operator s and the differential equations written in matrix form as

$$\begin{bmatrix} v_d \\ 0 \end{bmatrix} = \begin{bmatrix} R + sL_d & skM_f \\ skM_f & sL_f \end{bmatrix} \begin{bmatrix} i_d \\ \Delta i_f \end{bmatrix} \tag{10.35}$$

This equation* can be solved for v_d by eliminating Δi_f and writing

$$v_d = (R + sL'_d)i_d \tag{10.36}$$

---

* The general matrix equation

$$\begin{bmatrix} \mathbf{v}_1 \\ \mathbf{0} \end{bmatrix} = \begin{bmatrix} \mathbf{z}_{11} & \mathbf{z}_{12} \\ \mathbf{z}_{21} & \mathbf{z}_{22} \end{bmatrix} \begin{bmatrix} \mathbf{i}_1 \\ \mathbf{i}_2 \end{bmatrix}$$

has the solution $\mathbf{v}_1 = \mathbf{z}_{eq}\mathbf{i}_1$ where $\mathbf{z}_{eq} = [\mathbf{z}_{11} - \mathbf{z}_{12}\mathbf{z}_{22}^{-1}\mathbf{z}_{21}]$

where the d-axis transient inductance

$$L'_d = L_d - \frac{k^2 M_f^2}{L_f}, \quad X'_d = \omega L'_d \tag{10.37}$$

As there is no field winding in the quadrature axis

$$L'_q = L_q, \quad X'_q = \omega L'_q = X_q \tag{10.38}$$

However in many cases it is convenient to represent the rotor body of a turbo-generator by an additional q-axis rotor coil when an equation for L'_q, similar to that for L'_d with appropriate parameter changes, will result. This important point will arise again later when generator models are considered.

It is also useful to define the decay time constant of the induced field current. This time constant will depend on whether the d-axis armature coil is open-circuit or short-circuit, Figure 10.5b. The circuit situation is very similar to that used to establish the relationship for the transient inductance, but now viewed from the field winding, so that the same equation can be used for the effective field winding inductance, with appropriate symbol changes. When the direct axis armature circuit is open-circuit the *d-axis transient open-circuit time constant* T'_{do} is obtained as

$$T'_{do} = \frac{L_f}{R_f} \tag{10.39}$$

and, when the armature is short-circuit, the time constant becomes the *d-axis transient short-circuit time constant* T'_d

$$T'_d = \left(L_f - \frac{k^2 M_f^2}{L_d}\right)\frac{1}{R_f} = T'_{do}\frac{L'_d}{L_d} \tag{10.40}$$

As there is no quadrature axis field circuit there are no q-axis transient time constants.

Subtransient-state

In the subtransient-state the armature flux is deflected round the damper winding screening the field winding from the armature flux. The circuit configurations that reflect this flux condition are shown in Figure 10.6a. All the rotor circuits are now represented by short-circuited coils with zero resistance.

In the d-axis the matrix equation for the coupled circuits becomes

$$\begin{bmatrix} v_d \\ 0 \\ 0 \end{bmatrix} = \begin{bmatrix} R + sL_d & skM_f & skM_D \\ skM_f & sL_f & sL_{fD} \\ skM_D & sL_{fD} & L_D \end{bmatrix} \begin{bmatrix} i_d \\ \Delta i_f \\ i_D \end{bmatrix} \tag{10.41}$$

with

$$v_d = (R + sL''_d)i_d \tag{10.42}$$

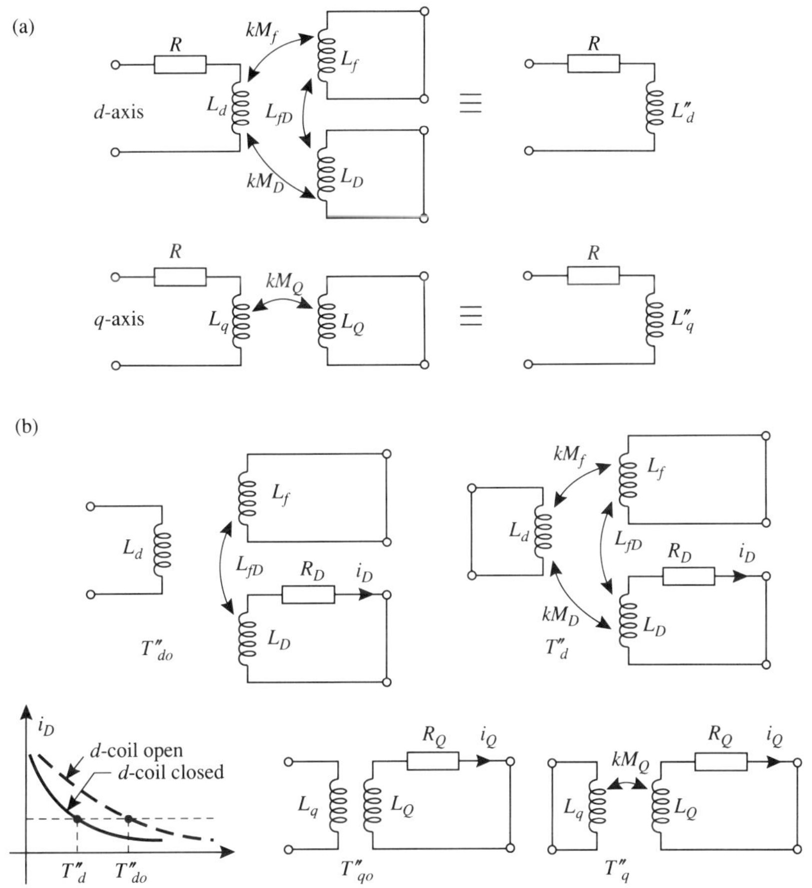

Figure 10.6 The d and q axis coupled circuits in the subtransient-state: (a) for determining subtransient inductance; (b) for determining the damper winding time constants

Eliminating the second two rows and columns using the same matrix procedure as before gives

$$L_d'' = L_d - \left[\frac{k^2M_f^2L_D + k^2M_D^2L_f - 2kM_f kM_D L_{fD}}{L_D L_f - L_{fD}^2}\right] \tag{10.43}$$

$$X_d'' = \omega L_d''$$

while in the quadrature axis an equation similar to equation 10.37 for the d-axis transient reactance gives

$$L_q'' = L_q - \frac{k^2M_Q^2}{L_Q}, \quad X_q'' = \omega L_q'' \tag{10.44}$$

As for the transient-state direct-axis time constants can be established for the decay of current in the damper winding, using the equivalent circuits of Figure 10.6b, as

$$T''_{do} = \left(L_D - \frac{L_{fD}^2}{L_f}\right)\frac{1}{R_D}$$
$$T''_d = \left[L_D - \left(\frac{L_{fD}^2 L_d + k^2 M_D^2 L_f - 2L_{fD} kM_D kM_f}{L_d L_f - k^2 M_f^2}\right)\right]\frac{1}{R_D} = T''_{do}\frac{L''_d}{L'_d} \tag{10.45}$$

where T''_{do} is the *d-axis subtransient open-circuit time constant* and T''_d the *d-axis subtransient short-circuit time constant.*

If there are no rotor body screening effects in the q-axis then the equivalent quadrature axis time constants are

$$T''_{qo} = \frac{L_Q}{R_Q},\ T''_q = \left(L_Q - \frac{k^2 M_Q^2}{L_q}\right)\frac{1}{R_Q} = T''_{qo}\frac{L''_q}{L'_q} \tag{10.46}$$

Recall that similar relations between open- and short-circuit time constants have been derived in a less precise way in Chapter 4, equation 4.16.

10.1.5 Synchronous generator equations

Having established how the parameters of the coupled circuits are related to the generator reactances and time constants the set of equations that constitute the full generator model, with the armature transformer emfs neglected, can now be examined more closely with a view to establishing more meaningful expressions.

Steady-state operation

In the steady-state the field winding current is constant and the damper winding currents $i_D = i_Q = 0$ so that the armature flux linkages Ψ_d and Ψ_q in equations 10.18 and 10.19 become

$$\Psi_d = L_d i_d + kM_f i_f, \quad \Psi_q = L_q i_q \tag{10.47}$$

Substituting for these flux linkages into the armature voltage equations 10.32 gives

$$v_d = -Ri_d - X_q i_q,\ v_q = -Ri_q + X_d i_d + e_q \tag{10.48}$$

where $e_q = \omega kM_f i_f$ is the open-circuit armature voltage induced by the field current i_f. When on open-circuit the armature current is zero and the field current can be related to the self-flux linkage $\Psi_{f(i_d=0)}$ from the flux equation 10.18 to give

$$e_q = \omega kM_f i_f = \omega\frac{kM_f}{L_f}\Psi_{f(i_d=0)} \tag{10.49}$$

Transient operation

When the generator is in the transient-state the armature flux has penetrated the damper coils and the damper currents have decayed to a relatively small value. This allows the circuits representing the damper windings to be removed from the equation set so that the flux equations become

$$\begin{bmatrix} \Psi_d \\ \Psi_f \end{bmatrix} = \begin{bmatrix} L_d & kM_f \\ kM_f & L_f \end{bmatrix} \begin{bmatrix} i_d \\ i_f \end{bmatrix}, \quad \Psi_q = L_q i_q \tag{10.50}$$

whilst from equation 10.33

$$\dot{\Psi}_f = v_f - R_f i_f \tag{10.51}$$

and from equation 10.32

$$v_d = -Ri_d - \omega\Psi_q$$

$$v_q = -Ri_q + \omega\Psi_d \tag{10.52}$$

These equations can be considered in two parts. Firstly the way in which the armature voltage equations 10.52 are influenced by the presence of the field winding and secondly how the differential equation 10.51 determines the way in which the armature flux penetrates the field winding. Consider first the armature voltage equations and in particular the voltage on the quadrature axis.

The flux linkage equation allows Ψ_d to be written in terms of i_d and Ψ_f so that when substituted into the quadrature axis voltage equation

$$v_q = -Ri_q + \omega \left[i_d \left(L_d - \frac{k^2 M_f^2}{L_f} \right) + \frac{kM_f}{L_f} \Psi_f \right] \tag{10.53}$$

This equation is readily simplified by noting that the coefficient of the first term in the square brackets is L_d' whilst the second term represents a voltage proportional to the field flux linkage Ψ_f. This voltage is given the symbol e_q' and is called the *quadrature axis transient emf* where

$$e_q' = \omega \left(\frac{kM_f}{L_f} \right) \Psi_f \tag{10.54}$$

This emf can be usefully compared with the q-axis steady-state emf, $e_q = \omega \frac{kM_f}{L_f} \Psi_{f(i_d=0)}$. Here $\Psi_{f(i_d=0)}$ is the field winding self-flux linkages in the steady-state and e_q the emf that the corresponding field current would induce in the armature. This emf is equal to the armature open-circuit voltage. In contrast in the transient-state Ψ_f is the field flux linkages that includes the effect of armature reaction. The voltage e_q' is the equivalent armature emf that would be induced by a field current proportional to these flux linkages. As these field flux linkages must remain constant in the short period after the disturbance Ψ_f only changes its value slowly. Making the substitutions for the transient inductance and transient emf, and assuming that $\omega \approx \omega_s$, gives

$$v_q = -Ri_q + X_d' i_d + e_q' \tag{10.55}$$

As there is no field winding on the quadrature axis $X_q' = X_q$ and

$$v_d = -Ri_d - X_q' i_q \tag{10.56}$$

Although equation 10.56 is correct for the rotor model assumed many generators, and in particular turbo-generators, have a solid steel rotor body which acts as a screen in the

q-axis. It is convenient to represent this by an additional q-axis, short-circuited coil, given the symbol g, when the q-axis flux equation becomes

$$\begin{bmatrix} \Psi_q \\ \Psi_g \end{bmatrix} = \begin{bmatrix} L_q & kM_g \\ kM_g & L_g \end{bmatrix} \begin{bmatrix} i_q \\ i_g \end{bmatrix} \tag{10.57}$$

with the change of flux linking this coil defined by the additional differential equation

$$\dot{\Psi}_g = v_g - R_g i_g = -R_g i_g \quad (v_g = 0) \tag{10.58}$$

The similarity with the d-axis rotor coils is immediately apparent and the voltage in the d-axis armature coil becomes

$$v_d = -Ri_d - X'_q i_q + e'_d \tag{10.59}$$

where $X'_q \neq X_q$ and

$$e'_d = -\omega \left(\frac{kM_g}{L_g} \right) \Psi_g \tag{10.60}$$

The flux linking the field winding Ψ_f does not remain constant during the entire period but changes slowly as the armature flux penetrates through the winding. This change in the field flux linkage is determined from the differential equation 10.51. Although this equation, along with equation 10.54 for e'_q, can be used directly to evaluate how e'_q changes with time it is usually more convenient to rephrase the flux linkage differential equation so that it can be more easily related to the armature. This modification can be accomplished by substituting into equation 10.51 for i_f obtained from the flux linkage equation 10.50 to give

$$v_f = \dot{\Psi}_f + \frac{R_f}{L_f}\Psi_f - R_f \frac{kM_f}{L_f} i_d \tag{10.61}$$

Differentiating equation 10.54 gives

$$\dot{e}'_q = \omega \frac{kM_f}{L_f} \dot{\Psi}_f \tag{10.62}$$

which, when rearranged and substituted into equation 10.61, gives, after some simplification,

$$e_f = \dot{e}'_q T'_{do} + e'_q - (X_d - X'_d) i_d \tag{10.63}$$

where e_f is the field voltage v_f referred to the armature given by

$$e_f = \omega k M_f v_f / R_f \tag{10.64}$$

e_f is also the output voltage of the exciter referred to the armature. Rearranging equation 10.63 gives

$$\dot{e}'_q = \frac{e_f - e'_q + i_d (X_d - X'_d)}{T'_{do}} \tag{10.65}$$

This analysis can be repeated for the quadrature axis when, assuming an additional rotor coil to represent the rotor body,

$$\dot{e}'_d = \frac{-i_q(X_q - X'_q) - e'_d}{T'_{qo}}, \quad X'_q \neq X_q \tag{10.66}$$

If no additional coil is present $X'_q = X_q$ and $e'_d = 0$.

Subtransient operation

During the subtransient period the rotor damper coils screen both the field winding and the rotor body from changes in the armature flux. The field flux linkages Ψ_f remain constant during this period whilst the damper winding flux linkages are constant immediately after the fault or disturbance but then decay with time as the generator moves towards the transient-state. These changes can be quantified using a similar procedure as for the transient period. Now the full equation set for the flux linkages, equations 10.18 and 10.19, and the flux decay equation 10.33 apply.

The armature voltage equations 10.32 are now modified due to their coupling with the rotor circuits in both the d and the q axes. The d axis flux linkage equations 10.18 allow the armature flux Ψ_d to be written in terms of i_d, Ψ_D and Ψ_f as

$$\Psi_d = L''_d i_d + (k_1 \Psi_f + k_2 \Psi_D) \tag{10.67}$$

where

$$k_1 = \frac{kM_f L_D - kM_D L_{fD}}{L_f L_D - L_{fD}^2}, \qquad k_2 = \frac{kM_D L_f - kM_f L_{fD}}{L_f L_D - L_{fD}^2} \tag{10.68}$$

which, when substituted into the armature voltage equation 10.32, gives

$$v_q = -Ri_q + X''_d i_d + e''_q \tag{10.69}$$

where

$$e''_q = \omega(k_1 \Psi_f + k_2 \Psi_D) \tag{10.70}$$

and represents an armature voltage proportional to the d-axis rotor flux linkages. These flux linkages remain constant immediately after the fault and only change as Ψ_D changes.

A similar analysis for the direct axis armature voltage yields

$$v_d = -Ri_d - X''_q i_q + e''_d \tag{10.71}$$

The way in which the subtransient voltage decays can be found using a similar approach as for the transient period. The differential equation governing the decay of the flux through the d-axis damper is given in equation 10.33 as

$$\dot{\Psi}_D = -R_D i_D \tag{10.72}$$

From the d-axis flux linkage equations 10.18, i_D can be written in terms of i_d, Ψ_D and Ψ_f when

$$\dot{\Psi}_D = k_2 i_d + \frac{1}{T''_{d0}} \frac{L_{fD}}{L_f} \Psi_f - \frac{1}{T''_{d0}} \Psi_D \tag{10.73}$$

Differentiating e_q'', equation 10.70 gives

$$\dot{e}_q'' = \omega k_2 \dot{\Psi}_D \tag{10.74}$$

as the field flux linkages Ψ_f are constant during the subtransient period. The relationships for e_q'' and $\dot{e}_q''$ can now be substituted into equation 10.73 to give, after some simplification,

$$\dot{e}_q'' = \frac{e_q' + (X_d' - X_d'')i_d - e_q''}{T_{d0}''} \tag{10.75}$$

A similar analysis for the q-axis flux linkage gives

$$\dot{e}_d'' = \frac{e_d' - (X_q' - X_q'')i_q - e_d''}{T_{q0}''} \tag{10.76}$$

The generator (d, q) reference frame and the system (a, b) reference frame

All the generator equations so far developed have been expressed in the (d, q) axis reference frame whereas the network equations developed in Chapter 3 are expressed in phase quantities in the system (a, b) reference frame. It is now necessary to examine how the two reference frames can be linked together. This is most conveniently done by considering the steady-state operation of the generator when the instantaneous phase voltages and currents constitute a balanced set given by

$$\begin{aligned} v_A &= \sqrt{2}V_g \sin(\omega t), & i_A &= \sqrt{2}I_g \sin(\omega t + \phi) \\ v_B &= \sqrt{2}V_g \sin(\omega t - 2\pi/3), & i_B &= \sqrt{2}I_g \sin(\omega t - 2\pi/3 + \phi) \\ v_C &= \sqrt{2}V_g \sin(\omega t - 4\pi/3), & i_C &= \sqrt{2}I_g \sin(\omega t - 4\pi/3 + \phi) \end{aligned} \tag{10.77}$$

At time $t = 0$ the terminal voltage of phase A is zero and the angle between the axis of phase A and the rotor d-axis is the rotor angle δ_g. As the rotational speed of the generator is ω the position of the rotor, relative to the axis of phase A, at any instant in time is given by $\gamma = \omega t + \delta_g$. Applying the transformation of equation 10.8 to the phase voltages in equation 10.77 then gives

$$v_d = -\sqrt{3}V_g \sin\delta_g, \qquad v_q = \sqrt{3}V_g \cos\delta_g \tag{10.78}$$

The phasor diagram in Figure 10.7 is similar to that in Figure 3.30 and shows how the rms terminal phase voltage V_g can be resolved into two orthogonal components, V_d and V_q, along the (d, q) axis where

$$V_q = V_g \cos\delta_g, \quad V_d = -V_g \sin\delta_g \text{ and } \overline{V}_g = V_q + jV_d \tag{10.79}$$

Substituting equation 10.79 into 10.78 allows the instantaneous generator voltages v_d and v_q to be related to the orthogonal rms terminal voltage components V_d and V_q by the relationship

$$v_d = \sqrt{3}V_d, \quad v_q = \sqrt{3}V_q \tag{10.80}$$

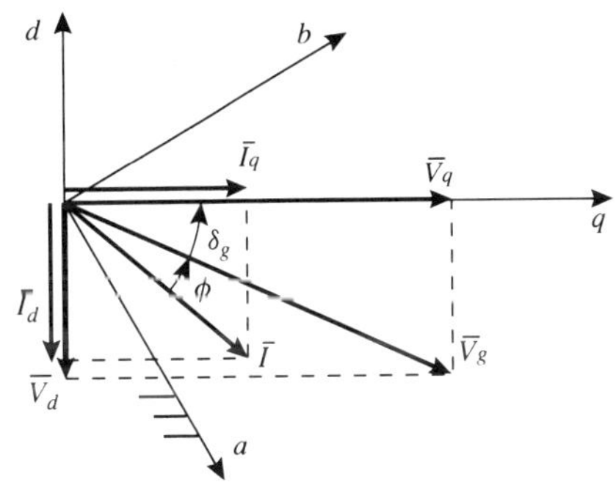

Figure 10.7 Phasor diagram in the (d, q) reference frame showing how the terminal voltage and current can be resolved into two components in quadrature (Also shown is the relative position of the system (a,b) reference frame)

This same process can also be applied to the currents to give

$$I_q = I_g \cos(\delta_g + \phi), \qquad I_d = -I_g \sin(\delta_g + \phi) \tag{10.81}$$

with the instantaneous currents being

$$i_d = -\sqrt{3} I_g \sin(\delta_g + \phi) = \sqrt{3} I_d, \quad i_q = \sqrt{3} I_g \cos(\delta_g + \phi) = \sqrt{3} I_q \tag{10.82}$$

The identities in equations 10.80 and 10.82 show that the instantaneous (d, q) currents and voltages are dc variables that are proportional to the orthogonal components of the rms phase currents and voltages. Consequently the instantaneous voltages and currents in the steady-state armature voltage equations 10.48 can be replaced by the orthogonal components of the phase currents and voltages to give

$$\begin{aligned} V_d &= -RI_d - X_q I_q \\ V_q &= -RI_q + X_d I_d + E_q \end{aligned} \tag{10.83}$$

which can be expressed in matrix form as

$$\begin{bmatrix} V_d \\ V_q \end{bmatrix} = \begin{bmatrix} 0 \\ E_q \end{bmatrix} - \begin{bmatrix} R & X_q \\ -X_d & R \end{bmatrix} \begin{bmatrix} I_d \\ I_q \end{bmatrix} \tag{10.84}$$

and

$$E_q = \omega M_f i_f / \sqrt{2} \tag{10.85}$$

The meaning of E_q is now much clearer as it is the rms voltage that would be induced in each of the armature phases by the field current i_f when the generator is on open-circuit. Equation 10.83 is identical to equation 3.65 derived in Chapter 3 and the equivalent circuit and phasor diagram corresponding to this equation were shown in Figures 3.17 and 3.18.

This concept can now be extended to the generator in the transient and subtransient state by recognising that by neglecting the transformer emf terms in the armature voltage equations the effect of asymmetric armature currents (dc offset) has been omitted so that the armature currents are always ac quantities of varying magnitude. The effect of the transformation of equation 10.8 is therefore to produce (d, q) currents that are solely dc quantities; they have no ac component. The dc (d, q) currents and voltages (i_d, i_q, v_d, v_q)

are related to the armature rms values (I_g, V_g) by the identities in equations 10.80 and 10.82. This allows the generator to be represented in the (d, q) reference frame by slowly changing dc quantities in all three characteristic states. All the emf equations and armature voltage equations developed in the previous section maintain the same form except that the instantaneous values i_d, i_q, e_d, e_q, etc. are replaced by orthogonal phase values I_d, I_q, E_d, E_q etc. This then allows equivalent circuits and phasor diagrams similar to those developed in Section 4.2 to be used to model the generator in the different operating states.

As the network equations are all expressed in the system (a, b) reference frame the current and voltage equations for each generator must be transformed from its own (d, q) reference frame to the system reference. The two reference frames are linked through the terminal voltage which appears explicitly in the generator equations, (d, q) reference frame, and the system equations, (a, b) reference frame. Although Figure 10.7 correctly shows the relationship between the generator (d, q) and the system (a, b) reference frames the necessary transformation is more clearly seen by reference to Figure 3.30. A phasor $\overline{E}$ can be defined in either the (d, q) or (a, b) reference frame with the two related by the **T** transformation defined by equations 3.121 and 3.122.

Power, torque and the swing equation

To complete the set of equations necessary to describe the generator, expressions are required for the three-phase terminal electrical power and the air-gap power. With transformer emfs neglected the terminal power can be obtained by substituting for v_d, v_q, i_d, i_q from equations 10.80 and 10.82 into the instantaneous terminal power equation 10.21 to obtain

$$P_g = 3(V_d I_d + V_q I_q) \qquad \text{Watts} \tag{10.86}$$

This equation is consistent with the normal three-phase power output expression as $V_d I_d + V_q I_q = VI\cos\phi$ and the generator terminal power is, as would be expected, three times the phase power.

The air-gap power is obtained from the terminal power, equation 10.86, by adding the resistance loss

$$P_e = 3(V_d I_d + V_q I_q + (I_d^2 + I_q^2)R) \qquad \text{Watts} \tag{10.87}$$

As $P = \omega\tau$ the air-gap torque is given by

$$\tau = \frac{3}{\omega}(V_d I_d + V_q I_q + (I_d^2 + I_q^2)R) \qquad \text{Nm} \tag{10.88}$$

The final equation required to complete the equation set is the swing equation derived in Chapter 5, equation 5.15, as

$$\frac{\mathrm{d}\Delta\omega}{\mathrm{d}t} = \frac{1}{M}(P_m - P_e - D\Delta\omega), \quad \Delta\omega = \omega - \omega_s = \frac{\mathrm{d}\delta}{\mathrm{d}t} \tag{10.89}$$

where P_e is the air-gap electrical power, P_m is the mechanical turbine power, D is the damping power coefficient, ω is the generator rotational speed, ω_s is the synchronous speed and $\Delta\omega$ is the speed deviation.

Per unit notation

It is standard practise in power system analysis to normalise all quantities to a common MVA base and, in the per unit system used here (see Appendix A), this is taken as the generator rated MVA/phase, $S_{1\phi}$. A base voltage V_b is also defined equal to the rated generator phase voltage. This allows the base current and the base impedance to be defined as

$$I_b = \frac{S_{1\phi}}{V_b}, \quad Z_b = \frac{V_b}{I_b} \tag{10.90}$$

Every parameter and equation can now be normalised by dividing by the appropriate base value as described in Appendix A. Of particular importance is the impact that normalisation has on power and torque as these are normalised to the total three-phase rated generator output $S_{3\phi}$. As shown in Appendix A when currents, voltages and flux linkages are all expressed in per unit the three-phase power expressions in equations 10.86 and 10.87 normalised to the three-phase MVA base become

$$\begin{aligned} P_g &= (V_d I_d + V_q I_q) \qquad \text{per unit} \\ P_e &= [V_d I_d + V_q I_q + (I_d^2 + I_q^2)R] \end{aligned} \tag{10.91}$$

whilst the per unit notation adopted requires $\tau_{pu} = P_{pu}$ so that in per unit, equation 10.88 is

$$\tau = \frac{\omega_s}{\omega}[V_d I_d + V_q I_q + (I_d^2 + I_q^2)R] \qquad \text{per unit} \tag{10.92}$$

A full explanation of the per unit system is given in Appendix A. In general all voltage, current and flux linkage equations retain the same form whether or not they are expressed in MKSA or in per unit notation. It is only the power and torque equations that change their form with the introduction of a 1/3 factor into the power expressions and a $\omega_s/3$ factor into the torque expressions. The introduction of the 1/3 factor into the total power expression carries the implication that the total generator power output, normalised to the $S_{3\phi}$ base, is the same as the power output per phase, normalised to $S_{1\phi}$, where $S_{1\phi} = S_{3\phi}/3$.

One of the advantages of referring all the rotor quantities to the armature winding, as in Section 10.1.5, is that only base quantities for the armature circuit need be considered. If the rotor equations had not been referred to the armature then additional base quantities for the rotor circuits would be necessary. Such a per unit system based on the concept of *equal mutual flux linkages* (Anderson & Fouad, 1977) is described in Appendix A and ensures that all the equations developed earlier in this chapter retain exactly the same form whether in per unit or MKSA. This per unit system also has the effect that the per unit value of all the mutual inductances on one axis are equal and, in per unit, is given the symbol L_{ad} on the d-axis and L_{aq} on the q-axis, that is

$$kM_f = kM_D = L_{fD} \equiv L_{ad}, \quad kM_Q \equiv L_{aq} \tag{10.93}$$

The self-inductances are also expressed as being equal to the sum of a magnetising inductance and a leakage component so that

$$\begin{aligned} L_d &= L_{md} + l_l, \quad L_f = L_{mf} + l_f, \quad L_D = L_{mD} + l_D \\ L_q &= L_{mq} + l_l, \quad L_Q = L_{mQ} + l_Q \end{aligned} \tag{10.94}$$

and, in per unit,

$$L_{md} = L_{mf} = L_{mD} \equiv L_{ad}, \quad L_{mq} = L_{mQ} \equiv L_{aq} \tag{10.95}$$

It is common to use these substitutions when expressing all the reactance and time constant expressions etc. derived in section 10.1.4 in per unit.

10.1.6 Synchronous generator models

The equations derived in the previous sub-section can now be used to model the behaviour of a synchronous generator. A number of different models will be developed where the generator is modelled by either its subtransient or transient emfs acting behind appropriate reactances. The way the armature flux gradually penetrates into the rotor during a fault and affects these emfs is quantified by the differential equations 10.75, 10.76, 10.65 and 10.66. These differential equations are gathered here and expressed in orthogonal phase quantities as

$$T''_{d0}\dot{E}''_q = E'_q - E''_q + I_d(X'_d - X''_d) \tag{10.96}$$

$$T''_{q0}\dot{E}''_d = E'_d - E''_d - I_q(X'_q - X''_q) \tag{10.97}$$

$$T'_{d0}\dot{E}'_q = E_f - E'_q + I_d(X_d - X'_d) \tag{10.98}$$

$$T'_{q0}\dot{E}'_d = \quad -E'_d - I_q(X_q - X'_q) \tag{10.99}$$

It is worth noting the similar structure of these equations. On the left-hand-side is the time derivative of the emf multiplied by the relevant time constant whilst the right-hand-side relates to the equivalent d or q axis armature circuit, with resistance neglected, shown in Figure 10.8. (These armature circuits were initially introduced in Chapter 4, Figure 4.14.) The first component of the right-hand-side in the equations constitutes a driving voltage whilst the final component constitutes a voltage drop in the relevant reactance.

The right-hand-side of equation 10.96 constitutes Kirchhoff's Voltage Law for the middle part of the circuit shown in Figure 10.8a, that is for the driving voltage E'_q, the voltage drop $I_d(X'_d - X''_d)$ and the emf E''_q. Similarly, the right-hand-side of equation 10.97 constitutes Kirchhoff's Voltage Law for the middle part of Figure 10.8b while equation 10.98 corresponds to the part of Figure 10.8a with the driving voltage E_f, the voltage drop $I_d(X_d - X'_d)$, and the emf E'_q. In equation 10.99, corresponding to the left-hand part of Figure 10.8b, there is no driving voltage due to the lack of excitation in the q-axis.

Equations 10.96, 10.97, 10.98, and 10.99 allow five different generator models of decreasing complexity and accuracy to be developed. Each model is given a model number

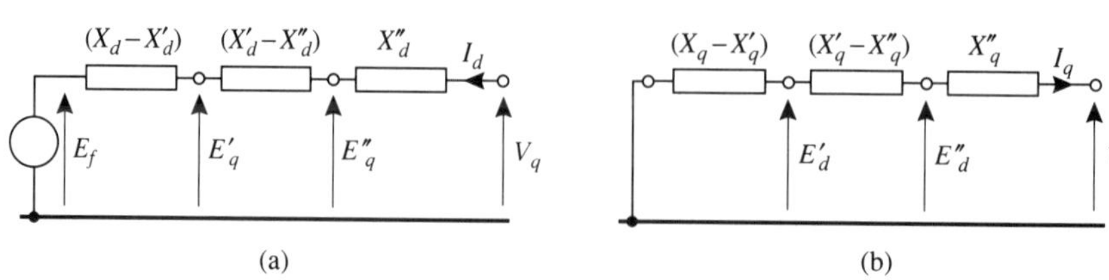

Figure 10.8 Generator equivalent circuits with resistances neglected: (a) d-axis; (b) q-axis

that indicates the number of differential equations required in the model. The larger the number the greater the model complexity and the greater would be the time required to solve the differential equations. The model number is then followed by a number of terms enclosed in brackets which define the differential equations used by the model. In developing the generator models it is assumed that all quantities are expressed in per unit.

Model 6 — ($\dot{E}''_d$, $\dot{E}''_q$, $\dot{E}'_d$, $\dot{E}'_q$, $\dot{\omega}$, $\dot{\delta}$)

In this model the generator is represented by the subtransient emfs E''_q and E''_d behind the subtransient reactances X''_d and X''_q as defined by the modified armature voltage equations 10.69 and 10.71

$$\begin{bmatrix} V_d \\ V_q \end{bmatrix} = \begin{bmatrix} E''_d \\ E''_q \end{bmatrix} - \begin{bmatrix} R & X''_q \\ -X''_d & R \end{bmatrix} \begin{bmatrix} I_d \\ I_q \end{bmatrix} \tag{10.100}$$

This equation corresponds to equation 4.17 derived in Chapter 4. The corresponding equivalent circuit and phasor diagrams are shown in Figure 4.12.

The differential equations 10.96, 10.97, 10.98 and 10.99 describe the change in these emfs as the flux linking the rotor circuits decays. To these must be added equations 10.89 in order to include the speed deviation and angle change of the rotor. As the differential equations 10.96 and 10.97 include thc influence of the damper windings, the damping coefficient in the swing equation need only quantify the mechanical damping due to windage and friction and, as this is usually small, it may be neglected ($D \approx 0$). The air-gap power of the generator can be calculated using equation 10.91, which, after substituting for the armature voltages from equation 10.100, gives

$$P_e = (E''_d I_d + E''_q I_q) + (X''_d - X''_q) I_d I_q \tag{10.101}$$

The second term in this power equation defines the subtransient saliency power discussed in Chapter 4.

Model 5 — ($\dot{E}''_d$, $\dot{E}''_q$, $\dot{E}'_q$, $\dot{\omega}$, $\dot{\delta}$)

In this model the screening effect of the rotor body eddy-currents in the q-axis are neglected so that $X'_q = X_q$ and $E'_d = 0$. This model reverts to the classical 5 winding model with armature transformer emfs neglected. *Model 5* eliminates equation 10.99 from the equation set of *Model 6*.

Model 4 — ($\dot{E}'_d$, $\dot{E}'_q$, $\dot{\omega}$, $\dot{\delta}$)

In this model the effect of the damper windings in *Model 6* are neglected and equations 10.96 and 10.97 are removed from the model. The generator is now represented by the transient emfs E'_q and E'_d behind the transient reactances X'_d and X'_q as defined by the equation

$$\begin{bmatrix} V_d \\ V_q \end{bmatrix} = \begin{bmatrix} E'_d \\ E'_q \end{bmatrix} - \begin{bmatrix} R & X'_q \\ -X'_d & R \end{bmatrix} \begin{bmatrix} I_d \\ I_q \end{bmatrix} \tag{10.102}$$

This equation is identical to equation 4.19 derived in Chapter 4. The corresponding equivalent circuit and phasor diagrams were shown in Figure 4.13.

The changes in the emfs E'_q and E'_d are determined from the differential equations 10.98 and 10.99 whilst the electrical air-gap power is

$$P_e = E'_q I_q + E'_d I_d + (X'_d - X'_q) I_d I_q \tag{10.103}$$

with the second part of the equation defining the transient saliency power.

As the damper windings are ignored the air-gap power calculated by this equation neglects the asynchronous torque produced by the damper windings. Because of this the damping coefficient in the swing equation should be increased by an amount corresponding to the average asynchronous torque, or power, calculated using the simplified formulae in equation 5.24.

This simplified model of the synchronous generator is widely considered to be sufficiently accurate to analyse electromechanical dynamics (Stott, 1979). Experience has shown that including the second differential equation to account for rotor body effects in the q-axis improves the accuracy of the model. The main disadvantage of this model is that the equivalent damping coefficient, appearing in the swing equation, can only be calculated approximately.

Model 3 — ($\dot{E}'_q$, $\dot{\omega}$, $\dot{\delta}$)

This model is similar to *Model 4* except that the d-axis transient emf E'_d is assumed to remain constant allowing equation 10.99 to be removed from the equation set. The generator is described only by equations 10.96 and 10.89 with the real power again given by equation 10.103. Besides neglecting the effect of the damper windings, by assuming E'_d to be constant this model also neglects the damping produced by the rotor body eddy-currents, even if an additional coil is used to represent the rotor body. If there is no winding in the quadrature axis to represent the rotor body $E'_d = 0, X'_q = X_q$ and equation 10.103 reduces to

$$P = E'_q I_q + (X'_d - X_q) I_d I_q \tag{10.104}$$

As in *Model 4* damper winding effects are neglected and so their effect must be included by increasing the value of the damping coefficient in the swing equation.

Model 2 — The Classical Model ($\dot{\omega}$, $\dot{\delta}$)

The classical synchronous generator model, widely used in all the previous chapters for a simplified analysis of power system dynamics, assumes that neither the d-axis armature current I_d nor the internal emf E_f representing the excitation voltage change very much during the transient-state. In this model the generator is represented by the swing equation and a constant emf E' behind the transient reactance X'_d.

The justification of the classical model is that the time constant T'_{d0}, appearing in equation 10.98, is relatively long, the order of a few seconds, so that E'_q does not change much provided that changes in E_f and I_d are small. This means that $E'_q \approx$ constant and, because it has already been assumed that $E'_d =$ constant, both the magnitude of the transient emf E' and its position α with respect to the rotor may be assumed to be constant. If rotor transient saliency is neglected $X'_q = X'_d$ and the two equivalent circuits in Figure 4.13 may be replaced by the one equivalent circuit shown in Figure 5.8 for a generator connected to an infinite busbar. As

$$\overline{I} = \overline{I}_q + j\overline{I}_d, \quad \overline{V} = \overline{V}_q + j\overline{V}_d, \quad \overline{E'} = \overline{E'_q} + j\overline{E'_d} \tag{10.105}$$

the two algebraic equations describing the armature voltage in equations 10.102 can now be replaced by one equation

$$\overline{V} = (\overline{E'_q} + j\overline{E'_d}) - jX'_d(\overline{I}_q + j\overline{I}_d) = \overline{E'} - jX'_d\overline{I} \tag{10.106}$$

The assumption of small changes in the direct component of the generator current, and in the internal emf, means that only generators located a long way from the point of disturbance should be represented by the classical model.

Summary

The number of reactances and time constants representing the generator depend on the number of equivalent windings used in a particular model. The five winding model, *Model 5*, has two equivalent rotor windings and time constants (T''_{d0}, T'_{d0}) on the d-axis and three armature reactances (X''_d, X'_d, X_d). In the q-axis there is one equivalent rotor winding, with a time constant (T''_{q0}), and two armature reactances (X''_q, X_q). Although generators with rotors constructed from ferromagnetic, electrically insulated, sheets of steel are well characterised by these parameters this is not the case for generators whose rotors are made from solid steel. In such generators the rotor eddy-currents play a significant role in the q-axis damping. This damping may be modelled approximately by introducing an additional q-axis winding as in *Model 6*. This expands the model by an additional reactance $X'_q \neq X'_d$ and a time constant T'_{q0} to represent the flux decay through this circuit. If these parameters are not specified by the manufacturer it is typical to assume that $X'_q = 2X''_q$ and $T'_q = 10T''_q$. Due to the screening effect of the field winding the influence of eddy-currents in the d-axis rotor body is small and there is no need to introduce an additional winding to account for this. For slower changing disturbances, and to enable faster solution in complex systems, the generator model can be simplified by omitting the damper windings from the electrical equations, *Model 4* and *Model 3*, and representing their damping effect by an increased damping coefficient in the swing equation. The simplest model, *Model 2*, is the model traditionally used in quantitative power system analysis. It is simple, but very approximate, and only really suitable for representing generators well away from the fault. It can also be useful for evaluating generator behaviour during the first rotor swing.

In developing all the generator models it has been assumed that $\omega \approx \omega_s$ and that the transformer emfs in the armature voltage equations can be neglected. Speed changes can be accounted for in all the models by introducing a factor ω/ω_s in front of every reactance. Accounting for the armature transformer emfs is significantly more difficult and is only justified if the detailed modelling of the electromagnetic transients immediately after the fault is required, for example in computing short-circuit currents.

10.1.7 Saturation effects

The generator equations derived in the previous sub-section ignored the effect of magnetic saturation in the stator and rotor iron and the effect this would have on the generator parameters and the operating conditions. Saturation effects are highly non-linear and depend on the generator loading conditions so that trying to account for them accurately in the generator model is nigh an impossible process. What is required is a relatively simple saturation model that produces acceptable results, is linked to the physical process and uses easily obtainable data.

At this stage it is interesting to consider the flux paths in Figure 4.8 that are associated with the different generator reactances. As it is only the iron paths that saturate one might expect those generator reactances where the flux path is mainly in air, that is X'_d, X'_q, X''_d, X''_q, to be less susceptible to saturation effects than X_d and X_q where a substantial part of the flux path is through iron (at least for the round-rotor generator). This concept of parameter sensitivity will be investigated later in this section but first it is necessary to investigate the general effects of saturation itself. This is more easily accomplished and understood if all generator parameters are expressed in per unit.

Saturation characteristic

For any magnetic circuit that comprises an iron-path and an air-path the mmf and flux will be related by the general curve shown in Figure 10.9a. When the iron path is unsaturated the relationship between the mmf and the flux is linear and is represented by the *air-gap*

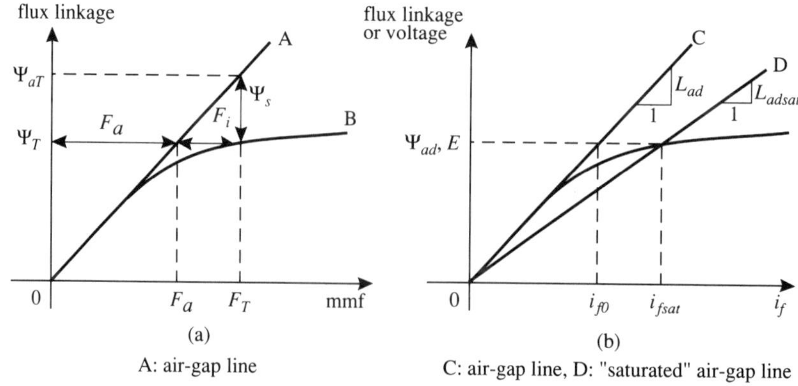

Figure 10.9 Saturation: (a) typical saturation curve for a magnetic circuit whose flux path is in iron and air; (b) generator open-circuit saturation characteristic

line, OA. In this situation the reluctance of the magnetic circuit is dominated by the reluctance of the air-gap. As the mmf is increased the iron saturates and the mmf/flux relationship ceases to be linear and follows the *saturation curve* OB. Consequently for some flux linkage Ψ_T the mmf required to produce this flux can be considered to have two components, the air-gap mmf F_a and the iron mmf F_i, so that the total mmf is given by

$$F_T = F_a + F_i \tag{10.107}$$

A *saturation factor* S can now be defined as

$$S = \frac{\text{air gap mmf}}{\text{air gap mmf} + \text{iron mmf}} = \frac{F_a}{F_a + F_i} = \frac{F_a}{F_T} \tag{10.108}$$

and, by similar triangles,

$$S = \frac{F_a}{F_a + F_i} = \frac{\Psi_T}{\Psi_T + \Psi_s} \tag{10.109}$$

Having defined a saturation factor it is now necessary to provide a simple method of finding its value for any generator loading condition and then be able to take this into account when computing the generator parameters. To achieve these objectives it is usually assumed that:

(i) the open-circuit generator saturation curve can be used under load conditions. On no-load the mmf in the d-axis is due to the field current, i_f, whilst on-load it is due to $(i_f + i_D + i_d)$. As Kundur (1994) points out this is usually the only saturation data available

(ii) as the leakage flux path is mainly in air the leakage inductances are assumed to be independent of saturation. This implies that it is only the mutual inductances L_{ad} and L_{aq} that are affected by saturation (and of course the self inductance $L_d = L_{ad} + l_l$, $L_f = L_{ad} + l_f$ etc. as defined in equation 10.94 and Appendix A.4)

(iii) saturation on the d-axis and the q-axis are independent of each other

Calculation of the saturation factor

When the generator is on no-load equation 10.85 shows that the voltage induced in one of the armature phases V_{g0} is

$$V_{g0} = E = E_q = \frac{1}{\sqrt{2}}\omega_s M_f i_{f0} = \frac{1}{\sqrt{3}}\omega_s k M_f i_{f0} = \left[\frac{1}{\sqrt{3}}\omega_s\right] L_{ad} i_{f0}$$

$$= \left[\frac{1}{\sqrt{3}}\omega_s\right] \Psi_{ad} \tag{10.110}$$

as $\Psi_{ad} = L_{ad} i_{f0}$ is the mutual flux linkage defined in equation A.34. This equation can now be related to the open-circuit characteristic in Figure 10.9b where the slope of the air-gap line is proportional to L_{ad}. It should also be recognised that as the flux linkage and the voltage are related by equation 10.110 the vertical axis can be interpreted in *either* voltage or flux linkage.

If at some open-circuit voltage E the required flux linkage is Ψ_{ad} then, if there is no saturation of the iron paths, the field current required to set up this flux linkage would be i_{f0}; however if saturation is present then the required field current would be i_{fsat}. From the definition of saturation factor in equation 10.109

$$i_{fsat} = \frac{i_{f0}}{S} \tag{10.111}$$

whilst

$$L_{ad} = \frac{\Psi_{ad}}{i_{f0}} \quad \text{and} \quad L_{adsat} = \frac{\Psi_{ad}}{i_{fsat}} \tag{10.112}$$

Substituting equation 10.111 into equation 10.112 gives

$$L_{adsat} = SL_{ad} \tag{10.113}$$

where L_{adsat} is the slope of the "saturated" air-gap line in Figure 10.9b. Consequently if Ψ_{ad} is known then the saturation factor S can be calculated from equation 10.109 by recognising that Ψ_{ad} and Ψ_T are equivalent. The saturated value of L_{ad} is then found from equation 10.113 and used to account for saturation in all the generator parameter equations developed in sub-Section 10.1.4.

The flux linkage, Ψ_{ad}, can be calculated provided that the armature voltage proportional to these flux linkages is known. As shown in Figure 10.10 this voltage is the voltage behind the armature leakage reactance X_l and, as X_l is not affected by saturation,

$$\overline{E}_l = \overline{V}_g + (R + jX_l)\overline{I} \tag{10.114}$$

so that provided $\overline{V}_g$ and $\overline{I}$ are known E_l and Ψ_{ad} can be computed for any loading condition. In some instances the voltage E_p behind the Potier reactance X_p is used instead of E_l (Arrillaga & Arnold, 1990).

The final piece of the jig-saw is to have some method of conveniently storing the saturation curve in computer memory so as to facilitate the calculation of the saturation factor. A number of methods are available with perhaps the simplest being to fit the saturation curve by the function

$$I_f = V + C_n V^n \tag{10.115}$$

where n is 7 or 9 and C_n is a constant for the particular curve (Arrillaga & Arnold, 1990, Hammons & Winning, 1971). Alternatively a polynomial can be used.

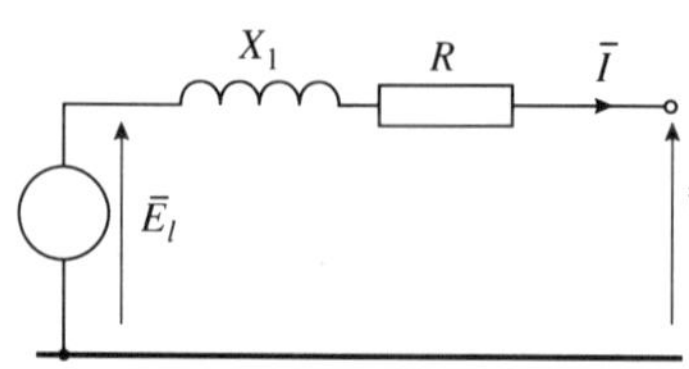

Figure 10.10 Equivalent circuit to calculate the armature voltage behind leakage reactance

A slightly different approach is to use a two stage exponential process (Kundur, 1994, Anderson & Fouad, 1977, Anderson, Agrawal & Van Ness, 1990) where

$$\begin{aligned} \Psi_{ad} &< 0.8 \quad \Psi_s = 0 \\ \Psi_{ad} &\geq 0.8 \quad \Psi_s = A_{sat} e^{B_{sat}(\Psi_{ad} - 0.8)} \end{aligned} \tag{10.116}$$

where A_{sat} and B_{sat} are constants that are easily calculated by taking two points from the known saturation curve. When using this method there is usually a slight discontinuity at $\Psi_{ad} = 0.8$ but it is usually of no consequence (Anderson & Fouad, 1977).

The procedure for accounting for saturation in the generator equations can now be summarised as:

Step 1: Knowing $\overline{V}_g$ and $\overline{I}$ use equation 10.114 to calculate E_l

Step 2: Divide E_l by $\dfrac{\omega_s}{\sqrt{3}}$ to obtain Ψ_{ad}

Step 3: Use equation 10.116 to compute Ψ_s

Step 4: Use equation 10.109 to calculate the saturation factor S

Step 5: Modify L_{ad} to account for saturation using equation 10.113

Step 6: Modify all the generator parameters that are a function of L_{ad}

The procedure described above defines saturation in the d-axis and, although a similar procedure can be used in the q-axis, it is usual to assume:

(i) for round-rotor generators the saturation factor is the same in both axes when $S_d = S_q$

(ii) for salient-pole generators as the q-axis reluctance is dominated by air-paths L_{aq} does not vary as much with saturation as L_{ad}, and $S_q = 1$ is assumed under all loading conditions

Although the saturation factor can be varied during a dynamic simulation it is quite common to calculate it at the beginning of the simulation and assume that it remains constant during the simulation period. This ensures that the initial conditions are calculated correctly. If the saturation factor is varied during a dynamic simulation then it will be necessary to iterate round steps 1 to 6 at each integration step in order to find a solution for L_{adsat}. This is because $\overline{V}_g$ and $\overline{I}$ depend on the degree of saturation whilst the saturation factor itself is dependent on $\overline{V}_g$ and $\overline{I}$. Other variations to this method of accounting for saturation can be found in Arrillaga & Arnold. (1990) and Pavella & Murthy (1994).

Sensitivity of parameters to saturation

It was explained at the beginning of this section how generator reactances associated with flux paths mainly in air are less susceptible to saturation effects than are those associated with flux paths in iron. Such parameter sensitivity can be quantified by assessing the sensitivity of the particular parameter to variations in L_{ad} (or L_{aq}). For example, assume that the general parameter X is a function of L_{ad} so that for a small change in X

$$X = X_0 + \Delta X \tag{10.117}$$

where ΔX, the change in X, is given by

$$\Delta X = \left.\frac{\partial X}{\partial L_{ad}}\right|_0 \Delta L_{ad} \tag{10.118}$$

and the suffix "0" indicates an initial value.

In the context of the present discussion Table 10.1 summarises the per unit d-axis steady-state and transient parameters defined in section 10.1.4 with the sensitivity parameter determined using equation 10.117, for example the synchronous inductance $L_d = L_{ad} + l_l$. In this example $\Delta L_d = \left.\frac{\partial L_d}{\partial L_{ad}}\right|_0 \Delta L_{ad}$ but, as $\left.\frac{\partial L_d}{\partial L_{ad}}\right|_0 = 1$, $\Delta L_d = \Delta L_{ad}$. Similarly for the d-axis transient inductance

$$L'_d = L_d - \frac{L_{ad}^2}{L_f} = L_{ad} + l_l - \frac{L_{ad}^2}{L_f} \tag{10.119}$$

and differentiating gives

$$\Delta L'_d = \Delta L_{ad} - \left[\frac{2L_{ad0}\Delta L_{ad}}{L_{f0}} - \frac{L_{ad0}^2 \Delta L_f}{L_{f0}^2}\right] \tag{10.120}$$

but $L_f = L_{ad} + l_l$ so that $\Delta L_f = \Delta L_{ad}$ and

$$\Delta L'_d = \Delta L_{ad} - \left[\frac{2L_{ad0}\Delta L_{ad}}{L_{f0}} - \frac{L_{ad0}^2 \Delta L_{ad}}{L_{f0}^2}\right] \tag{10.121}$$

Rearranging and writing $L_{f0} = L_{ad0} + l_f$ gives

$$\Delta L'_d = \left[\frac{l_f}{L_{f0}}\right]^2 \Delta L_{ad} \tag{10.122}$$

Anderson, Agrawal & Van Ness (1990) calculated typical values for these sensitivity parameters assuming a 10% change in the mutual flux linkage Ψ_{ad}; these values are

Table 10.1 Sensitivity of d-axis parameters to saturation

Parameter	Equation	Equation Number	Δ Parameter	Typical sensitivity to a 10% change in mutual flux linkage
L_d	$L_{ad} + l_l$	10.94	ΔL_{ad}	21%
L'_d	$L_d - \frac{L_{ad}^2}{L_f}$	10.37	$\left[\frac{l_f}{L_{f0}}\right]^2 \Delta L_{ad}$	1.5%
T'_{d0}	$\frac{L_f}{R_f}$	10.39	$\frac{1}{R_f}\Delta L_{ad}$	20%
E_f	$\omega_s L_{ad}\frac{v_f}{R_f}$	10.64	$\left[\frac{\omega_s v_f}{R_f}\right]\Delta L_{ad}$	21%

shown in Table 10.1. As expected the synchronous inductance is particularly sensitive to saturation but the transient inductance, associated with flux paths mainly in air, shows little change. Indeed if the analysis is extended to include sub-transient inductance the change in this parameter is negligible.

10.2 Excitation Systems

In Chapter 2 the different types of excitation system and AVR used to supply, and control, the field current to the generator were described and the complete exciter and AVR subsystem were shown in block diagram form in Figure 2.4. Generally the exciter is equipped with both an automatic voltage regulator, AVR, and a manual regulator. In some cases a back up AVR will also be provided. It is the automatic voltage regulator that is of particular interest here. As the power system industry uses a wide range of exciters and AVRs, with the details varying from one manufacturer to another, the aim of this section is to develop generic models of some of the most common types of excitation system. With careful parameter selection these models can then be used to represent different exciter systems from different manufacturers. A comprehensive study of excitation system models can be found in the IEEE Committee Reports on excitation systems (IEEE, 1968, 1973a, 1981) and the interested reader should consult these excellent sources.

Ultimately the exciter and the AVR model must be interfaced with the generator model developed in Section 10.1.6. This interface is through the variable E_f, which represents the generator main field voltage v_f referred to the generator armature winding (v_f is also the exciter output voltage v_{ex}). Initially exciter subsystem models will be developed in terms of the currents and voltages at the field winding but will finally be referred to the armature, and the values converted to per unit. This is equivalent to expressing the exciter variables in a per unit system where 1 per unit exciter output voltage corresponds to the field voltage required to produce 1 per unit armature voltage on the air-gap line, the per unit system used in the armature. This exciter per unit system is commonly used by manufacturers.

10.2.1 Transducer and comparator model

Figure 10.11 shows the block diagram of the transducer and comparator together with the load compensation element. The first block in this model corresponds to the load compensation element, shown previously in Figure 2.5, which corrects the generator terminal voltage according to the value of the compensation impedance. The second block represents the delay introduced by the signal transducers. Normally the equivalent time constant of the transducers is small so that this block is often neglected. The third element

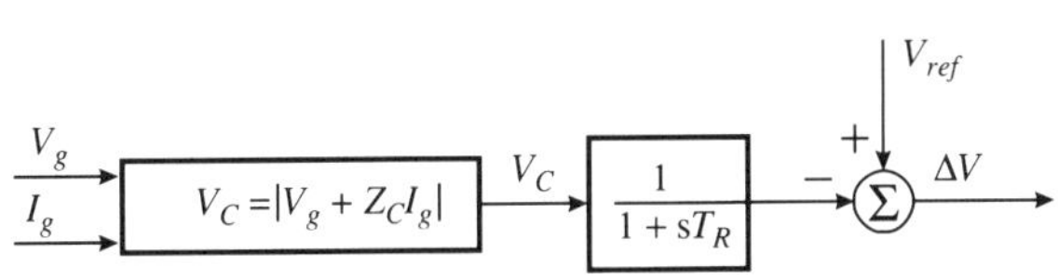

Figure 10.11 Block diagram of the transducer and comparator

represents the comparator at which the corrected generator voltage is added algebraically to the reference voltage V_{ref}.

10.2.2 Exciters and regulators

Sub-Section 2.2.2 discussed different types of exciters. Modern excitation systems usually employ either a brushless ac exciter or a static exciter with either an analogue or digital AVR, although digital AVRs are becoming more common. However many older generators are fitted with dc excitation systems and consequently the models developed in this section are typical of the excitation systems in use today.

dc exciters

Two different dc exciters are shown in Figure 10.12; the first being separately-excited and the second self-excited. In order to develop a mathematical model of these two exciters consider first the separately-excited case in Figure 10.12a. A change in the exciter field current i_{exf} can be described by the following equation

$$v_R = R_{exf} i_{exf} + L_{exf} \frac{di_{exf}}{dt} \tag{10.123}$$

where L_{exf} depends on saturation and is an incremental inductance. The relationship between the exciter field current i_{exf} and the emf e_{ex} induced in the exciter armature is non-linear because of magnetic saturation in the exciter core, whilst the exciter output voltage v_{ex} depends on both the saturation characteristic and the armature loading. Both of these effects can be included in the modelling process by using the constant resistance load saturation curve shown in Figure 10.13.

The first step in the modelling process is to observe that the slope of the air-gap line in Figure 10.13 is tangential to the linear part of the no-load saturation curve and can be represented by a resistance with the value $R = v_A/i_C$. The exciter field current that corresponds to the non-linear part of the constant resistance load characteristic can then be expressed using similar triangles as

$$i_A = i_C \frac{v_B}{v_A} = i_C(1 + S_e) = \frac{v_A}{R}(1 + S_e) \tag{10.124}$$

where S_e is a saturation coefficient defined in Figure 10.13 with reference to the constant resistance load saturation curve. This means that any point on the constant resistance load

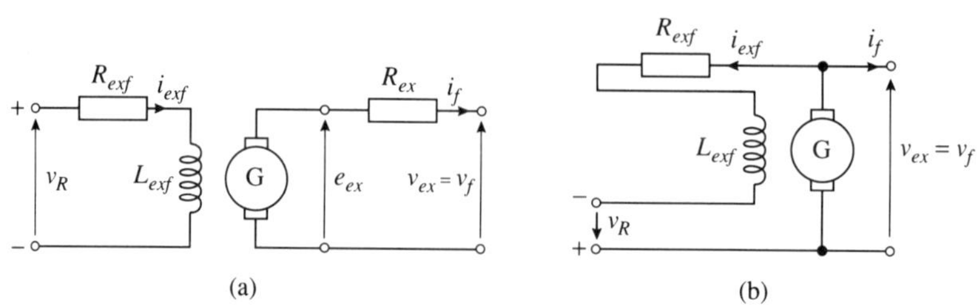

Figure 10.12 Equivalent circuit diagrams of dc exciters: (a) separately excited; (b) self-excited

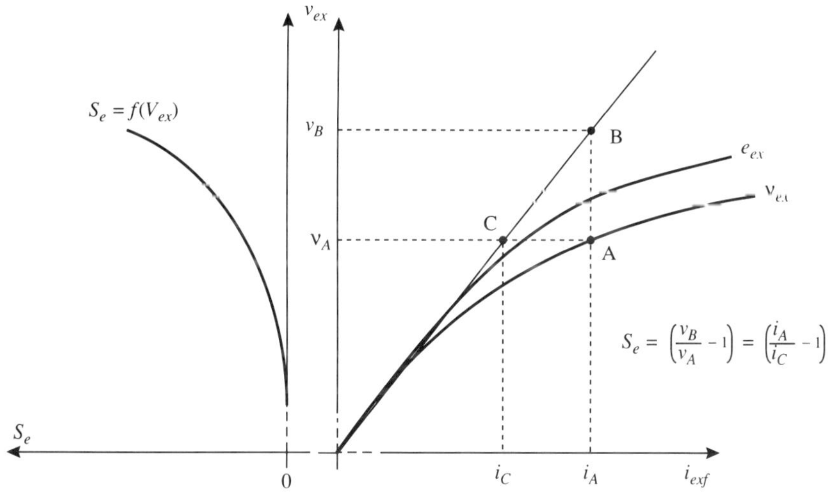

Figure 10.13 Illustration to define the saturation coefficient: e_{ex} is the no-load saturation curve; v_{ex} the constant resistance load saturation curve. The saturation characteristic is shown only around its saturation knee and has been exaggerated for clarity

saturation characteristic is defined by

$$i_{exf} = \frac{v_{ex}}{R}(1 + S_e) \tag{10.125}$$

Substituting equation 10.125 into equation 10.123 and writing

$$L_{exf}\frac{\mathrm{d}i_{exf}}{\mathrm{d}t} = L_{exf}\left[\frac{\mathrm{d}i_{exf}}{\mathrm{d}v_{ex}}\right]_{v_{ex0}}\frac{\mathrm{d}v_{ex}}{\mathrm{d}t}$$

gives

$$v_R = \frac{R_{exf}}{R}(1 + S_e)v_{ex} + L_{exf}\left[\frac{\mathrm{d}i_{exf}}{\mathrm{d}v_{ex}}\right]_{v_{ex0}}\frac{\mathrm{d}v_{ex}}{\mathrm{d}t} \tag{10.126}$$

where the derivative term in square brackets is simply the slope of the saturation characteristic at the initial operating point.

To enable the exciter to be easily interfaced with the generator model the exciter base quantities are defined with the base exciter voltage E_{exb} being that voltage which gives rated open-circuit generator voltage $V_{go/c}$ on the generator air-gap line.

This means that E_{exb} and the generator voltage are related by $V_{go/c} = \frac{1}{\sqrt{2}}\frac{\omega M_f}{R_f}E_{exb}$.

The base exciter resistance is defined as $R_b = R$ with base current $I_{exfb} = E_{exb}/R_b$. Equation 10.126 can now be expressed in per unit by dividing through by E_{exb}. It is also convenient to normalise the saturation curve to the base quantities E_{exb} and I_{exfb} when $L_{exf}\left[\frac{\mathrm{d}i_{exf}}{\mathrm{d}v_{ex}}\right]_{v_{ex0}} = \frac{L_{exf}}{R}\left[\frac{\mathrm{d}I_{exf}}{\mathrm{d}V_{ex}}\right]_{V_{ex0}}$ and equation 10.126 becomes

$$V_R = \frac{R_{exf}}{R}(1 + S_e)V_{ex} + \frac{L_{exf}}{R}\left[\frac{\mathrm{d}I_{exf}}{\mathrm{d}V_{ex}}\right]_{V_{ex0}}\frac{\mathrm{d}V_{ex}}{\mathrm{d}t} \tag{10.127}$$

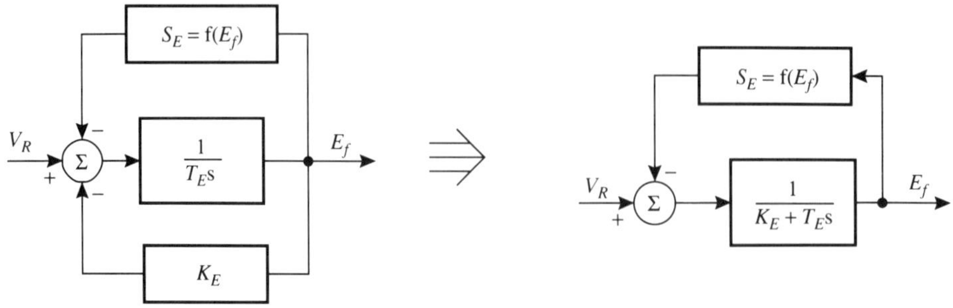

Figure 10.14 Block diagram of the regulated dc exciter

Noting that $E_f = V_{ex}$ and writing $\frac{L_{exf}}{R}\left[\frac{\mathrm{d}I_{exf}}{\mathrm{d}V_{ex}}\right]_{V_{ex0}} = T_E$ gives equation 10.127 in normalised form as

$$V_R = (K_E + S_E)E_f + T_E\frac{\mathrm{d}E_f}{\mathrm{d}t} \tag{10.128}$$

where $K_E = R_{exf}/R$ and $S_E = (R_{exf}/R)S_e$. This equation is shown in block diagram form in Figure 10.14 and comprises an integrating element, with integration time T_E, and two negative feedback loops with gains K_E and S_E respectively*. The negative feedback loop with gain S_E models the saturation in the exciter iron. As the saturation increases so does the value S_E, the magnitude of the negative feedback signal increases, and the influence of the regulator on the exciter voltage E_f is reduced. A dc exciter that is separately excited usually operates with $R_{exf} < R$, so that $K_E = 0.8 - 0.95$. Often approximated values of $K_E = 1$ and $S_E = S_e$ are assumed in the exciter model. The constant T_E is under one second and is often taken to be $T_E \approx 0.5$ s.

If the exciter is self-excited as shown in Figure 10.12b then the voltage v_R is the difference between the exciter internal emf and the excitation voltage, v_{ex}. Including this in equation 10.123 would give a differential equation that is identical to equation 10.128 except that $K_E = (R_{exf}/R - 1)$. Consequently the exciter block diagram in Figure 10.14 is also valid for self-excited exciters. Typically R_{exf} is slightly less than R so that K_E assumes a small negative value in the range -0.05 to -0.2.

The block diagram of the main part of the excitation system can now be formulated by combining the block diagram of the exciter with that of the regulator and the stabilising feedback signal as shown in Figure 10.15. The regulator is represented by a first-order transfer function with a time constant T_A and gain K_A. Typical values of these parameters are: $T_A = 0.05$–0.2 s and $K_A = 20$–400. The high regulator gain is necessary to ensure small voltage regulation of the order 0.5%. Unfortunately, although this high gain ensures low steady-state error, when coupled with the length of the time constants the transient performance of the exciter is unsatisfactory. To achieve acceptable transient performance the system must be stabilised in some way that reduces the transient (high frequency) gain.

* The constant T_E is sometimes called the exciter time constant. It is not a time constant. As shown in Figure 10.14 the time constant would depend on K_E and S_E. If saturation is neglected then the effective exciter time constant would be T_E/K_E.

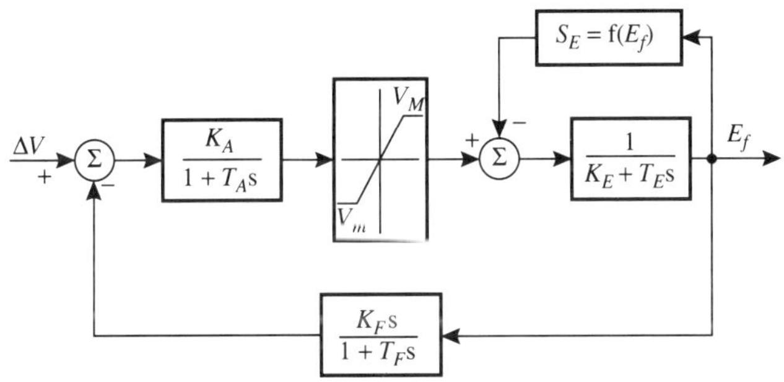

Figure 10.15 Block diagram of the excitation system with a dc exciter (Based on IEEE, 1968)

This is achieved by a feedback stabilisation signal represented by the first-order differentiating element with gain K_F and time constant T_F. Typical values of the parameters in this element are: $T_F = 0.35–1$ s and $K_F = 0.01–0.1$ s.

Although the saturation function $S_E = f(E_f)$ can be approximated by any non-linear function an exponential function of the form $S_E = A_{ex}e^{B_{ex}E_f}$ is commonly used. As this function must model the saturation characteristic over a wide range of exciter operating conditions the parameters A_{ex} and B_{ex} of the exponential function are determined by considering the heavily saturated region of the characteristic corresponding to high excitation voltage and high exciter field current. For example, Anderson *et al.* (1977) recommend that points on the characteristic corresponding to 100% and 75% of the maximum excitation voltage be used.

It is important to note that the limits on E_f are linked to the regulator limit and the saturation function such that the maximum value E_{fM} is obtained from

$$V_M - (K_E E_{fM} + S_{EM}) = 0 \tag{10.129}$$

ac rotating exciters

These exciters usually use a three-phase bridge rectifier consisting of six diode modules as shown in Figure 10.16a. The rectifier is fed from a three phase ac voltage source of emf V_E and reactance X_E. As with any rectifier system the output voltage depends on the commutation characteristics of the rectifier as determined by the degree of commutation overlap. As the effect of commutation overlap depends on the current through the rectifiers and the commutating reactance X_E three main operating states can be identified as a function of field current as shown in Figure 10.16b. These three states, denoted as I, II, III on the characteristic, depend on the commutation behaviour of the rectifier and allow the voltage at the rectifier terminals to be determined for a given field current I_f and voltage source emf V_E (Witzke, Kresser & Dillard, 1953, IEEE, 1981).

State I refers to the case where commutation in one branch of the bridge finishes before commutation in another branch starts. During this state the relationship between the rectifier terminal voltage and the field current is linear and is described by

$$\frac{E_f}{V_E} = 1 - \frac{1}{\sqrt{3}}\frac{I_f}{I_{SHC}} \tag{10.130}$$

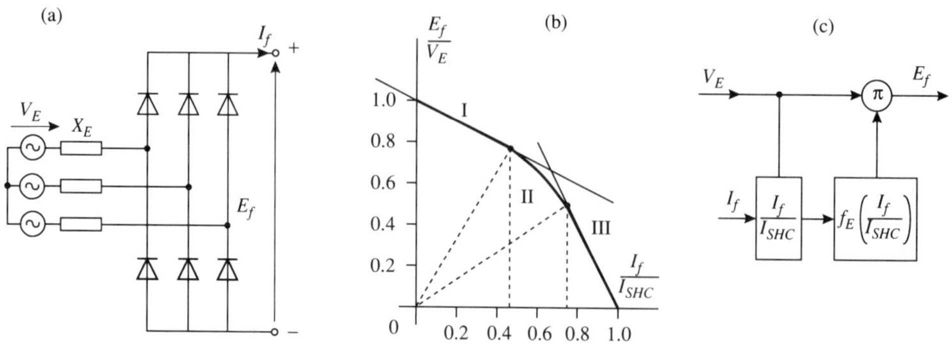

V_E, X_E are the emf and the reactance of the voltage source, E_f, I_f are the internal emf and the field current, and I_{SHC} is the rectifier short-circuit current

Figure 10.16 Three-phase uncontrolled bridge rectifier: (a) equivalent circuit; (b) voltage-current characteristic; (c) block diagram (Based on IEEE, 1981)

where $I_{SHC} = \sqrt{2}V_E/X_E$ is the rectifier short-circuit current conveniently chosen here to be one per unit. This relationship is valid for field currents $I_f < 2I_{SHC}/\sqrt{3}$.

As the field current increases commutation overlap also increases and the rectifier reaches state *II* when each diode can only conduct current when the counter-connected diode of the same phase has finished its conduction. In this state the relationship between the rectifier voltage and the field current is non-linear and corresponds to a circle with radius $\sqrt{3}/2$ as described by the equation

$$\left(\frac{E_f}{V_E}\right)^2 + \left(\frac{I_f}{I_{SHC}}\right)^2 = \left(\frac{\sqrt{3}}{2}\right)^2 \tag{10.131}$$

This relationship is valid for current $\frac{2}{\sqrt{3}}I_{SHC} \le I_f \le \frac{2}{4}I_{SHC}$.

As the field current increases further the rectifier reaches state *III* where commutation overlap is such that four diodes all conduct at the same time, two in each of the upper and lower arms. In this state the relationship between the rectifier voltage and the field current is linear and is described by the equation

$$\frac{E_f}{V_E} = \sqrt{3}\left(1 - \frac{I_f}{I_{SHC}}\right) \tag{10.132}$$

This relationship is valid for currents $\frac{3}{4}I_{SHC} \le I_f \le I_{SHC}$.

The block diagram that models the rectifier is shown in Figure 10.16c with all values in per unit. In this diagram the first block on the left calculates the value of the field current relative to the short circuit current I_{SHC}. The second block calculates the value of the function $f_E(I_f/I_{SHC})$ before passing it to the multiplying element to find the voltage at the terminals of the rectifier.

Figures 2.3b and 2.3c show how the rectifier can be fed from either an inductor generator or an inside-out ac generator with the latter being most common as it eliminates the need for slip rings. Although the alternator can be modelled using one of the generator

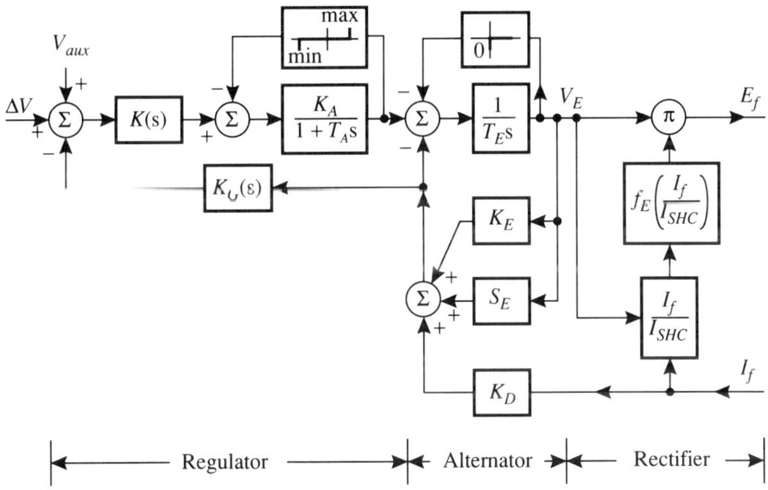

Figure 10.17 Block diagram of the excitation system with ac alternator and the uncontrolled rectifier

models described in Section 10.1 it is usually sufficiently accurate to simplify the alternator model to be similar to that used to represent the dc exciter. This simplification allows the complete excitation system to be modelled as shown in Figure 10.17.

As for the dc exciter the alternator is modelled by an integrating element with three feedback loops. The feedback loops with gain K_E and S_E play the same role as in the dc excitation system. However the effect of the alternator resistance is now included in the voltage/current characteristics of the rectifier when S_E is determined from the no-load saturation characteristic rather than the load saturation line as used for the dc exciter. As the armature current in the ac exciter is proportional to the current in the main generator field winding the third feedback loop, with gain K_D, uses this current to model the demagnetising effect of the armature reaction in the ac exciter. As the output voltage of a diode rectifier cannot drop below zero this is modelled by the negative feedback loop containing the signal limiter. If the excitation voltage drops below zero a large negative signal is fedback to the summing point to prevent a further drop in the voltage thereby keeping it at zero.

The system is stabilised by the feedback loop with transfer function $K_G(s) = K_F s/(1 + T_F s)$. In this case the stabilising loop is supplied by a signal that is proportional to the exciter field current. Alternatively the system could be stabilised by supplying this block directly from the output of the voltage regulator or from the excitation voltage E_f. In this diagram the feedback stabilisation is supplemented by an additional block with the transfer function $K(\mathrm{s})$ in the forward path preceding the regulator block. Both $K_G(\mathrm{s})$ and $K(\mathrm{s})$ depend on the specific excitation system and can be implemented by either analogue or digital techniques. Normally $K(\mathrm{s})$ will have a PI or PID type of structure and is often represented by the transfer function $(1 + T_C s)/(1 + T_B s)$. A major simplification to the model can be made by neglecting the variable effect of the field current on the rectifier voltage. In this case the model reduces to one that is very similar to that shown in Figure 10.15 for the dc exciter.

Static exciters

In static excitation systems the source of the dc current is a controlled three-phase bridge rectifier consisting of six thyristor modules as shown in Figure 10.18. The output characteristic of the rectifier depends on both the thyristor firing angle α and the system commutation characteristic. In the limiting case $\alpha = 0$ and the output characteristic is similar to that of the uncontrolled rectifier shown in Figure 10.16b. One very important characteristic of the controlled rectifier is the ability to provide a negative exciter output voltage so providing the exciter with *exciter buck* capability. Although the exciter output voltage may go negative the current cannot and must always flow in the same direction. As the firing angle increases the output voltage of the controlled rectifier decreases proportionally to $\cos\alpha$ (Lander, 1987) to give the family of characteristics shown in Figure 10.18.

The firing angle is set by the voltage regulator. The cosinusoidal dependency between the firing angle and the rectifier output voltage can be negated by introducing an inverted cosine function at the regulator output so as to produce a linear relationship. As the slight delays caused by discretisation of the firing sequence in each phase are much smaller than the power system time constants the rectifier can be regarded as a current source with no time delay.

The complete excitation system can then be modelled by the block diagram shown in Figure 10.19. In this figure the regulator and the stabilisation element are shown in the upper part of the diagram whilst the static characteristic of the rectifier is shown in the bottom part of the diagram. The rectifier supply voltage V_E is proportional to both the generator armature voltage and armature current as determined by the constants K_v and K_i. The value of these constants depend on how the rectifier is fed, Figure 2.3d-f. When $K_i = 0$ and $K_v = 1$ the system has no load compensation and corresponds to the rectifier being supplied directly from the terminals of the generator.

The way in which the main generator field current affects the rectifier output voltage is modelled in a similar way as for the uncontrolled rectifier. The regulator and the firing

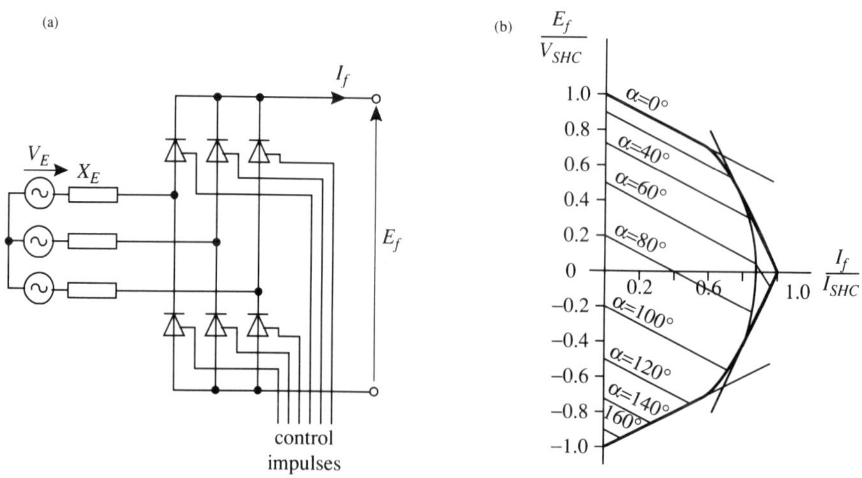

Figure 10.18 Three-phase controlled bridge rectifier: (a) circuit diagram; (b) voltage-current characteristic

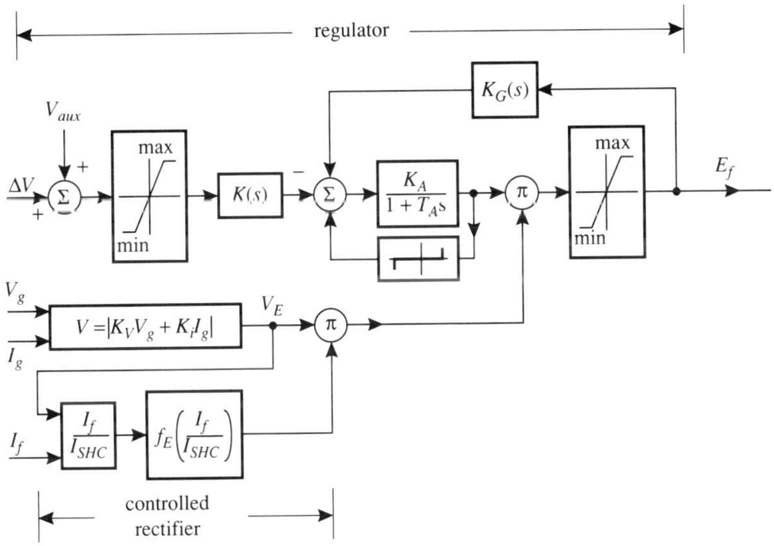

Figure 10.19 Block diagram of the excitation system with a static exciter (Based on IEEE, 1981)

circuits are modelled by a first order transfer function with gain K_A and time constant T_A. If the system does not contain cosinusoidal compensation of the voltage-firing angle dependency, the gain K_A will not be constant and should be modelled as a cosine function of the regulator signal. The system is stabilised by the transfer function $K(s)$ in the forward path and by feedback of the exciter output voltage through the block $K_G(s)$. For example the transfer function $K(s) = K_1(1 + T_C s)/(1 + T_B s)$ and a constant gain $K_G(s) = K_G$ may be used. Although $K(s)$ and $K_G(s)$ can be implemented using digital or analogue methods, digital AVRs are becoming common as they allow more sophisticated functions to be built into the AVR whilst only software changes are needed between different generators. The exciter output voltage E_f is given by the product of the supply voltage and the regulator output signal which represents the firing angle. If the influence of the generator field current on the rectifier output voltage is neglected then the exciter block diagram may be simplified to that consisting of the transfer function of the regulator and its stabilisation element.

10.2.3 Power system stabiliser (PSS)

Section 9.1 explained how a PSS can help to damp generator rotor oscillations by providing an additional input signal that produces a torque component in phase with the rotor speed deviation. The general structure of a PSS was shown in Figure 9.2 where the main elements are: the signal sensor, a low-pass filter, a high-pass filter, at least one lead compensation element, an amplifier and a limiter. The block diagram of the PSS model corresponding to the functional diagram of Figure 9.2 is shown in Figure 10.20. The input signal q can be rotor speed, real power, frequency, or some other signal as described in Section 9.1. The output signal is shown in Figure 10.20 as V_{PSS} and is passed to the AVR as the auxiliary signal V_{aux}. The parameters within the PSS are carefully selected for each PSS depending on its input signals and location in the system.

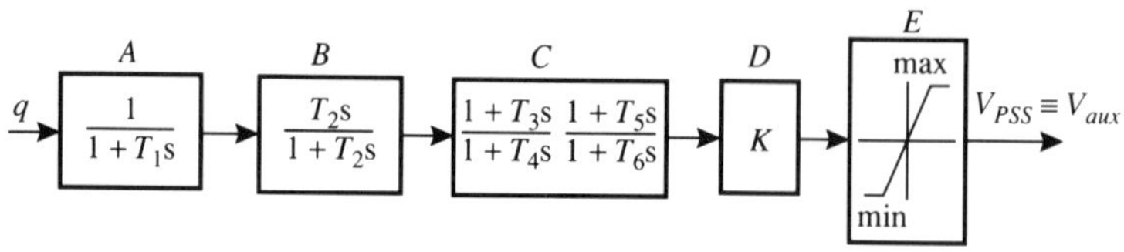

A - signal sensor, *B* - high-pass filter, *C* - lead/lag compensation element, *D* - amplifier gain, *E* - limiter

Figure 10.20 Block diagram of the power system stabiliser

10.3 Turbines and Turbine Governors

Chapter 2 explained how turbines, of one form or another, are used almost exclusively to provide the mechanical input power to the generator. It also explained how the basic governing systems on these turbines work. In this section models of the turbine and its governor will be developed so that the input power into the generator model can be regulated in the same way as in the actual system. The turbine models developed are simplified models designed solely to be used for power system analysis rather than for detailed mechanical modelling of the turbine itself. Models of both the mechanical-hydraulic and the electro-hydraulic governor will be developed. As with the AVR modern generators tend to be fitted with sophisticated digital governors that allow a high degree of functionality. As the structure of these vary between manufacturers the models developed here are typical of the turbines and governors currently in use. Although the turbine control system allows for both start up of the turbine and for its control when in operation it is the latter that is of interest here. A comprehensive discussion on the modelling of both steam and water turbines and their governors can be found in IEEE 1973b, 1991, 1992.

10.3.1 Steam turbines

Steam turbines are used extensively throughout the world to provide mechanical power to the generator. The mathematical model of the turbine will be derived using a simple model of a steam flow through a steam vessel shown in Figure 10.21.

The vessel introduces a time delay into the system as changes in the steam flow at the input take a finite time to appear at the output. This time delay can be quantified by considering a steam vessel of volume V as shown in Figure 10.21a. In this diagram m is the mass of steam in the vessel, p the steam pressure and $\dot{m}_1$ and $\dot{m}_2$ the steam mass flowrate at the input and the output. The mass of the steam in the vessel is constant when $\dot{m}_1 = \dot{m}_2$. When the steam input flowrate changes due to a change in the valve

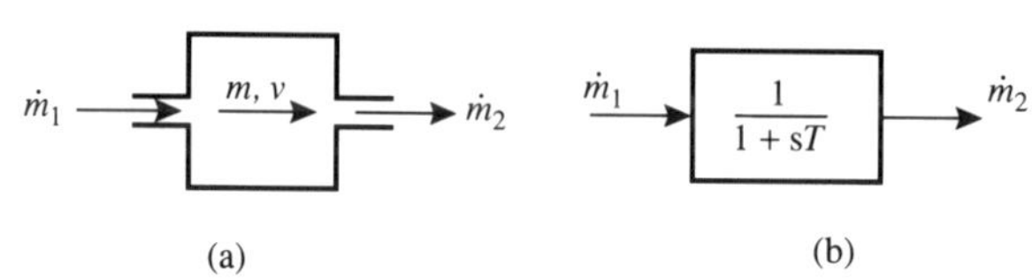

Figure 10.21 Model of the steam flow through a steam vessel: (a) vessel of capacity V; (b) block diagram

position then the mass of steam in the vessel will change at a rate proportional to the difference between the input flowrate and the output flowrate, that is $\mathrm{d}m/\mathrm{d}t = (\dot{m}_1 - \dot{m}_2)$. If the steam temperature is constant then the change of mass in the vessel must result in a pressure change, and this equation can be written as

$$\dot{m}_1 - \dot{m}_2 = \frac{\mathrm{d}m}{\mathrm{d}t} = \frac{\partial m}{\partial p}\frac{\mathrm{d}p}{\mathrm{d}t} = V\frac{\partial}{\partial p}\left(\frac{1}{v}\right)\frac{\mathrm{d}p}{\mathrm{d}t} \tag{10.133}$$

where v is the steam specific volume at a given pressure (volume divided by mass). Assuming that the outflow of steam is proportional to the pressure in the vessel

$$\dot{m}_2 = \dot{m}_0\frac{p}{p_0} \quad \text{or} \quad \frac{\mathrm{d}p}{\mathrm{d}t} = \frac{p_0}{\dot{m}_0}\frac{\mathrm{d}\dot{m}_2}{\mathrm{d}t} \tag{10.134}$$

where $\dot{m}_0 = \dot{m}_1(t = 0) = \dot{m}_2(t = 0)$ and $p_0 = p(t = 0)$. Substituting equation 10.134 into equation 10.133 gives

$$\dot{m}_1 - \dot{m}_2 = T\frac{\mathrm{d}\dot{m}_2}{\mathrm{d}t} \tag{10.135}$$

where $T = V\dfrac{p_0}{\dot{m}_0}\dfrac{\partial}{\partial p}\left(\dfrac{1}{v}\right)$ is a time constant corresponding to the inertia of the mass of steam in the vessel. Applying Laplace transforms and writing in transfer function form gives

$$\frac{\dot{m}_2(s)}{\dot{m}_1(s)} = \frac{1}{(1 + Ts)} \tag{10.136}$$

and corresponds to the inertia block in Figure 10.21b.

Figure 10.22 shows how the above equation can be used to model the tandem compound single-reheat turbine described in sub-Section 2.2.3. The schematic diagram of this arrangement, Figure 10.22a, shows how the steam passes through the governor control valves and the inlet piping to the HP steam chest (*A*). On leaving the steam chest it passes through the HP turbine before entering the reheat stage between the HP and IP steam turbines. After being reheated the steam passes through the intercept valves to the IP turbine and then through the crossover piping to the LP turbine. The mathematical model of such a system is shown in Figure 10.22b and can be divided into two parts. Firstly as the power extracted by a turbine is proportional to the steam mass flowrate $\dot{m}_s$ each turbine stage can be modelled by a constant α, β, or γ which corresponds to the portion of the total turbine power developed in the different turbine stage with $\alpha + \beta + \gamma = 1$. Typically, for the single-reheat turbine shown here $\alpha \approx 0.3$, $\beta \approx 0.4$ and $\gamma \approx 0.3$ defined on the turbine MW output base. If the generator MVA base is used then these values must be modified in the ratio of the two bases.

The second part of the turbine model concerns the storage of the steam chests and associated piping and corresponds to Figure 10.21b. Typical values of the parameters are: $T_A = 0.1$ to 0.4 s, $T_B = 4$ to 11s, $T_C = 0.3$ to 0.5 s, $\alpha = 0.3$, $\beta = 0.4$, $\gamma = 0.3$.

Assuming that there is no control of the intercept valves the block diagram of Figure 10.22b can be simplified to that shown in Figure 10.22c by combining the three inertia blocks to give one equivalent third-order block with the combined parameters:

$$a_1 = T_A + T_B + T_C, \quad a_2 = T_AT_B + T_AT_C + T_BT_C, \quad a_3 = T_AT_BT_C$$

$$b_1 = \alpha(T_B + T_C) + \beta T_C, \quad b_2 = \alpha T_BT_C$$

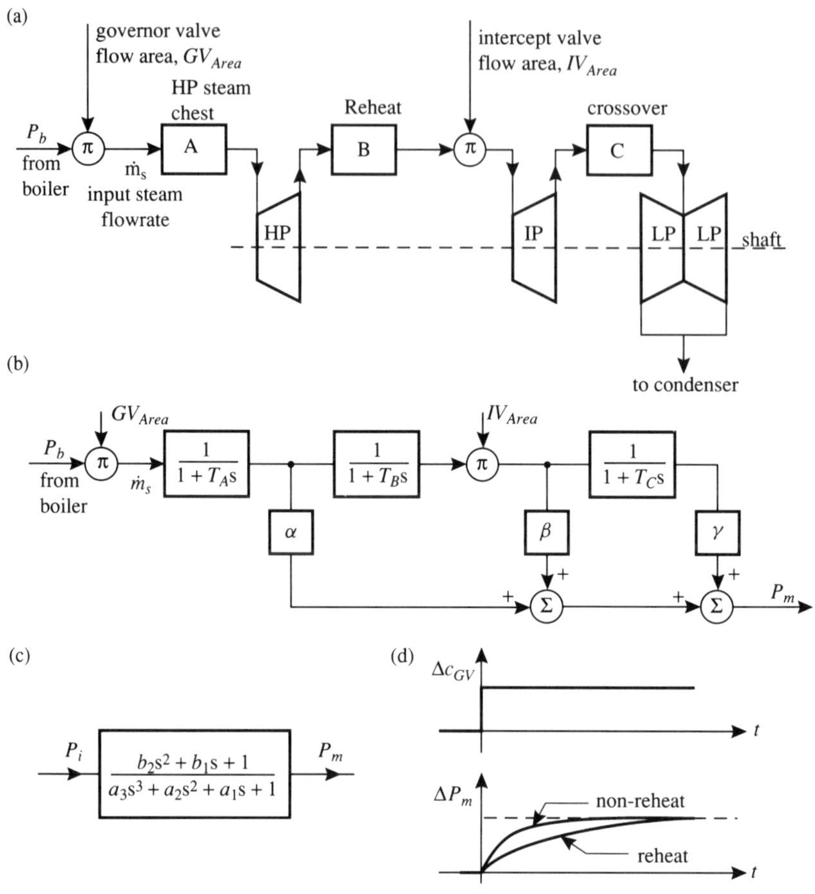

P_i, P_m - power of the inlet steam and of the turbine

Figure 10.22 Tandem compound single-reheat turbine: (a) schematic diagram; (b) block diagram; (c) transformed block diagram; (d) response of the linear turbine model to a step change in valve position

As turbine power is proportional to the steam mass flowrate the input flowrate is changed in this diagram to input power to further simplify the diagram. As the reheat time constant T_B is several times higher than either T_A or T_C the turbine model can be further simplified by assuming $T_A \approx 0$ or even $T_A \approx T_C \approx 0$ reducing the turbine model to a second and first order block respectively.

A steam turbine without reheat can be modelled by a first-order block with time constant $T_A = 0.2$ to 0.5 s and with $\alpha = 1$, $\beta = \gamma = 0$, $T_B = T_C = 0$.

A comparison of the time response of a turbine with and without reheat to an incremental step increase in the opening of the governor valve Δc_{GV} is shown in Figure 10.22d. When the valve opening increases so does the valve flow area GV_{Area} so that for a constant boiler pressure the steam flowrate also increases. However steam storage effects introduce a delay before the increased steam flowrate can reach the turbine blades and increase the output power. This effect is particular noticeable for a turbine with reheat when the rise time may be as long as 10 seconds as shown in Figure 10.22d. As this delay between activating the valve and the power changing can be excessively long in reheat turbines coordinated control of both the governor and the interceptor valves is necessary as explained in the next section.

Governing system for steam turbines

Chapter 2 explained that the reheat turbine shown in Figure 10.22a is fitted with two sets of control valves and two sets of emergency stop valves. Normally turbine control is accomplished by regulating the position of the governor valves and the intercept valves whilst the emergency stop valves are kept fully open and only used in an emergency to stop the turbine. As both sets of stop valves are normally kept fully open they can be neglected for modelling purposes.

Although the way in which the control of the governor valves and the intercept valves is co-ordinated depends on the purpose of the control action, the type of governor and the manufacturer concerned, the generic features of the governor can be included in a general model capable of representing both the mechanical-hydraulic and the electro-hydraulic governor. This model allows for overspeed control and load/frequency control as described in Chapter 8. The fast valving features described in Chapter 9 can be readily added if necessary.

In order to develop this governor model the functional diagram of the mechanical-hydraulic governor shown in Figure 10.23a will be used. This diagram shows the main elements of a mechanical-hydraulic governor to be the speed transducer, regulator, speed relay, servomotor and the steam control valves. Compared to Figure 2.11 an additional element, the speed relay, is shown. This develops an output proportional to the load reference signal less any contribution due to speed deviation. This additional element is required as on some larger generators the force required to adjust the position of the main steam valves is very large and an additional spring loaded servomotor, the speed relay, is required between the speeder linkage and the main servomotor. On some smaller machines the spring loaded servomotor of the speed relay is used to adjust the position of the control valves rather than the servomotor arrangement shown in Figure 2.11.

The elements in this functional diagram are modelled as shown in the block diagram of Figure 10.23b. The regulator is modelled by a gain $K = 1/\rho$, where ρ is the droop of the regulator static characteristic, whilst the speed relay is modelled by a first order lag with time constant T_2. The effect of the load reference has been modelled as the required opening of the main governor control valves c_{0GV}. The main servomotor, which alters the position of the control valves, is modelled by an integrator with direct feedback of valve position along with two limiters. The first limiter is necessary to protect the turbine from too rapid opening or closing of the steam valves. The second limiter limits the valve position between fully open and fully closed, or to some set position if a load limiter is present. The final, non-linear, block models the valve characteristics effectively converting the valve position to normalised flow area. The non-linearity of the valve characteristic may be compensated in the governor by a linearising cam located between the speed relay and the servomotor. These two, mutually compensating, non-linearities are often neglected in the model. The output of the model is the normalised valve flow area and appears as the main input to the turbine model in Figure 10.22b.

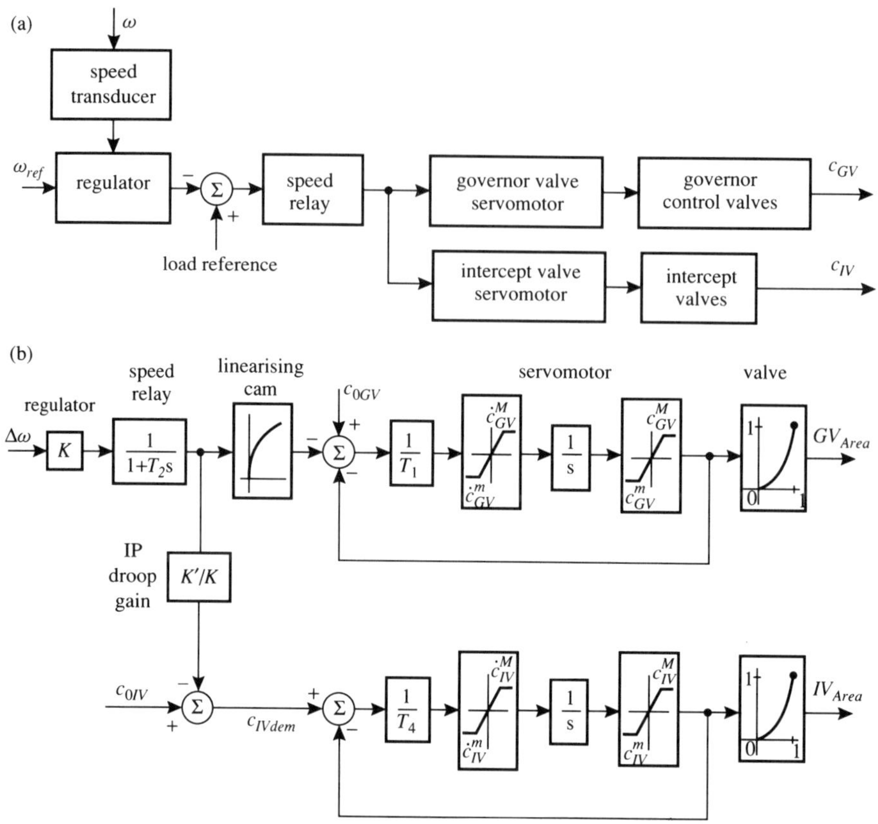

c - steam valve position, GV - governor valves, IV - intercept valves

Figure 10.23 Mechanical-hydraulic steam turbine governor (b) block diagram;

One way in which the intercept valves can be controlled to limit overspeed is also shown in Figure 10.23b. In this example the position of the intercept valves is controlled to follow a demand position signal c_{IVdem} generated according to the equation

$$c_{IVdem} = c_{0IV} - K'\Delta\omega \tag{10.137}$$

where K' is the IP gain and is the inverse of the IP droop. $K' > K$.

To examine in further detail how this control loop works assume $K = 25$ (main droop 4%) and $K' = 50$ (IP droop 2%) and that it is required to keep the intercept valves fully open until the overspeed is 4%. At this speed the main governor valves will just have fully closed. The reference signal c_{0IV} is set to 3. With $0.04 > \Delta\omega > 0$, the demand $c_{IVdem} \geq 1$ and the intercept valves are kept fully open. When $\Delta\omega = 0.04$ the intercept valves start to close and are fully closed when $\Delta\omega = 0.06$.

In some cases an auxiliary governor is used that acts in parallel with the normal regulator. This auxiliary governor only comes into operation when large speed deviations are registered and has the effect of increasing the regulator gain K. In this way the regulator output is increased to quickly close both the interceptor valves and the governor control valves.

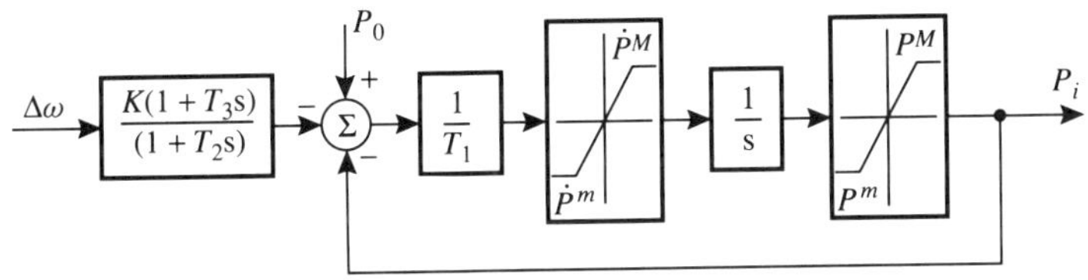

Figure 10.24 Simplified governor model

The governor model can also be used to model the electro-hydraulic governor as the main power components are similar except that now the mechanical components have been replaced by electronic circuits. Such changes provide a greater degree of flexibility and functionality than can be achieved in the traditional mechanical-hydraulic governor. A further advance is the digital-hydraulic governor where the control functions are implemented in software thus providing even greater flexibility. From a modelling point of view this increased flexibility means that the actual governor control used is manufacturer dependent particularly with regard to the more advanced features such as fast valving. Nevertheless the governor model developed in Figure 10.23b with the speed relay replaced by the general transfer function $(1 + T_3s)/(1 + T_2s)$ to provide any phase compensation forms a good basis for modelling an electro-hydraulic governor. If a particular control logic is to be implemented then it is relatively straightforward to modify Figure 10.23b accordingly.

When analyzing electromechanical dynamics in a time interval of around 10 seconds the boiler pressure can be assumed to be constant and can be equated to the initial power P_0 so that $P_b = P_m(t = 0) = P_0$ = constant. With this assumption the governor block diagram can be simplified to that shown in Figure 10.24 by multiplying the limits, and the initial conditions, in the servomotor model by P_0. In this model the regulator and the speed relay transfer functions have been replaced by a general control block $K(1 + T_3s)/(1 + T_2s)$ to enable this model to be used for both mechanical-hydraulic and electro-mechanical governors. The additional time constant now allows both types of governor to be modelled by the correct choice of values. Intercept valve control is not shown on this diagram as it assumes that they are kept fully open but could be added if required. Other forms of electro-hydraulic and mechanical-hydraulic governors are described in the references (IEEE, 1973b,1991 and Kundur 1994).

Boiler control

In the modelling process the boiler output is linked to the turbine via the boiler pressure parameter P_b. It is not the purpose here to describe boiler models but a brief discussion of boiler/turbine interaction is necessary as throughout this book it has been assumed that P_b effectively remains constant. Traditionally conventional turbine control operated in a *boiler follows turbine* mode where all load changes are carried out by control of the turbine valves with the boiler controls operating to maintain constant steam conditions, i.e. pressure and flow. Although this control mode allows rapid turbine response and good frequency control by utilizing the stored energy in the boiler, the changes in the boiler pressure and other variables can be quite large. To prevent this an alternative *turbine follows boiler* control mode can be used when all load changes are made via the boiler controls. The turbine valves are controlled to regulate the boiler pressure. As the speed of response of the turbine valves is very fast almost perfect pressure control can be achieved. However load changes are now very slow with time constants of 1-2 minutes, depending on the type of boiler and the fuel, as now no use is made of the stored energy in the boiler. As a compromise *integrated boiler and turbine* control may be used to achieve a quick turbine response whilst limiting the changes in the boiler variables.

10.3.2 Hydraulic turbines

If water flows from a high level to a low level through a hydraulic turbine the potential energy of the water stored in a high reservoir is converted into mechanical work on the turbine shaft as described in sub-Section 2.2.3. The turbine may be either an impulse turbine or a reaction turbine and, although the way in which they operate hydraulically differs, the work done by both types is entirely due to the conversion of kinetic energy. In impulse turbines, such as the Pelton wheel, all the available energy in the water is converted into kinetic energy as the water passes through the nozzle. The water forms a free jet as it leaves the nozzle and strikes the runner where the kinetic energy is converted into mechanical work. In the reaction turbine, such as the Francis turbine, only a part of the energy in the water is converted into kinetic energy as the water passes through the adjustable wicket gates with the remaining conversion taking place inside the runner itself. All passages are filled with water including the draft tube. In both turbines the power is controlled by regulating the flow into the turbine by wicket gates on the reaction turbine and by the needle, or spear, on the impulse turbine. What is required is a mathematical description of how the turbine power changes as the position of the regulating device is changed.

Figure 10.25a shows a schematic diagram of a turbine installation where water flows down the penstock and through the turbine before exiting into the tail water. The penstock is modelled by assuming that the flow is incompressible, then the rate of change of flow in the penstock is obtained by equating the rate of change of momentum of the water in the penstock to the net force on the water in the penstock when

$$\rho L \frac{\mathrm{d}Q}{\mathrm{d}t} = F_{net} \tag{10.138}$$

where Q is the volumetric flowrate, L the penstock length and ρ the mass density of water.

The net force on the water can be obtained by considering the pressure head at either end of the conduit. On entry to the penstock the force on the water is simply proportional to the static head H_s whilst at the wicket gate it is proportional to the head H across the turbine. Due to friction effects in the conduit there is also a friction force on the water represented by the head loss H_l so that the net force on the water in the penstock is

$$F_{net} = (H_s - H_l - H)A\rho g \tag{10.139}$$

where A is the penstock cross-sectional area and g the acceleration due to gravity. Substituting the net force into equation 10.138 gives

$$\rho L \frac{\mathrm{d}Q}{\mathrm{d}t} = (H_s - H_l - H)A\rho g \tag{10.140}$$

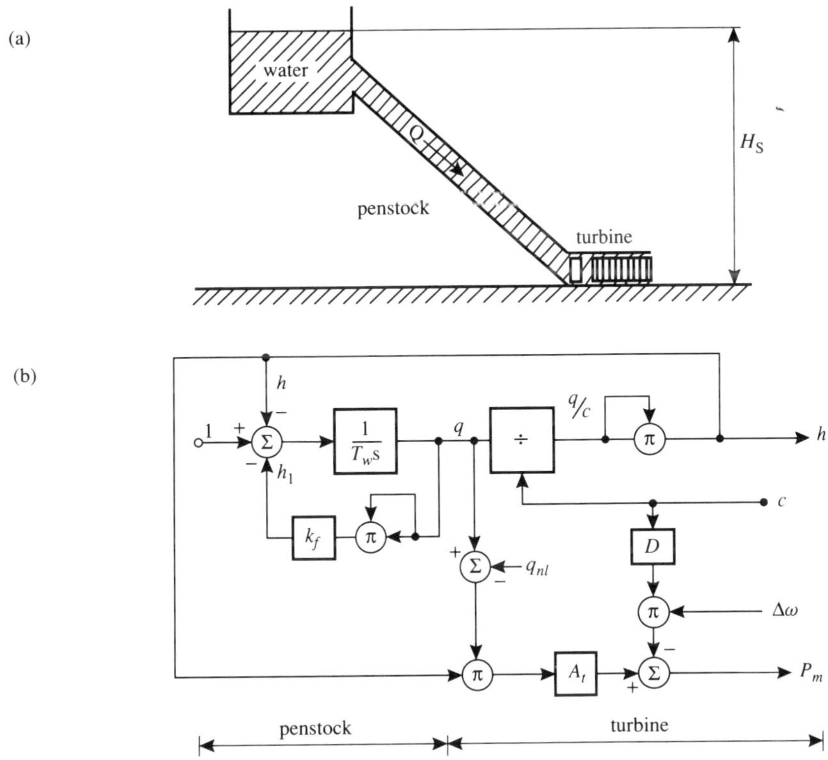

Figure 10.25 Hydraulic turbine: (a) schematic diagram; (b) non-linear model (Based on IEEE 1992)

It is usual to normalise this equation to a convenient base. Although this base system is arbitrary the base head h_{base} is taken as the static head *above* the turbine, in this case H_s, whilst the base flowrate q_{base} is taken as the flowrate through the turbine with the gates fully open and the head at the turbine equal to h_{base} (IEEE, 1992). Dividing both sides of equation 10.140 by $h_{base}q_{base}$ gives

$$\frac{dq}{dt} = \frac{1}{T_w}(1 - h_l - h) \tag{10.141}$$

where $q = Q/q_{base}$ and $h = H/h_{base}$ are the normalised flowrates and pressure heads respectively and $T_w = \dfrac{Lq_{base}}{Agh_{base}}$ is the *water starting time*. Theoretically T_w is defined as the time taken for the flowrate in the penstock to change by a value equal to q_{base} when the head term in brackets changes by a value equal to h_{base}. The head loss h_l is proportional to the flowrate squared and depends on the conduit dimensions and friction factor. It suffices here to assume that $h_l = k_f q^2$ and can often be neglected. This equation defines the penstock model and is shown in block diagram form in the left-hand part of Figure 10.25b.

In modelling the turbine itself both its hydraulic characteristics and mechanical power output must be modelled. Firstly the pressure head across the turbine is related to the flowrate by assuming the turbine can be represented by the valve characteristic

$$Q = kc\sqrt{H} \tag{10.142}$$

where c is the gate position between *0-1* and k is a constant. With the gate fully open c = *1* and this equation can be normalised by dividing both sides by $q_{base} = k\sqrt{h_{base}}$ to give, in per unit,

$$q = c\sqrt{h} \tag{10.143}$$

Secondly the power developed by the turbine is proportional to the product of the flowrate and the head and depends on the efficiency. To account for the turbine not being 100% efficient the no-load flow q_{nl} is subtracted from the actual flow to give, in normalised parameters,

$$P_m = h(q - q_{nl}) \tag{10.144}$$

Unfortunately this expression is in a different per unit system from that used for the generator whose parameters are normalised to the generator MVA base so that equation 10.144 is written as

$$P_m = A_t h(q - q_{nl}) \tag{10.145}$$

where the factor A_t is introduced to account for the difference in the bases. The value of the factor A_t can be obtained by considering the operation of the turbine at rated load when

$$P_m = A_t h_r(q_r - q_{nl}) = \frac{\text{turbine power (MW)}}{\text{generator MVA rating}} \tag{10.146}$$

and the suffix "r" indicates the value of the parameters at rated load. Rearranging equation 10.146 gives

$$A_t = \frac{\text{turbine power (MW)}}{\text{generator MVA rating}} \frac{1}{h_r(q_r - q_{nl})} \tag{10.147}$$

Also present is a damping effect that is dependent on gate opening, so that at any load condition the turbine power can be expressed by

$$P_m = A_t h(q - q_{nl}) - Dc\Delta\omega \tag{10.148}$$

where D is the damping coefficient. Equations 10.143 and 10.148 constitute the turbine non-linear model shown in the right-hand side of Figure 10.25b where the wicket gate position is the control variable.

Linear turbine model

The classical model of the water turbine (IEEE, 1973b) uses a linearised version of the non-linear model. Such a model is valid for small changes of mechanical power and can be obtained by linearising equations 10.141, 10.143 and 10.145 about an initial operating point to give

$$\frac{d\Delta q}{dt} = -\frac{\Delta h}{T_w}, \; \Delta q = \frac{\partial q}{\partial c}\Delta c + \frac{\partial q}{\partial h}\Delta h, \; \Delta P_m = \frac{\partial P_m}{\partial h}\Delta h + \frac{\partial P_m}{\partial q}\Delta q \tag{10.149}$$

Introducing the Laplace operator s and eliminating Δh and Δq from the equations gives

$$\frac{\Delta P_m}{\Delta c} = \frac{\left[\dfrac{\partial q}{\partial c}\dfrac{\partial P_m}{\partial q} - T_w s\,\dfrac{\partial P_m}{\partial h}\dfrac{\partial q}{\partial c}\right]}{1 + T_w s\,\dfrac{\partial q}{\partial h}} \tag{10.150}$$

where the partial derivatives are

$$\begin{aligned} \frac{\partial q}{\partial h} &= \frac{1}{2}\frac{c_0}{\sqrt{h_0}}, & \frac{\partial q}{\partial c} &= \sqrt{h_0} \\ \frac{\partial P_m}{\partial q} &= A_t h_0, & \frac{\partial P_m}{\partial h} &= A_t(q_0 - q_{nl}) \approx A_t(q_0) \end{aligned} \tag{10.151}$$

and the suffix "0" indicates an initial value. Substituting into equation 10.150 and noting that $q_0 = c_0\sqrt{h_0}$ gives

$$\frac{\Delta P_m}{\Delta c} = A_t h_0^{3/2}\frac{1 - T'_w s}{1 + \dfrac{T'_w}{2}s} \tag{10.152}$$

where $T'_w = T_w\dfrac{q_0}{h_0} = \dfrac{L}{Ag}\dfrac{Q_0}{H_0}$. Typically T'_w is between 0.5 and 5 s.

This is the classic definition of water starting time but is dependent on the values of the head and flowrate at the linearisation point. It therefore varies with load. If required, the constant A_t can be absorbed into the gate position when it effectively converts the gate opening to per unit turbine power on the generator MVA base. The block diagram of this linear turbine model is shown in Figure 10.26a.

Equation 10.152 describes an interesting and important characteristic of water turbines. For example, suppose that the position of the gate is suddenly closed slightly so as to reduce the turbine power output. The flowrate in the penstock cannot change instantaneously so the velocity of the flow through the turbine will initially increase. This increase in water velocity will produce an initial increase in the turbine power until, after a short delay, the flowrate in the penstock has time to reduce when the power will also reduce.

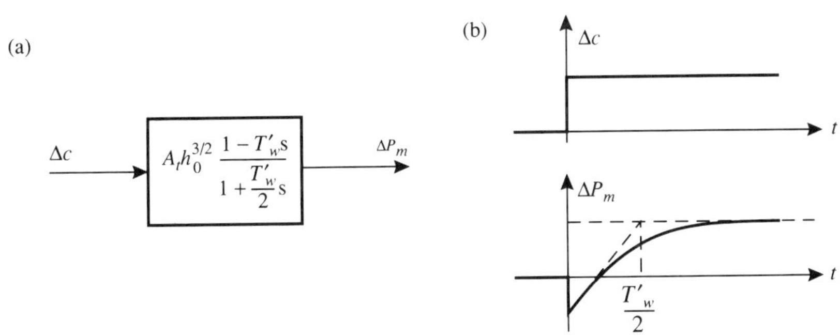

Figure 10.26 Hydraulic turbine: (a) linear model; (b) response of the linear turbine model to a step change in gate position

This effect is reflected in equation 10.152 by the minus sign in the numerator. This characteristic is shown in Figure 10.26b where a step increase in the gate position Δc initially produces a rapid drop in power output. As the flowrate in the penstock increases the power output increases.

Although the linearised model (equation 10.152) has been successfully used in both steady-state and transient stability studies, the IEEE Committee (1992) recommends the use of the non-linear turbine model in power system studies, as its implementation using computers is no more difficult than the approximate linear transfer function. Other, more detailed models, are discussed in the references (IEEE, 1973b, 1992).

Governing system for hydraulic turbines

Governing systems for hydraulic turbines differ from those used in steam turbines in two main ways. Firstly a very high force is required to move the control gate as it must overcome both high water pressure and high friction forces and, secondly, the peculiar response of the hydraulic turbine to changes in valve position must be adequately compensated. To provide the necessary force to move the gate, two servomotors are used as shown in the functional diagram in Figure 10.27a. In a similar way as for the steam turbine, the speed regulator acts through a system of levers on the *pilot valve* which controls the flow of hydraulic fluid into the *pilot servomotor*. The pilot servomotor then acts on the *relay valve* of the very high power *main servomotor* which controls the gate position. Just as in the servomotor on the steam turbine, negative feedback of the position of both the servomotors is necessary to achieve the required movement. To compensate for the peculiar response of the water turbine to changes in the gate position it is necessary to slow down the initial gate movement to allow the water flow in the penstock to catch up. This is achieved by the *transient droop* element which reduces the gain of the governor for fast changes in valve position and, in mechanical-hydraulic governor systems, is achieved by feeding back the position of the gate via a system of levers that includes a dashpot system. Similarly as for the steam turbine governor shown in Figure 2.14 direct feedback of the gate position through a series of levers controls the static droop.

The block diagram model of the governor, including the transient droop, is shown in Figure 10.27b. The main servomotor is modelled by an integrating element, with integration time T_g, and two limiters. The first limiter limits the gate position between fully open and fully closed whilst the second, rate limiter, limits the rate at which the gate can be moved. This is necessary as if the gate is closed too rapidly the resulting high pressure could damage the penstock. The pilot servomotor with its position feedback is modelled by a first order lag with a time constant T_p. The system has two main feedback loops. The proportional feedback loop provides the static droop characteristic equal to ρ whilst the feedback loop with the differentiating element corrects the transient droop to a value δ. Typical values of the parameters recommended by Ramey & Skooglund (1970) are: $T_p = 0.04$ s, $T_g = 0.2$ s, $T_r = 5T'_w$, $\delta = 2.5T'_w/T_m$ and $\rho = 0.03$ to 0.06, where T'_w is the water starting time and T_m is the mechanical time constant of the unit.

If the non-linearities introduced by the limiters are, for the moment, neglected the system may be described by the third-order transfer function

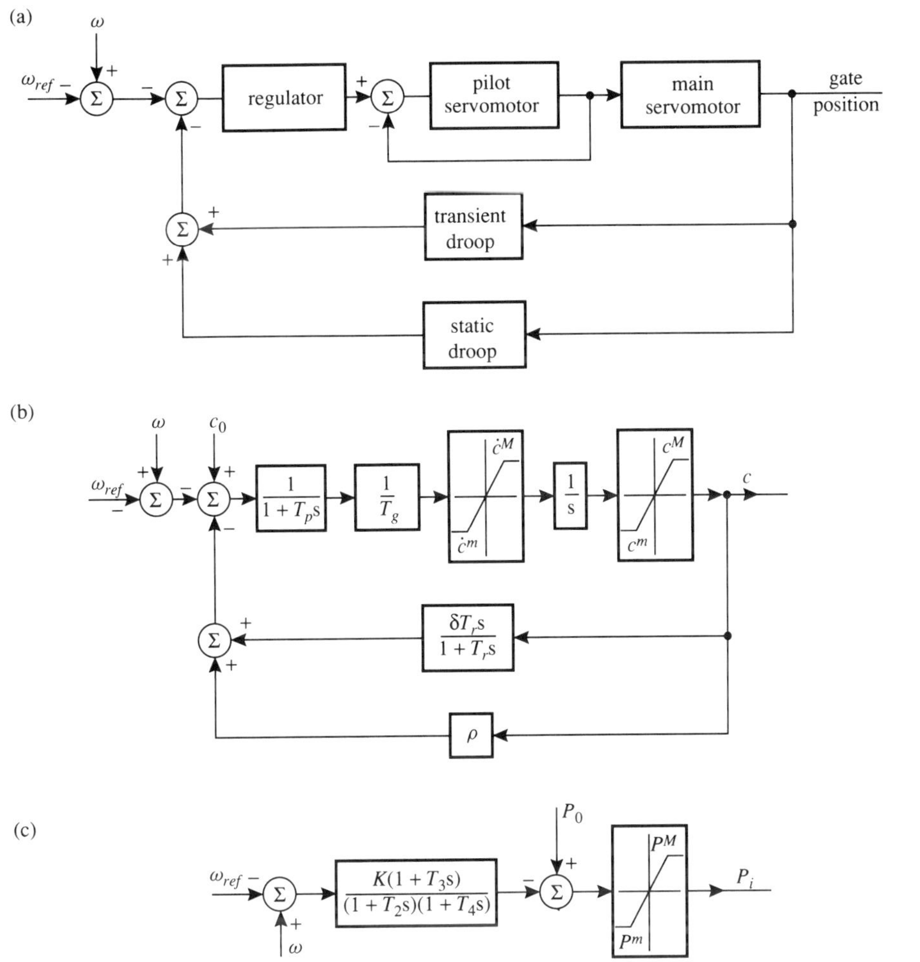

Figure 10.27 Block diagram of hydraulic turbine governing system: (a) functional diagram; (b) full diagram; (c) simplified diagram

$$\frac{\Delta c}{\Delta \omega} = \frac{\dfrac{(1 + T_r s)}{\rho}}{\dfrac{T_p T_r T_g}{\rho} s^3 + \dfrac{(T_p + T_r) T_g}{\rho} s^2 + \dfrac{T_g + T_r(\rho + \delta)}{\rho} s + 1} \tag{10.153}$$

As the time constant T_p is several times smaller than the time constants T_g and T_r it may be neglected to give the second-order transfer function

$$\frac{\Delta c}{\Delta \omega} = \frac{(1 + T_3 s) K}{(1 + T_2 s)(1 + T_4 s)} \tag{10.154}$$

where $K = 1/\rho$, $T_2 \approx T_r T_g / [T_g + T_r(\rho + \delta)]$, $T_3 = T_r$ and $T_4 = (T_g + T_r(\rho + \delta))/\rho$. Usually $T_4 >> T_2$, then $T_4 + T_2 \approx T_4$.

Table 10.2 Typical values of parameters of turbine governing systems

Type of turbine	Parameters				
	ρ	T_1	T_2	T_3	T_4
		s	s	s	s
steam	0.02-0.07	0.1	0.2-0.3	0	—
hydraulic	0.02-0.04	—	0.5	5	50

If the gate limiters are now added to the transfer function the simplified governor block diagram shown in Figure 10.27c is obtained and is similar to the simplified system used to represent the steam turbine governor.

Typical parameter values for use in the simplified block diagram of the steam and hydraulic turbine governors are tabulated in Table 10.2.

10.4 Dynamic Load Models

Section 3.4 explained how power system loads, at the transmission and the sub-transmission level, can be represented by a static load characteristic which describes how the real and the reactive power at a busbar changes with both voltage and frequency. Although this is an adequate method of representing modest changes of voltage and/or frequency for a composite load there are cases when it is necessary to account for the dynamics of the load components themselves.

Typically, motors consume 60 to 70% of the total power system energy and their dynamics are important for studies of inter-area oscillations, voltage stability and long-term stability. In this section the dynamic model of an induction motor will be presented. The dynamic model of the synchronous motor is not discussed as it is identical to that of the synchronous generator presented in Section 10.1, except for a sign change on the armature current necessary to reflect the motoring action.

The induction motor can be modelled in a similar way as the synchronous generator but with three important differences. Firstly, as there is no field winding in the induction motor, the cage rotor is modelled by two coils in quadrature (as for the damper windings in the synchronous machine) whilst the armature is modelled in the normal way by a d- and a q-axis armature coil. Secondly, as the induction motor does not rotate at synchronous speed, both the stator armature coils and the two rotor coils are transformed into a reference frame that rotates at synchronous speed. This means that the two rotor equations now include a rotational emf term proportional to the rotor slip speed $s\omega_s$ and are of the form

$$\begin{aligned} v_{dR} &= 0 = R_R i_{dR} + \dot{\Psi}_{dR} - s\omega_s \Psi_{qR} \\ v_{qR} &= 0 = R_R i_{qR} + \dot{\Psi}_{qR} + s\omega_s \Psi_{dR} \end{aligned} \tag{10.155}$$

where the suffix "R" signifies rotor quantities and the slip $s = (\omega_s - \omega)/\omega_s$.

The two rotational voltages are proportional to the rotor slip speed and to the flux linkage of the other rotor coil. During a disturbance the flux linkages Ψ_{qR} and Ψ_{dR} cannot change instantaneously and, just as in equations 10.54 and 10.60, can be equated to d and q axis armature emfs E'_d and E'_q. If equation 10.155 is further compared with equation

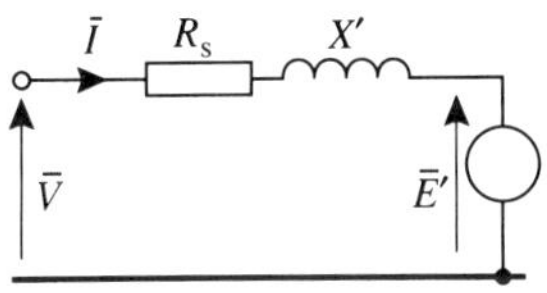

Figure 10.28 Transient state representation of induction motor

10.51 the rotational voltages $s\omega_s\Psi_{qR}$ and $s\omega_s\Psi_{dR}$ can be seen to play a similar role as the excitation voltage E_f and will therefore appear in the corresponding flux decay equations.

The third important difference is that positive current is now defined for motoring action requiring a sign change in the appropriate equations.

With these points in mind Figure 10.28 shows how the induction motor can be modelled by a transient emf E' behind a transient impedance X' in a similar way as the synchronous generator, *Model 4* (Section 10.1.6). However as the reactance is unaffected by rotor position, and the model is in the synchronously rotating reference frame, the necessary equations are more conveniently expressed in the network (a, b) coordinates (Arrillaga & Arnold, 1990) as

$$\begin{bmatrix} V_b \\ V_a \end{bmatrix} = \begin{bmatrix} E'_b \\ E'_a \end{bmatrix} + \begin{bmatrix} R_s & X' \\ -X' & R_s \end{bmatrix} \begin{bmatrix} I_b \\ I_a \end{bmatrix} \tag{10.156}$$

with the change in the emfs E'_b and E'_a given by

$$\begin{aligned} \dot{E}'_b &= -s\omega_s E'_a - \frac{E'_b - (X - X')I_a}{T'_0} \\ \dot{E}'_a &= s\omega_s E'_b - \frac{E'_a + (X - X')I_b}{T'_0} \end{aligned} \tag{10.157}$$

where $s = (\omega_s - \omega)/\omega_s$ is the rotor slip, $X' = X_s + X_\mu X_R/(X_\mu + X_R)$ is the transient reactance and is equal to the blocked-rotor (short-circuit) reactance, $X = X_S + X_\mu$ is the motor no-load (open-circuit) reactance and $T'_0 = (X_R + X_\mu)/(\omega_s R_R)$ is the transient open-circuit time constant. The other reactances take the meaning explained in the induction motor steady-state equivalent circuit in Figure 3.22. The stator current can be calculated from equation 10.156. Alternatively equations 10.156 and 10.157 can be written directly in phasor notation as

$$\begin{aligned} \overline{V} &= \overline{E}' + \mathrm{j}X'\overline{I} + \overline{I}R_s \\ \frac{\mathrm{d}}{\mathrm{d}t}\overline{E}' &= -s\omega_s \mathrm{j}\overline{E}' - \frac{[\overline{E}' - \mathrm{j}\overline{I}(X - X')]}{T'_0} \end{aligned} \tag{10.158}$$

To evaluate the slip $s = (\omega_s - \omega)/\omega_s$ it is necessary to calculate the rotor speed ω from the equation of motion

$$I\frac{\mathrm{d}\omega}{\mathrm{d}t} = \tau_e - \tau_m \tag{10.159}$$

where I is the inertia of the motor and load, τ_m is the mechanical load torque and τ_e is the electromagnetic torque that can be converted into useful work. Using the notation shown in the equivalent circuit of Figure 3.22 the induction motor air-gap power is

$$P_{ag} = I^2 \frac{R}{s} \tag{10.160}$$

whilst the electromagnetic power P_e that is converted into mechanical power is given by

$$P_e = I^2 R \frac{(1-s)}{s} = P_{ag}(1-s) \tag{10.161}$$

so that the torque

$$\tau_e = \frac{P_e}{\omega} = \frac{P_{ag}}{\omega_s} \tag{10.162}$$

Normalising by dividing by $T_{base} = \dfrac{MVA}{\omega_s}$ gives

$$\tau_e = P_{ag} = \text{Re } (\overline{E'I}^*) = E'_b I_b + E'_a I_a \tag{10.163}$$

whilst the normalised version of the equation of motion is

$$M \frac{d\omega}{dt} = \tau_e - \tau_m \tag{10.164}$$

where $M = 2H/\omega_s$. Normally τ_m will vary with speed and is commonly represented by a quadratic equation of the form

$$\tau_m = \tau_{m0}(A\omega^2 + B\omega + C) \tag{10.165}$$

where τ_{m0} is the rated load torque, $A\omega_0^2 + B\omega_0 + C = 1$, and ω_0 is the rated speed. For example, a simple pump load, the torque of which is proportional to speed squared, can be represented by setting $B = C = 0$.

The above analysis is derived for single-cage induction motors but can be readily extended to double-cage and deep-bar rotors by using slip dependent rotor parameters $X_R(s)$ and $R_R(s)$ in the calculation of the X' and T'_0, as explained by Say (1976) and Arrillaga & Arnold (1990). However in many cases it is sufficient to model the induction motor by its steady-state equivalent circuit; then the electrical torque is obtained from the static torque-speed characteristic and the slip calculated via equation 10.164. As the format of the induction motor dynamic model is similar to that of the synchronous generator, its inclusion in a simulation program is relatively straightforward.

11 Power System Model Reduction — Equivalents

Because contemporary power systems are so large, power system analysis programs do not usually model the complete system in detail. This problem of modelling a large system arises for a number of reasons including:

- practical limitations on the size of computer memory
- the excessive computing time required by large power systems, particularly when running dynamic simulation and stability programs
- parts of the system far away from a disturbance have little effect on the system dynamics and it is therefore unnecessary to model them with great accuracy
- often parts of large, interconnected, systems belong to different utilities, each having their own control centres which treat the other parts of the system as external subsystems
- in some countries private utilities compete with each other and are reluctant to disclose detailed information about their business. This means that vital data regarding the whole system may not be available
- even assuming that the full system data is available, maintaining the relevant databases would be very difficult and expensive

To avoid all these problems only a part of the system, called the *internal subsystem*, is modelled in detail. The remainder of the system, called the *external subsystem*, is represented by simple models referred to as the *equivalent system* or simply as the *equivalent*.

11.1 Types of Equivalents

The methods by which the equivalent of an external subsystem can be produced can be broadly divided into two groups depending on whether or not they require any knowledge of the configuration and the parameters of the external subsystem itself. Methods that do not require any knowledge of the external subsystem are used for on-line security assessment and will not be considered further here but details of these methods can be found in Deville & Schweppe (1972), Debs (1975), Dopazo et al. (1977) and Contaxis &

Debs (1977). Typically these methods use the measurement of certain electrical quantities taken inside the internal subsystem and at the border nodes and tie-lines to form the equivalent. Methods that do require a knowledge of the sub-system are called *model reduction methods*. These methods are used for off-line system analysis and are the subject of this chapter.

Model reduction methods can be further subdivided into the following three subgroups:

- *physical reduction* which consists of choosing appropriate models for the system elements (generators, loads etc.) depending on how influential an individual element is in determining the system response to a particular disturbance. Generally elements electrically close to the disturbance are modelled more accurately whilst elements further away are modelled more simply
- *topological reduction* which consists of eliminating and/or aggregating selected nodes in order to reduce the size of the equivalent network and the number of generating units modelled
- *modal reduction* techniques which use linearised models of the external subsystem to eliminate, or neglect, the unexcited system modes

The equivalent model obtained using modal reduction is in the form of a reduced set of linear differential equations (Undrill & Turner, 1971). This requires extending the standard power system software to take into account the special form of the equivalent. As standardisation of the software is difficult to achieve with modal reduction, this type of equivalent is rarely used in practice.

Topological reduction, used together with physical reduction, gives an equivalent model that comprises of standard system elements such as equivalent generating units, equivalent lines, equivalent nodes etc. Consequently topological equivalents are easy to attach to the internal subsystem model and allow the whole system to be analysed using standard software.

If the topological reduction is performed using one of the methods described in this chapter then the reduced model obtained will generally be a good representation of both the system static performance and the system dynamic performance for the first few seconds following a disturbance. The reduced model can therefore be used for load flow analysis and transient stability analysis when disturbances occur in the internal subsystem.

The division of the whole system into external and internal components is illustrated in Figure 11.1. A reduced model of the external subsystem is created assuming that

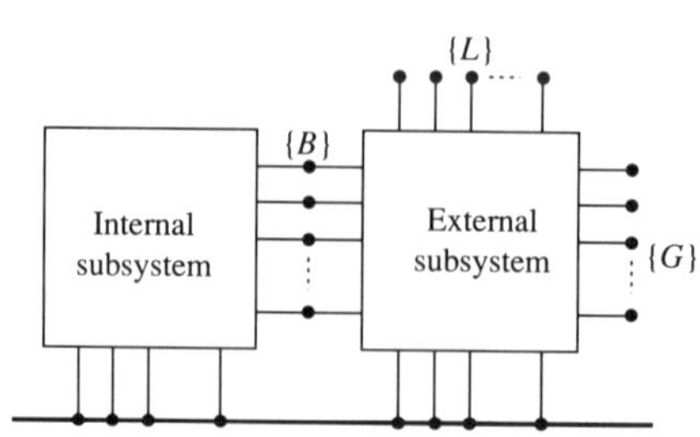

{B} – boundary nodes, {L} – load nodes,
{G} – generator nodes of the external subsystem

Figure 11.1 Internal and external subsystems

the disturbance occurs only inside the internal subsystem. The *border nodes* between the internal and external subsystems are sometimes referred to as the *boundary*, or *tearing*, nodes. Topological reduction consists of transforming a large external network, that consists of load nodes and/or generation nodes, into a smaller network by eliminating and/or aggregating the nodes. Eliminated nodes are removed completely from the network while every group of aggregated nodes is replaced by one equivalent node.

11.2 Network Transformation

Topological reduction is achieved by transforming a large network into a smaller equivalent network by either elimination or aggregation of nodes.

11.2.1 Elimination of nodes

Figure 11.2 illustrates that when nodes are eliminated from the network model, set $\{E\}$, they must be removed in such a way that the currents and nodal voltages at the retained nodes, set $\{R\}$, are unchanged.

Before any nodes are eliminated the network is described by the following nodal equation (see Section 3.5)

$$\begin{bmatrix} \underline{\mathbf{I}}_R \\ \underline{\mathbf{I}}_E \end{bmatrix} = \begin{bmatrix} \underline{\mathbf{Y}}_{RR} & \underline{\mathbf{Y}}_{RE} \\ \underline{\mathbf{Y}}_{ER} & \underline{\mathbf{Y}}_{EE} \end{bmatrix} \begin{bmatrix} \underline{\mathbf{V}}_R \\ \underline{\mathbf{V}}_E \end{bmatrix} \tag{11.1}$$

where the subscripts refer to the eliminated $\{E\}$ and retained $\{R\}$ sets of nodes. The eliminated voltages and currents can be swapped using simple matrix algebra to give

$$\begin{bmatrix} \underline{\mathbf{I}}_R \\ \underline{\mathbf{V}}_E \end{bmatrix} = \begin{bmatrix} \underline{\mathbf{Y}}_R & \underline{\mathbf{K}}_I \\ \underline{\mathbf{K}}_V & \underline{\mathbf{Y}}_{EE}^{-1} \end{bmatrix} \begin{bmatrix} \underline{\mathbf{V}}_R \\ \underline{\mathbf{I}}_E \end{bmatrix} \tag{11.2}$$

where

$$\underline{\mathbf{Y}}_R = \underline{\mathbf{Y}}_{RR} - \underline{\mathbf{Y}}_{RE}\underline{\mathbf{Y}}_{EE}^{-1}\underline{\mathbf{Y}}_{ER}, \quad \underline{\mathbf{K}}_I = \underline{\mathbf{Y}}_{RE}\underline{\mathbf{Y}}_{EE}^{-1}, \quad \underline{\mathbf{K}}_V = -\underline{\mathbf{Y}}_{EE}^{-1}\underline{\mathbf{Y}}_{ER} \tag{11.3}$$

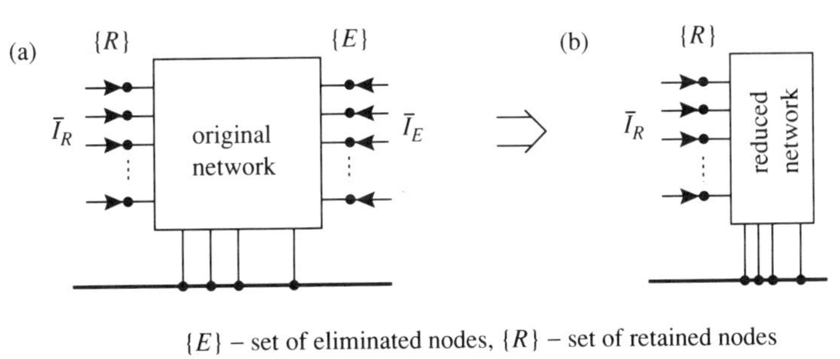

Figure 11.2 Elimination of nodes: (a) network before elimination; (b) network after elimination

The square matrix in equation 11.2 is the *partial inversion of the admittance matrix* and is described in detail in sub-Section 13.2.1. The nodal currents in the set $\{R\}$ are

$$\overline{\mathbf{I}}_R = \overline{\mathbf{Y}}_R \overline{\mathbf{V}}_R + \Delta \overline{\mathbf{I}}_R \tag{11.4}$$

where $\Delta \overline{\mathbf{I}}_R = \overline{\mathbf{K}}_I \overline{\mathbf{I}}_E$.

Equation 11.4 describes the relationship between the currents and voltages of the retained nodes in the reduced network. As any electrical network is uniquely described by its admittance matrix the matrix $\overline{\mathbf{Y}}_R$ corresponds to a reduced equivalent network that consists of the retained nodes and equivalent branches linking them. This network is often referred to as the *transfer network* and the matrix describing it as the *transfer admittance matrix*. Matrix $\overline{\mathbf{K}}_I$ passes the nodal currents from the eliminated nodes to the retained nodes and is referred to as the *distribution matrix*. Each equivalent current is a combination of the eliminated currents.

Another form of equation 11.4 can be obtained by replacing the nodal power injection at each eliminated node by a constant shunt admittance $\overline{Y}_{Ei} = \overline{S}_i^*/V_i^2$ added, with an appropriate sign, to the diagonal elements of the submatrix $\overline{\mathbf{Y}}_{EE}$ (and to the network diagram as a shunt connection). The nodal injections at the eliminated nodes then become zero ($\overline{\mathbf{I}}_E = \mathbf{0}$) and the reduced model does not contain any equivalent currents ($\Delta \overline{\mathbf{I}}_R = 0$). This is quite convenient but has a drawback. The equivalent shunt branches have large conductance values, corresponding to the real power injections, which become part of the equivalent branches in the reduced model. Consequently, the branches of the equivalent network may have a poor X/R ratio causing convergence problems for some load flow computer programs.

Different authors give different names to the elimination of the network nodes using equations 11.2 and 11.4. Edelmann (1974) refers to it as Gauss-Rutishauser elimination while Brown (1975) and Grainger & Stevenson (1994) call the reduced circuit a *Ward equivalent*.

Sparse matrix techniques

Equation 11.4 describes formally the elimination algorithm. In practice sparse matrix techniques are used and the nodes are processed one at a time in order to minimise the complexity and memory requirements of the elimination process (Stott & Hobson, 1971, Tinney & Bright, 1986, Machowski & Bernas, 1989). This is equivalent to Gaussian elimination of a corresponding row and column from the admittance matrix.

Consider one elimination step, that of eliminating node k of set $\{E\}$. Matrix $\overline{\mathbf{Y}}_{EE} = \overline{Y}_{kk}$ is a scalar, $\overline{\mathbf{Y}}_{RE}$ is a column and $\overline{\mathbf{Y}}_{ER}$ is a row. The second component of matrix $\overline{\mathbf{Y}}_R$ becomes

$$\overline{\mathbf{Y}}_{RE}\overline{\mathbf{Y}}_{EE}^{-1}\overline{\mathbf{Y}}_{ER} = \frac{1}{\overline{Y}_{kk}} \begin{bmatrix} \overline{Y}_{1k} \\ \vdots \\ \overline{Y}_{ik} \\ \vdots \\ \overline{Y}_{nk} \end{bmatrix} \begin{bmatrix} \overline{Y}_{k1} \cdots \overline{Y}_{kj} \cdots \overline{Y}_{kn} \end{bmatrix} = \frac{1}{\overline{Y}_{kk}} \begin{bmatrix} & \vdots & \\ \cdots & \overline{Y}_{ik}\overline{Y}_{kj} & \cdots \\ & \vdots & \\ & j & \end{bmatrix} i \tag{11.5}$$

where n is the number of nodes in set $\{R\}$. Assume now that $\overline{Y}_{ij}^{old}$ is an element of matrix $\overline{\mathbf{Y}}_{RR}$ while $\overline{Y}_{ij}^{new}$ is an element of matrix $\overline{\mathbf{Y}}_R$. Equations 11.3 and 11.5 show that elimination of node k modifies each element of the "new" matrix $\overline{\mathbf{Y}}_R$ to

$$\overline{\boldsymbol{Y}}_{ij}^{new} = \overline{\boldsymbol{Y}}_{ij}^{old} - \frac{\overline{\boldsymbol{Y}}_{ik}\overline{\boldsymbol{Y}}_{kj}}{\overline{Y}_{kk}} \quad \text{for} \quad i \neq k, j \neq k \tag{11.6}$$

If node i is directly connected to the eliminated node k then it is called its *neighbour* and the corresponding mutual admittance is $\overline{\boldsymbol{Y}}_{ik} \neq \mathbf{0}$. The mutual admittances $\overline{\boldsymbol{Y}}_{ik}$ of a node i which is not a neighbour of node k are all zero. Equation 11.6 shows that:

- if nodes i and j are not neighbours of node k then elimination of k does not modify the admittance $\overline{\boldsymbol{Y}}_{ij}$
- elimination of node k modifies the admittance between all its neighbours. This creates additional connections between the neighbours replacing the original connections of node k
- the self-admittance of all the neighbours of node k are also modified according to equation (11.6) when $i = j$.

Both situations are illustrated in Figure 11.3. Nodes $\{1, 2, 3\}$ are the neighbours of node k so that its elimination creates additional connections between the nodes. Nodes $\{4, 5\}$ are not neighbours of node k and their connections do not change.

When using sparse matrix techniques, the order in which the rows/columns of a matrix (or nodes of the network) are processed is important from the point of view of preserving the sparsity of the resultant matrix and minimising the number of algebraic manipulations required. Although it is not possible to devise a general optimal elimination ordering strategy, simple heuristic schemes usually work well (Tinney & Walker, 1967, Stott & Hobson, 1971, Brameller, 1976). Typically these node elimination schemes adopt one of the following procedures at each elimination step:

- eliminate the node which has the least number of neighbours
- eliminate the node which introduces the least number of new connections

It is worth noting that the successive elimination defined by equation 11.6 is a generalisation of the well-known star-delta transformation. This can be proved by letting $\overline{Z}_A$, $\overline{Z}_B$ and $\overline{Z}_C$ be the impedances of a star-connected network. Equation 11.6 can then be used

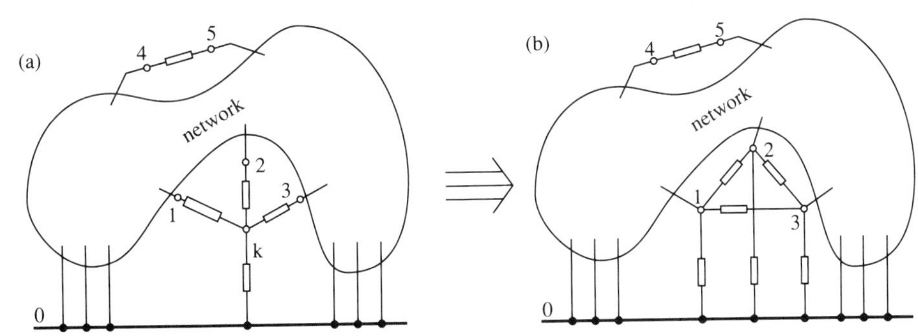

Figure 11.3 Elimination of a single node: (a) situation before elimination; (b) situation after elimination

to derive a set of equations that define the admittances of the equivalent delta connection. The transformation of these admittances to impedances then gives the equivalent impedances in the form $\overline{Z}_{AB} = \overline{Z}_A + \overline{Z}_B + \overline{Z}_A\overline{Z}_B/\overline{Z}_C$.

11.2.2 Aggregation of nodes using Dimo's method

This method is illustrated in Figure 11.4 and consists of replacing a group of nodes $\{A\}$ by an equivalent node a. As before $\{R\}$ is the set of retained nodes.

In the first step of the transformation (Dimo, 1971), some fictitious branches are added to the aggregated nodes, set $\{A\}$. Each branch admittance is chosen in such a way as to make the terminal voltage of all the added branches equal. The terminal equipotential nodes can then be connected together to form a fictitious auxiliary node f. The admittance of each of the fictitious branches can be chosen freely provided that they all give the same terminal voltage. Usually these admittances are made to correspond to the nodal injections (at a given voltage) in the aggregated nodes

$$\overline{Y}_{fi} = \frac{\overline{S}_i^*}{V_i^2} \qquad \text{for } i \in \{A\} \tag{11.7}$$

and then the voltage at the fictitious node f is zero. As it is inconvenient to have an equivalent node operating at zero voltage an extra fictitious branch with negative admittance is usually added to node f. This branch raises the voltage at its terminal node a to a value close to the rated network voltage. A typical choice of the negative admittance is

$$\overline{Y}_{fa} = -\frac{\overline{S}_a^*}{V_a^2} \qquad \text{where } \overline{S}_a = \sum_{i\in\{A\}} \overline{S}_i \tag{11.8}$$

This makes the voltage $\overline{V}_a$ at the equivalent node equal to the weighted average of the voltages at the aggregated nodes:

$$\overline{V}_a = \frac{\sum_{i\in\{A\}} \overline{S}_i}{\sum_{i\in\{A\}} \left(\frac{\overline{S}_i}{\overline{V}_i}\right)^*} \tag{11.9}$$

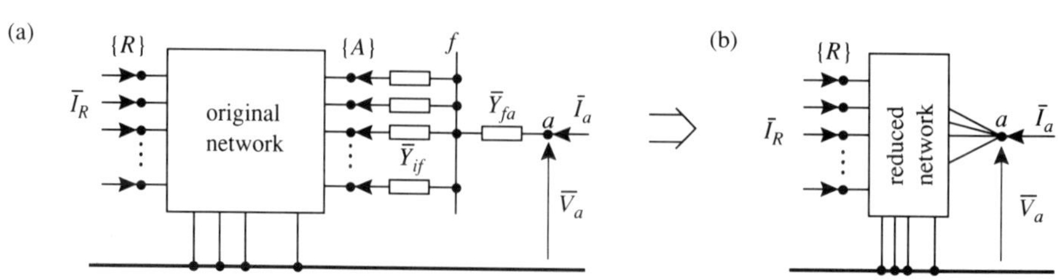

Figure 11.4 Node aggregation using Dimo's method: (a) network with fictitious branches; (b) network after elimination of the nodes and fictitious branches

The auxiliary node f is eliminated together with the nodes belonging to set $\{A\}$ giving an equivalent network, referred to as the *radial equivalent independent (REI)* circuit, connecting the equivalent node a with the retained nodes $\{R\}$. As well as the REI circuit, the elimination process also creates additional connections between the retained nodes.

If the operating conditions are different from the ones for which the reduction was performed then the obtained equivalent will only imitate the external network accurately if the admittances of the fictitious branches, equation 11.7, can be assumed to remain constant. For load nodes this is equivalent to assuming that the loads can be modelled as constant admittances and is only valid for loads with a power-voltage characteristic of the form $\overline{S}_i = V_i^2 \overline{Y}^*_{fi}$, $\overline{Y}_{fi}$ = constant. The generation nodes operate at a constant voltage and the condition $\overline{Y}_{fi} = \overline{S}^*_i / V_i^2$ = constant is only satisfied by those generators where the real and reactive power output can be assumed to be constant.

Dimo's method produces a large number of fictitious branches due to the elimination of node f and nodes $\{A\}$. As aggregation introduces a branch with negative admittance, equation 11.8, the branches in the final network model may have negative admittances. Moreover, large nodal injections in the aggregated nodes produce large resistance values in the equivalent branches, equation 11.7. Negative branch admittances combined with large resistances may cause convergence problems for some load flow programs.

11.2.3 Aggregation of nodes using Zhukov's method

This method of aggregation was first formulated by Zhukov (Жуков, 1964) although a similar approach was also taken by Di Caprio & Marconato (1975) and by De Mello, Podmore & Stanton (1975). The matrix formulation described below was developed by Bernas (1971) but in view of Zhukov's early publication it is referred to here as Zhukov's method.

Aggregation consists of replacing a set of nodes $\{A\}$ by a single equivalent node a as shown in Figure 11.5. $\{R\}$ denotes the set of retained nodes. Aggregation must satisfy the following conditions:

1. It does not change the currents and voltages, $\overline{\mathbf{I}}_R$ and $\overline{\mathbf{V}}_R$, at the retained nodes
2. The real and reactive power injection at the equivalent node must be equal to the sum of injections at the aggregated nodes, $\overline{S}_a = \Sigma_{i \in \{A\}} \overline{S}_i$

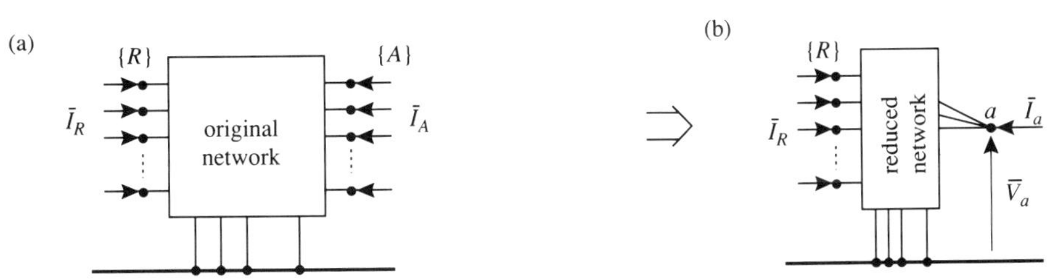

Figure 11.5 Node aggregation using Zhukov's method: (a) network before aggregation; (b) network after aggregation

The transformation of the network can then be described by

$$\begin{bmatrix} \overline{\mathbf{I}}_R \\ \overline{\mathbf{I}}_A \end{bmatrix} = \begin{bmatrix} \overline{\mathbf{Y}}_{RR} & \overline{\mathbf{Y}}_{RA} \\ \overline{\mathbf{Y}}_{AR} & \overline{\mathbf{Y}}_{AA} \end{bmatrix} \begin{bmatrix} \overline{\mathbf{V}}_R \\ \overline{\mathbf{V}}_A \end{bmatrix} \Rightarrow \begin{bmatrix} \overline{\mathbf{I}}_R \\ \overline{I}_a \end{bmatrix} = \begin{bmatrix} \overline{\mathbf{Y}}_{RR} & \overline{\mathbf{Y}}_{Ra} \\ \overline{\mathbf{Y}}_{aR} & \overline{Y}_{aa} \end{bmatrix} \begin{bmatrix} \overline{\mathbf{V}}_R \\ \overline{V}_a \end{bmatrix} \tag{11.10}$$

where the subscripts refer to the appropriate sets. As a is a single node, $\overline{\mathbf{Y}}_{Ra}$ is a column, $\overline{\mathbf{Y}}_{aR}$ is a row and $\overline{Y}_{aa}$ is a scalar.

The first condition is satisfied when

$$\overline{\mathbf{Y}}_{RR}\overline{\mathbf{V}}_R + \overline{\mathbf{Y}}_{RA}\overline{\mathbf{V}}_A = \overline{\mathbf{Y}}_{RR}\overline{\mathbf{V}}_R + \overline{\mathbf{Y}}_{Ra}\overline{V}_a \quad \text{or} \quad \overline{\mathbf{Y}}_{RA}\overline{\mathbf{V}}_A = \overline{\mathbf{Y}}_{Ra}\overline{V}_a \tag{11.11}$$

If this condition is to be satisfied for any vector $\overline{\mathbf{V}}_A$, it must hold that

$$\overline{\mathbf{Y}}_{Ra} = \overline{\mathbf{Y}}_{RA}\overline{\boldsymbol{\vartheta}} \tag{11.12}$$

where $\overline{\boldsymbol{\vartheta}} = \overline{V}_a^{-1}\overline{\mathbf{V}}_A$ is the vector of voltage transformation ratios between the aggregated nodes and the equivalent node.

The second condition is satisfied when

$$\overline{V}_a\overline{I}_a^* = \overline{\mathbf{V}}_A^{\mathrm{T}}\,\overline{\mathbf{I}}_A^* \tag{11.13}$$

where the left-hand-side expresses the injection at the equivalent node and the right-hand-side expresses the sum of all the aggregated injections. Substituting into equation 11.13 for $\overline{I}_a$ and $\overline{\mathbf{I}}_A$ calculated from equation 11.10 gives

$$\overline{V}_a\overline{\mathbf{Y}}_{aR}^*\overline{\mathbf{V}}_R^* + \overline{V}_a\overline{Y}_{aa}^*\overline{V}_a^* = \overline{\mathbf{V}}_A^{\mathrm{T}}\,\overline{\mathbf{Y}}_{AR}^*\overline{\mathbf{V}}_R^* + \overline{\mathbf{V}}_A^{\mathrm{T}}\,\overline{\mathbf{Y}}_{AA}^*\overline{\mathbf{V}}_A^* \tag{11.14}$$

If this equation is to be satisfied for any vector $\overline{\mathbf{V}}_A$ the following two conditions must hold

$$\overline{\mathbf{Y}}_{aR} = \overline{\boldsymbol{\vartheta}}^{*\mathrm{T}}\,\overline{\mathbf{Y}}_{AR} \tag{11.15}$$

$$\overline{\mathbf{Y}}_{aa} = \overline{\boldsymbol{\vartheta}}^{*\mathrm{T}}\,\overline{\mathbf{Y}}_{AA}\overline{\boldsymbol{\vartheta}} \tag{11.16}$$

Equations 11.12, 11.15 and 11.16 describe the admittances of the equivalent network. The admittances of the equivalent branches linking the equivalent node with the retained nodes depend on the vector of transformation ratios $\overline{\boldsymbol{\vartheta}}$, and hence on the voltage angle at the equivalent node. As it is convenient to have equivalent branches of low resistances the voltage angle δ_a at the equivalent node is assumed to be equal to the weighted average of voltage angles at the aggregated nodes

$$\delta_a = \frac{\sum\limits_{i\in\{A\}} S_i\delta_i}{\sum\limits_{i\in\{A\}} S_i} \quad \text{or} \quad \delta'_a = \frac{\sum\limits_{i\in\{A\}} M_i\delta'_i}{\sum\limits_{i\in\{A\}} M_i} \tag{11.17}$$

where S_i is the apparent power injection at the aggregated node i and $M_i = T_{mi}S_{ni}/\omega_s$ is the inertia coefficient of the unit installed at the i-th aggregated node. The first formula can be used for forming the equivalent to be used for steady-state analysis while the second formula can be applied for aggregation of a group of generators represented by the classic transient stability model (constant transient emfs E'_i).

When compared to Dimo's method, the advantage of Zhukov's method is that it does not introduce fictitious branches between the retained nodes $\{R\}$. This is because submatrix $\overline{\mathbf{Y}}_{RR}$ is unchanged by aggregation. It does, however, introduce fictitious shunt branches at the retained nodes. To understand this examine the i-th diagonal element of $\overline{\mathbf{Y}}_{RR}$ which is equal to the sum of admittances of all the series and shunt branches connected to i:

$$(\overline{\mathbf{Y}}_{RR})_{i,i} = \overline{y}_{i0} + \sum_{j \in \{R\}} \overline{y}_{ij} + \sum_{k \in \{A\}} \overline{y}_{ik} \tag{11.18}$$

where $\overline{y}_{i0}$ is the sum of admittances of all shunt branches connected to i and $\overline{y}_{ij}$ is the admittance of a branch linking nodes i and j. During aggregation, all the branches of admittance $\overline{y}_{ik}$ that link node $i \in \{R\}$ with the aggregated nodes $k \in \{A\}$ are replaced by a single branch with admittance $\overline{y}_{ia}$ generally not equal to $\Sigma_{k \in \{A\}} \overline{y}_{ik}$. As $(\overline{\mathbf{Y}}_{RR})_{i,i}$ and $\Sigma_{j \in \{R\}} \overline{y}_{ij}$ must remain unchanged, replacing $\Sigma_{k \in \{A\}} \overline{y}_{ik}$ by $\overline{y}_{ia}$ must be compensated by a change in the value $\overline{y}_{i0}$. The interpretation of this in network terms is that Zhukov's aggregation introduces some equivalent shunt admittances at the retained nodes $\{R\}$.

Electrical interpretation of Zhukov's aggregation

De Mello, Podmore & Stanton (1975) proposed an aggregation method whereby all the nodes to be aggregated are connected together through ideal transformers with transformation ratios that give a common secondary voltage $\overline{V}_a$. This is illustrated in Figure 11.6. With this aggregation method it can easily be shown that the vector of transformation ratios must be equal to $\overline{\boldsymbol{\vartheta}}$ and that the transformation is equivalent to Zhukov's method. This can be proved by constructing the nodal matrix equation for the network shown in Figure 11.6 but without ideal transformers. All the currents and voltages are then transformed to the secondaries of the ideal transformers using the vector of the transformation ratios $\overline{\boldsymbol{\vartheta}}$. The transformers' secondaries are then connected together to create the equivalent node a, and the nodal matrix equation for the reduced network can be derived. Comparing the resulting parameters with equations 11.12, 11.15 and 11.16 completes the proof. The advantage of De Mello's approach is that it gives an electrical interpretation of the mathematical transformations.

Symmetry of the equivalent admittance matrix

If the vector $\overline{\boldsymbol{\vartheta}}$ is complex then Zhukov's equivalent admittance matrix is not generally symmetric ($\overline{\mathbf{Y}}_{aR} \neq \overline{\mathbf{Y}}_{Ra}^{\mathrm{T}}$). This means that if $\overline{Y}_{ia} \neq \overline{Y}_{ai}$ for $i \in \{R\}$ then the value of the

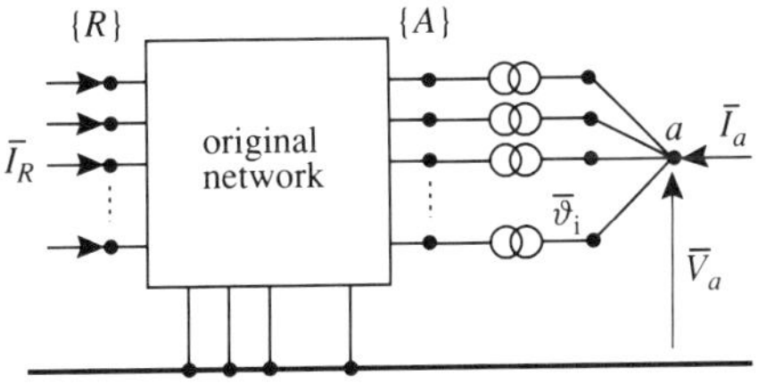

Figure 11.6 Electrical interpretation of Zhukov's aggregation

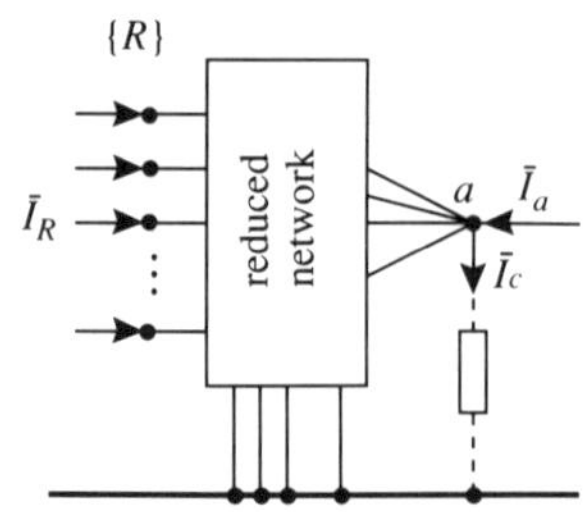

Figure 11.7 Symmetry of the equivalent network

admittances in the equivalent branches obtained after aggregation are direction dependent. From a computational point of view asymmetry of the admittance matrix is inconvenient and Figure 11.7 shows how this asymmetry can be removed by inserting a correction current $\overline{I}_c$ at the equivalent node a. The nodal equation of the system then takes the form

$$\begin{bmatrix} \overline{\mathbf{I}}_R \\ \overline{I}_a \end{bmatrix} = \begin{bmatrix} \overline{\mathbf{Y}}_{RR} & \overline{\mathbf{Y}}_{RA} \\ \overline{\mathbf{Y}}_{Ra}^T & \overline{Y}_{aa} + \dfrac{\overline{I}_c}{\overline{V}_a} \end{bmatrix} \begin{bmatrix} \overline{\mathbf{V}}_R \\ \overline{V}_a \end{bmatrix} \tag{11.19}$$

where $\overline{I}_c = \left[(\overline{\boldsymbol{\vartheta}}^* - \overline{\boldsymbol{\vartheta}})^{\mathrm{T}}\overline{\mathbf{Y}}_{AR}\right]\overline{\mathbf{V}}_R$ is the correction current. This current is not constant as it depends on the voltages in set $\{R\}$. The correction current is small (negligible when compared to $\overline{I}_a$) when the difference $(\overline{\boldsymbol{\vartheta}}^* - \overline{\boldsymbol{\vartheta}})$ is small, that is when the imaginary parts of the transformation ratios are small. This condition is usually satisfied because the angle of the equivalent voltage, equation 11.17, is averaged over the aggregated nodes. Consequently, variations of the correction current can be neglected and a constant current replaced by a constant admittance $(\overline{I}_c/\overline{V}_a)$ added to the self-admittance of the equivalent node a. This has been shown using the dashed line in Figure 11.7.

Coherency

The admittances in the equivalent Zhukov network depend on the transformation ratios $\overline{\vartheta}_i = \overline{V}_i/\overline{V}_a$ between the aggregated nodes $i \in \{A\}$ and the equivalent node a. This means that an equivalent network obtained for an initial (pre-fault) state is only valid for other states (transient or steady-state) if the transformation ratios can be assumed to remain constant for all nodes $i \in \{A\}$ in a given group:

$$\frac{\overline{V}_i(t)}{\overline{V}_a(t)} = \frac{\overline{V}_{i0}}{\overline{V}_{a0}} = \overline{\vartheta}_i = \text{constant} \qquad \text{for } i \in \{A\} \tag{11.20}$$

where subscript "0" denotes the initial state. For any two nodes $i, j \in \{A\}$ this condition is equivalent to

$$\frac{\overline{V}_i(t)}{\overline{V}_j(t)} = \frac{V_i(t)}{V_j(t)} e^{\mathrm{j}[\delta_i(t) - \delta_j(t)]} = \frac{V_{i0}}{V_{j0}} e^{\mathrm{j}[\delta_{i0} - \delta_{j0}]} = \text{constant} \qquad \text{for } i, j \in \{A\} \tag{11.21}$$

Nodes satisfying this condition are referred to as being *electrically coherent*, or simply *coherent*. If the voltage magnitude of the aggregated node can be assumed to be constant (as for PV nodes in the steady-state power flow problem or generators modelled by the classic model in the dynamic analysis), the coherency condition simplifies to

$$\delta_i(t) - \delta_j(t) = \delta_{ij0} \tag{11.22}$$

where $\delta_{ij0} = \delta_{i0} - \delta_{j0}$ and $\delta_{i0} = \delta_i(t = 0)$ are the initial values for which the parameters of the reduced model have been calculated.

Practical experience with power system simulation shows that load nodes are almost never electrically coherent. Only the load nodes very far away from a disturbance maintain constant voltage magnitude and angle. On the other hand it is usually possible to find groups of coherent generation nodes because some groups of generators in the system have a natural tendency to swing together. This means that Zhukov's method is well suited for aggregation of groups of electrically coherent generation nodes.

11.3 Aggregation of Generating Units

The elimination and aggregation of nodes considered so far will produce a reduced network model for use in steady-state analysis. If the reduced model is to be used for dynamic analysis then equivalent generating units must be added to the equivalent nodes. The simplest solution to this problem is to assume that all the generators installed at the aggregated nodes can be modelled by the classic generator model, equation 5.40, that is by $E' =$ constant behind the transient reactance X'_d, and by the rotor swing equation 5.15. As with this model each generator emf behind its respective transient reactance is assumed to be constant, the generator nodes can be aggregated if the coherency condition in equation 11.22 is satisfied by the transient power angles, that is $\delta'_i(t) - \delta'_j(t) = \delta'_{ij0}$. Because these angles define the space position of the generator rotors this condition is referred to as *electromechanical coherency*. Obviously if a group of generators is electromechanically coherent then the generators will remain in synchronism with each other.

An example of the rotor angle variation for three generators is shown in Figure 11.8. Generators i and j are electromechanically coherent as the difference between their rotor angles is almost constant despite both angles undergoing quite deep oscillations. Generator k is not coherent with the other two as its rotor angle variations are completely different.

From the mechanical view point, the rotors of electromechanically coherent generators can be treated as if they rotated on one common rigid shaft, Figure 11.9. A group of such

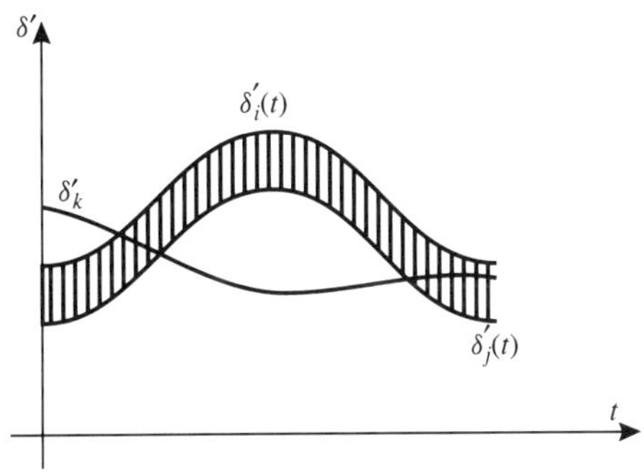

Figure 11.8 Example of variation of rotor angles for three generators

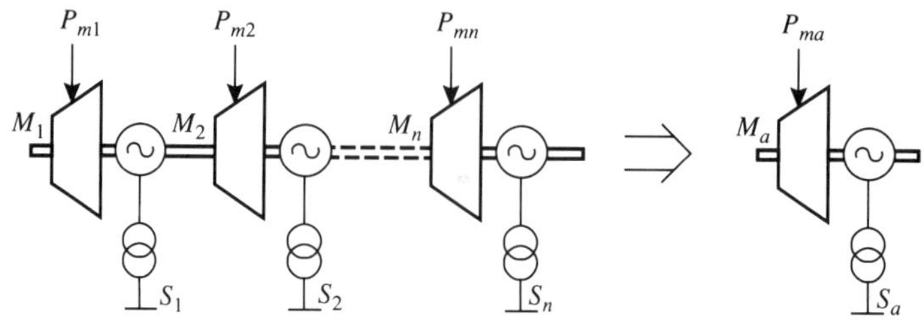

Figure 11.9 Mechanical aggregation of coherent rotors

generators can be replaced by one equivalent generator with inertia coefficient M_a and mechanical power input P_{ma} given by:

$$M_a = \sum_{i \in \{A\}} M_i, \quad P_{ma} = \sum_{i \in \{A\}} P_{mi} \tag{11.23}$$

where M_i is the inertia coefficient and P_{mi} is the mechanical power input of the i-th aggregated generator. This is consistent with Zhukov's aggregation which sets the power injection at the equivalent node equal to the sum of power injections to all the aggregated nodes, equation 11.13.

The equivalent model of a group of electromechanically coherent generation units is therefore created by Zhukov's aggregation of the generation nodes and by replacing the aggregated generators by one equivalent generator with inertia coefficient and mechanical power given by equation 11.23. The equivalent generator is represented by the classic model with constant equivalent transient emf and by the swing equation.

If more detailed models are used then parameters of the equivalent unit can be found by matching the frequency response characteristics of the equivalent unit to the characteristics of the aggregated units. Details can be found in Garmond & Podmore (1978) and Cai Yun-Quing & Wu Ci-Shun (1986).

11.4 Equivalent Model of the External Subsystem

The node elimination and aggregation methods described so far can be used to reduce the model of the external system. Figure 11.1 showed that this can be done by separating the external subsystem from the internal subsystem and by making the injected currents at the boundary nodes equal to the tie line currents. Reduction of the external subsystem consists of three steps: (1) elimination and/or aggregation of the load nodes in the external subsystem; (2) identification of coherent groups of generators; (3) aggregation of the coherent groups. All three steps are briefly described in the following sub-sections.

11.4.1 Elimination/aggregation of load nodes

All the load nodes in the external subsystem can be eliminated using the method described in sub-Section 11.2.1. The resulting external equivalent network is referred to as the *PV equivalent* because, with the exception of the border nodes, it contains only generation nodes (the terminology used with respect to load flow calculations in Section 3.6 referred to such nodes as PV nodes). The power demand of the external system is then distributed among the border nodes. All the generation nodes, and the border nodes, are connected by the equivalent network which is much more dense than the original.

For some power system analysis problems it may be more convenient not to eliminate the load nodes altogether but to replace a few of them by equivalent load nodes using Dimo's aggregation method. These equivalent nodes can then be used to change the power demand of the external subsystem if a change in tie-line flows is required.

11.4.2 Coherency recognition

The topological network equivalent obtained by aggregating nodes will only give valid results if, following a disturbance inside the internal subsystem, the generators within each aggregated group are coherent. The problem is therefore how to assess generator coherency without first completing a detailed dynamic simulation of the complete system for the particular disturbance. Fortunately methods are available that will assess generator coherency without the need for such detailed simulation and this is referred to as *coherency recognition*.

Several methods of coherency recognition have been reported in the literature. The approach described here is based on analysing the parameters of the transfer network that links the border nodes with the generator nodes of the external system. The relevant coherency criteria were originally derived by Machowski (1985) using the theory of non-linear dynamic systems, and confirmed using the modal analysis of the linearised external subsystem (Machowski et al., 1988). As both proofs are rather long and complicated the idea of the method will be described using a single disturbance when the voltage angle at one of the nodes $k \in \{B\}$ is changed. All the aggregated generators are represented by the classic model.

Provided that all the load nodes of the external subsystem have been eliminated, any disturbance inside the internal subsystem influences the generators in the external subsystem through the equivalent branches of the transfer network. According to equation 3.113 if the mutual conductances G_{ij} in the transfer admittance matrix are neglected the real power produced by generator $i \in \{G\}$ in the external subsystem can be expressed as

$$P_i = V_i^2 G_{ii} + \sum_{k \in \{B\}} V_i V_k B_{ik} \sin \delta'_{ik} + \sum_{l \in \{G\}} V_i V_l B_{il} \sin \delta'_{il} \tag{11.24}$$

where V_i for $i \in \{G\}$ is the generator transient emf (that is E'), V_k for $k \in \{B\}$ is the voltage at the border node, $\delta'_{ik} = \delta'_i - \delta_k$, $\delta'_{il} = \delta'_i - \delta_l$ and G_{ii}, B_{ik} and B_{il} are the appropriate elements of the transfer admittance matrix.

Assume that the disturbance is caused by a change in the voltage angle of the border node k from the initial value δ_{k0} to a value $\delta_k = \delta_{k0} + \Delta\delta_k$. Assuming that the voltages at the other nodes are constant, this change in angle will cause a change in the power generation at node $i \in \{G\}$ equal to

$$\Delta P_i(\Delta\delta_k) = b_{ik} \left[\sin(\delta'_{ik0} + \Delta\delta_k) - \sin \delta'_{ik0}\right] \tag{11.25}$$

where $b_{ik} = V_i V_k B_{ik}$ is the maximum power transfer in the equivalent branch linking a generator node $i \in \{G\}$ with a border node $k \in \{B\}$. As $\Delta\delta_k$ is small, it holds that

$\cos\Delta\delta_k \approx 1$ and $\sin\Delta\delta_k \approx \Delta\delta_k$. Expanding the sine term in equation 11.25 gives

$$\Delta P_i(\Delta\delta_k) \approx h_{ik}\Delta\delta_k \tag{11.26}$$

where $h_{ik} = b_{ik}\cos\delta_{ik0}$ is the synchronising power between a given generator $i \in \{G\}$ and a given border node $k \in \{B\}$. The considered disturbance causes the rotor acceleration:

$$\varepsilon_i = \frac{\Delta P_i(\Delta\delta_k)}{M_i} = \frac{h_{ik}}{M_i}\Delta\delta_k \quad \text{for} \quad i \in \{G\}, k \in \{B\} \tag{11.27}$$

where M_i is the inertia coefficient. A similar expression for acceleration can be written for another generator in the external subsystem

$$\varepsilon_j = \frac{\Delta P_j(\Delta\delta_k)}{M_j} = \frac{h_{jk}}{M_j}\Delta\delta_k \quad \text{for} \quad j \in \{G\}, k \in \{B\} \tag{11.28}$$

Generators i and j are electromechanically coherent if their rotor accelerations ε_i and ε_j caused by the disturbance are the same, that is when

$$\frac{h_{ik}}{M_i} = \frac{h_{jk}}{M_j} \quad \text{for} \quad i, j \in \{G\}, k \in \{B\} \tag{11.29}$$

Equation 11.29 constitutes the coherency condition during the post-fault state and means that the synchronising powers divided by the inertia constants must be identical.

In practice, apart from the trivial case of identical generating units operating in parallel on the same busbar, the coherency condition is never exactly satisfied. This is not a significant problem as the simulation of the internal system gives results of satisfactory accuracy if the external subsystem can be replaced by an approximate equivalent. For practical purposes the equality in equation 11.29 can be replaced by the following inequality:

$$\max_{i\in\{G\}} \frac{h_{ik}}{M_i} - \min_{i\in\{G\}} \frac{h_{jk}}{M_j} \le \rho_h \quad \text{for} \quad i, j \in \{G\}, k \in \{B\} \tag{11.30}$$

where ρ_h is a small number determining the admissible error. As generators close to the internal system have a greater influence on the dynamics than do distant ones, ρ_h can be made dependent on the distance a given generator is from the border nodes.

The coherency recognition algorithm works as follows:

1. Determine the transfer admittance matrix for the border nodes $\{B\}$ and all the generation nodes in the external subsystem
2. Mark all the generators of the external subsystem as eligible generators for grouping
3. Order all the equivalent lines in ascending order according to the values of synchronising power. This puts those generators with the strongest inter-connection, and therefore with the greatest probability of coherency, higher in the order
4. Take the next equivalent line with terminal nodes i and j. If there are no lines left, stop the algorithm
5. If the generator i or j is non-eligible for grouping, go to step 4
6. If the criterion defined in equation 11.30 is not satisfied for pair $\{i, j\}$, then go to step 4. Otherwise create group $\{g\}$ consisting of generators $\{i, j\}$

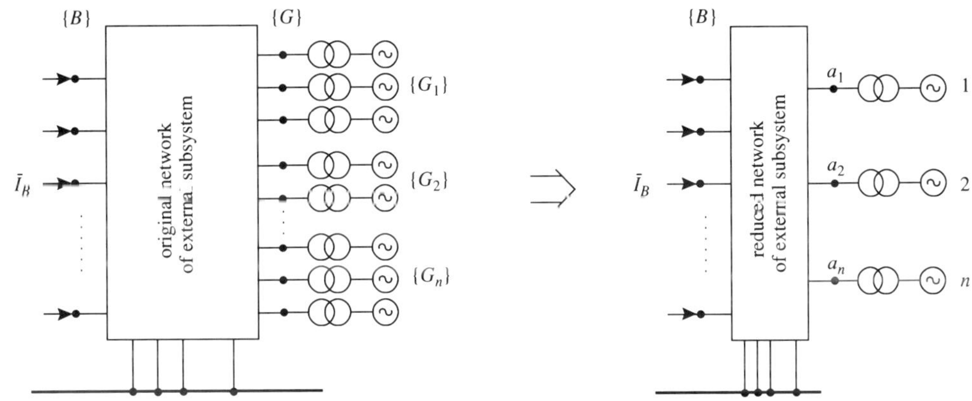

Figure 11.10 Model reduction of the external system

7. Search all the eligible generators for a new generator x which satisfies the criterion in equation 11.30 for the extended group $\{g, x\}$ and gives a minimum value for the left-hand-side of the criterion in equation 11.30. If such a generator cannot be found, store group $\{g\}$ as a new group and go to step 4. Otherwise go to step 8
8. Mark generator x as non-eligible and add it to group $\{g\}$. Go to step 7

The coherency recognition algorithm, test results and recommended values of the error ρ_h can be found in Machowski, 1985 and Machowski et al., 1986 and 1988.

11.4.3 Equivalent generation nodes and generation units

When all the groups of coherent generators in the external subsystems have been identified the next step is to use Zhukov's method to aggregate the nodes in these groups.

Figure 11.10 illustrates the whole process of forming an equivalent model of the external subsystem. The original model of the subsystem contains a large number of load nodes and a large number of generator nodes, set $\{G\} = \{G_1\} + \{G_2\} + \cdots + \{G_n\}$. The load nodes are either completely eliminated or aggregated into a few equivalent nodes using Dimo's method. The generator nodes are divided into the groups of approximately coherent nodes, sets $\{G_1\}, \{G_2\}, \ldots, \{G_n\}$, and each of the groups is replaced by one equivalent node with an equivalent generating unit.

12
Steady-state Stability of Multi-machine System

The *steady-state*, or *small disturbance*, stability of a power system is the ability of the power system to maintain synchronism when subjected to a small disturbance. In Chapter 5 the steady-state stability of the generator-infinite busbar system was discussed and the system seen to be steady-state stable about an equilibrium point if, following a small disturbance, the system remained within a small region surrounding the equilibrium point. In such cases the system is said to be *locally stable*. Furthermore if, as time progresses, the system returns to the equilibrium point, then it is also said to be *asymptotically stable*. These concepts, and those introduced in Chapter 5, will be expanded in this chapter to assess the steady-state stability of a multi-machine power system in which the generators are described by the mathematical models introduced in Chapter 10.

The chapter starts by giving a brief overview of the mathematics required to assess the steady-state stability of a non-linear dynamic system. The eigenvalue analysis was already used in Section 6.6 to analyse torsional shaft oscillations and it will be now formally derived.

12.1 Mathematical Background

Generally a non-linear dynamic system can be described by the differential matrix equation:

$$\dot{\mathbf{x}} = \mathbf{F}(\mathbf{x}) \tag{12.1}$$

where $\mathbf{x}$ is the vector of the n state variables. The equilibrium points $\hat{\mathbf{x}}$ are those points where the system is at rest, that is where all the state variables are constant and their values do not change with time, so that $\mathbf{F}(\hat{\mathbf{x}}) = \mathbf{0}$. Expanding the function $\mathbf{F}(\mathbf{x})$ in the vicinity of $\hat{\mathbf{x}}$ by a Taylor series, and neglecting the non-linear part of the expansion, gives the *linear approximation* of the non-linear equation 12.1 as:

$$\Delta\dot{\mathbf{x}} = \mathbf{A}\Delta\mathbf{x} \tag{12.2}$$

where $\Delta\mathbf{x} = \mathbf{x} - \hat{\mathbf{x}}$ and $\mathbf{A} = \partial\mathbf{F}(\mathbf{x})/\partial\mathbf{x}$ is the Jacobian calculated at the point $\hat{\mathbf{x}}$. Equation 12.2 is known as the *state equation* and the matrix $\mathbf{A}$ as the *plant* or *state matrix*.

Lyapunov's First Method now defines the stability of the non-linear system based on its linear approximation by:

> A non-linear system is steady-state stable in the vicinity of the equilibrium point $\hat{\mathbf{x}}$ if its linear approximation is asymptotically stable. If the linear approximation is unstable, then the non-linear system is also unstable. If the linear approximation is stable, but not asymptotically stable, then it is not possible to assess the system stability based on its linear approximation.

The stability of the linear system is determined by the eigenvalues of the matrix **A**. Let λ_i and $\mathbf{w}_i$ be the *eigenvalues* and *eigenvectors* of the matrix **A**. Then for every pair of eigenvalues and eigenvectors it holds that $\mathbf{A}\mathbf{w}_i = \mathbf{w}_i\lambda_i$ and

$$\mathbf{A}[\mathbf{w_1}, \mathbf{w_2}, \ldots, \mathbf{w_n}] = [\mathbf{w_1}, \mathbf{w_2}, \ldots, \mathbf{w_n}] \begin{bmatrix} \lambda_1 & 0 & \ldots & 0 \\ 0 & \lambda_2 & \ldots & 0 \\ \vdots & \vdots & \ddots & \vdots \\ 0 & 0 & \ldots & \lambda_n \end{bmatrix} \quad \text{or } \mathbf{AW} = \mathbf{W\Lambda} \tag{12.3}$$

where $\mathbf{W} = [\mathbf{w_1}, \mathbf{w_2}, \ldots \mathbf{w_n}]$ is a square matrix whose columns are the eigenvectors of the matrix **A** and $\mathbf{\Lambda} = \text{diag}\lambda_i$ is a diagonal matrix of the corresponding eigenvalues. If all the eigenvalues λ_i are distinct then the corresponding eigenvectors are linearly independent and the matrix **W** is non-singular. Pre-multiplying both sides of equation 12.3 by $\mathbf{W}^{-1}$ gives

$$\mathbf{\Lambda} = \text{diag}\ \lambda_i = \mathbf{W}^{-1}\mathbf{AW} \tag{12.4}$$

To eliminate the cross-coupling between the state variables in the state equation 12.2 the state vector $\Delta\mathbf{x}$ can be transformed into a new state vector **z** using the linear transformation:

$$\Delta\mathbf{x} = \mathbf{Wz}, \qquad \mathbf{z} = \mathbf{W}^{-1}\Delta\mathbf{x} \tag{12.5}$$

Substituting these two equations into equation 12.2 gives $\mathbf{W}\dot{\mathbf{z}} = \mathbf{AWz}$, or $\dot{\mathbf{z}} = \mathbf{W}^{-1}\mathbf{AWz}$ which, on substituting for $\mathbf{\Lambda}$, gives

$$\dot{\mathbf{z}} = \mathbf{\Lambda z} \tag{12.6}$$

Because the matrix Λ is diagonal the matrix equation 12.6 describes a set of uncoupled scalar differential equations

$$\dot{z}_i = \lambda_i z_i \qquad \text{for } i = 1, 2, \ldots, n \tag{12.7}$$

The functions $z_i(t)$ are the solutions to this equation and are referred to as the *mode functions* or *modes* of the linear equation 12.2. Equation 12.6 is the *modal form* of equation 12.2.

Each non-zero initial condition $\Delta x_0 = \Delta x(t = 0^+)$ results in a solution $\Delta x(t)$ which forms a trajectory in the system state space. The transformation in equation 12.5 allows the initial conditions of each of the modes to be found as $\mathbf{z}_0 = \mathbf{W}^{-1}\Delta\mathbf{x}_0$ when the solution of each of the scalar equations 12.7 is then given by

$$z_i(t) = e^{\lambda_i t} z_{i0} \tag{12.8}$$

where z_{i0} is the initial condition of a given mode. When all the modal equations are solved, the solution of equation 12.2 can be found as $\Delta\mathbf{x}(t) = \mathbf{Wz}(t)$. The solution for each state variable is given by a linear combination of the dynamic modes as:

$$\Delta x_i(t) = \sum_{j=1}^{n} w_{ij} e^{\lambda_i t} z_{j0} \tag{12.9}$$

where w_{ij} is the (i, j) element of the matrix $\mathbf{W}$. Consequently if any of the eigenvalues λ_i has a positive real part then there is always a variable $\Delta x_i(t)$ which will tend to infinity and the linear system 12.2 is unstable.

The modal equation 12.6, and the solution of the state equation 12.9, have been derived assuming that all the eigenvalues of matrix $\mathbf{A}$ are distinct, $\lambda_1 \neq \lambda_2 \neq \ldots \neq \lambda_n$. If the matrix $\mathbf{A}$ has multiple eigenvalues then the situation is more complicated. Nevertheless, it can be shown (Willems, 1970) that

> The linear equation $\Delta\dot{\mathbf{x}} = \mathbf{A}\Delta\mathbf{x}$ is stable if, and only if, all the eigenvalues of the matrix $\mathbf{A}$ have non-positive real parts (Re $(\lambda_i) \leq 0$). The system is asymptotically stable if, and only if, all the eigenvalues of the matrix A have negative real parts (Re $(\lambda_i) < 0$).

The stability of a linear equation does not depend on the initial condition but only on the eigenvalues of the matrix $\mathbf{A}$.

The above theorem, together with Lyapunov's first theorem, leads to the conclusion that if the non-linear equation $\dot{\mathbf{x}} = \mathbf{F}(\mathbf{x})$ can be approximated by the linear equation $\Delta\dot{\mathbf{x}} = \mathbf{A}\Delta\mathbf{x}$ then the non-linear system is asymptotically stable if all the eigenvalues of the state matrix $\mathbf{A}$ are negative (Re $(\lambda_i) < 0$). If any of the eigenvalues have a positive real part then the system is unstable. If any of the eigenvalues are zero then no conclusion can be reached with regards the system stability and some other method, such as Lyapunov's second method, must be used. Algorithms for determining the eigenvalues and eigenvectors of a matrix can be found in Press *et al.* (1992).

12.2 Steady-state Stability of Unregulated System

The *inherent steady-state stability* of a power system is concerned with analysing the response of the generators when the system is subjected to a small disturbance and when the effect of the voltage regulators is neglected. This form of stability was first discussed in Section 5.4 for the generator-infinite busbar system where the steady-state stability limit (critical power) was determined by the steady-state model of the generator; that is constant emfs $E_f = E_q$ and $E_d = 0$ behind the synchronous reactances X_d and X_q in conjunction with the swing equation 5.15. Assuming small angle variations this analysis can be expanded for a system of n generators and equation 5.15 can be written for generator i as:

$$\begin{aligned} \frac{\mathrm{d}\Delta\delta_i}{\mathrm{d}t} &= \Delta\omega_i \\ \frac{\mathrm{d}\Delta\omega_i}{\mathrm{d}t} &= -\frac{\Delta P_i}{M_i} - \frac{D_i}{M_i}\Delta\omega_i \end{aligned} \qquad \text{for } i = 1, 2, \ldots, n \tag{12.10}$$

where ΔP_i is the change in the generator real power determined using the incremental network equations 3.114. Equations 3.114 and 12.10 constitute the basic linearised system model suitable for assessing small-signal stability. However the final form of the state equation will depend on additional assumptions regarding the value of the damping coefficients and the load model used.

It should be emphasised that it is the synchronous reactances X_d and X_q that are attached to the network model and not the transient reactances. Chapter 5, Figure 5.6, showed that the steady-state power-angle characteristic P_{E_q} (obtained using reactances X_d and X_q) should be used to check the steady-state stability conditions, while the transient power-angle characteristic should be used to analyse the dynamic response of the generator near the equilibrium point. In Section 12.3 differential equations that describe the transient state will be added to the system equations which will require the transient reactances to be included in the network model.

12.2.1 Constant impedance loads

Computations are considerably simplified if all the system loads are modelled as constant-impedance loads. Further simplifications can be introduced by neglecting steady-state saliency so that for all the generators $X_d = X_q$. The next step is to eliminate all the load nodes, including the generator terminal nodes, in the network model of Figure 12.1 using the method described in Section 11.2. The only retained nodes are the fictitious generator nodes behind the synchronous reactances. The equivalent transfer network directly links all the generator nodes. For this network, equation 3.111 and 3.114 can be used with the voltages V_i replaced by the synchronous emfs E_i. As $\Delta E_i = 0$ equation 3.114 can be simplified to

$$\Delta \mathbf{P} = \mathbf{H} \Delta \delta \tag{12.11}$$

where, according to equation 3.117, the elements of the Jacobian matrix are:

$$\begin{aligned} H_{ij} &= \frac{\partial P_i}{\partial \delta_j} = E_i E_j(-B_{ij} \cos \delta_{ij} + G_{ij} \sin \delta_{ij}) \\ H_{ii} &= \frac{\partial P_i}{\partial \delta_i} = \sum_{j=1}^{n} E_i E_j (B_{ij} \cos \delta_{ij} - G_{ij} \sin \delta_{ij}) \end{aligned} \tag{12.12}$$

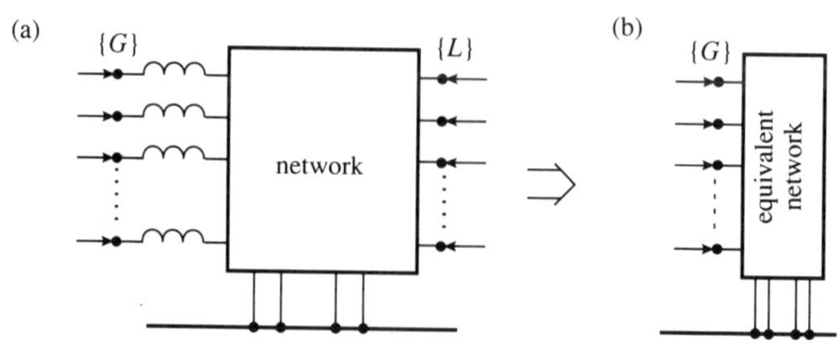

$\{G\}$ is the set of generator nodes behind the synchronous reactances,
$\{L\}$ is the set of load nodes including generator terminals

Figure 12.1 Network model for the steady-state stability analysis: (a) before elimination of load nodes; (b) after elimination

The matrix **H** is singular as the sum of the elements in each of its rows is zero:

$$\left[\sum_{j=1}^{n} H_{ij} = H_{ii} + \sum_{j \neq i}^{n} H_{ij} = 0\right] \tag{12.13}$$

At this stage it is tempting to define the increments in the power angles as the state variables but, as a loss of synchronism does not correspond to a simultaneous increase in all the power angles, this is not a valid choice. Rather, as a loss of synchronism is determined by the relative angles $\Delta\delta_{in}$ calculated with respect to a *reference generator*, it is the increment in these relative angles that must be used as the state variables. This is illustrated in Figure 12.2.

The power change in all the generators must now be related to the relative angles $\Delta\delta_{in}$. Assuming that the last generator, numbered n, acts as the reference, equation 12.13 gives

$$H_{in} = -\sum_{j \neq n}^{n} H_{ij} \tag{12.14}$$

The power change in any of the generators, equation 12.11, can be expressed as

$$\begin{aligned}\Delta P_i &= \sum_{j=1}^{n} H_{ij}\Delta\delta_j = \sum_{j \neq n} H_{ij}\Delta\delta_j + H_{in}\Delta\delta_n = \sum_{j \neq n} H_{ij}\Delta\delta_j - \sum_{j \neq n} H_{ij}\Delta\delta_n \\ &= \sum_{j \neq n} H_{ij}(\Delta\delta_j - \Delta\delta_n) = \sum_{j \neq n} H_{ij}\Delta\delta_{jn}\end{aligned} \tag{12.15}$$

Equation 12.11 can now be expressed as a function of the relative angles by removing the last column of **H**, that is

$$\begin{bmatrix} \Delta P_1 \\ \Delta P_2 \\ \vdots \\ \Delta P_n \end{bmatrix} = \begin{bmatrix} H_{11} & H_{12} & \dots & H_{1,n-1} \\ H_{21} & H_{22} & \dots & H_{2,n-1} \\ \vdots & \vdots & \ddots & \vdots \\ H_{n1} & H_{n2} & \dots & H_{n,n-1} \end{bmatrix} \begin{bmatrix} \Delta\delta_{1n} \\ \Delta\delta_{2n} \\ \vdots \\ \Delta\delta_{n-1,n} \end{bmatrix} \quad \text{or } \Delta\mathbf{P} = \mathbf{H_n}\Delta\boldsymbol{\delta}_\mathbf{n} \tag{12.16}$$

where the matrix $\mathbf{H_n}$ is rectangular and is of dimension $n \times (n-1)$ while the vector $\Delta\boldsymbol{\delta}_\mathbf{n}$ contains $(n-1)$ relative rotor angles calculated with respect to the n-th reference generator.

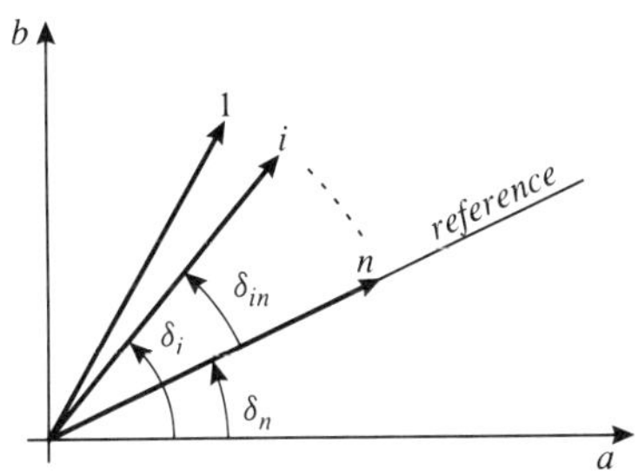

Figure 12.2 Emfs in the complex plane. Finding the relative angles $\delta_{in} = \delta_i - \delta_n$ where n is the reference generator and (a, b) are the rectangular co-ordinates of network equations

Substituting equation 12.16 into the differential equation 12.10 gives

$$\begin{bmatrix} \Delta\dot{\delta}_{1n} \\ \Delta\dot{\delta}_{2n} \\ \vdots \\ \Delta\dot{\delta}_{n-1,n} \\ \hline \Delta\dot{\omega}_1 \\ \Delta\dot{\omega}_2 \\ \vdots \\ \Delta\dot{\omega}_{n-1} \\ \hline \Delta\dot{\omega}_n \end{bmatrix} = \left[\begin{array}{cccc|cccc|c} 0 & 0 & \dots & 0 & 1 & 0 & \dots & 0 & -1 \\ 0 & 0 & \dots & 0 & 0 & 1 & \dots & 0 & -1 \\ \vdots & \vdots & \ddots & \vdots & \vdots & \vdots & \ddots & \vdots & \vdots \\ 0 & 0 & \dots & 0 & 0 & 0 & \dots & 1 & -1 \\ \hline -\frac{H_{11}}{M_1} & -\frac{H_{12}}{M_1} & \dots & -\frac{H_{1,n-1}}{M_1} & -\frac{D_1}{M_1} & 0 & \dots & 0 & 0 \\ -\frac{H_{21}}{M_2} & -\frac{H_{22}}{M_2} & \dots & -\frac{H_{2,n-1}}{M_2} & 0 & -\frac{D_2}{M_2} & \dots & 0 & 0 \\ \vdots & \vdots & \ddots & \vdots & \vdots & \vdots & \ddots & \vdots & \vdots \\ -\frac{H_{n-1,1}}{M_{n-1}} & -\frac{H_{n-1,2}}{M_{n-1}} & \dots & -\frac{H_{n-1,n-1}}{M_{n-1}} & 0 & 0 & \dots & -\frac{D_{n-1}}{M_{n-1}} & 0 \\ \hline -\frac{H_{n1}}{M_n} & -\frac{H_{n2}}{M_n} & \dots & -\frac{H_{n,n-1}}{M_n} & 0 & 0 & \dots & 0 & -\frac{D_n}{M_n} \end{array}\right] \begin{bmatrix} \Delta\delta_{1n} \\ \Delta\delta_{2n} \\ \vdots \\ \Delta\delta_{n-1,n} \\ \hline \Delta\omega_1 \\ \Delta\omega_2 \\ \vdots \\ \Delta\omega_{n-1} \\ \hline \Delta\omega_n \end{bmatrix} \tag{12.17}$$

Equation 12.17 can be expressed more compactly as

$$\begin{bmatrix} \Delta\dot{\boldsymbol{\delta}}_{\mathbf{n}} \\ \hline \Delta\dot{\boldsymbol{\omega}} \end{bmatrix} = \left[\begin{array}{c|c} \mathbf{0} & \mathbf{1}_{-1} \\ \hline -\mathbf{M}^{-1}\mathbf{H}_{\mathbf{n}} & -\mathbf{M}^{-1}\mathbf{D} \end{array}\right] \begin{bmatrix} \Delta\boldsymbol{\delta}_{\mathbf{n}} \\ \hline \Delta\boldsymbol{\omega} \end{bmatrix} \quad \text{or } \Delta\dot{\mathbf{x}} = \mathbf{A}\Delta\mathbf{x} \tag{12.18}$$

where $\mathbf{1}_{-1}$ denotes a diagonal unity matrix extended by a column whose elements are equal to (-1). In this equation the state vector has $(2n-1)$ elements consisting of $(n-1)$ angle changes and n speed deviations. The matrix $\mathbf{A}$ has a rank equal to $(2n-1)$.

For the generator-infinite busbar system equation 12.17 takes the form

$$\begin{bmatrix} \Delta\dot{\delta} \\ \Delta\dot{\omega} \end{bmatrix} = \begin{bmatrix} 0 & 1 \\ -\dfrac{H}{M} & -\dfrac{D}{M} \end{bmatrix} \begin{bmatrix} \Delta\delta \\ \Delta\omega \end{bmatrix} \tag{12.19}$$

The characteristic equation of the state matrix is

$$\lambda^2 + \frac{D}{M}\lambda + \frac{H}{M} = 0 \tag{12.20}$$

and results in two eigenvalues. This equation is identical with the characteristic equation 5.56 but with $H = K_{E_q}$ used instead of the transient synchronising power coefficient $K_{E'}$. This is a consequence of using equation 12.17 or 12.19 to assess the steady-state stability, not the frequency of rotor oscillations. If the classical transient model was used with the generator represented by the transient reactance X'_d then equation 12.17 or 12.19 could be used to determine the frequency of oscillations but not assess the steady-state stability. As shown in Section 5.4, Figure 5.16, the steady-state stability limit is determined by the steady-state characteristic obtained when the generator is represented by the emf E_q acting behind the synchronous reactance $X_d(X_q)$.

Assuming *uniform damping*, that is

$$\frac{D_1}{M_1} = \frac{D_2}{M_2} = \ldots = \frac{D_n}{M_n} = d \tag{12.21}$$

allows one state variable to be eliminated from equation 12.17 as explained below. With uniform damping the swing equation for any machine, and for the reference machine, can be written as:

$$\frac{\mathrm{d}\Delta\omega_i}{\mathrm{d}t} = -\frac{\Delta P_i}{M_i} - d\Delta\omega_i, \qquad \frac{\mathrm{d}\Delta\omega_n}{\mathrm{d}t} = -\frac{\Delta P_n}{M_n} - d\Delta\omega_n \tag{12.22}$$

Subtracting the last equation from the first gives:

$$\frac{\mathrm{d}\Delta\omega_{in}}{\mathrm{d}t} = -\left(\frac{\Delta P_i}{M_i} - \frac{\Delta P_n}{M_n}\right) - d\Delta\omega_{in} \qquad \text{for } i = 1, 2, \ldots, (n-1) \tag{12.23}$$

where $\Delta\omega_{in} = \Delta\omega_i - \Delta\omega_n = \mathrm{d}\delta_{in}/\mathrm{d}t$ and is the rotor speed deviation of the i-th generator relative to the reference machine. Substituting the power changes calculated from equation 12.16 into equation 12.23 gives

$$\begin{bmatrix} \Delta\dot{\boldsymbol{\delta}}_\mathbf{n} \\ \Delta\dot{\boldsymbol{\omega}} \end{bmatrix} = \begin{bmatrix} \mathbf{0} & \mathbf{1} \\ \mathbf{a} & -\mathbf{d} \end{bmatrix} \begin{bmatrix} \Delta\boldsymbol{\delta}_\mathbf{n} \\ \Delta\boldsymbol{\omega}_\mathbf{n} \end{bmatrix} \quad \text{or } \Delta\dot{\mathbf{x}} = \mathbf{A}\Delta\mathbf{x} \tag{12.24}$$

where $\mathbf{1}$ is the unity matrix, $\Delta\boldsymbol{\delta}_\mathbf{n}$ and $\Delta\boldsymbol{\omega}_\mathbf{n}$ are the vectors of the $(n-1)$ relative changes in the power angles and speed deviations, and $\mathbf{d}$ is a diagonal matrix in which all the diagonal elements are identical and equal to d given by equation 12.21. The matrix $\mathbf{a}$ is equal to:

$$\mathbf{a} = \begin{bmatrix} \dfrac{H_{11}}{M_1} - \dfrac{H_{n1}}{M_n} & \dfrac{H_{12}}{M_1} - \dfrac{H_{n2}}{M_n} & \cdots & \dfrac{H_{1,n-1}}{M_1} - \dfrac{H_{n,n-1}}{M_n} \\ \dfrac{H_{21}}{M_2} - \dfrac{H_{n1}}{M_n} & \dfrac{H_{22}}{M_2} - \dfrac{H_{n2}}{M_n} & \cdots & \dfrac{H_{2,n-1}}{M_2} - \dfrac{H_{n,n-1}}{M_n} \\ \vdots & \vdots & \ddots & \vdots \\ \dfrac{H_{n-1,1}}{M_{n-1}} - \dfrac{H_{n1}}{M_n} & \dfrac{H_{n-1,2}}{M_{n-1}} - \dfrac{H_{n2}}{M_n} & \cdots & \dfrac{H_{n-1,n-1}}{M_{n-1}} - \dfrac{H_{n,n-1}}{M_n} \end{bmatrix} \tag{12.25}$$

Comparing the matrix $\mathbf{a}$ with the square state matrix in equation 12.17 shows that the matrix $\mathbf{a}$ is created by subtracting the last row in equation 12.17 from all the upper rows corresponding to the elements H_{ij}. The state vector in equation 12.24 has $(2n-2)$ elements, including $(n-1)$ relative angles $\Delta\delta_{in}$ and $(n-1)$ relative speed deviations $\Delta\omega_{in}$. The problem of finding the eigenvalues of the state matrix in equation 12.24 can now be simplified to that of determining the eigenvalues of the matrix $\mathbf{a}$ which, with rank $(n-1)$, is half the size. This can be explained as follows.

Let λ_i and w_i be the eigenvalues and eigenvectors of the matrix **A** in equation 12.24. The definition of the eigenvalues and eigenvectors gives:

$$\begin{bmatrix} \mathbf{0} & \mathbf{1} \\ \mathbf{a} & -\mathbf{d} \end{bmatrix} \begin{bmatrix} \mathbf{w}'_i \\ \mathbf{w}''_i \end{bmatrix} = \lambda_i \begin{bmatrix} \mathbf{w}'_i \\ \mathbf{w}''_i \end{bmatrix} \tag{12.26}$$

or, after multiplying, $\mathbf{w}''_i = \lambda_i \mathbf{w}'_i$ and $(\mathbf{a}\mathbf{w}'_i - \mathbf{d}\mathbf{w}''_i = \lambda_i \mathbf{w}''_i)$. Substituting $\mathbf{w}''_i$ calculated from the first equation into the second gives $\mathbf{a}\mathbf{w}'_i = (\lambda_i^2 + d\lambda_i)\mathbf{w}'_i$ or $\mathbf{a}\mathbf{w}'_i = \mu_i \mathbf{w}'_i$ where $\mu_i = \lambda_i^2 + d\lambda_i$. Hence

$$\lambda_i^2 + d\lambda_i - \mu_i = 0 \tag{12.27}$$

Obviously μ_i and $\mathbf{w}'_i$ also satisfy the definition of eigenvalues and eigenvectors of the matrix **a**. The eigenvalue λ_i can be found by solving equation 12.27 to give

$$\lambda_i = -\frac{d}{2} \pm \sqrt{\frac{d^2}{4} + \mu_i} \tag{12.28}$$

Thus, knowing the eigenvalues μ_i of the matrix **a** it is possible to determine the eigenvalues λ_i. With this information it is possible to evaluate the system stability knowing the eigenvalues μ_i and the damping coefficient d. Obviously determining the eigenvalues μ_i of the matrix **a**, of rank $(n - 1)$, is less computationally intensive than determining the eigenvalues λ_i of the matrix **A**, of rank $(2n - 2)$. Thus the assumption of uniform damping, if true, considerably simplifies the stability calculations.

12.2.2 Including the voltage characteristics of the loads

The linear equation 12.17 was derived in two stages. First the load nodes in the network admittance model were eliminated, Figure 12.1a, and then the power equations of the equivalent transfer matrix were linearised. By reversing the order of these steps the voltage characteristics of the loads can be included in the model. To accomplish this it is first necessary to linearise the original network model and replace the power increments of the loads by the increments resulting from the voltage characteristics before eliminating the load nodes in the linearised network equation.

As the voltage magnitude at all the generator nodes $\{G\}$ is constant (steady-state representation), all the increments $\Delta E_i = 0$ for $i \in \{G\}$. In this case the incremental network equation 3.114 has the following form for the original network model (Figure 12.1a):

$$\begin{bmatrix} \Delta \mathbf{P}_{\mathrm{G}} \\ \hline \Delta \mathbf{P}_{\mathrm{L}} \\ \Delta \mathbf{Q}_{\mathrm{L}} \end{bmatrix} = \left[\begin{array}{c|cc} \mathbf{H}_{\mathrm{GG}} & \mathbf{H}_{\mathrm{GL}} & \mathbf{M}_{\mathrm{GL}} \\ \hline \mathbf{H}_{\mathrm{LG}} & \mathbf{H}_{\mathrm{LL}} & \mathbf{M}_{\mathrm{LL}} \\ \mathbf{N}_{\mathrm{LG}} & \mathbf{N}_{\mathrm{LL}} & \mathbf{K}_{\mathrm{LL}} \end{array}\right] \begin{bmatrix} \Delta \boldsymbol{\delta}_{\mathrm{G}} \\ \hline \Delta \boldsymbol{\delta}_{\mathrm{L}} \\ \Delta \mathbf{V}_{\mathrm{L}} \end{bmatrix} \tag{12.29}$$

where the submatrices, forming the square matrix, are the partial derivatives of the power with respect to the nodal voltage angles and magnitudes. The indices show which set of nodes, $\{G\}$ or $\{L\}$, a particular submatrix refers to.

Section 3.4 introduced the voltage sensitivities of the loads k_{PV} and k_{QV} so that for a small change in the voltage at each of the load nodes $j \in \{L\}$ it holds that $\Delta P_j = k_{PVj} \Delta V_j$

and $\Delta Q_j = k_{QVj}\Delta V_j$. This can be written for the whole set $\{L\}$ as

$$\Delta \mathbf{P_L} = -\mathbf{k_p}\Delta \mathbf{V_L}, \qquad \Delta Q_L = -\mathbf{k_Q}\Delta \mathbf{V_L} \tag{12.30}$$

where $\mathbf{k_P} =$ diag (k_{PVi}) and $\mathbf{k_Q} =$ diag (k_{QVi}) are diagonal matrices of the voltage sensitivities. The minus sign is present as loads in the network equations represent negative injections, that is outflow from the node.

Substituting equation 12.30 into equation 12.29 gives, after some algebra,

$$\begin{bmatrix} \Delta \mathbf{P_G} \\ \hdashline \mathbf{0} \\ \mathbf{0} \end{bmatrix} = \left[\begin{array}{c:cc} \mathbf{H_{GG}} & \mathbf{H_{GL}} & \mathbf{M_{GL}} \\ \hdashline \mathbf{H_{LG}} & \mathbf{H_{LL}} & \mathbf{M_{LL}} + \mathbf{k_P} \\ \mathbf{N_{LG}} & \mathbf{N_{LL}} & \mathbf{K_{LL}} + \mathbf{k_Q} \end{array}\right] \begin{bmatrix} \Delta \boldsymbol{\delta}_\mathbf{G} \\ \hdashline \Delta \boldsymbol{\delta}_\mathbf{L} \\ \Delta \mathbf{V_L} \end{bmatrix} \tag{12.31}$$

Partial inversion of this equation allows the submatrices $\Delta\boldsymbol{\delta}_\mathbf{L}$, $\Delta \mathbf{V_L}$ to be eliminated to obtain an equation of the same form as equation 12.11 where

$$\mathbf{H} = \mathbf{H_{GG}} - [\mathbf{H_{GL}} \quad \mathbf{M_{GL}}] \begin{bmatrix} \mathbf{H_{LL}} & \mathbf{M_{LL}} + \mathbf{k_P} \\ \mathbf{N_{LL}} & \mathbf{K_{LL}} + \mathbf{k_Q} \end{bmatrix}^{-1} \begin{bmatrix} \mathbf{H_{LG}} \\ \mathbf{N_{LG}} \end{bmatrix} \tag{12.32}$$

The remaining calculations follow the same pattern as described in the previous section with equations 12.17 and 12.26 being equally valid but now with the elements H_{ij} calculated from the matrix defined in equation 12.32.

12.3 Steady-state Stability of the Regulated System

Section 5.5 discussed the steady-state stability of the regulated generator-infinite busbar system. In this section this analysis will be extended to multi-machine systems.

In order to assess the steady-state stability of a regulated power system detailed models of the generators, the exciters, the turbine governors and the power system stabilisers must be used. This leads to a large number of state variables being required to describe each generating unit. As a full description of the full linearised system model would therefore be complicated, and take a lot of space, the approach adopted here is to describe the overall methodology of creating the system model using relatively simple component models. The creation of a more extensive system models would then follow similar lines.

12.3.1 Generator and network

To simplify considerations the fourth-order generator model described in sub-Section 10.1.6 as *Model 4* will be used. For this model it is convenient to neglect transient saliency and attach the generator transient reactance, or the average transient reactance, to the network model. The procedure is similar to that shown in Figure 12.1 with the only difference being that for the purpose of assessing the inherent steady-state stability the generator was represented by the synchronous reactances and synchronous emf $E_q =$ constant. As now variations in the transient emfs have to be considered, the generator reactances attached to the network are the transient, not steady-state, reactances.

After all the load nodes, including the generator terminal nodes, have been eliminated, the generator currents can be expressed in terms of the transient emfs. Working in rectangular co-ordinates the currents can be expressed in the same way as in equation (3.110) to give

$$I_{ai} = \sum_{j=1}^{n}(G_{ij}E'_{aj} - B_{ij}E'_{bj}), \quad I_{bi} = \sum_{j=1}^{n}(G_{ij}E'_{bj} + B_{ij}E'_{aj}) \tag{12.33}$$

The emfs E'_d and E'_q in the individual generator (d, q) coordinate system can be transformed to the (a, b) coordinate system using equation (3.121) to give

$$E'_{aj} = -E'_{dj}\sin\delta_j + E'_{qj}\cos\delta_j, \quad E'_{bj} = E'_{dj}\cos\delta_j + E'_{qj}\sin\delta_j \tag{12.34}$$

whilst the reverse transformation in equation (3.122) can be used for the generator currents

$$I_{di} = -I_{ai}\sin\delta_i + I_{bi}\cos\delta_i, \quad I_{qi} = I_{ai}\cos\delta_i + I_{bi}\sin\delta_i \tag{12.35}$$

It should be noted that the rotor angle δ is the angle of the rotor q-axis measured with respect to the reference frame and not the phase angle of the transient emf E'.

Substituting equations 12.33 and 12.34 into equation 12.35 gives, after some simple but arduous algebra,

$$\begin{aligned} I_{di} &= \sum_{j=1}^{n}\{(B_{ij}\cos\delta_{ij} - G_{ij}\sin\delta_{ij})E'_{qj} + (B_{ij}\sin\delta_{ij} + G_{ij}\cos\delta_{ij})E'_{dj}\} \\ I_{qi} &= \sum_{j=1}^{n}\{(B_{ij}\sin\delta_{ij} + G_{ij}\cos\delta_{ij})E'_{qj} - (B_{ij}\cos\delta_{ij} - G_{ij}\sin\delta_{ij})E'_{dj}\} \end{aligned} \tag{12.36}$$

These are the transfer network equations in the (d, q) co-ordinates of the individual generators.

Similarly, as for the power increment in equation 12.15, equation 12.36 allows the current increments to be expressed in terms of the increments in the relative angles $\Delta\delta_{jn}$ and the increments of the component emfs as

$$\begin{bmatrix} \Delta\mathbf{I_q} \\ \Delta\mathbf{I_d} \end{bmatrix} = \begin{bmatrix} \dfrac{\partial\mathbf{I_q}}{\partial\mathbf{E'_q}} & \dfrac{\partial\mathbf{I_q}}{\partial\mathbf{E'_d}} & \dfrac{\partial\mathbf{I_q}}{\partial\boldsymbol{\delta}_\mathbf{n}} \\ \dfrac{\partial\mathbf{I_d}}{\partial\mathbf{E'_q}} & \dfrac{\partial\mathbf{I_d}}{\partial\mathbf{E'_d}} & \dfrac{\partial\mathbf{I_d}}{\partial\boldsymbol{\delta}_\mathbf{n}} \end{bmatrix} \begin{bmatrix} \Delta\mathbf{E'_q} \\ \Delta\mathbf{E'_d} \\ \Delta\boldsymbol{\delta}_\mathbf{n} \end{bmatrix} \tag{12.37}$$

where $\Delta\mathbf{I_q}$, $\Delta\mathbf{I_d}$, $\Delta\mathbf{E'_q}$, $\Delta\mathbf{E'_d}$, and $\Delta\boldsymbol{\delta}_\mathbf{n}$ are the appropriate generator increment vectors. The elements of the Jacobian matrix can be found by differentiating the right-hand-side of equations 12.36 and will not be considered further here. However it should be remembered that vector $\Delta\boldsymbol{\delta}_\mathbf{n}$ is of size $(n-1)$ so that the sub-matrices $\partial\mathbf{I_q}/\partial\boldsymbol{\delta}_\mathbf{n}$ and $\partial\mathbf{I_d}/\partial\boldsymbol{\delta}_\mathbf{n}$ are rectangular with the dimensions $n \times (n-1)$.

If transient saliency is neglected $X'_q \approx X'_d$ and equation 10.103 simplifies to $P_i = E'_{qi}I_{qi} + E'_{di}I_{di}$ which, after substituting for I_{di} and I_{qi} from equation 12.36, yields

$$
\begin{aligned}
P_i = E'_{di} \sum_{j=1}^{n} \{(B_{ij}\cos\delta_{ij} - G_{ij}\sin\delta_{ij})E'_{qj} + (B_{ij}\sin\delta_{ij} + G_{ij}\cos\delta_{ij})E'_{dj}\} \\
+ E'_{qi} \sum_{j=1}^{n} \{(B_{ij}\sin\delta_{ij} + G_{ij}\cos\delta_{ij})E'_{qj} - (B_{ij}\cos\delta_{ij} - G_{ij}\sin\delta_{ij})E'_{dj}\}
\end{aligned}
\tag{12.38}
$$

Linearising and expressing this equation in matrix form gives

$$
[\Delta \mathbf{P}] = \begin{bmatrix} \dfrac{\partial \mathbf{P}}{\partial \mathbf{E'_q}} & \dfrac{\partial \mathbf{P}}{\partial \mathbf{E'_d}} & \dfrac{\partial \mathbf{P}}{\partial \boldsymbol{\delta}_\mathbf{n}} \end{bmatrix} \begin{bmatrix} \Delta \mathbf{E'_q} \\ \Delta \mathbf{E'_d} \\ \Delta \boldsymbol{\delta}_\mathbf{n} \end{bmatrix}
\tag{12.39}
$$

where $\Delta\mathbf{P}$ is the vector of the power increments in all the generators and $\partial\mathbf{P}/\partial\boldsymbol{\delta}_\mathbf{n}$ is a rectangular Jacobian matrix of dimension $n \times (n-1)$. The Jacobian submatrices are calculated by differentiating the right-hand-side of equation 12.38.

The terminal voltage equation 10.102 can be treated in a similar way. Substituting equation 12.36 into equation 10.102 gives

$$
\begin{aligned}
V_{qi} = E'_{qi} + X'_{di} \sum_{j=1}^{n} \{(B_{ij}\cos\delta_{ij} - G_{ij}\sin\delta_{ij})E'_{qj} + (B_{ij}\sin\delta_{ij} + G_{ij}\cos\delta_{ij})E'_{dj}\} \\
V_{di} = E'_{di} - X'_{qi} \sum_{j=1}^{n} \{(B_{ij}\sin\delta_{ij} + G_{ij}\cos\delta_{ij})E'_{qj} - (B_{ij}\cos\delta_{ij} - G_{ij}\sin\delta_{ij})E'_{dj}\}
\end{aligned}
\tag{12.40}
$$

The quantity of interest with regard to voltage regulator action is the voltage magnitude $V_i = \sqrt{V_{qi}^2 + V_{di}^2}$ so that

$$
[\Delta \mathbf{V}] = \begin{bmatrix} \dfrac{\partial \mathbf{V}}{\partial \mathbf{E'_q}} & \dfrac{\partial \mathbf{V}}{\partial \mathbf{E'_d}} & \dfrac{\partial \mathbf{V}}{\partial \boldsymbol{\delta}_\mathbf{n}} \end{bmatrix} \begin{bmatrix} \Delta \mathbf{E'_q} \\ \Delta \mathbf{E'_d} \\ \Delta \boldsymbol{\delta}_\mathbf{n} \end{bmatrix}
\tag{12.41}
$$

where $\Delta\mathbf{V}$ is the vector of the magnitude increments in the terminal voltages of all the generators. The Jacobian submatrices can be calculated by taking the partial derivative of the voltage magnitude with respect to a variable α as

$$
\frac{\partial V_i}{\partial \alpha} = \frac{1}{V_i}\left(V_{qi}\frac{\partial V_{qi}}{\partial \alpha} + V_{di}\frac{\partial V_{di}}{\partial \alpha} \right)
\tag{12.42}
$$

where α signifies E'_{qi}, E'_{qj}, E'_{di}, E'_{dj}, δ_i, or δ_j. The derivatives $\partial V_{qi}/\partial\alpha$ and $\partial V_{di}/\partial\alpha$ are obtained by differentiating the right-hand-side of equations 12.40.

The linearised generator differential equations can now be obtained from equations 10.98, 10.99, and 10.89 as:

$$
\begin{aligned}
\mathbf{T'_{d0}}\Delta\dot{\mathbf{E}}'_\mathbf{q} &= -\Delta\mathbf{E'_q} + \Delta\mathbf{X}\Delta\mathbf{I_d} + \Delta\mathbf{E_f} \\
\mathbf{T'_{q0}}\Delta\dot{\mathbf{E}}'_\mathbf{d} &= -\Delta\mathbf{E'_d} + \Delta\mathbf{X}\Delta\mathbf{I_q} \\
\mathbf{M}\Delta\dot{\boldsymbol{\omega}} &= -\Delta\mathbf{P} - \mathbf{D}\Delta\boldsymbol{\omega} \\
\Delta\dot{\boldsymbol{\delta}}_\mathbf{n} &= \mathbf{1}_{-1}\Delta\boldsymbol{\omega}
\end{aligned}
\tag{12.43}
$$

where $\Delta \mathbf{E}'_\mathbf{q}$, $\Delta \mathbf{E}'_\mathbf{d}$, $\Delta \mathbf{I_q}$, $\Delta \mathbf{I_d}$, $\Delta \mathbf{P}$, and $\Delta \boldsymbol{\omega}$ are the generator increment vectors of size n whilst $\Delta \boldsymbol{\delta}_\mathbf{n}$ is the increment vector of the relative angles of size $(n-1)$. The matrices $\mathbf{T}'_{\mathbf{q0}}$, $\mathbf{T}'_{\mathbf{d0}}$, $\mathbf{M}$, and $\mathbf{D}$ are the respective diagonal matrices of the generator time constants, inertia coefficients and damping coefficients. The matrix $\Delta \mathbf{X}$ is also diagonal with values $\Delta X_i = (X_{di} - X'_{di}) \approx (X_{qi} - X'_{qi})$. Substituting equations 12.37, 12.39 and 12.41 into equation 12.43 gives:

$$\begin{bmatrix} \Delta \dot{\mathbf{E}}'_\mathbf{q} \\ \Delta \dot{\mathbf{E}}'_\mathbf{d} \\ \Delta \dot{\boldsymbol{\delta}}_\mathbf{n} \\ \Delta \dot{\boldsymbol{\omega}} \end{bmatrix} = \begin{bmatrix} (\mathbf{T}'_{\mathbf{d0}})^{-1}\left[\Delta\mathbf{X}\left(\dfrac{\partial \mathbf{I_d}}{\partial \mathbf{E}'_\mathbf{q}}\right) - \mathbf{1}\right] & (\mathbf{T}'_{\mathbf{d0}})^{-1}\Delta\mathbf{X}\left(\dfrac{\partial \mathbf{I_d}}{\partial \mathbf{E}'_\mathbf{d}}\right) & (\mathbf{T}'_{\mathbf{d0}})^{-1}\Delta\mathbf{X}\left(\dfrac{\partial \mathbf{I_d}}{\partial \boldsymbol{\delta}_\mathbf{n}}\right) & \mathbf{0} \\ -(\mathbf{T}'_{\mathbf{q0}})^{-1}\Delta\mathbf{X}\left(\dfrac{\partial \mathbf{I_q}}{\partial \mathbf{E}'_\mathbf{q}}\right) & -(\mathbf{T}'_{\mathbf{q0}})^{-1}\left[\Delta\mathbf{X}\left(\dfrac{\partial \mathbf{I_q}}{\partial \mathbf{E}'_\mathbf{d}}\right) - \mathbf{1}\right] & (\mathbf{T}'_{\mathbf{q0}})^{-1}\Delta\mathbf{X}\left(\dfrac{\partial \mathbf{I_q}}{\partial \boldsymbol{\delta}_\mathbf{n}}\right) & \mathbf{0} \\ \mathbf{0} & \mathbf{0} & \mathbf{0} & \mathbf{1}_{-1} \\ -\mathbf{M}^{-1}\left(\dfrac{\partial \mathbf{P}}{\partial \mathbf{E}'_\mathbf{q}}\right) & -\mathbf{M}^{-1}\left(\dfrac{\partial \mathbf{P}}{\partial \mathbf{E}'_\mathbf{d}}\right) & -\mathbf{M}^{-1}\left(\dfrac{\partial \mathbf{P}}{\partial \boldsymbol{\delta}\mathbf{n}}\right) & -\mathbf{M}^{-1}\mathbf{D} \end{bmatrix} \begin{bmatrix} \Delta \mathbf{E}'_\mathbf{q} \\ \Delta \mathbf{E}'_\mathbf{d} \\ \Delta \boldsymbol{\delta}_\mathbf{n} \\ \Delta \boldsymbol{\omega} \end{bmatrix} + \begin{bmatrix} (\mathbf{T}'_{\mathbf{d0}})^{-1}\Delta \mathbf{E_f} \\ \mathbf{0} \\ \mathbf{0} \\ \mathbf{0} \end{bmatrix} \tag{12.44}$$

The four submatrices in the bottom right hand corner of the state matrix have the same structure as the state matrix used in equation 12.19 to assess the steady-state inherent stability. The changes in E_f, appearing in the last component of equation 12.44, are due to the action of the generator exciters and AVR systems.

12.3.2 Including the exciter, AVR and the PSS

Section 10.2 derived models for the different types of excitation systems. For steady-state stability purposes non-linearity effects do not play a major role and may be neglected allowing the excitation systems to be represented by a high-order transfer function. To show how the linearised equations of the excitation system can be included into the complete linearised system model a second-order transfer function will be used to represent the exciter and the AVR while a third-order transfer function will be used to represent the PSS. The block diagram corresponding to such an excitation system is shown in Figure 12.3. In this system the PSS will react to a signal proportional to rotor speed deviation, when $K_P = 0$, or to a signal proportional to the generator real power, when $K_\omega = 0$.

The differential equation describing the excitation system can be obtained from the transfer function in Figure 12.3 by introducing two auxiliary variables, x_1 and x_2, to give:

$$\begin{aligned} a_{2i}\dot{x}_{2i} &= \Delta z_i - a_{1i}x_{2i} - a_{0i}x_{1i} \\ \dot{x}_{1i} &= x_{2i} \end{aligned} \tag{12.45}$$

$$\Delta E_{fi} = b_{1i}x_{2i} + b_{0i}x_{1i} \tag{12.46}$$

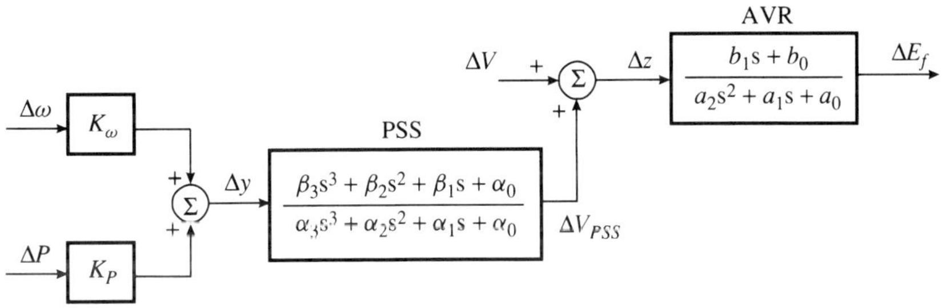

Figure 12.3 Simplified block diagram of the exciter with AVR and PSS

where i is the generator number. Expressing this equation in matrix form, and substituting equation 12.41, gives

$$\begin{bmatrix} \dot{\mathbf{x}}_1 \\ \dot{\mathbf{x}}_2 \end{bmatrix} = \begin{bmatrix} \mathbf{0} & \mathbf{1} \\ -\mathbf{a}_2^{-1}\mathbf{a}_0 & -\mathbf{a}_2^{-1}\mathbf{a}_1 \end{bmatrix} \begin{bmatrix} \mathbf{x}_1 \\ \mathbf{x}_2 \end{bmatrix}$$

$$+ \begin{bmatrix} \mathbf{0} & \mathbf{0} & \mathbf{0} \\ -\mathbf{a}_2^{-1}\left(\dfrac{\partial \mathbf{V}}{\partial \mathbf{E}'_\mathbf{q}}\right) & -\mathbf{a}_2^{-1}\left(\dfrac{\partial \mathbf{V}}{\partial \mathbf{E}'_\mathbf{d}}\right) & -\mathbf{a}_2^{-1}\left(\dfrac{\partial \mathbf{V}}{\partial \boldsymbol{\delta}_\mathbf{n}}\right) \end{bmatrix} \begin{bmatrix} \Delta\mathbf{E}'_\mathbf{q} \\ \Delta\mathbf{E}'_\mathbf{d} \\ \Delta\boldsymbol{\delta}_\mathbf{n} \end{bmatrix} + \begin{bmatrix} 0 \\ -\mathbf{a}_2^{-1}\Delta\mathbf{V}_\mathbf{PSS} \end{bmatrix}$$

$$[\Delta\mathbf{E}_\mathbf{f}] = [\,\mathbf{b}_0 \quad \mathbf{b}_1\,] \begin{bmatrix} \mathbf{x}_1 \\ \mathbf{x}_2 \end{bmatrix} \tag{12.47}$$

where $\mathbf{a}_0, \mathbf{a}_1, \mathbf{a}_2, \mathbf{b}_0, \mathbf{b}_1$ are diagonal submatrices of the coefficients in the excitation system transfer functions. The vector $\Delta\mathbf{V}_\mathbf{PSS}$ can be obtained from the stabiliser equations which, as the stabiliser transfer function is third order, can be constructed using three auxiliary variables x_3, x_4 and x_5. This results in

$$\begin{aligned} \alpha_{3i}\dot{x}_{5i} &= \Delta y_i - \alpha_{2i}x_{5i} - \alpha_{1i}x_{4i} - \alpha_{0i}x_{3i} \\ \dot{x}_{4i} &= x_{5i} \\ \Delta V_{PSS} &= \frac{\beta_{3i}}{\alpha_{3i}}\Delta y_i - \left(\frac{\alpha_{2i}}{\alpha_{3i}}\beta_{3i} - \beta_{2i}\right)x_{5i} - \left(\frac{\alpha_{1i}}{\alpha_{3i}}\beta_{3i} - \beta_{1i}\right)x_{4i} - \frac{\beta_{3i}}{\alpha_{3i}}x_{3i} \end{aligned} \tag{12.48}$$

where $\Delta y_i = K_{\omega i}\Delta\omega_i + K_{Pi}\Delta P_i$ is the input signal and i is the generator number. Representing the equations in the matrix form and taking into account equation 12.39 gives

$$\begin{bmatrix} \dot{\mathbf{x}}_3 \\ \dot{\mathbf{x}}_4 \\ \dot{\mathbf{x}}_5 \end{bmatrix} = \begin{bmatrix} \mathbf{0} & \mathbf{1} & \mathbf{0} \\ \mathbf{0} & \mathbf{0} & \mathbf{1} \\ -\boldsymbol{\alpha}_3^{-1}\boldsymbol{\alpha}_0 & -\boldsymbol{\alpha}_3^{-1}\boldsymbol{\alpha}_1 & -\boldsymbol{\alpha}_3^{-1}\boldsymbol{\alpha}_2 \end{bmatrix} \begin{bmatrix} \mathbf{x}_3 \\ \mathbf{x}_4 \\ \mathbf{x}_5 \end{bmatrix}$$

$$+ \begin{bmatrix} \mathbf{0} & \mathbf{0} & \mathbf{0} & \mathbf{0} \\ \mathbf{0} & \mathbf{0} & \mathbf{0} & \mathbf{0} \\ -\boldsymbol{\alpha}_3^{-1}\mathbf{K}_\mathbf{P}\left(\dfrac{\partial \mathbf{P}}{\partial \mathbf{E}'_\mathbf{q}}\right) & -\boldsymbol{\alpha}_3^{-1}\mathbf{K}_\mathbf{P}\left(\dfrac{\partial \mathbf{P}}{\partial \mathbf{E}'_\mathbf{d}}\right) & -\boldsymbol{\alpha}_3^{-1}\mathbf{K}_\mathbf{P}\left(\dfrac{\partial \mathbf{P}}{\partial \boldsymbol{\delta}_\mathbf{n}}\right) & -\boldsymbol{\alpha}_3^{-1}\mathbf{K}_\omega \end{bmatrix} \begin{bmatrix} \Delta\mathbf{E}'_\mathbf{q} \\ \Delta\mathbf{E}'_\mathbf{d} \\ \Delta\boldsymbol{\delta}_\mathbf{n} \\ \Delta\boldsymbol{\omega} \end{bmatrix}$$

$$[\Delta \mathbf{V}_{PSS}] = \left[\mathbf{K}_P \boldsymbol{\alpha}_3^{-1} \boldsymbol{\beta}_3 \left(\frac{\partial \mathbf{P}}{\partial \mathbf{E}'_q} \right) \middle| \mathbf{K}_P \boldsymbol{\alpha}_3^{-1} \boldsymbol{\beta}_3 \left(\frac{\partial \mathbf{P}}{\partial \mathbf{E}'_d} \right) \middle| \mathbf{K}_P \boldsymbol{\alpha}_3^{-1} \boldsymbol{\beta}_3 \left(\frac{\partial \mathbf{P}}{\partial \boldsymbol{\delta}_n} \right) \middle| \mathbf{K}_\omega \boldsymbol{\alpha}_3^{-1} \boldsymbol{\beta}_3 \middle| -\boldsymbol{\alpha}_3^{-1} \boldsymbol{\beta}_3 \middle| \boldsymbol{\beta}_1 - \boldsymbol{\alpha}_3^{-1} \boldsymbol{\alpha}_1 \boldsymbol{\beta}_3 \middle| \boldsymbol{\beta}_2 - \boldsymbol{\alpha}_3^{-1} \boldsymbol{\alpha}_2 \boldsymbol{\beta}_3 \right] \begin{bmatrix} \Delta \mathbf{E}'_q \\ \Delta \mathbf{E}'_d \\ \Delta \boldsymbol{\delta}_n \\ \Delta \boldsymbol{\omega} \\ \mathbf{x}_3 \\ \mathbf{x}_4 \\ \mathbf{x}_5 \end{bmatrix} \tag{12.49}$$

where $\boldsymbol{\alpha}_0$, $\boldsymbol{\alpha}_1$, $\boldsymbol{\alpha}_2$, $\boldsymbol{\alpha}_3$, $\boldsymbol{\beta}_1$, $\boldsymbol{\beta}_2$, $\boldsymbol{\beta}_3$, $\mathbf{K}_P$, and $\mathbf{K}_\omega$ are diagonal matrices of the stabiliser coefficients.

12.3.3 Linear state equation of the system

As the linear equations 12.44, 12.47, and 12.49 have common state variables they can be combined to form one large state equation that includes the effect of all the generators (fourth-order model), exciters, AVRs and PSSs. This results in the matrix equation 12.50. So far the turbine power has been assumed to be constant. If this is not the case then additional equations that describe the turbine and its governor must be added to the state equation further increasing its size. The size of the state equation will be increased yet further if the effect of control devices, such as SVC or other FACTS devices are included. Provided a suitable model is known then including the effect of such elements is straightforward and follows similar lines as for the exciter and AVR and for this reason is not considered further in this book.

As the state matrix in equation 12.50 is sparse it is advantageous to use sparse matrix techniques when calculating the eigenvalues and eigenvectors even if a small system is under investigation. In the case of a large interconnected system, when all the control devices influencing the steady-state stability and damping are included, then the dimension of the state matrix may be very large indeed and well outside the range of conventional eigenvalue analysis methods. In such cases special solution techniques, which evaluate a selected subset of the eigenvalues associated with the complete system response, are necessary (Kundur, 1994).

12.4 Sensitivity Analysis

By evaluating the eigenvalues of the state matrix Lyapunov's First Method allows the stability of a system to small disturbances to be established. If the eigenvectors are also evaluated then equation 12.5 allows the influence of each mode on a particular state variable to be predicted and vice versa. This latter technique has important applications with regard to establishing the sensitivity of the system to parameter changes as will now be discussed.

The first of equations 12.5, $\Delta \mathbf{x}(t) = \mathbf{W}\mathbf{z}(t)$, shows that each state variable is a linear combination of all the modes

$$\Delta x_i(t) = w_{i1} z_1(t) + w_{i2} z_2(t) + \ldots + w_{ij} z_j(t) + \ldots + w_{in} z_n(t) \tag{12.51}$$

$$
\begin{bmatrix} \Delta\dot{\mathbf{E}}'_q \\ \Delta\dot{\mathbf{E}}'_d \\ \Delta\dot{\delta}_n \\ \Delta\dot{\omega} \\ \dot{x}_1 \\ \dot{x}_2 \\ \dot{x}_3 \\ \dot{x}_4 \\ \dot{x}_5 \end{bmatrix}
=
\begin{bmatrix}
(\mathbf{T}'_{d0})^{-1}\left[\Delta\mathbf{X}\left(\frac{\partial \mathbf{I}_d}{\partial \mathbf{E}'_q}\right)-\mathbf{1}\right] & (\mathbf{T}'_{d0})^{-1}\Delta\mathbf{X}\left(\frac{\partial \mathbf{I}_d}{\partial \mathbf{E}'_d}\right) & (\mathbf{T}'_{d0})^{-1}\left(\frac{\partial \mathbf{I}_d}{\partial \delta_n}\right) & 0 & (\mathbf{T}'_{d0})^{-1}\mathbf{b}_0 & (\mathbf{T}'_{d0})^{-1}\mathbf{b}_1 & 0 & 0 & 0 \\
-(\mathbf{T}'_{q0})^{-1}\Delta\mathbf{X}\left(\frac{\partial \mathbf{I}_q}{\partial \mathbf{E}'_q}\right) & -(\mathbf{T}'_{q0})^{-1}\left[\Delta\mathbf{X}\left(\frac{\partial \mathbf{I}_q}{\partial \mathbf{E}'_d}\right)-\mathbf{1}\right] & -(\mathbf{T}'_{q0})^{-1}\left(\frac{\partial \mathbf{I}_q}{\partial \delta_n}\right) & 0 & 0 & 0 & 0 & 0 & 0 \\
0 & 0 & 0 & \mathbf{1}_{-1} & 0 & 0 & 0 & 0 & 0 \\
-\mathbf{M}^{-1}\left(\frac{\partial \mathbf{P}}{\partial \mathbf{E}'_q}\right) & -\mathbf{M}^{-1}\left(\frac{\partial \mathbf{P}}{\partial \mathbf{E}'_d}\right) & -\mathbf{M}^{-1}\left(\frac{\partial \mathbf{P}}{\partial \delta_n}\right) & -\mathbf{M}^{-1}\mathbf{D} & 0 & 0 & 0 & 0 & 0 \\
0 & 0 & 0 & 0 & 1 & 0 & 0 & 0 & 0 \\
a_2^{-1}\left[\mathbf{K}_P\alpha_3^{-1}\beta_3\left(\frac{\partial \mathbf{P}}{\partial \mathbf{E}'_q}\right)-\frac{\partial \mathbf{V}}{\partial \mathbf{E}'_q}\right] & a_2^{-1}\left[\mathbf{K}_P\alpha_3^{-1}\beta_3\left(\frac{\partial \mathbf{P}}{\partial \mathbf{E}'_d}\right)-\frac{\partial \mathbf{V}}{\partial \mathbf{E}'_d}\right] & a_2^{-1}\left[\mathbf{K}_P\alpha_3^{-1}\beta_3\left(\frac{\partial \mathbf{P}}{\partial \delta_n}\right)-\frac{\partial \mathbf{V}}{\partial \delta_n}\right] & a_2^{-1}\mathbf{K}_\omega\alpha_3^{-1}\beta_3 & -a_2^{-1}a_0 & -a_2^{-1}a_1 & -a_2^{-1}\alpha_3^{-1}\beta_3 & a_2^{-1}(\beta_1-\alpha_1\alpha_3^{-1}\beta_3) & a_2^{-1}(\beta_2-\alpha_2\alpha_3^{-1}\beta_3) \\
0 & 0 & 0 & 0 & 0 & 0 & 0 & 1 & 0 \\
0 & 0 & 0 & 0 & 0 & 0 & 0 & 0 & 1 \\
\alpha_3^{-1}\mathbf{K}_P\left(\frac{\partial \mathbf{P}}{\partial \mathbf{E}'_q}\right) & \alpha_3^{-1}\mathbf{K}_P\left(\frac{\partial \mathbf{P}}{\partial \mathbf{E}'_d}\right) & \alpha_3^{-1}\mathbf{K}_P\left(\frac{\partial \mathbf{P}}{\partial \delta_n}\right) & \alpha_3^{-1}\mathbf{K}_\omega & 0 & 0 & -\alpha_3^{-1}\alpha_0 & -\alpha_3^{-1}\alpha_1 & -\alpha_3^{-1}\alpha_2
\end{bmatrix}
\begin{bmatrix} \Delta\mathbf{E}'_q \\ \Delta\mathbf{E}'_d \\ \Delta\delta_n \\ \Delta\omega \\ x_1 \\ x_2 \\ x_3 \\ x_4 \\ x_5 \end{bmatrix}
\tag{12.50}
$$

If the eigenvectors are normalised then the factors w_{ij} determine the relative activity of the state variable $\Delta x_i(t)$ when a particular mode $z_j(t)$ is excited. This is referred to as the *mode shape*.

Denoting $\mathbf{U} = \mathbf{W}^{-1}$ in equation 12.5 gives $\mathbf{z}(t) = \mathbf{U}\Delta\mathbf{x}(t)$ or, for individual modes:

$$z_i(t) = u_{i1}\Delta x_1(t) + u_{i2}\Delta x_2(t) + \ldots + u_{ij}\Delta x_j(t) + \ldots + u_{in}\Delta x_n(t) \tag{12.52}$$

In this case the factors u_{ij} measure the amount of activity a particular state variable $\Delta x_j(t)$ contributes to a particular mode $z_i(t)$.

A knowledge of the matrices **W** and **U** allows the sensitivity of a particular mode $z_i(t)$ to the changes in a particular system parameter to be determined. This is of particular importance when establishing the parameter settings of any control device installed in the system. Such a *sensitivity analysis* is accomplished in the following way. Let $\mathbf{w}_i$ be a column of the matrix **W** and $\mathbf{u}_i^T$ be a row of the matrix **u** then it holds from equation 12.3 that $\mathbf{A}\mathbf{w}_i = \lambda_i\mathbf{w}_i$ and $\mathbf{u}_i^T\mathbf{A} = \lambda_i\mathbf{u}_i^T$. The vector $\mathbf{w}_i$ is referred to as the *right, or right-hand, eigenvector* while the vector $\mathbf{u}_i$ is referred to as the *left, or left-hand, eigenvector*. It results from these two equations, or equation 12.4, that

$$\mathbf{u}_i^T\mathbf{A}\mathbf{w}_i = \lambda_i\mathbf{u}_i^T\mathbf{w}_i \text{ or } \lambda_i = \frac{\mathbf{u}_i^T\mathbf{A}\mathbf{w}_i}{\mathbf{u}_i^T\mathbf{w}_i} \tag{12.53}$$

If the vectors are normalised then, from $\mathbf{U} = \mathbf{W}^{-1}$, it results that $\mathbf{u}_i^T\mathbf{w}_i = \mathbf{1}$ and $\mathbf{u}_i^T\mathbf{w}_j = \mathbf{0}$.

Now let ξ be a system parameter when, using equation 12.53, it can be shown that

$$\frac{\partial\lambda_i}{\partial\xi} = \frac{\mathbf{u}_i^T\left(\dfrac{\partial\mathbf{A}}{\partial\xi}\mathbf{w}_i\right)}{\mathbf{u}_i^T\mathbf{w}_i} \tag{12.54}$$

For many system parameters the elements of the state matrix **A**, given by equation 12.50, are simple functions and the determination of the matrix of derivatives $\partial\mathbf{A}/\partial\xi$ presents little problem. This is especially true with regard to the parameters of the excitation and voltage control systems. Knowing the value of the derivative $\partial\mathbf{A}/\partial\xi$ in equation 12.54 it is possible to determine which parameters may improve the system stability by observing if the system eigenvalues move to the left on the complex plane.

13
Power System Dynamic Simulation

The simulation of power system behaviour is a highly useful tool for power system planning, stability analysis and for operator training. The individual models of the generator, AVR, turbine-governor and the system loads are given by the differential and algebraic equations developed in Chapter 10, whilst the network is modelled by the algebraic equations developed in Chapter 3. Together these equations form a complete mathematical model of the system which can be solved numerically to simulate the system behaviour.

To develop a dynamic power system simulation the equations used to model the different elements are collected together to form a set of differential equations

$$\dot{\mathbf{x}} = \mathbf{f}(\mathbf{x}, \mathbf{y}) \tag{13.1}$$

and a set of algebraic equations

$$\mathbf{0} = \mathbf{g}(\mathbf{x}, \mathbf{y}) \tag{13.2}$$

The differential equations describe the system dynamics and are primarily contributed by the generating units and the dynamic loads whilst the algebraic equations describe the network, the static loads and the generator algebraic equations. The solution of these two equations defines the electromechanical state of the power system at any instant in time. A disturbance in the network usually requires a change to both the network configuration and the boundary conditions. These are modelled by changing the coefficients in the functions appearing on the right-hand side of equations 13.1 and 13.2. The dynamic power system simulation computer program must then solve the differential and algebraic equations over a period of time for a given sequence of network disturbances.

Equations 13.1 and 13.2 can be solved using either a partitioned-solution or a simultaneous-solution. The partitioned-solution is sometimes referred to as the *alternating-solution* and the simultaneous-solution as the *combined-solution*. In the *partitioned-solution* the differential equations are solved using a standard explicit numerical integration method with the algebraic equations 13.2 being solved separately at each time step. The *simultaneous-solution* uses implicit integration methods to convert the differential equations 13.1 into a set of algebraic equations which are then combined with the algebraic network equations 13.2 to be solved as one set of simultaneous algebraic equations. The effectiveness of these two solutions depends both on the generator model used and the method of numerical integration.

In order to select the most appropriate integration method to use it is necessary to understand the time scale of the dynamics included in the model of the generating unit. As explained in Chapter 12 the solution of any set of linear differential equations is in the form of a linear combination of exponential functions each of which describe the individual system modes. These modes are themselves defined by the system eigenvalues which are linked to the time scale of the different dynamics in the model. When the eigenvalues have a range of values that are widely distributed on the complex plane the solution will consist of the sum of fast-changing dynamics, corresponding to large eigenvalues, and slow-changing dynamics, corresponding to small eigenvalues. In this instance the system of differential equations is referred to as a *stiff system*. A non-linear system is referred to as stiff if its linear approximation is stiff. Among the power system electromechanical models developed in Chapter 10 all the models that include both the subtransient equations, with their very small time constants, and the relatively slow rotor dynamics constitute stiff models. The model stiffness is further aggravated if the AVR equations, with their small time constants, and the turbine equations, with long time constants, are included in the model. Consequently if the power system model includes AVR systems and high order generator models then the solution method should use integration methods well suited to stiff systems. In contrast the classical power system model does not constitute a stiff set of differential equations as it only includes slow rotor dynamics. Transient stability programs using this model can use simpler integration formulae.

13.1 Numerical Integration Methods

The analytical solution of non-linear differential equations is not generally possible and a numerical solution consisting of a series of values $(x_1, x_2, \ldots, x_k, \ldots)$ that satisfies the equation $\dot{x} = f(x, t)$ at the time instants $(t_1, t_2, \ldots, t_k, \ldots)$ must be found. This requires the use of *numerical integration formulae* that calculate the value x_{k+1} knowing all the previous values $(\ldots, x_{k-2}, x_{k-1}, x_k)$. These formulae fall into two general categories: the single-step *Runge-Kutta* methods and the multi-step *predictor-corrector* methods. Both are used in power system simulation programs and the interested reader is referred to Chua & Lin (1975) or Press et al. (1992) where these methods, and in particular the Runge-Kutta method, are discussed in detail along with examples of the necessary computer code. In this section discussion is mainly concentrated on implicit integration methods as they can be effectively used in both the partitioned-solution and the simultaneous-solution of stiff differential equations. Such equations are generally used to describe power system behaviour. The standard Runge-Kutta methods are restricted to the partitioned-solution of non-stiff systems.

When numerically solving differential equations each calculated value of the solution will differ from the accurate solution by an amount called the *local error*. This error comprises a *round-off error*, which depends on the computational accuracy of the particular computer used, and the *method error*, which depends on the type, order and step length of the integration method used. As the local error propagates to subsequent steps the total error at any one step consists of the local error made at that step plus the local errors transmitted from previous steps. The way in which the error propagates to subsequent steps determines the practical usefulness of a method. If the error does not increase from step to step then the formulae are *numerically stable*, otherwise the formulae are *numerically unstable*. In the latter case the cumulative effect of the errors may cause the solution x_k to be drastically different from the accurate solution.

For the differential equation $\dot{x} = f(x, t)$ the value of x_{k+1} can be found by either integrating the function $f(x, t)$ along its time path between t_k and t_{k+1} or alternatively by integrating $x(t)$ along its path from x_k to x_{k+1}. Each method will result in a different set of formulae known as the *Adams formulae* and the *Gear formulae* respectively.

The *Adams difference formulae* used in predictor-corrector schemes are obtained by approximating the function $f(x, t)$ by a power series $w(t)$ in the time interval over which the function $f(x, t)$ must be integrated. This gives

$$x_{k+1} = x_k + \int_{t_k}^{t_{k+1}} f(x, t)dt \approx x_k + \int_{t_k}^{t_{k+1}} w(t)dt \tag{13.3}$$

where the power series $w(t)$ is based on r values of $f(x, t)$. The coefficients in this power series depend on the values of $f(x, t)$ at the individual points so that the integration formula becomes

$$x_{k+1} = x_k + h\left(\sum_{j=1}^{r} b_j f_{k+1-j} + b_0 f_{k+1}\right) \tag{13.4}$$

where the function $f_i = f(x(t_i))$ is the value of the function at a given point t_i in time and h is the integration step length. The number of points r used in the power series is referred to as the *order* of the formula.

If $b_0 = 0$ then the resulting formulae are referred to as the *explicit*, or *Adams-Bashforth formulae*. In these formulae the approximating polynomial $w(t)$ is calculated using the known values $(\ldots, f_{k-2}, f_{k-1}, f_k)$ and used to extrapolate the function $f(x, t)$ in the new interval from t_k to t_{k+1}. If $b_0 \neq 0$, then the resulting formulae are referred to as the *implicit*, or *Adams-Moulton formulae*. In these formulae the approximating polynomial $w(t)$ is calculated using the known values $(\ldots, f_{k-2}, f_{k-1}, f_k)$ and the unknown value of f_{k+1} in order to interpolate function $f(x, t)$ in the interval from t_k to t_{k+1}. Table 13.1 contains Adams-Bashforth-Moulton formulae up to the third order.

The first order formulae in the Adams family is the *Euler formulae* and can have either an explicit or implicit form. The second order interpolation (implicit) formula is the *trapezoidal rule* when the polynomial $w(t)$ corresponds to the area of a trapezium below a line linking the points f_k and f_{k+1}.

The error in the Adams difference formulae depends on the order and is the integral of the error between the function $f(x, t)$ and the polynomial approximation w(t). This error may be expressed as

$$\varepsilon_{k+1} = \varepsilon_0 h^{r+1} x_k^{r+1}(\tau) \tag{13.5}$$

where $x_k^{(r+1)}(\tau)$ is the $(r+1)$th derivative at a point τ lying in the interval from t_{k-r} to t_{k+1}, while ε_0 is a constant. Table 13.1 shows that for $r > 1$ the size of the error introduced by the implicit formulae is much smaller than that introduced by the explicit formulae.

An additional advantage of the implicit formulae is their better numerical stability. Figure 13.1 shows the area of numerical stability on the complex λh plane where h

Table 13.1 Examples of Adams-Bashforth-Moulton formulae

Type	Order	Formula	ε_0
Adams-Bashforth (explicit) formulae	1	$x_{k+1} = x_k + hf_k$	1/2
	2	$x_{k+1} = x_k + \frac{h}{2}(3f_k - f_{k-1})$	5/12
	3	$x_{k+1} = x_k + \frac{h}{12}(23f_k - 16f_{k-1} + 5f_{k-2})$	9/24
Adams-Moulton (implicit) formulae	1	$x_{k+1} = x_k + hf_{k+1}$	−1/2
	2	$x_{k+1} = x_k + \frac{h}{2}(f_{k+1} + f_k)$	−1/12
	3	$x_{k+1} = x_k + \frac{h}{12}(5f_{k+1} + 8f_k - f_{k-1})$	−1/24

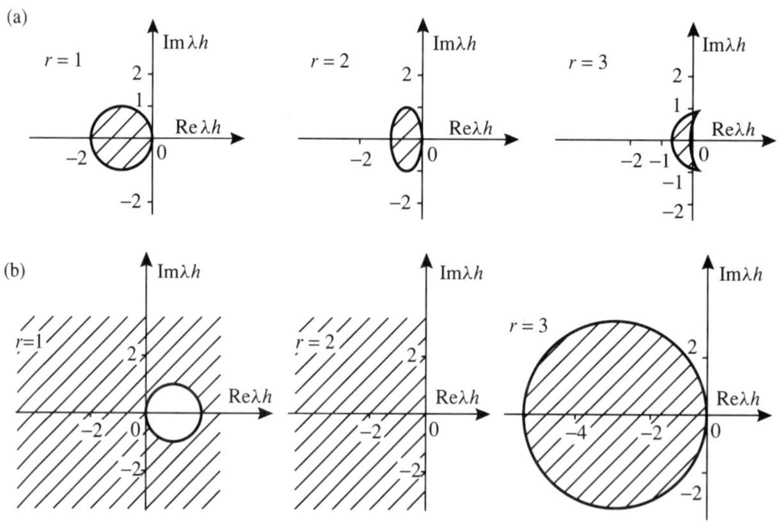

Figure 13.1 Areas of numerical stability for the Adams formulae : (a) explicit; (b) implicit

is the integration step length and λ is the largest system eigenvalue. The integration process is numerically stable if the integration step length is small enough so that for all the system eigenvalues λ_i the product $h\lambda_i$ lies inside the stability area. The figure shows that the stability area of the explicit formulae is much smaller than that of their implicit counterparts. The first and second order implicit formulae are stable over the whole left half-plane and are therefore numerically *absolutely stable*. This stability issue is particularly important for stiff systems where large eigenvalues force the use of a small integration step h. The low order implicit formulae are numerically stable over a large part of the complex plane and potentially allow the use of a longer integration step. When using these formulae the length of the integration step is limited only by the accuracy required in the calculation and the error as determined by equation 13.5.

A disadvantage of the implicit formulae is that the value of x_{k+1} cannot be calculated directly. When $b \neq 0$ equation 13.4 can be expressed as

$$x_{k+1} = \beta_k + hb_0 f(x_{k+1}) \tag{13.6}$$

where $\beta_k = x_k + \sum_{j=1}^{r} b_j f_{k+1-j}$. The unknown value of x_{k+1} now lies on both sides of the equation which means that, if the function $f(x)$ is non-linear, it must be found iteratively.

The simplest method of solving equation 13.6, referred to as *functional iteration*, consists of a series of substitutions according to the following iterative formula

$$x_{k+1}^{(l+1)} = \beta_k + hb_0 f(x_{k+1}^{(l)}) \tag{13.7}$$

where the upper index in brackets denotes the iteration number. If the first value $x_{k+1}^{(0)}$ used in the iteration is calculated using an explicit Adams formula, then the whole procedure is known as the *prediction-correction method*. The explicit formula is used as a *predictor*, while the implicit formulae is the *corrector*. The iterative process (equation 13.7) converges if

$$hb_0 L < 1 \tag{13.8}$$

where L is the *Lipschitz constant*, $L = \sqrt{\alpha_M}$, and α_M is the maximum eigenvalue of the matrix product $(\mathbf{A}^T\mathbf{A})$, where $\mathbf{A} = \left[\dfrac{\partial f}{\partial x}\right]$ is the Jacobian matrix calculated at point $\hat{x}_{k+1}$, being a solution of equation 13.6. The smaller the value of the product (hb_0L) the faster the convergence.

Because of the presence of large eigenvalues in stiff systems the convergence condition of equation 13.8 may force a limitation on the integration step length that is more restrictive than that required either by the required accuracy or by numerical stability. In such a case equation 13.6 may be solved using *Newton's method* instead of by functional iteration. For any equation $F(x) = 0$ Newton's formula is

$$x^{(l+1)} = x^{(l)} - \left[\frac{\partial F}{\partial x}\right]_{(l)}^{-1} F(x^{(l)}) \tag{13.9}$$

where the upper, bracketed, index denotes the iteration number. Applying equation 13.9 to the implicit formula of equation 13.6 gives

$$x_{k+1}^{(l+1)} = x_{k+1}^{(l)} - [1 - hb_0 A_{k+1}^{(l)}]^{-1}[x_{k+1}^{(l)} - \beta_k - hb_0 f(x_{k+1}^{(l)})] \tag{13.10}$$

where $A_{k+1}^{(l)}$ is the Jacobian matrix calculated for the l-th iteration. Newton's method allows the integration step length to be increased to a much higher value than that defined by equation 13.8 but, because the matrix is inverted at each step, the complexity of the method is much greater than functional iteration. However if the integration step length is sufficiently large then the added complexity of Newton's method may be justified.

In developing the Adams formulae the function $f(x, t)$ was approximated by the power series $w(t)$ in order to find x_{k+1}. An alternative approach is to approximate $x(t)$, rather than the function $f(x, t)$, by the power series $w(t)$. In this case $w(t) \approx x(t)$ and the coefficients in the approximating polynomial $w(t)$ are functions of consecutive values of $(\ldots, x_{k-2},$

$x_{k-1}, \ldots$). Taking the time derivative gives $\dot{x} = \dot{w}(t)$, or $\dot{w}(t) = f(x, t)$. When $w(t)$ is used as the extrapolation formula this leads to the following explicit integration formula

$$x_{k+1} = \sum_{j=0}^{r} a_j x_{k-j} + b_0 h f_k \tag{13.11}$$

If $w(t)$ is used as the interpolation formula the following implicit integration formula is obtained

$$x_{k+1} = \sum_{j=0}^{r} a_j x_{k-j} + b_0 h f_{k+1} \tag{13.12}$$

These formulae, known as the *Gear formulae*, are shown in Table 13.2 (Variant I) up to the third order. The first order formula is the Euler formula whilst the second order formula is referred to as the *intermediate point formula*.

The solution of equation 13.12 may be obtained using either functional iteration or Newton's method, just as for the Adams formulae. If equation 13.12 is solved by functional iteration then equation 13.11 is used as the predictor and equation 13.12 as the corrector.

The main advantage of the Gear formulae over the Adams formulae is that they have a larger area of numerical stability as shown in Figure 13.2. They are therefore better suited to stiff systems. If a large integration step length is used on a stiff system the predictor in equation 13.11 may not give a good approximation and instead $x(t)$ may be extrapolated directly using the values $(\ldots, x_{k-2}, x_{k-1}, x_k)$. The Lagrange extrapolation equations then give

$$x_{k+1} = a_0 x_k + \sum_{j=1}^{r} a_j x_{k-1} \tag{13.13}$$

Values of the coefficients for these formulae up to third order are given in Table 13.2, Variant II.

When a set of differential equations are non-linear the Jacobian matrix and its eigenvalues are not constant and the criteria that limit the integration step length constantly change with time. For example the numerical stability depends on the eigenvalues and

Table 13.2 Examples of Gear formulae

Type	Order	Formula	ε_0
explicit (Gear) Variant I	1	$x_{k+1} = x_k + hf_k$	
	2	$x_{k+1} = x_{k-1} + 2hf_k$	
	3	$x_{k+1} = -\frac{3}{2}x_k + 3x_{k-1} - \frac{1}{2}x_{k-2} + 3f_k h$	
extrapolation (Lagrange) Variant II	1	$x_{k+1} = 2x_k - x_{k-1}$	
	2	$x_{k+1} = 3x_k - 3x_{k-1} + x_{k-2}$	
	3	$x_{k+1} = 4x_k - 6x_{k-1} + 4x_{k-2} - x_{k-3}$	
implicit (Gear)	1	$x_{k+1} = x_k + hf_{k+1}$	-1/2
	2	$x_{k+1} = \frac{4}{3}x_k - \frac{1}{3}x_{k-1} + \frac{2}{3}hf_{k+1}$	-2/9
	3	$x_{k+1} = \frac{18}{11}x_k - \frac{9}{11}x_{k-1} + \frac{2}{11}x_{k-2} + \frac{6}{12}hf_{k+1}$	-3/22

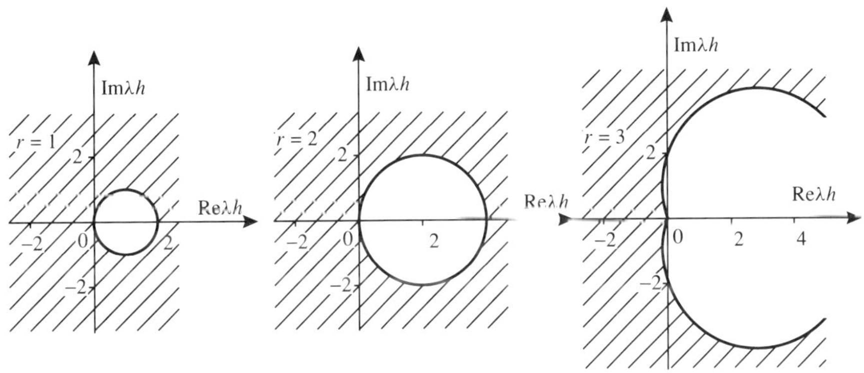

Figure 13.2 Areas of numerical stability of the Gear formulae

the step length, the convergence properties on the Lipschitz constant and the step length, while the local error is determined by the derivative $x^{(r+1)}(\tau)$ and the step length so that the correct choice of integration step length is of fundamental importance.

To combat these problems there are two extreme ways in which the difference formulae can be used. The first is to use a low-order formula, which is absolutely stable numerically, together with a constant integration step length that is limited to a value that will guarantee good convergence of the iterative corrector and limited local errors for the whole of the simulation period. Alternativcly the order of the formulae can be automatically varied during the simulation process so that the integration step length can be maximised without compromising either the accuracy or the convergence properties.

Because higher-order formulae have a low area of numerical stability the highest order formula that is normally used in practice is the sixth. Simple programs use second or third order formulae in order to avoid problems with numerical stability. It is also possible to use a constant-order, variable-step procedure. Automatic changes in either the integration step length or the order of the formula used requires additional calculation. In addition a change in the integration step length is not easy as if the step length is changed then the coefficients in most of the formulae also change as they depend on the distance between the interpolation nodes. Problems with the use of variable-order, constant-step formulae is alleviated if the equations are written in the *canonical form*, details of which can be found in Chua & Lin (1975).

13.2 The Partitioned-Solution

In each step of the numerical integration procedure the partitioned-solution alternates the solution of the differential equations 13.1 and the algebraic equations 13.2. In order to match the values of the variables $\mathbf{y}(t)$ to the values of the variables $\boldsymbol{x}(t)$ it is necessary to solve the algebraic equations before numerically integrating the differential equations. A general solution algorithm for the partitioned-solution, using a predictor-corrector numerical integration method, is shown in Figure 13.3. In this algorithm the algebraic equations are solved at three stages. The first solution is in stage 3, and occurs whenever there is a change in the network configuration. This change in the network configuration alters the coefficients in the algebraic equations so that for a given set of state variables $\mathbf{x}_k$

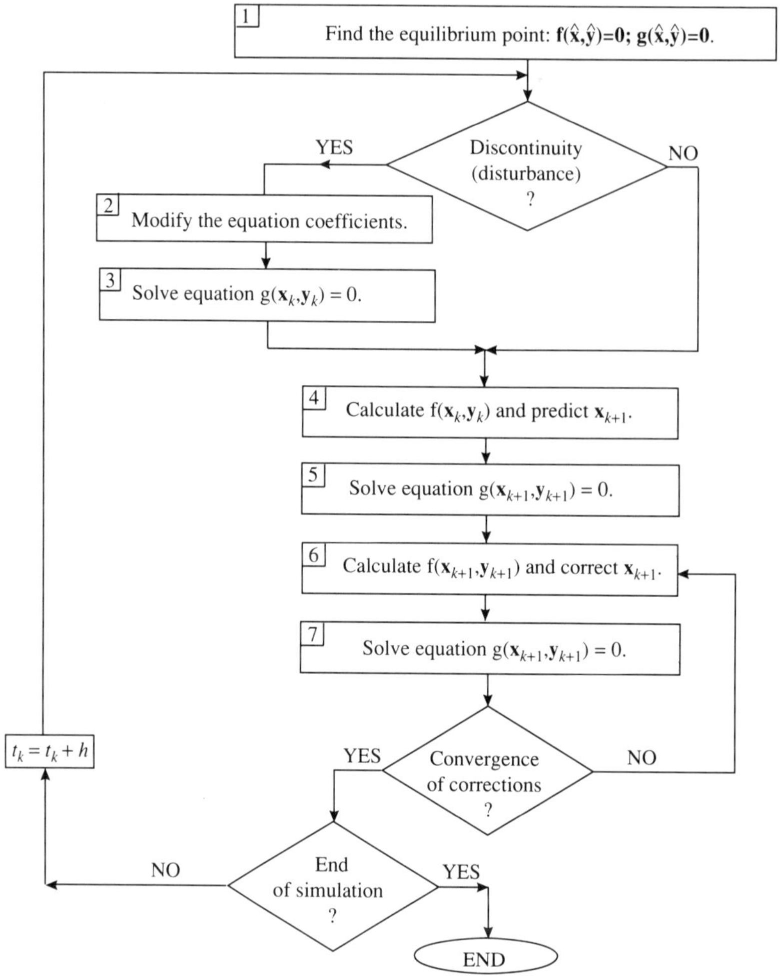

Figure 13.3 Simplified algorithm of the partitioned-solution using the predictor-corrector method

the value of the dependent variables $\mathbf{y}_k$ change. The second solution of the algebraic equations takes place in stage 5, after predicting the new values of the variables $\mathbf{x}_{k+1}$. The third, and final solution, is in stage 7 after correcting x_{k+1} in stage 6. The solution in stage 7 is repeated each integration step as many times as there are corrections of $\mathbf{x}_{k+1}$.

In the algorithm shown in Figure 13.3 the solution of the algebraic equations takes a significant proportion of the total simulation time. It is therefore important to consider the methods available for effectively realising the solution. In the following discussion the network equations will be solved assuming the network to be represented as shown in Figure 13.4. In this diagram additional generator nodes i' have been added to allow each generator to be represented by an equivalent fictitious emf $\overline{E}_i^f$ behind a fictitious reactance X_i^f with $\overline{I}_i$ being the injected generator current. If rotor saliency is neglected then, depending on the generator model used, the fictitious emf and the fictitious reactance

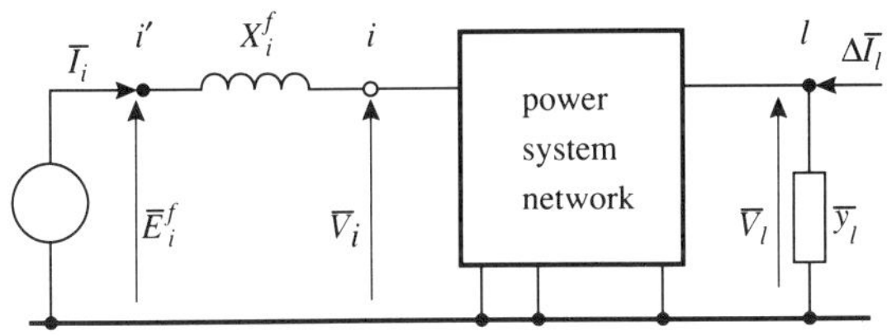

i' - a generator node; l - a load node

Figure 13.4 Network model with each generator replaced by a Thevenin voltage source

will correspond to the transient or subtransient values. If, on the other hand, rotor saliency is included then both the reactance and the emf will have some fictitious value.

With regard to the load model, each load is represented by an equivalent admittance y_l and a correction current $\Delta\bar{I}_l$ that depends on the voltage at the load. This correction current is used to match the power drawn by the admittance to the value resulting from the static load characteristic.

Under the above assumptions the network is described by the following nodal equation

$$\begin{bmatrix} \mathbf{I}_G \\ \Delta\mathbf{I}_L \end{bmatrix} = \begin{bmatrix} \mathbf{Y}_{GG} & \mathbf{Y}_{GL} \\ \mathbf{Y}_{LG} & \mathbf{Y}_{LL} \end{bmatrix} \begin{bmatrix} \mathbf{E}_G \\ \mathbf{V}_L \end{bmatrix} \tag{13.14}$$

where $\{G\}$ is the set of fictitious generator nodes, $\{L\}$ is the set of all the other nodes (called the load nodes) that includes the generator terminal nodes, $\mathbf{I}_G$ is a vector of the generator currents, $\Delta\mathbf{I}_L$ a vector of the load corrective currents, $\mathbf{E}_G$ a vector of the fictitious emfs and $\mathbf{V}_L$ a vector of the voltages at the load nodes. The matrix $\mathbf{Y}_{GG}$ is a diagonal matrix of generator admittance $y_i = 1/X_i^f$. $\mathbf{Y}_{GL}$ is a rectangular matrix comprising a diagonal submatrix with elements equal to $-y_i$ and all other elements equal to zero. $\mathbf{Y}_{LG}$ is the transpose of $\mathbf{Y}_{GL}$. The matrix $\mathbf{Y}_{LL}$ is a modified version of the nodal admittance matrix, introduced in Chapter 3, whose diagonal terms now include the load and generator admittances at the rows corresponding to the relevant load nodes and generator terminal nodes respectively. If each load is represented by a constant admittance then the correction currents $\Delta\mathbf{I}_L = 0$.

13.2.1 Partial matrix inversion

Equation 13.14 can be expanded as

$$\mathbf{I}_G = \mathbf{Y}_{GG}\mathbf{E}_G + \mathbf{Y}_{GL}\mathbf{V}_L \tag{13.15}$$

$$\Delta\mathbf{I}_L = \mathbf{Y}_{LG}\mathbf{E}_G + \mathbf{Y}_{LL}\mathbf{V}_L \tag{13.16}$$

Rearranging equation 13.16

$$\mathbf{V}_L = -\mathbf{Y}_{LL}^{-1}\mathbf{Y}_{LG}\mathbf{E}_G + \mathbf{Y}_{LL}^{-1}\Delta\mathbf{I}_L \tag{13.17}$$

and substituting into equation 13.15 gives

$$\mathbf{I}_G = (\mathbf{Y}_{GG} - \mathbf{Y}_{GL}\mathbf{Y}_{LL}^{-1}\mathbf{Y}_{LG})\mathbf{E}_G + \mathbf{Y}_{GL}\mathbf{Y}_{LL}^{-1}\Delta\mathbf{I}_L \tag{13.18}$$

The last two equations 13.17 and 13.18 can be rewritten in matrix form as

$$\begin{bmatrix} \mathbf{I}_G \\ \mathbf{V}_L \end{bmatrix} = \begin{bmatrix} \mathbf{Y}_G & \mathbf{K}_I \\ \mathbf{K}_V & \mathbf{Z}_{LL} \end{bmatrix} \begin{bmatrix} \mathbf{E}_G \\ \Delta\mathbf{I}_L \end{bmatrix} \tag{13.19}$$

where $\mathbf{Y}_G = \mathbf{Y}_{GG} - \mathbf{Y}_{GL}\mathbf{Y}_{LL}^{-1}\mathbf{Y}_{LG}$, $\mathbf{K}_I = \mathbf{Y}_{GL}\mathbf{Y}_{LL}^{-1}$, $\mathbf{K}_V = -\mathbf{Y}_{LL}^{-1}\mathbf{Y}_{LG}$ and $\mathbf{Z}_{LL} = \mathbf{Y}_{LL}^{-1}$. The square matrix in equation 13.19 is called the *partial inversion matrix* and refers to the fact that only the submatrix $\mathbf{Y}_{LL}$ is explicitly inverted to obtain $\mathbf{Z}_{LL} = \mathbf{Y}_{LL}^{-1}$.

If rotor saliency is neglected then the fictitious emfs $\overline{E}_i^f$ in the network model are equal to the generator emfs and are calculated during the numerical integration of the differential equations. If, in addition, each load is represented by a constant admittance then the correction currents $\Delta\mathbf{I}_L$ are equal to zero and the generator currents and the load voltages can be calculated directly from

$$\begin{aligned} \mathbf{I}_G &= \mathbf{Y}_G\mathbf{E}_G \\ \mathbf{V}_L &= \mathbf{K}_V\mathbf{E}_G \end{aligned} \tag{13.20}$$

without the need for an iterative solution. However if the loads are non-linear the correction currents are non-zero and depend on the load voltage according to the function $\Delta\mathbf{I}_L(\mathbf{V}_L)$. As the unknown load voltages now appear on both sides of equation 13.19 they must be calculated iteratively. The lower of equations 13.19 can be used to formulate the iterative formula

$$\mathbf{V}_L^{(l+1)} = \mathbf{K}_V\mathbf{E}_G + \mathbf{Z}_{LL}\Delta\mathbf{I}_L(\mathbf{V}_L^{(l)}) \tag{13.21}$$

where the upper, bracketed, index denotes the iteration number. When the iteration process is complete the voltage $\mathbf{V}_L = \mathbf{V}_L^{(l+1)}$ and the generator currents may be calculated from the upper of equations 13.19 as

$$\mathbf{I}_G = \mathbf{Y}_G\mathbf{E}_G + \mathbf{K}_I\Delta\mathbf{I}_L \tag{13.22}$$

where $\Delta\mathbf{I}_L$ are the correction currents corresponding to the calculated values of the voltages.

If rotor saliency is included then the fictitious emfs representing the generators in the network equations must also be calculated iteratively together with the generator currents. To explain how this is done the generator fourth order transient model ($\dot{E}'_d$, $\dot{E}'_q$, $\dot{\delta}$, $\dot{\omega}$), *Model 4*, will be used but the technique can equally well be applied to the sixth order subtransient model ($\dot{E}''_d$, $\dot{E}''_q$, $\dot{E}'_d$, $\dot{E}'_q$, $\dot{\delta}$, $\dot{\omega}$), *Model 6*, or the fifth order subtransient model ($\dot{E}''_d$, $\dot{E}''_q$, $\dot{E}'_q$, $\dot{\delta}$, $\dot{\omega}$), *Model 5*, defined in Chapter 10.

Figure 13.5 shows three circuit diagrams. The first two diagrams correspond to the generator in the transient state, while the third diagram corresponds to the fictitious Thevenin source used in the network model to replace the generator. The emf of the fictitious generator voltage source $\overline{E}^f$ must satisfy the equation $\overline{E}^f = \overline{V} + jX^f\overline{I}$, or $(E_a^f + jE_b^f) = (V_a + jV_b) + jX^f(I_a + jI_b)$, which expressed in matrix form is

$$\begin{bmatrix} E_a^f \\ E_b^f \end{bmatrix} = \begin{bmatrix} V_a \\ V_b \end{bmatrix} - \begin{bmatrix} 0 & -X^f \\ X^f & 0 \end{bmatrix} \begin{bmatrix} I_a \\ I_b \end{bmatrix} \quad \text{or} \quad \mathbf{E}_{ab}^f = \mathbf{V}_{ab} - \mathbf{Z}_{ab}^f\mathbf{I}_{ab} \tag{13.23}$$

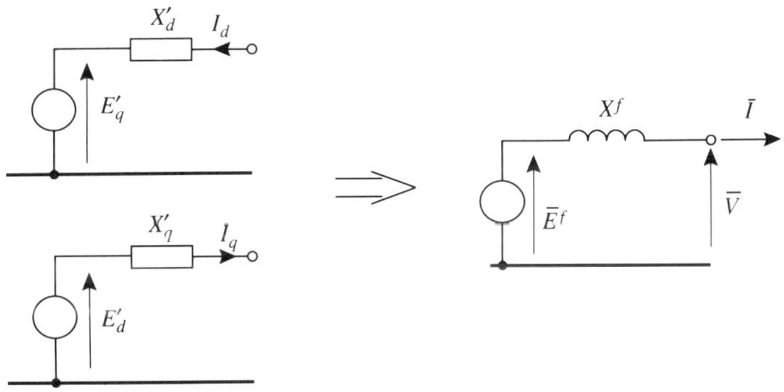

Figure 13.5 Replacing the equivalent d-axis and q-axis circuit of the generator by one equivalent circuit with fictitious reactance X^f and fictitious emf $\overline{E}^f$

This equation can be transformed from the system (a,b) reference frame to the individual generator (d,q) reference frame using the $\mathbf{T}$ transformation defined in equation 3.122 to give

$$\begin{bmatrix} E_d^f \\ E_q^f \end{bmatrix} = \begin{bmatrix} V_d \\ V_q \end{bmatrix} - \begin{bmatrix} 0 & -X^f \\ X^f & 0 \end{bmatrix} \begin{bmatrix} I_d \\ I_q \end{bmatrix} \quad \text{or} \quad \mathbf{E}_{dq}^f = \mathbf{V}_{dq} - \mathbf{Z}_{dq}^f \mathbf{I}_{dq} \tag{13.24}$$

where $\mathbf{Z}_{dq}^f = \mathbf{T}^{-1}\mathbf{Z}_{ab}^f\mathbf{T} = \mathbf{Z}_{ab}^f$ and $\mathbf{T}^{-1} = \mathbf{T}$. On the other hand the armature voltage equation for the generator fourth order model ($\dot{E}'_d$, $\dot{E}'_q$, $\dot{\delta}$, $\dot{\omega}$), *Model 4*, is given by equation 10.102, as

$$\begin{bmatrix} V_d \\ V_q \end{bmatrix} = \begin{bmatrix} E'_d \\ E'_q \end{bmatrix} - \begin{bmatrix} 0 & X'_q \\ -X'_d & 0 \end{bmatrix} \begin{bmatrix} I_d \\ I_q \end{bmatrix} \quad \text{or} \quad \mathbf{V}_{dq} = \mathbf{E}'_{dq} - \mathbf{Z}_{dq}\mathbf{I}_{dq} \tag{13.25}$$

which, when substituted into equation 13.24, gives

$$\begin{bmatrix} E_d^f \\ E_q^f \end{bmatrix} = \begin{bmatrix} E'_d \\ E'_q \end{bmatrix} - \begin{bmatrix} 0 & -\Delta X_q \\ \Delta X_d & 0 \end{bmatrix} \begin{bmatrix} I_d \\ I_q \end{bmatrix} \quad \text{or} \quad \mathbf{E}_{dq}^f = \mathbf{E}'_{dq} - \Delta\mathbf{Z}\mathbf{I}_{dq} \tag{13.26}$$

where $\Delta X_q = X'_q - X^f$ and $\Delta X_d = X'_d - X^f$. This equation determines the emf of the fictitious voltage source in terms of the generator current.

Equation 13.26 can be solved iteratively together with the network equations 13.19. To show how this is done equation 13.26 is re-written as

$$\begin{bmatrix} E_d^{f(l+1)} \\ E_q^{f(l+1)} \end{bmatrix} = \begin{bmatrix} E'_d \\ E'_q \end{bmatrix} - \begin{bmatrix} 0 & -\Delta X_q \\ \Delta X_d & 0 \end{bmatrix} \begin{bmatrix} I_d^{(l)} \\ I_q^{(l)} \end{bmatrix} \quad \text{or} \quad \mathbf{E}_{dq}^{f(l+1)} = \mathbf{E}'_{dq} - \Delta\mathbf{Z}\mathbf{I}_{dq}^{(l)} \tag{13.27}$$

where l is the iteration number. The iterative solution algorithm is then:

1. Estimate the values of E_d^f and E_q^f for all the generators and transform them to the system (a, b) reference frame
2. Solve the network equations 13.19, calculate the generator currents I_a and I_b and transform their values to the generator (d,q) reference frame

3. Use equation 13.27 to correct the values of E_d^f and E_q^f
4. Compare the result with the previous iteration; if they differ, transform the new values of E_d^f and E_q^f to the system (a,b) reference frame and repeat from step 2 until the voltages converge.

The number of iterations necessary to solve the generator equations and the network equations depends on the value of the reactance chosen for the fictitious Thevenin source. Let $\hat{E}_d^f$, $\hat{E}_q^f$ and $\hat{I}_d$, $\hat{I}_q$ be the solutions to the equations. According to equation 13.27 the solution must satisfy

$$\begin{bmatrix} \hat{E}_d^f \\ \hat{E}_q^f \end{bmatrix} = \begin{bmatrix} E'_d \\ E'_q \end{bmatrix} - \begin{bmatrix} 0 & -\Delta X_q \\ \Delta X_d & 0 \end{bmatrix} \begin{bmatrix} \hat{I}_d \\ \hat{I}_q \end{bmatrix} \quad \text{or} \quad \hat{\mathbf{E}}_{dq}^f = \mathbf{E}'_{dq} - \Delta\mathbf{Z}\hat{\mathbf{I}}_{dq} \tag{13.28}$$

Subtracting equation 13.27 from equation 13.28 gives

$$(\mathbf{E}_{dq}^{f(l+1)} - \hat{\mathbf{E}}_{dq}^f) = \Delta\mathbf{Z}(\mathbf{I}_{dq}^{(l)} - \hat{\mathbf{I}}_{dq}) \tag{13.29}$$

This equation is important as it means that for a given error $\Delta\mathbf{I}_{dq}^{(l)} = (\mathbf{I}_{dq}^{(l)} - \hat{\mathbf{I}}_{dq})$ in the current estimation then the smaller the elements $\Delta X_q = X'_q - X^f$ and $\Delta X_d = X'_d = X^f$ in the matrix $\Delta\mathbf{Z}$ the closer the next estimation of the fictitious emf will be to the final solution. Consequently the iterative process will converge quickly if the equivalent generator reactance is chosen to have a value equal to an average value of X'_d and X'_q when the fictitious emf $\overline{E}^f$ then has a value close to the generator transient emf $\overline{E}'$. Generally one of the following "average" reactance values is used for the fictitious reactance:

$$X^f = \frac{1}{2}(X'_d + X'_q), \quad X^f = 2\frac{X'_d X'_q}{X'_d + X'_q}, \quad X^f = \sqrt{X'_d X'_q} \tag{13.30}$$

The use of one of these fictitious reactance values guarantees a small number of iterations in the solution of equation 13.27. If X'_d is close to X'_q then one or two iterations will suffice. If transient saliency is neglected then $X'_d = X'_q = X_f$ and $\hat{\mathbf{E}}_{dq}^f = \mathbf{E}'_{dq}$ and the solution is obtained without any iterations. When $X'_d \neq X'_q$, then the number of iterations depends on the initial choice of the fictitious emfs. If the value of X^f is chosen according to one of the formulae in equation 13.30 then the magnitude and the angle of the fictitious emf are close to the magnitude and the angle of the generator transient emf. In order to achieve a good estimation of the final value the fictitious emfs should be changed in each integration step in proportion to the generator emfs.

When rotor saliency and load non-linearity are included, the algorithm will involve both of the iterative processes described above. A simplified algorithm is shown in Figure 13.6.

A disadvantage of the partial matrix inversion technique used in equation 13.19 to solve the network equation is that all the submatrices $\mathbf{Y}_G$, $\mathbf{K}_I$, $\mathbf{K}_V$ and $\mathbf{Z}_{LL}$ are dense and therefore have a large computer memory requirement whilst also requiring a large number of arithmetic operations to calculate the currents and the voltages. If the number of loads for which the voltage must be computed is small, in comparison to the total number of nodes, the partial matrix inversion technique is worth using as then only part of the submatrices $\mathbf{K}_I$, $\mathbf{K}_V$ and $\mathbf{Z}_{LL}$ need be stored, and the number of arithmetic operations is

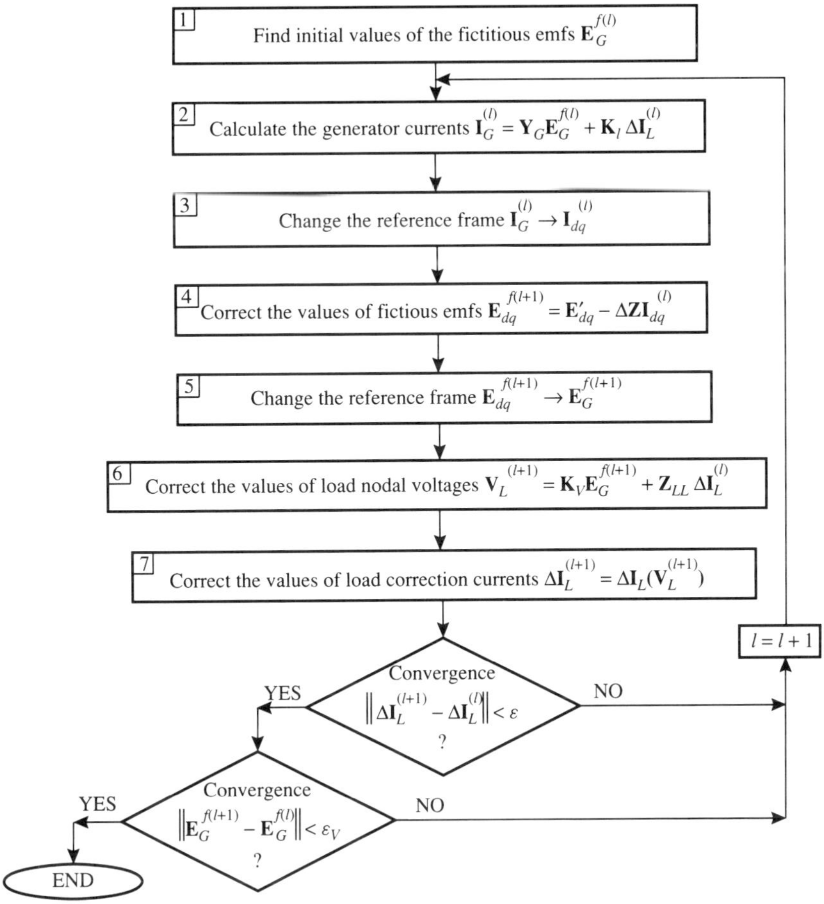

Figure 13.6 A simplified flow chart for the solution of the network equations using partial matrix inversion with both load non-linearity and rotor saliency included.

also reduced. If the voltage must be computed for the majority of the load nodes then the partial matrix inversion method is not recommended as the matrix factorisation technique described in the next section is more efficient.

13.2.2 Matrix factorisation

The lower of equations 13.14 gives

$$\mathbf{Y}_{LL}\mathbf{V}_L = (\Delta\mathbf{I}_L + \mathbf{I}_N) \tag{13.31}$$

where $\mathbf{I}_N = -\mathbf{Y}_{LG}\mathbf{E}_G$ is a vector of the currents having non-zero elements only at the nodes where the generators are connected to the system. Equation 13.31 corresponds to the network shown in Figure 13.7.

Using *triangular factorisation* (Press *et al.*, 1992) the square matrix $\mathbf{Y}_{LL}$ can be replaced by the product of an upper, or right, triangular matrix $\mathbf{R}_{LL}$ and lower, or left,

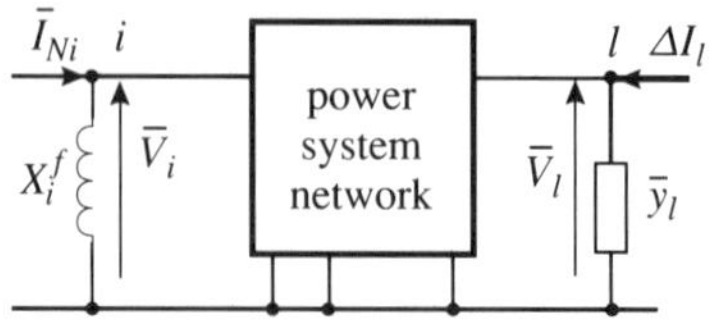

Figure 13.7 Network model with each generator replaced by a Norton source

triangular matrix $\mathbf{L}_{LL}$ to give

$$\mathbf{Y}_{LL} = \mathbf{L}_{LL}\mathbf{R}_{LL} \tag{13.32}$$

Equation 13.31 can then be re-written as two equations

$$\mathbf{L}_{LL}\mathbf{b}_L = (\Delta\mathbf{I}_L + \mathbf{I}_N), \quad \mathbf{R}_{LL}\mathbf{V}_L = \mathbf{b}_L \tag{13.33}$$

where $\mathbf{b}_L$ is an unknown vector.

Assuming that the loads are represented by constant admittances then, for a given set of generator emfs $\mathbf{E}_G$, equations 13.33 can be solved non-iteratively. As the vector $\mathbf{I}_N = -\mathbf{Y}_{LG}\mathbf{E}_G$ and the lower triangular matrix $\mathbf{L}_{LL}$ are both known and $\Delta\mathbf{I}_L = 0$ the unknown vector $\mathbf{b}_L$ can be found using a series of *forward substitutions* starting from the upper left hand corner of $\mathbf{L}_{LL}$. After calculating $\mathbf{b}_L$ the values of $\mathbf{V}_L$ can be found by a series of *back substitutions* starting from the lower right hand corner of the upper triangular matrix $\mathbf{R}_{LL}$. The advantage of this method is that if the matrix $\mathbf{Y}_{LL}$ is sparse then both the factor submatrices $\mathbf{R}_{LL}$ and $\mathbf{L}_{LL}$ are also sparse allowing sparse matrix techniques to be used to save both on computer memory and the number of arithmetic operations needed to effect a solution (Tewerson, 1973, Brameller, 1976, Pissanetzky, 1984, Duff, Erisman & Reid, 1986).

If non-linearity of the loads is included equations 13.33 must be solved iteratively as the correction currents $\Delta\mathbf{I}_L$ depend on the voltages. The iteration formulae are

$$\mathbf{L}_{LL}\mathbf{b}_L^{(l+1)} = (\Delta\mathbf{I}_L^{(l)} + \mathbf{I}_N), \quad \mathbf{R}_{LL}\mathbf{V}_L^{(l+1)} = \mathbf{b}_L^{(l+1)} \tag{13.34}$$

where $\Delta\mathbf{I}_L^{(l)} = \Delta\mathbf{I}_L(\mathbf{V}_L^{(l)})$ is a vector of the load correction currents. The upper index l denotes the iteration number.

If rotor saliency is included then the generator Norton current $\mathbf{I}_N$ must also be calculated iteratively, in a similar way as for the equivalent emfs in equation 13.27. These iterations can be executed together with the iterations necessary to calculate the voltages and the correction currents at the load nodes. A simplified algorithm of the method is shown in Figure 13.8.

13.2.3 Newton's method

Newton's method was introduced in Chapter 3 as a way to solve the network power-voltage equations. In system simulation Newton's method is also used but must now solve a set of current-voltage equations so that the solution algorithm is different from that used in the steady-state load-flow. A particularly attractive feature of Newton's method is that if the network nodal current-voltage equations are solved in rectangular co-ordinates, in

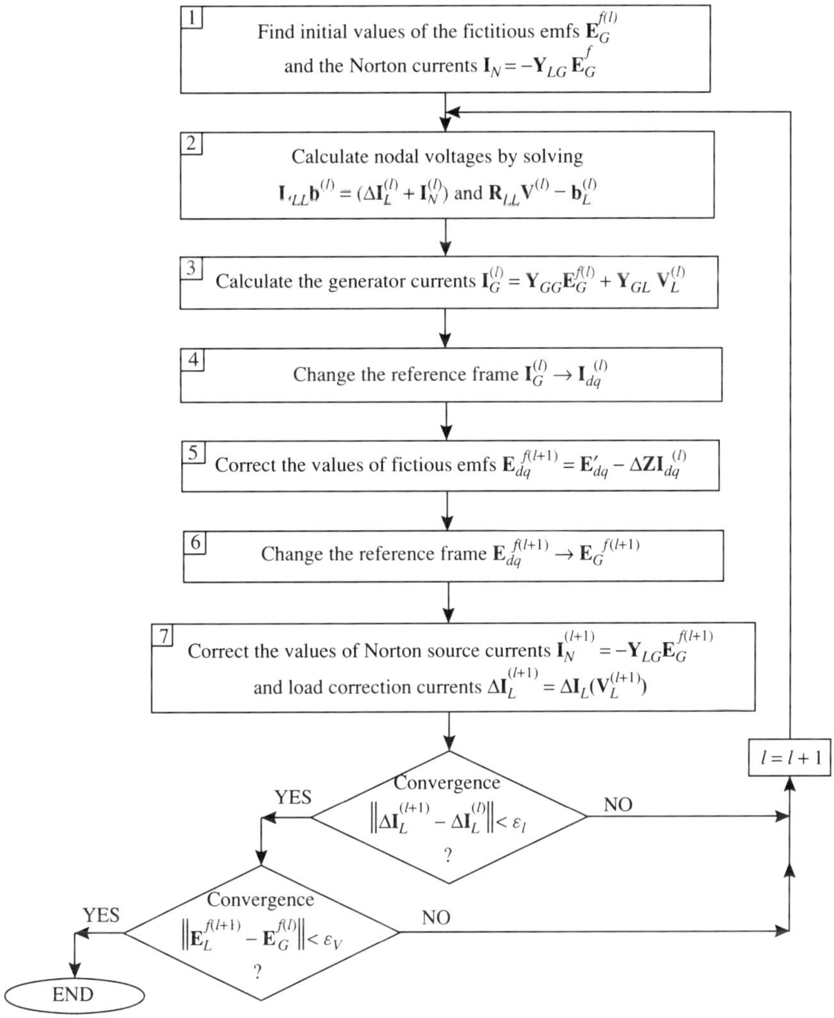

Figure 13.8 Simplified network solution flowchart using triangular factorisation with load non-linearity and rotor saliency included

the system (a, b) reference frame, then rotor saliency can conveniently be included in the solution. To account for saliency the generator equations are added to the network nodal admittance equation expressed in real numbers as in equation 3.111.

Assuming that the generators are represented by the fourth order transient model ($\dot{E}'_d$, $\dot{E}'_q$, $\dot{\delta}$, $\dot{\omega}$), *Model 4*, the generator armature voltage equations 10.102 can be written as

$$\begin{bmatrix} E'_d - V_d \\ E'_q - V_q \end{bmatrix} = \begin{bmatrix} 0 & -X'_q \\ X'_d & 0 \end{bmatrix} \begin{bmatrix} I_d \\ I_q \end{bmatrix} \tag{13.35}$$

which, when inverted, gives

$$\begin{bmatrix} I_d \\ I_q \end{bmatrix} = \frac{1}{X'_d X'_q} \begin{bmatrix} 0 & X'_q \\ -X'_d & 0 \end{bmatrix} \begin{bmatrix} E'_d - V_d \\ E'_q - V_q \end{bmatrix} \quad \text{or} \quad \mathbf{I}_{dq} = \mathbf{Y}_{dq}(\mathbf{E}'_{dq} - \mathbf{V}_{dq}) \tag{13.36}$$

These voltages and currents are in the generator (d,q) reference frame and are transformed into the system (a, b) reference frame using the transformation matrix $\mathbf{T}$ in equation 3.121 to give

$$\begin{bmatrix} I_a \\ I_b \end{bmatrix} = (\mathbf{T}^{-1}\mathbf{Y}_{dq}\mathbf{T}) \begin{bmatrix} E'_a - V_a \\ E'_b - V_b \end{bmatrix} \quad \text{or} \quad \mathbf{I}_{ab} = \mathbf{Y}_{ab}(\mathbf{E}'_{ab} - \mathbf{V}_{ab}) \tag{13.37}$$

where

$$\begin{aligned} \mathbf{Y}_{ab} = \mathbf{T}^{-1}\mathbf{Y}_{dq}\mathbf{T} &= \frac{1}{X'_d X'_q} \begin{bmatrix} -\sin\delta & \cos\delta \\ \cos\delta & \sin\delta \end{bmatrix} \begin{bmatrix} 0 & X'_q \\ -X'_d & 0 \end{bmatrix} \begin{bmatrix} -\sin\delta & \cos\delta \\ \cos\delta & \sin\delta \end{bmatrix} \\ &= \frac{1}{X'_d X'_q} \begin{bmatrix} \frac{1}{2}(X'_d - X'_q)\sin 2\delta & -X'_q \sin^2\delta - X'_d \cos^2\delta \\ X'_q \cos^2\delta + X'_d \sin^2\delta & -\frac{1}{2}(X'_d - X'_q)\sin 2\delta \end{bmatrix} = \begin{bmatrix} g_a & -b_{ab} \\ b_{ba} & g_b \end{bmatrix} \end{aligned} \tag{13.38}$$

is a submatrix similar in form to $\mathbf{Y}_{ij}$ in equation 3.111. In the nodal admittance technique the submatrix $\mathbf{Y}_{ab}$ describes a generator with transient saliency. Generally as $g_a \neq g_b$ and $b_{ab} \neq b_{ba}$ there is no one equivalent branch with admittance $\overline{Y}_{ab}$ so that an equivalent circuit for the salient-pole machine cannot be drawn. In such a case equation 13.37 cannot be written in complex form $\overline{I} = \overline{Y}_{ab}(\overline{E}' - \overline{V})$ as the equivalent admittance $\overline{Y}_{ab}$ does not exist. If transient saliency is neglected, $X'_d = X'_q$, then the submatrix $\mathbf{Y}_{ab}$ is skew symmetric as $g_a = g_b = 0$ and $b_{ab} = b_{ba} = 1/X'_d$ and an equivalent branch with admittance $\overline{Y}_{ab} = 0 + jb_{ab} = j\dfrac{1}{X'_d}$ now exists to represent the generator as shown in Figure 13.4 for the Thevenin source or Figure 13.7 for the Norton source with, in both cases, $X_i^f = X'_d$.

Because of saliency the generator current-voltage equation can only be written in real numbers in the (a,b) co-ordinate system; complex notation cannot be used. In order to include such an equation with the network equation 13.14 or 13.31 these must also be written in real numbers in the same way as equation 3.112. When using such equations each generator will be represented by a submatrix as in equation 13.38 which is added to the respective elements of the submatrices $\mathbf{Y}_{GG}$, $\mathbf{Y}_{GL}$, $\mathbf{Y}_{LG}$ and $\mathbf{Y}_{LL}$. In the case of the Norton source, equation 13.31 can be rewritten as

$$\mathbf{Y}_{LL}(\delta)\mathbf{V}_L = \Delta\mathbf{I}_L(\mathbf{V}_L) + \mathbf{I}_N \tag{13.39}$$

where the matrix $\mathbf{Y}_{LL}(\delta)$ is shown as a function of δ to emphasise that the diagonal elements of this matrix that refer to the generator depend on the power angle in the way defined by the submatrix in equation 13.38. The elements of these submatrices change with time as the generator rotor angles change. The Norton source currents $\mathbf{I}_N$ are given by

$$\mathbf{I}_N = -\mathbf{Y}_{LG}(\delta)\mathbf{E}_G \tag{13.40}$$

where matrix $\mathbf{Y}_{LG}(\delta)$ consists of the submatrices in equation 13.38 while $\mathbf{E}_G$ is a vector comprising of the individual generator emfs $\mathbf{E}'_{ab}$. All variables are in the (a, b) system co-ordinates.

To solve equation 13.39 it can be re-written in the standard Newton form as

$$\mathbf{F}(\mathbf{V}_L) = [\mathbf{Y}_{LG}(\delta)\mathbf{E}_G - \Delta\mathbf{I}_L(\mathbf{V}_L)] + \mathbf{Y}_{LL}(\delta)\mathbf{V}_L = 0 \tag{13.41}$$

Using Newton's iterative formula gives

$$\mathbf{V}_L^{(l+1)} = \mathbf{V}_L^{(l)} - \left[\frac{\partial \mathbf{F}}{\partial \mathbf{V}_L}\right]_l^{-1} \mathbf{F}(\mathbf{V}_L^l) \tag{13.42}$$

where the Jacobian matrix

$$\left[\frac{\partial \mathbf{F}}{\partial \mathbf{V}_L}\right] = \mathbf{Y}_{LL}(\delta) - \left[\frac{\partial \Delta \mathbf{I}_L}{\partial \mathbf{V}_L}\right] \tag{13.43}$$

is equal to the nodal admittance matrix minus the matrix of the derivatives of the correction currents with respect to the voltages. This correction matrix is diagonal and its elements are of the form

$$\left[\frac{\partial \Delta I_i}{\partial V_i}\right] = \begin{bmatrix} \dfrac{\partial \Delta I_{ai}}{\partial V_{ai}} & \dfrac{\partial \Delta I_{ai}}{\partial V_{bi}} \\ \dfrac{\partial \Delta I_{bi}}{\partial V_{ai}} & \dfrac{\partial \Delta I_{bi}}{\partial V_{bi}} \end{bmatrix} \tag{13.44}$$

The way in which the derivatives in equation 13.44 are calculated needs some explanation. In rectangular co-ordinates the relationship between the correction powers and the correction currents can be expressed as

$$\begin{bmatrix} \Delta P_i \\ \Delta Q_i \end{bmatrix} = \begin{bmatrix} V_{ai} & V_{bi} \\ V_{bi} & -V_{ai} \end{bmatrix} \begin{bmatrix} \Delta I_{ai} \\ \Delta I_{bi} \end{bmatrix} \quad \text{or}$$

$$\begin{bmatrix} \Delta I_{ai} \\ \Delta I_{bi} \end{bmatrix} = \frac{1}{|V_i|^2} \begin{bmatrix} V_{ai} & V_{bi} \\ V_{bi} & -V_{ai} \end{bmatrix} \begin{bmatrix} \Delta P_i \\ \Delta Q_i \end{bmatrix} \tag{13.45}$$

Making the substitution

$$\Delta p_i = \frac{\Delta P_i}{|V_i|^2}, \quad \Delta q_i = \frac{\Delta Q_i}{|V_i|^2} \tag{13.46}$$

allows equation 13.45 to be rewritten as

$$\begin{bmatrix} \Delta I_{ai} \\ \Delta I_{bi} \end{bmatrix} = \begin{bmatrix} V_{ai} & V_{bi} \\ V_{bi} & -V_{ai} \end{bmatrix} \begin{bmatrix} \Delta p_i \\ \Delta q_i \end{bmatrix} \tag{13.47}$$

Differentiating equation 13.47 gives

$$\frac{\partial \Delta I_{ai}}{\partial V_{ai}} = V_{ai} \frac{\partial \Delta p_i}{\partial V_{ai}} + \Delta p_i + V_{bi} \frac{\partial \Delta q_i}{\partial V_{ai}}$$

$$\frac{\partial \Delta I_{ai}}{\partial V_{bi}} = V_{bi} \frac{\partial \Delta q_i}{\partial V_{bi}} + \Delta q_i + V_{ai} \frac{\partial \Delta p_i}{\partial V_{bi}}$$

$$\frac{\partial \Delta I_{bi}}{\partial V_{ai}} = -V_{ai} \frac{\partial \Delta q_i}{\partial V_{ai}} - \Delta q_i + V_{bi} \frac{\partial \Delta p_i}{\partial V_{ai}}$$

$$\frac{\partial \Delta I_{bi}}{\partial V_{bi}} = V_{bi} \frac{\partial \Delta p_i}{\partial V_{ai}} + \Delta p_i - V_{ai} \frac{\partial \Delta q_i}{\partial V_{bi}}$$

which, when substituted into equation 13.44 and the matrix rearranged, leads to

$$\left[\frac{\partial \Delta I_i}{\partial V_i}\right] = \begin{bmatrix} V_{ai} & V_{bi} \\ V_{bi} & -V_{ai} \end{bmatrix} \begin{bmatrix} \dfrac{\partial \Delta p_i}{\partial V_{ai}} & \dfrac{\partial \Delta p_i}{\partial V_{bi}} \\ \dfrac{\partial \Delta q_i}{\partial V_{ai}} & \dfrac{\partial \Delta q_i}{\partial V_{bi}} \end{bmatrix} + \begin{bmatrix} \Delta p_i & \Delta q_i \\ -\Delta q_i & \Delta p_i \end{bmatrix}$$

$$= \frac{1}{|V_i|} \begin{bmatrix} V_{ai} & V_{bi} \\ -V_{bi} & -V_{ai} \end{bmatrix} \begin{bmatrix} \dfrac{\partial \Delta p_i}{\partial |V_i|} & \dfrac{\partial \Delta p_i}{\partial |V_i|} \\ \dfrac{\partial \Delta q_i}{\partial |V_i|} & \dfrac{\partial \Delta q_i}{\partial |V_i|} \end{bmatrix} \begin{bmatrix} V_{ai} & 0 \\ 0 & V_{bi} \end{bmatrix} + \begin{bmatrix} \Delta p_i & \Delta q_i \\ -\Delta q_i & \Delta p_i \end{bmatrix} \tag{13.48}$$

If the derivatives in the second matrix are expressed in the form

$$\frac{\partial \Delta p_i}{\partial V_{ai}} = \frac{\partial \Delta p_i}{\partial |V_i|}\frac{\partial |V_i|}{\partial V_{ai}} = \frac{\partial \Delta p_i}{\partial |V_i|}\frac{\partial}{\partial V_{ai}}\sqrt{V_{ai}^2 + V_{bi}^2} = \frac{1}{|V_i|}\frac{\partial \Delta p_i}{\partial |V_i|}V_{ai}$$

then the partial derivatives $\partial \Delta p_i / \partial |V_i|$ and $\partial \Delta q_i / \partial |V_i|$ can be computed from the static load characteristics $P(V)$ and $Q(V)$.

The solution algorithm consists of the iterative equation 13.42, the Jacobian matrix equation 13.43 and the correction currents in equation 13.48. The algorithm may be simplified by using the *dishonest Newton method*, where the Jacobian matrix is calculated only once at the beginning of each integration step using the initial values of the correction currents. This simplification should not be used during those integration steps when a network disturbance is being simulated as the change in the correction currents may be large.

If rotor saliency, and the load correction currents, are neglected equation 13.42 of the Newton method is identical to the second of equations 13.20 in the partial matrix inversion method. This can be shown by substituting $\Delta \mathbf{I}_L = 0$, $\mathbf{Y}_{LL}(\delta) = \mathbf{Y}_{LL} =$ constant and $\mathbf{V}_L^{(l+1)} = \mathbf{V}_L^l$ into equations 13.42 and 13.41.

13.2.4 Ways of avoiding iterations and multiple network solutions

The basic algorithm of the partitioned-solution shown in Figure 13.3 attempts to match the values of the variables $\mathbf{y}(t)$, for given values of $\mathbf{x}(t)$, by solving the linear algebraic equation 13.2. This solution is repeated after each prediction and each correction of $\mathbf{x}(t)$. One way to speed-up the algorithm is to replace the solution of the algebraic equation by an extrapolation of the value of $\mathbf{y}(t)$ at some appropriate stage in the solution. As the extrapolated values of $\mathbf{y}(t)$ are only approximate an error is introduced into the right hand side of equation 13.1, called the *interface error*, which influences the accuracy of $\mathbf{x}(t)$.

There are three ways of introducing extrapolation into the algorithm:

- the algebraic equations are solved after each prediction, with extrapolated values being used after each correction
- the algebraic equations are solved after each prediction and after the last correction
- the algebraic equations are solved after each correction with the prediction of $\mathbf{y}(t)$ and $\mathbf{x}(t)$ being done together, by extrapolation

The first method is not recommended as it may introduce large interface errors which force the integration step length to shorten. The second and the third methods are much better with the third method being preferred as, with this method, interface errors generated at the prediction stage are eliminated during correction. In this case extrapolation eliminates one solution of the algebraic equations in each integration step. Obviously this method is beneficial only if the number of corrections required is small and the corrector converges quickly.

In most cases the variables in $\mathbf{y}(t)$ are extrapolated individually and independently of each other. Simple extrapolation formulae are normally employed that use the past values of the variable obtained from the previous two or three steps. Typical of these formulae are those listed as Variant II in Table 13.2 as

$$x_{k+1} = 2x_k - x_{k-1} \qquad \text{or} \qquad x_{k+1} = 3x_k - 3x_{k-1} + x_{k-2}$$

whilst Adibi, Hirsch & Jordan (1983) suggest updating the complex load voltages using

$$\overline{V}_{k+1} = \frac{\overline{V}_k^2}{\overline{V}_{k-1}} \qquad \text{or} \qquad \overline{V}_{k+1} = \frac{\overline{V}_k^3 \overline{V}_{k-2}}{\overline{V}_{k-1}^2} \tag{13.49}$$

After extrapolating the voltages at all the load nodes using equation 13.49 the generator currents are computed from the nodal equation 13.14 using the generator emfs obtained from the numerical integration. Extrapolation procedures can also be used to obtain the values of other variables such as generator real power, voltage error etc. (Stott, 1979). Generally high-order extrapolation formulae are not used to update $\mathbf{y}(t)$ as the improvement in the accuracy is small compared to the simple formulae. In addition complications occur following network disturbances because, at the instant of the discontinuity, all the previous, old, values of the variable are invalid and extrapolation must start at the step where the disturbance occurs. Sometimes no previous values are used in the extrapolation process but a linearised equation is formed that links the increments in $\mathbf{y}(t)$ with the increments in $\mathbf{x}(t)$ (Stott, 1979).

Besides reducing the number of times the algebraic equations need be solved each integration step it is also possible to avoid iterations in the network solution. These iterations result from non-linear load characteristics and from saliency in the generator rotors. The number of iterations necessary to account for generator saliency can be reduced by using either the sixth or the fifth order subtransient models rather than the fourth or third order transient models. Table 4.3 shows that transient saliency, $X'_q \neq X'_d$, is usually much larger than subtransient saliency, $X''_q \neq X''_d$, so that the iteration process for the subtransient model converges faster, and with fewer iterations, than that for the transient model. Unfortunately the reduction in computing time due to the faster convergence is partially offset by the shorter integration time step required by the subtransient model to account for the smaller time constants. As subtransient saliency effects are normally quite small Dandeno & Kundur (1973) suggest that in order to produce a fast, non-iterative, solution the subtransient model with subtransient saliency effects neglected should be used. By adopting this approach a solution is obtained that is more accurate than that produced by the transient model where the damper windings are neglected. In this non-iterative algorithm the iterations required to account for load non-linearity are performed only at discontinuities, that is only at the time of the disturbance. Except at disturbances the change of the voltage at the load nodes is smooth and slow so that the correction currents can be approximately calculated in each integration step by basing them on the voltages in the previous step, i.e.

$$\Delta \mathbf{I}_{L(k+1)} = \Delta \mathbf{I}_L(\mathbf{V}_{L(k)}) \tag{13.50}$$

With these assumptions the network equations can be solved non-iteratively, apart from the instants of discontinuity, using either partial matrix inversion or matrix factorisation. At discontinuities changes in the voltages may be large and it is necessary to execute a few iterations in order to accurately calculate the correction currents. Adibi, Hirsch & Jordan (1974) noted that small errors resulting from 13.50 can be partially eliminated by extrapolating the voltages at the load nodes thereby improving the estimated correction currents.

Based on these assumptions the solution algorithm in Figure 13.3 can be modified. When solving the algebraic equations at the discontinuity, stage 3, equations 13.21 or 13.34 can be used to introduce the iterations necessary to model the non-linear loads. Solution of the algebraic equations at stage 5, after the prediction, can be replaced by extrapolation of the algebraic variables $\mathbf{y}(t)$, while the solution at stage 7, after correction, can be executed using equations 13.20 or 13.33 to solve the network equations non-iteratively.

13.3 The Simultaneous Solution Methods

The concept of the simultaneous-solution methods is to use implicit integration formulae to change the differential equations 13.1 into algebraic form and then to solve these algebraic equations simultaneously with the algebraic network equations 13.2.

Any implicit integration formula can be written in the general form

$$x_{k+1} = \beta_k + hb_0 f(x_{k+1}) \tag{13.51}$$

where h is the integration step length, b_0 is a coefficient that depends on the actual integration method, $f(x_{k+1})$ is the right-hand side of the differential equations 13.1 calculated at the value x_{k+1} and

$$\beta_k = x_k + \sum_j b_j f_{k+1-j} \tag{13.52}$$

is a coefficient depending on all the previous steps. Using the formula in equation 13.51, equations 13.1 and 13.2 can be rewritten as

$$\begin{aligned} \mathbf{F}_1(\mathbf{x}_{k+1}, \mathbf{y}_{k+1}) &= \mathbf{f}(\mathbf{x}_{k+1}, \mathbf{y}_{k+1}) - \frac{1}{hb_0}\mathbf{x}_{k+1} - \boldsymbol{\beta}_k = \mathbf{0} \\ \mathbf{F}_2(\mathbf{x}_{k+1}, \mathbf{y}_{k+1}) &= \mathbf{g}(\mathbf{x}_{k+1}, \mathbf{y}_{k+1}) = \mathbf{0} \end{aligned} \tag{13.53}$$

where $\boldsymbol{\beta}_k$ is a column vector containing the values β_k.

Newton's method gives the iteration formula as

$$\begin{bmatrix} \mathbf{x}_{k+1}^{(l+1)} \\ \mathbf{y}_{k+1}^{(l+1)} \end{bmatrix} = \begin{bmatrix} \mathbf{x}_{k+1}^{(l)} \\ \mathbf{y}_{k+1}^{(l)} \end{bmatrix} - \begin{bmatrix} \mathbf{f}_x - \frac{1}{hb_0}\mathbf{1}, & \mathbf{f}_y \\ \mathbf{g}_x, & \mathbf{g}_y \end{bmatrix}^{-1} \begin{bmatrix} \mathbf{F}_1(\mathbf{x}_{k+1}^{(l)}, \mathbf{y}_{k+1}^{(l)}) \\ \mathbf{F}_2(\mathbf{x}_{k+1}^{(l)}, \mathbf{y}_{k+1}^{(l)}) \end{bmatrix} \tag{13.54}$$

where $\mathbf{1}$ is the unity diagonal matrix and $\mathbf{f}_x = \partial\mathbf{f}/\partial\mathbf{x}$, $\mathbf{f}_y = \partial\mathbf{f}/\partial\mathbf{y}$, $\mathbf{g}_x = \partial\mathbf{g}/\partial\mathbf{x}$, $\mathbf{g}_y = \partial\mathbf{g}/\partial\mathbf{y}$ are the Jacobian submatrices. The Jacobian matrix in equation 13.54 is sparse so that computer programs that simulate large systems do not generally explicitly invert this matrix. Instead equation 13.54 is solved using triangular factorisation and forward and back substitution. The network equations are expressed in rectangular form in the system (a,b) reference frame so that rotor saliency can be included without any difficulty. The non-linear load correction currents are modified during each iteration. The effectiveness of the method depends on both the choice of the variables to be iterated in the Newton method and on skilful use of sparse matrix techniques. A significant role is also played by appropriate grouping of the variables allowing block matrices to be used.

Vorley (1974) presents one of the variants of this method where equations 13.1 and 13.2 are arranged in such a way that the generator differential and algebraic equations are grouped together to produce an equation of the form

$$\left[\begin{array}{ccc|ccc} 1 & \dots & 0 & & & \\ \vdots & \ddots & \vdots & & & \\ 0 & \dots & 1 & & & \\ \hline & & & 0 & \dots & 0 \\ & & & \vdots & \ddots & \vdots \\ & & & 0 & \dots & 0 \end{array}\right] \left[\begin{array}{c} \dot{x}_1 \\ \vdots \\ \dot{x}_r \\ \hline \dot{x}_{r+1} \\ \vdots \\ \dot{x}_m \end{array}\right] = \left[\begin{array}{c} f_1 \\ \vdots \\ f_r \\ \hline f_{r+1} \\ \vdots \\ f_m \end{array}\right] \quad \text{or} \quad \mathbf{c}_i\dot{\mathbf{x}}_i = \mathbf{f}_i(\mathbf{x}_i, \mathbf{V}) \qquad (13.55)$$

where $(x_1, \dots, x_r)$ are the variables of the differential equations describing the i-th generating unit and $(x_{r+1}, \dots, x_m)$ are the variables of the algebraic equations describing this unit. As matrix $\mathbf{c}_i$ is singular this equation is singular. The whole system is described by

$$\begin{aligned} \mathbf{C}\dot{\mathbf{x}} &= \mathbf{F}(\mathbf{x}, \mathbf{V}) \\ \mathbf{0} &= \mathbf{G}(\mathbf{x}, \mathbf{V}) \end{aligned} \qquad (13.56)$$

where the first of these equations consists of equations 13.55 corresponding to individual generating units whilst the second, describing the network, is the equation of nodal voltages. Using the implicit integration formulae and Newton's equation (as for equation 13.54), gives

$$\begin{bmatrix} \mathbf{x}_{k+1}^{(l+1)} \\ \mathbf{V}_{k+1}^{(l+1)} \end{bmatrix} = \begin{bmatrix} \mathbf{x}_{k+1}^{(l)} \\ \mathbf{V}_{k+1}^{(l)} \end{bmatrix} - \begin{bmatrix} \mathbf{F}_x - \dfrac{1}{hb_0}\mathbf{C} & \mathbf{F}_v \\ \mathbf{G}_x & \mathbf{G}_v \end{bmatrix}^{-1} \begin{bmatrix} \mathbf{F}(\mathbf{x}_{k+1}^{(l)}, \mathbf{V}_{k+1}^{(l)}) - \dfrac{1}{hb_0}\mathbf{C}(\mathbf{x}_{k+1}^{(l)} - \boldsymbol{\beta}_k) \\ \mathbf{G}(\mathbf{x}_{k+1}^{(l)}, \mathbf{V}_{k+1}^{(l)}) \end{bmatrix} \qquad (13.57)$$

where $\mathbf{F}_x = \partial\mathbf{F}/\partial\mathbf{x}$, $\mathbf{F}_v = \partial\mathbf{F}/\partial\mathbf{V}$, $\mathbf{G}_x = \partial\mathbf{G}/\partial\mathbf{x}$ and $\mathbf{G}_v = \partial\mathbf{G}/\partial\mathbf{V}$ are the submatrices of the Jacobian matrix.

The Jacobian matrices of the individual generating units have a block structure which simplifies the factorisation. To speed-up calculations the dishonest Newton method is used where the iterations in each integration step are executed for a constant matrix calculated from predicted values. It is also possible to simplify the method further by modifying the

Jacobian matrix only after network disturbances and after a certain number of integration steps. The number of iterations necessary for convergence can be used as an indicator as to when to modify the Jacobian matrix; it is updated if the number of iteration exceeds a pre-set value, e.g. three.

Variable integration step length and variable order interpolation formulae are also used. As the differential and algebraic equations are solved together there is no interfacing problem, while the use of Newton's method ensures no convergence problems, even when a long integration step length is used with a stiff system. At the start of each integration step extrapolated initial values are used in the iterations. Description of other examples of simultaneous-solution method can be found in Adibi, Hirsch & Jordan, 1974, Harkopf, 1978, Stott 1979, Rafian, Sterling & Irving, 1987.

13.4 Comparison between the Methods

The simultaneous-solution methods allow rotor saliency and non-linear loads to be readily included and are especially attractive for simulations that cover a long time period. Newton's method, together with implicit integration formulae, allow the integration step length to be increased when the changes in the variables are not very steep. The dishonest Newton method can be used to speed up the calculations. Interfacing problems between the algebraic and differential equations do not exist.

In contrast the partitioned-solution methods are attractive for simulations that cover a shorter time interval. They are more flexible, easier to organise, and allow a number of simplifications to be introduced to speed-up the solution. However, unless care is taken, these simplifications may cause large interfacing errors. The majority of dynamic simulation programs described in the literature are based on partitioned-solution methods.

The main characteristic of the partitioned-solution methods relates to the way in which the network equations are solved. Partial matrix inversion is only attractive for simplified systems because the submatrices of the partially inverted nodal matrix are dense. If the nodal matrix is large these submatrices take a lot of computer memory. Additionally, because of the large number of non-zero elements in these submatrices, the number of arithmetic operations needed to solve the network equations is also large. The speed of solution can be improved by assuming that the loads are linear (constant admittances) and by calculating the voltages at only a small number of load nodes thereby limiting the size of the relevant inverted submatrices. This method becomes particularly attractive when model reduction is employed based on the aggregation of coherent generators as discussed in Chapter 11. In this case, when the algorithm is re-organised, the transfer matrix that is used to predict groups of coherent generators (after certain transformations corresponding to aggregation) can also be used to solve the equations of the reduced network.

If non-linear loads are included, or the voltage change at a certain number of loads is required, then triangular factorisation is superior to partial inversion as the factor matrices remain sparse after factorisation. For a typical power network the factor matrices only contain about 50% more elements than the original admittance matrix and the number of arithmetic operations required to solve the network is not very high. If additional modifications that limit the number of iterations due to rotor saliency and non-linear loads are included then triangular factorisation becomes by far the fastest solution method.

The properties of the computer algorithms that use Newton's method are similar to those for the simultaneous solution method. Compared with triangular factorisation Newton's method requires a larger computer memory and more arithmetic calculations per integration step. However, thanks to good convergence, Newton's method can use a longer integration step than the factorisation method which partially compensates for the greater number of computations per step. The use of the dishonest Newton method speeds up the calculations quite considerably. Moreover rotor saliency and non-linear loads can be included more easily than is the case with triangular factorisation.

It is worth adding that recently, with ever-increasing power of computers, there is a tendency to develop *real-time simulators* to train operators for dispatch and security monitoring and which can also be used as the core of an on-line dynamic security assessment system. To make these simulators operate in real-time, it is often required to split the program into independent task to be executed in parallel (Chai and Bose, 1993, Bialek, 1994).

Appendix: Per-unit System

A.1 Introduction

Perhaps the one area in power system analysis that causes more confusion than any other is that of the per unit system. This confusion is further compounded when a synchronous machine is included in the system. However the per unit system is well established and has a number of attractions. For example, by normalising the generator equations derived in Chapter 10, the parameters of generators of the same type, but different ratings, will fall within the same range thereby providing the engineer with an intuitive understanding of the generator's performance. Such a normalised set of parameters can also lead to computational efficiencies.

In the following sub-sections the per unit system used in this book is described. Firstly the base system used in the stator armature is described followed by a brief discussion on power invariance in both MKSA and per unit. The per unit system is then examined in more detail in order to derive base values for the different rotor circuits before finally explaining how the generator and the network per unit systems fit together.

A.2 Stator Base Quantities

The principal armature base values used are:

$$\text{Base voltage, } V_b = \text{generator line to neutral rms terminal voltage, } V_{L-N},$$
$$\text{(this will normally be the rated voltage)}$$
$$\text{Base power, } S_b = \text{the generator MVA rating/phase, } S_{1\phi}$$
$$\text{Base time, } t_b = 1\,\text{second}$$

These principal base values lead to the following derived base values:

$$\text{Base current,} \quad I_b = \frac{S_b}{V_b} = \frac{S_{1\phi}}{V_b}\,\text{A}$$

$$
\begin{aligned}
\text{Base impedance,}\quad Z_b &= \frac{V_b}{I_b}\,\Omega \\
\text{Base inductance,}\quad L_b &= \frac{V_b t_b}{I_b}\,\text{H} \\
\text{Base flux linkage,}\quad \Psi_b &= L_b I_b = V_b t_b \equiv V_b\ \text{Vs} \\
\text{Base electrical angle}\quad \theta_b &= 1\text{electrical radian} \\
\text{Base electrical speed}\quad \omega_b &= 1\text{electrical radian/s} \\
\text{Base mechanical angle}\quad \theta_{mb} &= 1\text{mechanical radian} \\
\text{Base mechanical speed}\quad \omega_{mb} &= 1\text{mechanical radian/s} \\
\text{Base machine power}\quad S_{3\phi} &= 3S_{1\phi}\ \text{VA} \\
\text{Base torque,}\quad T_b &= \frac{S_{3\phi}}{\omega_{sm}}\ \text{Nm}
\end{aligned}
$$

To use these base values any particular current, voltage etc. in MKSA is simply divided by the corresponding base value to obtain the per unit value (or vice versa)

$$
per\ unit\ value = \frac{actual\ value}{base\ value} \tag{A.1}
$$

It is important to note that with the voltage and current ABC/dq transformation equations introduced in Chapter 10 the same base values are used for the armature coils in both the A,B,C and the d,q reference frames. This is not the case with other values of transformation coefficient (Harris, Lawrenson & Stephenson 1970). As explained in Chapter 10 the ABC/dq transformation is power invariant such that

$$
v_a i_a + v_b i_b + v_c i_c = v_d i_d + v_q i_q \tag{A.2}
$$

With the base values defined above the transformations are power invariant in both MKSA and per unit notation as shown in Appendix A.3.

The following points should be noted:

1. With a base time of 1 second all time constants are expressed in seconds.
2. A per unit reactance is related to a per unit inductance by $X_{pu} = \omega L_{pu}$ so that the normal relationship between inductance and reactance is maintained. The per unit inductance is NOT EQUAL to the per unit reactance.
3. The definition of base torque is such that at synchronous speed per unit torque is equal to per unit power, e.g. a turbine torque of 0.8 pu corresponds to a turbine power of 0.8 pu. In general

$$
P = T\omega_m\ \text{(MKSA)} \tag{A.3}
$$

Dividing by $S_{3\phi}$

$$
\frac{P}{S_{3\phi}} = \frac{T\omega_m}{S_{3\phi}} = \frac{T\omega_m}{T_b\omega_{sm}}
$$

$$
P_{pu} = T_{pu}\frac{\omega_m}{\omega_{sm}} \tag{A.4}
$$

but as $\omega_m = 2\omega/p$ and $\omega_{sm} = 2\omega_s/p$

$$P_{pu} = T_{pu}\frac{\omega_m}{\omega_{sm}} = T_{pu}\frac{\omega}{\omega_s} \tag{A.5}$$

At Synchronous speed $\omega = \omega_s$ and

$$P_{pu} = T_{pu} \tag{A.6}$$

4. Under balanced operation the power output of a single phase, normalised to $S_{1\phi}$, is numerically the same in per unit as the generator power output normalised to $S_{3\phi}$. Under balanced operation

$$P_{1\phi} = V_{rms}I_{rms}\cos\phi, \quad P_{3\phi} = 3V_{rms}I_{rms}\cos\phi \tag{A.7}$$

Dividing by $S_{1\phi}$ and $S_{3\phi}$ respectively gives

$$P_{pu} = V_{pu}I_{pu}\cos\phi \tag{A.8}$$

This is an extremely useful identity particularly when balanced operation is being studied by means of a phasor diagram.

5. Because of the per unit notation adopted most of the equations developed in this book are the same whether the quantities are expressed in MKSA or in per unit. The two important exceptions to this are generator power and torque both of which must be normalised to the generator MVA base rather than the phase MVA base. Consequently

$$P_{pu} = \frac{P_{MKSA}}{S_{3\phi}} = \frac{1}{3}\left[\frac{P_{MKSA}}{V_bI_b}\right] \tag{A.9}$$

whilst

$$T_{pu} = \frac{T_{MKSA}}{T_b} = T_{MKSA}\frac{\omega_s}{S_{3\phi}} = \frac{\omega_s}{3}\left[\frac{T_{MKSA}}{V_bI_b}\right] \tag{A.10}$$

The implication of these two equations is that generator power or torque equations derived in MKSA can simply be converted to per unit by multiplying by 1/3 and $\omega_s/3$ respectively. See, for example, the torque expressions in Chapter 4.

6. Full load power (and torque) should not be confused with 1 pu power (and torque). They are not the same. In general

$$\text{Full load power} = S_{3\phi}\cos\phi_{rated}$$

$$\text{Full load torque} = T_b\cos\phi_{rated}$$

Mechanical engineers like to refer to a shaft rated, for example, as 4 times full load torque. This does not mean 4 times T_b - they are different by $\cos\phi_{rated}$!

7. Because of the base values used the relationships between v_d, v_q & V_d, V_q and i_d, i_q & I_d, I_q derived in Chapter 10 are valid in both MKSA and per unit, that is

$$v_{d\,pu} = \sqrt{3}V_{d\,pu} \qquad i_{d\,pu} = \sqrt{3}I_{d\,pu} \tag{A.11}$$

$$v_{q\,pu} = \sqrt{3}V_{q\,pu} \qquad i_{q\,pu} = \sqrt{3}I_{q\,pu} \tag{A.12}$$

A.3 Power Invariance

A check on power invariance in MKSA is useful. Under balanced conditions, and using the current and voltage identities of equations 10.78–10.82,

$$\begin{aligned} P_{3\phi} &= v_d i_d + v_q i_q = 3(V_d I_d + V_q I_q) \\ &= 3V_g I_g\ (\sin\delta_0 \sin(\delta_0 + \phi) + \cos\delta_0 \cos(\delta_0 + \phi)) \\ &= 3V_g I_g \cos\phi \quad W \end{aligned} \tag{A.13}$$

showing that power invariance is maintained.

As $P_{3\phi} = 3V_g I_g \cos\phi$, dividing both sides by $S_{3\phi}$ gives the generator power in per unit

$$\frac{P_{3\phi}}{S_{3\phi}} = \frac{v_d i_d + v_q i_q}{3V_b I_b} = \frac{3(V_d I_d + V_q I_q)}{3V_b I_b} = \frac{3V_g I_g \cos\phi}{3V_b I_b} = V_{gpu} I_{gpu} \cos\phi \tag{A.14}$$

$$P_{pu} = \tfrac{1}{3}(v_{dpu} i_{dpu} + v_{qpu} i_{qpu}) = (V_{dpu} I_{dpu} + V_{qpu} I_{qpu}) = V_{gpu} I_{gpu} \cos\phi \tag{A.15}$$

and power invariance is also maintained in the per unit system.

A.4 Rotor Base Quantities

Although a number of per unit systems are possible (Harris, Lawrenson & Stephenson, 1970) the system considered here is that of *equal mutual flux linkages* as expounded by Anderson & Fouad (1977) and also explained in depth by Pavella & Murthy (1994). In this system the base field current, or base d-axis damper current, are defined so that each will produce the same fundamental air-gap flux wave as that produced by the base armature current acting in the fictitious d-axis armature coil. As will be seen as a consequence of this choice of per unit system all the per unit mutual inductances on a particular axis are equal.

It is convenient at this stage to separate each individual winding self-inductance into a magnetising inductance and a leakage inductance so that

$$\begin{aligned} L_d &= L_{md} + l_l \qquad & L_q &= L_{mq} + l_l \\ L_D &= L_{mD} + l_D \qquad & L_Q &= L_{mQ} + l_Q \\ L_f &= L_{mf} + l_f \end{aligned} \tag{A.16}$$

where l represents the winding leakage inductance. The per unit system requires the mutual flux linkage in each winding to be equal, that is:

$$\begin{aligned} &d - coil: && L_{md} I_b = kM_f I_{fb} = kM_D I_{Db} \\ &f - coil: && kM_f I_b = L_{mf} I_{fb} = L_{fD} I_{Db} \\ &D - coil: && kM_D I_b = L_{fD} I_{fb} = L_{mD} I_{Db} \\ &q - coil: && L_{mq} I_b = kM_Q I_{Qb} \\ &Q - coil: && kM_Q I_b = L_{mQ} I_{Qb} \end{aligned} \tag{A.17}$$

Multiplying each of these winding mutual flux linkages by the coil base current gives the fundamental constraint between the base currents as

$$L_{md}I_b^2 = L_{mf}I_{fb}^2 = L_{mD}I_{Db}^2 = kM_f I_{fb} I_b = kM_D I_{Db} I_b = L_{fD} I_{fb} I_{Db}$$
$$L_{mq}I_b^2 = L_{mQ}I_{Qb}^2 = kM_Q I_b I_{Qb} \tag{A.18}$$

As the MVA base for each winding must be the same and equal to $S_b = V_b I_b$ this gives

$$\frac{V_{fb}}{V_b} = \frac{I_b}{I_{fb}} = \sqrt{\frac{L_{mf}}{L_{md}}} = \frac{kM_f}{L_{md}} = \frac{L_{mf}}{kM_f} = \frac{L_{fD}}{kM_D} \equiv k_f$$
$$\frac{V_{Db}}{V_b} = \frac{I_b}{I_{Db}} = \sqrt{\frac{L_{mD}}{L_{md}}} = \frac{kM_D}{L_{md}} = \frac{L_{mD}}{kM_D} = \frac{L_{fD}}{kM_f} \equiv k_D \tag{A.19}$$
$$\frac{V_{Qb}}{V_b} = \frac{I_b}{I_{Qb}} = \sqrt{\frac{L_{mQ}}{L_{mq}}} = \frac{kM_Q}{L_{mq}} = \frac{L_{mQ}}{kM_Q} \equiv k_Q$$

As equation A.19 defines the base currents and voltages in all the windings as a function of the stator base quantities V_b and I_b

$$Z_{fb} = \frac{V_{fb}}{I_{fb}} = k_f^2 Z_b \quad \Omega, \quad Z_{Db} = \frac{V_{Db}}{I_{Db}} = k_D^2 Z_b \quad \Omega, \quad Z_{Qb} = \frac{V_{Qb}}{I_{Qb}} = k_Q^2 Z_b \quad \Omega \tag{A.20}$$

and

$$L_{fb} = \frac{V_{fb}t_b}{I_{fb}} = k_f^2 L_b \text{ H}, \quad L_{Db} = \frac{V_{Db}t_b}{I_{Db}} = k_D^2 L_b \text{ H}, \quad L_{Qb} = k_Q^2 L_b \text{ H} \tag{A.21}$$

whilst the base mutual inductances are

$$M_{fb} = \frac{V_{fb}t_b}{I_b} = \frac{V_b t_b}{I_{fb}} = k_f L_b \text{ H}, M_{Db} = k_D L_b \text{ H}$$
$$M_{Qb} = k_Q L_b \text{ H}, L_{fDb} = k_f k_D L_b \text{ H} \tag{A.22}$$

With the base values now defined equations 10.18, 10.19, 10.30 and 10.31 can be normalised and expressed in per unit. As an example consider the normalisation of the field flux linkage Ψ_f in equation 10.18 where

$$\Psi_f = kM_f i_d + L_f i_f + L_{fD} i_D \tag{A.23}$$

Divide by $\Psi_{fb} = L_{fb} I_{fb}$ to give

$$\Psi_{f\,pu} = \frac{kM_f}{L_{fb}}\frac{i_d}{I_{fb}} + \frac{L_f}{L_{fb}}\frac{i_f}{I_{fb}} + \frac{L_{fD}}{L_{fb}}\frac{i_D}{I_{fb}} \tag{A.24}$$

when substituting for I_{fb} and L_{fb} from equations A.19 and A.20 gives

$$\Psi_{f\,pu} = \left[\frac{kM_f}{k_f L_b}\right]\left[\frac{i_d}{I_b}\right] + \left[\frac{L_f}{L_{fb}}\right]\left[\frac{i_f}{I_{fb}}\right] + \left[\frac{L_{fD}}{k_f k_D L_b}\right]\left[\frac{i_D}{I_{Db}}\right] \tag{A.25}$$

and

$$\Psi_{f\,pu} = kM_{f\,pu} i_{d\,pu} + L_{f\,pu} i_{f\,pu} + L_{fD\,pu} i_{D\,pu} \tag{A.26}$$

This normalised equation is of exactly the same form as the equation in MKSA and this is true for all other equations in section 10.1.4. In other words *all the voltage, current and flux equations in section 10.1.4 are of the same form whether in per unit or MKSA.*

One further interesting feature of the per-unit system is that the per-unit values of all the mutual inductances on one axis are equal, that is L_{md}, L_{mf}, L_{mD}, kM_f, kM_D and L_{fD} are all equal. For example

$$kM_{f\,pu} = \frac{kM_f}{M_{fb}} = \frac{kM_f}{k_f L_b} = \frac{k_f L_{md}}{k_f L_b} = L_{md\,pu}$$

$$L_{fD\,pu} = \frac{L_{fD}}{L_{fDb}} = \frac{L_{fD}}{k_f k_D L_b} = \frac{L_{fD}}{\dfrac{kM_f}{L_{md}}\dfrac{L_{fD}}{kM_f} L_b} = \frac{L_{md}}{L_b} = L_{md\,pu} \tag{A.27}$$

It is common practise to replace all these per unit mutual values by a per unit mutual inductance L_{ad} so that

$$L_{ad} \equiv L_{md} = L_{mf} = L_{mD} = kM_f = kM_D = L_{fD} \tag{A.28}$$

and, in the q-axis,

$$L_{aq} \equiv L_{mq} = L_{mQ} = kM_Q \tag{A.29}$$

All the equations in section 10.1.4 can now be written in per unit in terms of the mutual inductances L_{ad} and L_{aq}. For example equation 10.43 for the d-axis subtransient inductance would become, in per unit,

$$L''_d = L_d - \left[\frac{L_{ad}^2 L_D + L_{ad}^2 L_f - 2L_{ad}^3}{L_D L_f - L_{ad}^2}\right] \tag{A.30}$$

where $L_d = L_{ad} + l_1$, $L_D = L_{ad} + l_D$ and $L_f = L_{ad} + l_f$.

With this knowledge it is constructive to examine the per unit flux linking each winding. Using equation 10.18, and dropping the pu symbol for simplicity, the flux linkage of the d-axis coil is

$$\Psi_d = L_d i_d + kM_f i_f + kM_D i_D \tag{A.31}$$

when substituting for L_{ad} and introducing the winding leakage inductance gives

$$\Psi_d = L_{ad}(i_d + i_f + i_D) + l_l i_d \tag{A.32}$$

and similarly for the field and d-axis damper coil

$$\Psi_f = L_{ad}(i_d + i_f + i_D) + l_f i_f$$
$$\Psi_D = L_{ad}(i_d + i_f + i_D) + l_D i_D \tag{A.33}$$

Thus if the per-unit leakage flux linkage of a particular winding is subtracted from the total flux linkage the remaining mutual flux linkage in all the windings on each axis is equal. This mutual flux linkage is often given the symbol Ψ_{ad}, and for the q-axis Ψ_{aq}, where

$$\begin{aligned} \Psi_{ad} &= L_{ad}(i_d + i_f + i_D) \\ \Psi_{aq} &= L_{aq}(i_q + i_Q) \end{aligned} \tag{A.34}$$

A.5 Power System Base Quantities

It is customary in three-phase power system analysis to use rated line-to-line voltage as the base voltage and an arbitrary three-phase volt-amp base, typically 10 MVA, 100 MVA etc. Such a base system would, at first sight, seem to be totally inconsistent with the generator armature base defined in section A.1. In fact the two are entirely consistent. For the power system

$$\begin{aligned} V_{L-L,b} &= V_{L-L} = \sqrt{3}V_{L-N} \text{ V} \\ S_b &= S_{3\phi} \text{ VA} \\ I_b &= \frac{S_{3\phi}}{\sqrt{3}\, V_{L-L,b}} \text{ A} \\ Z_b &= \frac{V_{L-L,b}}{\sqrt{3}\, I_b} = \frac{V_{L-L,b}^2}{S_{3\phi}} \ \Omega \end{aligned} \tag{A.35}$$

Ignoring any transformer effects and assuming that the system MVA base $S_{3\phi}$ is equal to the generator MVA rating then $V_{L-N} = V_b$, where V_b is the generator base voltage, and

$$\begin{aligned} Z_b &= \frac{V_{L-L,b}^2}{S_{3\phi}} = \frac{3V_{L-N}^2}{S_{3\phi}} = \frac{V_b^2}{S_{1\phi}} \\ I_b &= \frac{S_{3\phi}}{\sqrt{3}V_{L-L,b}} = \frac{S_{1\phi}}{V_b} \end{aligned} \tag{A.36}$$

showing that the power system base and the generator base are totally consistent.

However there is one complication and that is $S_{3\phi}$ for the system is chosen arbitrarily whilst $S_{1\phi}$ for the generator is the rated MVA per phase. In fact for all the equipment making up the power system their per unit impedance values will be defined with respect to their individual MVA ratings. Consequently in the system analysis it will be necessary to either:

(i) convert all the generator parameters to be on the system base or
(ii) have a base conversion between the individual generator equations and the system equations in the computer software. This is easily achieved and has the advantage that not only do the generator per unit values retain their familiarity but they are exactly as provided by the equipment manufacture.

Both methods are used and converting per unit values from one system (base 1) to another (base 2) is readily achieved via equation A.1 to obtain

$$\textit{per unit value (base 2)} = \textit{per unit value (base 1)}\frac{\textit{Base 1 value}}{\textit{Base 2 value}} \tag{A.37}$$

A.6 Transformers

Section 3.2 showed how a transformer could be represented by either the primary or the secondary equivalent circuit shown in Figure A.1a and A.1b. In these equivalent circuits the primary equivalent impedance Z_1 and the secondary equivalent impedance Z_2 are related by

$$Z_1 = n^2 Z_2 \tag{A.38}$$

where n is the nominal turns ratio. However the primary and secondary base values are defined as

$$\begin{aligned} V_{pb} &= V_{1,L-N} & V_{sb} &= V_{2,L-N} \\ I_{pb} &= \frac{S_{3\phi}}{3V_{pb}} & I_{sb} &= \frac{S_{3\phi}}{3V_{sb}} \\ Z_{pb} &= \frac{V_{pb}}{I_{pb}} & Z_{sb} &= \frac{V_{sb}}{I_{sb}} \end{aligned} \tag{A.39}$$

where

$$V_{pb} = nV_{sb} \tag{A.40}$$

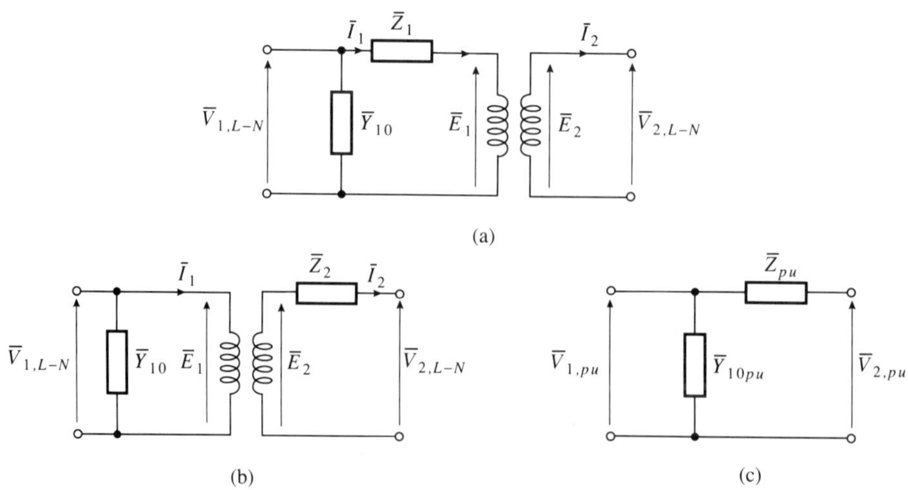

Figure A.1 Transformer equivalent circuit: (a) primary; (b) secondary; (c) per unit at nominal tap setting

and implies that

$$I_{pb} = \frac{I_{sb}}{n}, \qquad Z_{pb} = n^2 Z_{sb} \tag{A.41}$$

Consequently as 1 per unit voltage on the high voltage side of the transformer must be 1 per unit on the low voltage side and

$$Z_{pu} = \frac{Z_1}{Z_{pb}} = \frac{Z_2 n^2}{n^2 Z_{sb}} = \frac{Z_2}{Z_{sb}} = Z_{pu} \tag{A.42}$$

the per unit value of the primary and secondary equivalent impedance is the same, so that the transformer at nominal taps can be represented by the per unit equivalent shown in Figure A.1c. If the tap setting changes from the nominal the equivalent circuit is modified to that shown in Figure 3.8.

References

1. ABB "An introduction to ABB series capacitors" *ABB Information Publication*, September 9, 1991.
2. Abe, S., Isono, A. "Determination of power system voltage stability" *Electrical Engineering in Japan*, **103**, No. 3, 1983.
3. Adibi, M.M., Hirsch, P.M., Jordan, J.A. "Solution methods for transient and dynamic stability." *Proc. IEEE*, **62**, No. 7., pp 951–958, July 1974.
4. Adkins, B. *The general theory of electrical machines*. Chapman and Hall, 1957.
5. Ahlgren, L, Johansson, K.E. and Gadhammar, A. "Estimated life expenditure of turbine-generator shafts at network faults and risk for subsynchronous resonance in the Swedish 400 kV system", *IEEE Trans.*, **PAS-97**, No. 6, Nov/Dec 1978.
6. Ajjarapu, V., Christy, C. "The continuation power flow: a tool for steady-state voltage stability analysis" *IEEE Trans.*, **PWRS-7**, No. 1, pp. 416–423, 1992.
7. Alsac, O., Stott, B., Tinney, W.F. "Sparsity oriented compensation methods for modified network solutions" *IEEE Trans.*, **PAS-102**, pp. 1050–1060, 1983.
8. Anderson, P.M., Agrawal, B.L., Van Ness, J.E. *Subsynchronous resonance in power systems*. IEEE Press, 1990.
9. Anderson, P.M., Fouad, A.A. *Power System Control and Stability*. The Iowa State University Press, 1977.
10. Arrillaga, J., Arnold, C.P., Harker, B.J. *Computer modelling of electrical power systems*. John Wiley and Sons, Chichester 1983.
11. Arrillaga, J., Arnold, C.P. *Computer Analysis of Power Systems*. John Wiley & Sons, Chichester 1990.
12. Ashok Kumar, B.S.A., Parthasawathy, K., Prabhakara, F.S., Khincha, H. P. "Effectiveness of series capacitors in long distance transmission lines" *IEEE Trans.*, **PAS-89**, no. 4, pp. 941–950, 1970.
13. Balu, N.J. "Fast turbine valving and independent pole tripping breaker applications for plant stability" *IEEE Trans.*, **PAS-99**, No. 4, 1980.
14. Bellman, R. *Introduction to matyrix analysis*. McGraw-Hill, 2nd Ed., 1970.
15. Berg, G.L. "Power system load representation" *IEE Proc.*, **120**, pp. 344348, 1973.
16. Bernas, S. "Zastepowanie grupy generatorow przy badaniu stabilnosci systemu elektroenergetycznego" *Prace Naukowe PW*, Elektryka, **17**, 1971.
17. Bernas, S. *Systemy elektroenergetyczne*. WNT, Warsaw, 1982.
18. Bialek, J. "Tracing the flow of electricity" *IEE Proc. Gen. Transm. and Distrib.*, **143**, No. 4, July 1996, pp. 313–320.

19. Bialek, J, Grey, D.J. "Application of clustering and factorisation tree techniques for parallel solution of sparse network equations" *IEE Proc. Gen. Transm. and Distrib.*, **141**, No. 6, November 1996, pp. 609-616.
20. Boening, H.J., Hauer, J.F. "Commisioning tests of the Bonneville Power Administartion 30 MJ superconducting magnetic storage unit" *IEEE Trans.*, **PAS-104**, no. 2, pp. 302-312, 1985.
21. Bölderl, P., Kulig, T., Lambrecht, D. "Beurteilung der Torsionsbeanspruchung in den Wellen von Turbosätzen bei wiederholt auftrenden Störungen im Laufe der Betriebszeit" *ETZ-A*, **96**. H. 4, 1975.
22. Bourgin, F., Testud, G., Heilbronn, B., Verseille, J. "Present practices and trends on the French power system to prevent voltage collapse" *IEEE Trans.*, **PWRS-8**, no. 3, pp. 778-788, 1993.
23. Brameller, A., Allan, R.N., Hamam, Y.M. *Sparsity. Its practical application to system analysis*. Pitman Publishing, London, 1976.
24. British Electricity International *"Modern power station practice"* **D**, Electrical systems and equipment, Pergamon Press, 1991a.
25. British Electricity International *"Modern power station practice"* **F**, Control and Instrumentation, Pergamon Press, 1991b.
26. British Electricity International *"Modern power station practice"* **L**, System operation, Pergamon Press, 1991c.
27. Brown, H.E, Shipley, R.B., Coleman, D., Nied, R.B. "A study of stability equivalents" *IEEE Trans.*, **PAS-88**, No. 3, 1969.
28. Brown, P.G, de Mello, F.P., Lenfest, E.H., Mills, R.J. "Effects of excitation, turbine energy control and transmission on transient stability" *IEEE Trans.*, **PAS-89**, No. 6, 1970.
29. Brown, H.E. *Solution of large networks by matrix methods*. John Wiley and Sons, Chichester 1975.
30. Bumby, J.R. "Torsional natural frequencies in the superconducting turbogenerator" *IEE Proc.*, **129**, Pt.C, No. 4, pp. 141-151, July 1982.
31. Bumby, J.R. *Superconducting rotating electrical machines*. Clarendon Press, Oxford, 1983.
32. Bumby, J.R. and Wilson, J.M. 'Structural modes and undamped torsional natural frequencies of a superconducting turbogenerator rotor', *Journal of Sound and Vibration*, **87**, No. 4, April 1983, pp 589-602.
33. Cai, Y.Q., Wu, C.S. "A novel algorithm for aggregating coherent generating units" *IFAC Symp. on Power System and Power Plant Control*, Beijing, China, 1986.
34. Carpentier, J., Girard, R., Scano, E. "Voltage collapse proximity indicators computed from an optimal power flow" *Proc. of the 8th Power System Computing Conf.*, pp. 671-678, Helsinki, 1984.
35. Cegrell, T. *Power System Control Technology*. Prentice-Hall Intern. (UK), 1986.
36. Chai, J.S., Bose, A. "Bottlenecks in parallel algorithms for power system stability analysis" *IEEE Trans.*, **PWRS-8**, No. 1, pp. 9-15, 1993.
37. Ching, Y.K., Adkins, B.A. "Transient Theory of Synchronous Generators under Unbalanced Conditions" *IEE Proc.*, **101**, Pt. IV, pp 166-182, 1954.
38. Christie, R.D., Bose, A. "Load frequency control issues in power system operations after deregulation" *IEEE Trans.*, **PWRS-11**, No. 3, pp. 1191-1191, August 1996.
39. Chua, L.O. and Lin, P-M. *Computer-aided analysis of electronic circuits*. Prentice-Hall, 1975.
40. CIGRE Task Force 38-01-02 "Static VAR compensators" 1986.

41. CIGRE Working Group 38.01 "Planning against voltage collapse", *Electra*, pp. 55–75, March 1987.
42. CIGRE, paper 37/38-01, Working Group 38.04: Ultra High Voltage Technology, *CIGRE Session, 1994*.
43. Concordia, C. *Synchronous machines. Theory and performance*. Wiley, New York, 1951.
44. Concordia, C., Ihara, S. "Load Representation in Power System Stability Studies" *IEEE Trans.*, **PAS-101**, No 4, 1982.
45. Contaxis, G, Debs, A.S. "Identification of external equivalents for steady-state security assessment" *IEEE Power Summer Meeting*, Paper F 77-526-7, July 1977.
46. Crary S.B. *Power system stability*. Vol. **I, II**, John Wiley and Sons, New York, 1945, 1947.
47. Cushing, E.W., Drechsler, G.E., Killgoar, W.P., Marshall, H.G., Steward, H.R. "Fast valving as an aid to power system transient stability and prompt resynchronization and rapid reload after full load rejection" *IEEE Trans.*, **PAS-91**, no. 2, pp. 1624–1636, 1972.
48. Dahl, O.G.C. *Electric power circuits - theory and applications*, McGraw-Hill, New York, 1938.
49. Dandeno, P., Kundur, P. "Non-iterative transient stability program including the effects of variable load voltage characteristics" *IEEE Trans.*, **PAS-92**, no. 5, 1973.
50. De Mello, F.P., Concordia, C. "Concepts of synchronous machine stability as affected by excitation control" *IEEE Trans.*, **PAS-88**, No. 4, pp. 316–329, 1969.
51. De Mello, R.W., Podmore, R., Stanton, K.N. "Coherency based dynamic equivalents: applications in transient stability studies" *PICA Conf.*, 1975.
52. Debs, A.S. "Estimation of external network equivalents from internal system data" *IEEE Trans.*, **PAS-94**, No. 2, pp. 273–289, March/April 1975.
53. Debs, A.S. *Modern Power System Control and Operation*. Kluwer Academic Publishers, 1988.
54. Deckmann, S., Pizzolante, A., Monticelli, A., Stott, B. "Numerical testing of power system load-flow equivalencing" *IEEE Trans.*, **PAS-99**, pp. 2292–2300, No. 6, Nov. 1980.
55. Deville, T.G., Schweppe, F.C. "On line identification of interconnected network equivalents from operating data" *IEEE Power Summer Meeting*, Paper C 72-464-6, July 1972.
56. Di Caprio, U., Marconato, R. "A novel criterion for the development of multi-area simplified models oriented to on-line evaluation of power system dynamic security" *Proc. PSCC*, Cambridge, UK, 1975.
57. Dimo, P. *L'analyse nodale des reseaux d'energie*. Eyrolles Ed., Paris, 1971.
58. Dommel H.W., Sato, N. "Fast transient stability solution" *IEEE Trans.*, **PAS-91**, No. 4, pp. 1643–1650, 1972.
59. Dopazo, J.F., Dwarakonath, M.H., Li, J.J., Sasson, A.M. "An external system equivalent model using real-time measurements for system security evaluation" *IEEE Trans.*, **PAS-96**, March/April 1977.
60. Duff, I.S., Erisman, A.M. and Reid, J.K. *Direct methods for sparse matrices*. Oxford University Press, 1986.
61. Dunlop, R.D., Gutman, R., Marchenko, R.P. "Analytical development of loadability characteristic for EHV and UHV transmission lines" *IEEE Trans.*, **PAS-98**, pp 606–617, 1979.
62. Edelmann, H. *Berechung elektrischer Verbundnetze*. Berlin, Springer-Verlag, 1963.
63. Edelmann, H. "Direkte Verfahren der Netzanalyse mit sparlichen Matritzen" *Nachrichtechnische Zeitschrift*, Heft 2 und 3, 1974.

64. El-Abiad, A.H. *Power system analysis and planning*. Hemisphere Publishing Corporation, Washington-London, 1983.
65. Elgerd, O. *Electric Energy Systems Theory : An Introduction*. 2d ed. McGraw-Hill, New York, 1982.
66. El-Hawary, M.E. *Electric power systems - design and analysis*. Reston Publishing Company, Prentice-Hall Company, Reston, Virginia, 1983.
67. El-Sherbiny, M.K., Metha, D.M. "Dynamic system stability. Part 1: Investigation of the effect of different loading and excitation systems" *IEEE Trans.*, **PAS-92**, 1973.
68. El-Wakil, M.M. *Power Plant Technology*. McGraw-Hill, New York, 1984.
69. EPRI "Flexible AC transmission Systems (FACTS). Scoping study" *EPRI Report EL-6943*, Final Report on RP3022-02 by GE, September 1991.
70. Fahlen, N.T. "Series capacitors in power transmission: design and experience" *Int. Conf. High Voltage DC and/or AC Power Transmission*, London, IEE Conf. Publication No. 107, 19-23 November, 1973.
71. Fahlen, N.T. "EHV series capacitor equipment protection and control" *IEE Proc.*, **128**, Pt. C, 1981.
72. Fouad, A.A., Vittal, V. *Power System Transient Stability Analysis Using The Transient Energy Function Method*. Prentice Hall, Englewood Cliffs, New Jersey 07632, 1992.
73. Garmond, A.J., Podmore, R. "Dynamic aggregation of generating unit models" *IEEE Trans.*, **PAS-97**, No. 4, 1978.
74. Gavrilovic, M.M. "Optimal control of SMES systems for maximum power system stability and damping" *Electrical Power & Energy Systems*, **17**, No. 3, pp. 199–205, 1995.
75. Giles, R.L. *Layout of E.H.V. Substations*. Cambridge at the University Press, 1970.
76. Glover, J.D. Sarma M. *Power System Analysis and Design*. PWS Publishing Company, Second Edition, Boston 1994.
77. Gopal, R.I., Jones, C.Y. "Line-to -Line Short Circuit of Synchronous Machine: Ilustration of Computer-Aided Machine Analysis" *IEE Proc.*, No 1, pp 161–168, January 1971.
78. Grainger, J.J., Stevenson, W.D. *Power System Analysis*. McGraw-Hill, 1994.
79. Grebe, E., Handschin, E., Haubrich, H.J. and Traeder, G. "Dynamische Langezeitstabilitat von Netzen" Elektrizitatswirtschaft, Jg 78, Heft 19, 1979.
80. Greenwood, A. *Electrical Transients in Power Systems*. Wiley-Interscience, 1971.
81. Gross, C.A. *Power System Analysis*. 2nd. edition. Wiley, New York 1986.
82. Gubina, F., Omahen, P., Machowski, J. "Dynamic properties of a power system equivalent model for transient stability studies" *34th Congress on Electronics*, MELECON'87, Rome, Italy, pp 577–580, 1987
83. Gubina, F., Bakic, K, Omahen, Hrovatin, J., Jakl, F. "Economical, planning and operational aspects of East-West power transmission over the Slovenian power network" *35th CIGRE Session*, Paris,1994.
84. Guruprasada Rau, V., Krishnamoorthy, N. *Stability Analysis of Power Systems*. Omega Scientific Publishers, New Delhi, India, 1994.
85. Güth, G., Baker, R., Eglin, P. "Static thyristor-controlled regulating transformer for ac transmission " . *Int. Conf. Thyristor and Variable Static Equipment for AC and DC Transmission*, IEE Conf. Publication No. 205, 1981.
86. Hammons, T.J., Winning, D.J. "Comparisons of synchronous machine models in the study of the transient behaviour of electrical power systems" *Proc. IEE*, **118**, No. 10, Oct. 1971.

87. Handschin, E., Stephanblome, T. "New SMES strategies as a link between network and power plant control" *Int. IFAC Symp. Power Plants and Power System Control*, Munich, Germany, 9-11 March 1992.
88. Harkopf, T. "Simulation of power system dynamics using trapezoidal rule and Newton's method" *Proc. PSCC Conf.*, Darmstadt, 1978.
89. Harris, M.R., Lawrenson, P.J., Stephenson, J.M. *Per unit systems with special reference to electrical machines*. Cambridge University Press, 1970.
90. Hassenzahl, W.V. "Superconducting magnetic energy storage" *Proc. IEEE*, **71**, pp. 1089-1098, Sept. 1983.
91. Hicklin, J., Grace, A. *Simulink*. Math Works Inc., 1992.
92. Hill, D.J. "Nonlinear dynamic load models with recovery for voltage stability studies" *IEEE Trans.*, **PWRS-8**, No. 1, pp. 166-176, 1993.
93. Hingorani, N.G. "FACTS technology and opportunities" *IEE Colloquium "FACTS - the key to increased utilisation of power systems"*, 12 January 1994.
94. Horowitz, S.H., Phadke, A.G. *Power system relaying*. J. Wiley & Sons, 1992.
95. Humpage, W.D., Stott, B. "Predictor-corrector methods of numerical integration in digital computer analysis of power system transient stability" *IEE Proc.*, **112**, pp 1557-1565, Aug. 1965.
96. Humpage, W.D., Wong, K.P., Lee, Y.W. "Numerical integration algorithms in power-system dynamic analysis" *IEE Proc.*, **121**, pp 467-473, June 1974.
97. Huwer, R. *Robuste Power System Stabilizer für Mehrmaschinennetze*. Ph.D. Thesis, Universität Kaiserslautern, 1992.
98. IEE *"Power System Protection."* (4 volumes), Ed. by The Electricity Training Association, London, 1995.
99. IEEE Committee Report "Computer representation of excitation systems" *IEEE Trans.*, **PAS-87**, No. 6, pp. 1460-1464, 1968.
100. IEEE Committee Report "Recommended phasor diagrams for synchronous machines" *IEEE Trans.*, **PAS-88**, No. 11, pp. 1593-1610, 1969.
101. IEEE Committee Report "Excitation system dynamic characteristic" *IEEE Trans.*, **PAS-92**, No. 1, 1973a.
102. IEEE Committee Report "Dynamic models for steam and hydroturbines in power system studies" *IEEE Trans.*, **PAS-92**, No. 6, pp. 1904-1915, 1973b.
103. IEEE Committee Report "System load dynamics simulation effects and determination of load constants" *IEEE Trans.*, **PAS-92**, No. 2, pp. 600-609,1973c.
104. IEEE Power System Relaying Committee Report "Out-of-step relaying for generators" *IEEE Trans.*, **PAS-96**, No. 5, pp. 1556-1564, 1977.
105. IEEE Committee Report "Excitation system models for power system stability studies" *IEEE Trans.*, **PAS-100**, No. 2, pp. 494-509, 1981.
106. IEEE Working Group Report of panel discussion "Turine fast valving to aid system stability : benefits and other considerations" *IEEE Trans.*, **PWRS-1**, no. 2, pp. 143-153, 1986.
107. IEEE Committee Report "Dynamic models for fossil fuelled steam units in power system studies" *IEEE Trans.*, **PWRS-6**, No. 2, pp 753-761, 1991.
108. IEEE Committee Report "Hydraulic turbine and turbine control models for system dynamic studies" *IEEE Trans*, **PWRS-7**, No. 1, pp 167-179, 1992.
109. IEEE Task Force on Load Representation for Dynamic Performance "Load represenation for dynamic performance analysis" *IEEE Trans*, **PWRS-8**, No. 2, pp 472-482, 1993.

110. IEEE Committee Report "Static VAR compensator models for power flow and dynamic performance simulation" *IEEE Trans*, **PWRS-9**, No. 1, pp 229–240, 1994.
111. IEEE Task Force on Load Representation for Dynamic Performance "Standard load models for power flow and dynamic performance simulation" *IEEE Trans*, **PWRS-10**, No. 3, pp 1302–1312, 1995.
112. Iliceto, F., Cinieri, E. Comparative analysis of series and shunt compensation schemes for ac transmission systems" *IEEE Trans*, **PAS-96**, No. 1, pp 167–179, 1977.
113. Irvani, M.R., Maratukulam, D. "Review of semiconductor-controlled (static) phase shifters for power system applications" *IEEE PES Winter Meeting*, Paper No. PWRS 182-6, 1994.
114. Irving, M.R., Sterling, M.J.H. "Efficient Newton–Raphson algorithm for load-flow calculation in transmission and distribution networks" *IEE Proc.*, **134**, Pt. C, No. 5, pp. 325–328, 1987.
115. Ise, T., Murakami, Y., Tsuji, K. "Simultaneous active and reactive power control of superconducting magnetic storage using GTO converters" *IEEE Trans.*, **PWRD-1**, No. 1, pp. 143–150, 1986.
116. Jancke, G., Fahlen, N., Nerf, O. "Series capacitors in power systems" *IEEE Trans.*, **PAS-94**, pp. 915–925, 1975.
117. Jäger, J., Herold, G., Hosemann, G. "Current-source controlled phase-shifting transformer as a new equipment for high dynamical power flow control" Proc. APE'95 Conference, Gdansk, Poland, 12-14 June 1995.
118. Johns, A.T., Salman, S.K. *Digital Protection for Power Systems.* IEE Power Series, No. 15, 1994.
119. Jones, C.V. *The Unified Theory of Electrical Machines.* Butterworth 1967.
120. Kamwa, I, Grondin, R. "Fast adaptive scheme for tracking voltage phasor and local frequency in power transmission and distribution systems" *IEEE Trans.*, **PWRD-7**, No. 2, pp. 789–795, 1992.
121. Kehlhofer, R. *Combined-Cycle Gas and Steam Turbine Power Plants.* The Fairmont Press, Librun, GA 30247, 1991.
122. Kessel, P., Glavitsch, H. "Estimating the voltage stability of a power system" *IEEE Trans.*, **PWRD-1**, No. 3, pp 346–354, 1986.
123. Kimbark E.W. *Power system stability.* **I, II, III**, John Wiley and Sons, New York, 1948, 1950, 1956, re-printed by IEEE in 1995.
124. Kirchmayer, L.K. *Economic Control of Interconnected Systems.* Wiley, New York, 1959.
125. Kundur, P., Lee, D.C., Zein El-Din, H.M. "Power system satbilizers for thermal units: analytical techniques and on-site validation" *IEEE Trans.*, **PAS-100**, pp. 81–95, 1981.
126. Kundur, P. *Power System Stability and Control.* McGraw-Hill, New York, 1994.
127. Kulicke, B, Webs, A. "Elektromechanisches Verhalten von Turbosetzen bei Kurzschlüssen in Kraftwerksnähe" *ETZ-A*, Bd. 96, H. 4, 1975.
128. Kumano, S., Miwa, Y., Kokai, Y., Yatsu, M., Omata, K., Asano, T. "Evaluation of transient stability controller system model". *CIGRE Session 38-303*, 28 August-3 September 1994.
129. Lander, C.W. *Power Electronics.* McGraw-Hill, second edition, 1987.
130. Larsen, E.V., Bowler, C., Damsky, B., Nilsson, S. "Benefits of thyristor-controlled series compensation" *CIGRE Joint Session 14-37-38*, Paris, 1992.
131. Larsen, E.V., Swan, D.A. "Applying power system stabilizers, Parts I, II, and III" *IEEE Trans.*, **PAS-100**, pp. 3017–3046, 1981.
132. Laughton, M.A., Say, M.G. (ed) *Electrical Engineer's Reference Book.* Stoneham, Mass., Butterworth, 1985.

133. Läge, K., Lambrecht, D. "Die Auswirkung dreipoliger Netzkurz-schlüsse mit Kurzschlussfortschaltung auf die mechanische Beansprachung von Turbosätzen" *ETZ-A*, Bd. 95, H. 10, 1974.
134. Lee, S.T.Y, Schweppe, F.C. "Distance measures and coherency recognition for transient stability equivalents" *IEEE Trans.*, **PAS-92**, 1973.
135. Lee, D.C., Beaulieu, R.E., Service, J.R.R. "A power system stabilizer using speed and electrical power inputs - design and field experience" *IEEE Trans.*, **PAS-100**, pp. 4151–4167, 1981.
136. Lee, D.C., Kundur, P. "Advanced excitation control for power system stability enhancement" *CIGRE Session 38-01*, 1986.
137. Löf, P.A., Smed, T., Andersson, G., Hill, D.J. "Fast calculation of a voltage stability index" *IEEE Trans.*, **PWRS-7**, No. 1, pp. 54–64, 1992.
138. Löf, P.A., Hill, D.J., Arnborg, S., Andersson, G. "On the analysis of long term voltage stability" *Electric Power and Energy Systems*, **15**, No. 4, pp. 229–237, 1993.
139. Lüders, G.A. "Transient stability of multimachine power system via the direct method of Lyapunov" *IEEE Trans.*, **PAS-90**, No. 1, 1971.
140. MacDonald, J. "Present phase-angle regulating transformer technology" *IEE Proc. Colloquium "Facts — the key to increased utilisation of a power system"*, pp. 61–62, 12 January 1994.
141. Machowski, J. "Dynamic equivalents for transient stability studies of electrical power systems" *Int. J. on Electrical Power and Energy Systems*, **7**, No 4, pp. 215–223, 1985.
142. Machowski, J., Cichy, A., Gubina, F., Omahen, P. "External subsystem equivalent model for steady-state and dynamic security assessment" *IEEE Trans.*, **PWRS-3**, No 4, 1988.
143. Machowski, J., Cichy, A., Gubina, F., Omahen, P. "Modified algorithm for coherency recognition in large electrical power systems" *IFAC Symp. Power Systems and Power Plant Control*, Beijing, China, 1986.
144. Machowski, J., Gubina, F., Omahen, P. "Power system transient stability studies by Lyapunov method using coherency based aggregation" ibid.
145. Machowski, J., Bernas, S. *Stany nieustalone i stabilnosc systemu elektroenergetycznego.* Wydawnictwa Naukowo-Techniczne, Warszawa, 1989.
146. Machowski, J., Nelles, D. "Optimal control of superconducting magnetic energy storage unit" *Electric Machines and Power Systems*, **20**, No 6, 1992a.
147. Machowski, J., Nelles, D. "Power system transient stability enhancement by optimal control of static VAR compensators" *Int. J. on Electrical Power and Energy Systems*, **14**, No 5, 1992b.
148. Machowski, J., Nelles, D. "Simple robust adaptive control of static VAR compensator" *European Transaction on Electric Power Engineering*, **3**, No 6, 1993.
149. Machowski, J., Nelles, D. "Optimal modullation controller for superconducting magnetic energy storage" *Int. J. on Electrical Power and Energy Systems*, **16**, No 5, 1994.
150. McPherson, G., Laramore, R.D. *Introduction to Electric Machines and Transformers.* 2nd ed., Wiley, 1990.
151. Martin, H.F., Tapper, D.N., Alston, T.M. "Sustained fast valving applied to Tennessee Valley Authority's Watts Bar Nuclear Units" Trans of the ASME *Journal of Engineering for Power*. Paper 76-JPGC-Pwr55.
152. Meisel, J., Sen, A., Gilles, M.L. "Alleviation of a transient stability crisis using shunt braking resistors and series capacitors" *Int. J. Electrical Power and Energy Systems*, **2**, No. 1, 1981.

153. Moore, P., Ashmole, P. "Flexible AC transmission systems" *IEE Power Engineering Journal*, pp. 282–286, December 1995.
154. Moussa, H.A.M, Yu, Y.N. "Improving power system damping through supplementary governor control" *PES Summer Meeting*, paper C 72 470-3, 1972.
155. Miller, T.J.E. *Reactive Power Control in Electric Systems.* Wiley, New York, 1982.
156. The National Grid Company (NGC) "1994 Seven Year Statement", 1994.
157. Nagao, T. "Voltage collapse at load ends of power systems" *Electrical Engineering in Japan*, **95**, No. 4, 1975.
158. Nitta, T., Shirari, Y., Okada, T. "Power charging and discharging characteristics of SMES connected to artificial transmission line" *IEEE Trans. on Magnetics*, **21**, No. 2, March 1985.
159. Olas, A. "Synthesis of systems with prescribed trajectories" *Proc. Non-linear Vibrations*, **16**, 1975.
160. Omahen, P. "Fast transient stability assessment using corrective PEBS method" *Proc. 6-th IEEE Mediterranean Electrotechnical Conference*, **2**, pp. 1408–1411, 1991.
161. Omahen, P., Gubina, F. "Simulations and field tests of a reactor coolant pump emergency start-up by means of remote gas units" *IEEE Trans.*, EC-7, **7**, No. 4, pp. 691–697, 1992.
162. Omahen, P. *Unified approach to power system analysis in its multi-time scale dynamic response*. Ph.D. Thesis, Warsaw University of Tehnology, Faculty of Electrical Engineering, Warsaw, (in Polish), 1994.
163. Omahen, P., Gubina, F. "Experience with large power system dynamics model for security assessment" *CIGRE Power System Operation & Control Colloquium*, Johannesburg/Capetown, 1995.
164. Pai M.A. *Power System Stability. Analysis by the direct method of Lyapunov*. Amsterdam, North Holland Publishing Company 1981.
165. Pai, M.A. *Energy Function Analysis For Power System Stability.* Kluwer Academic Publishers, Boston, Dordrecht, London, 1989.
166. Pal, M.K. "Voltage stability conditions considering load characteristics" *IEEE Trans.*, **PWRS-7**, No. 1, pp. 243–249, 1992.
167. Park, R.H. "Fast turbine valving" *IEEE Trans.*, **PAS-92**, pp. 1065–1073, 1973.
168. Paszek, W. *Stany nieustalone maszyn elektrycznych pradu przemiennego*. Warsaw, WNT, 1986.
169. Pavella, M., Murthy, P.G. *Transient Stability of Power Systems. Theory and Practice*. John Wiley & Sons, 1994.
170. Pissanetzky, S. *Sparse matrix technology*. Academic Press, 1984.
171. Phadke, A.G., Thorap, J.S., Adamiak, M.G. "A new measurement technique for tracking voltage phasors, local system frequency and rate of change of frequency" *IEEE Trans.*, **PAS-102**, No. 5, 1983.
172. Phadke, A.G., Thorap, J.S. *Computer relaying for power systems.* J. Wiley & Sons, 1988.
173. Podmore, R. "Identification of coherent generators for dynamic equivalents" *IEEE Trans.*, **PAS-97**, pp. 1344–1354, 1978.
174. Press, W.H., Teukolsky, S.A., Vetterling, W.T., Flannery, B.P. *Numerical Recipes in C. The art of scientific computing.* Cambridge University Press, 2nd edition, 1992.
175. Racz, L.Z., Bokay, B. *Power System Stability*. Kluwer Academic Publishers, Boston, Dordrecht, London, 1988.

176. Rafian, M., Sterling, M.J.H., Irving, M.R. "Real time power system simulation" *IEE Proc.*, **134**, Pt.C, No. 3, pp. 206–223, 1987.

177. Rajagopalan, C., Lesiuetre, B., Sauer, P.W., Pai, M.A. "Dynamic aspects of voltage/power characteristics", *IEEE Trans.*, **PWRS-7**, No. 3, pp. 990–1000, 1992.

178. Ralls, K.J. "The growth of power electronics in electrical power transmission systems" *IEE Power Engineering Journal*, pp. 15–23, February 1995.

179. Ramey, D.G., Skooglund, J.W. "Detailed hydrogovernor representation for system stability studies" *IEEE Trans.*, **PAS-89**, 1970.

180. Riaz, M. "Hybrid-parameter models of synchronous machines" *IEEE Trans.*, **PAS-93**, pp. 849-858, 1974.

181. Rogers, J.D., Scermer, R.I., Miller, B.L., Hauer, J.F. "30 MJ superconducting energy storage system for electric utility transmission stabilization" *Proc.IEEE*, **71**, pp. 1099–11–7, Sept. 1983.

182. Rüdenberg, R. *Elektrische Schaltvorgänge und verwandte Störungserscheinungen in Starkstromanlagen*. Julius Springer, Berlin, 1923.

183. Rüdenberg, R. *Transient Performance of Electric Power Systems*. Mc Graw-Hill, New York, 1950.

184. Say, M.G. *Alternating current machines*. 4th ed., Pitman, 1976.

185. Schlueter, R.A., Ilu, T., Chang, J.C., Lo, J.C., Costi, A. "Methods for determining proximity to voltage collapse", *IEEE Trans.*, **PWRS-6**, No. 2, pp. 285–292, 1992.

186. Siodelski, A, Szczerba, Z., Zajczyk, R. "Protection de la centrale intervenant dans les réglages de la turbine et de l'alternateur lors de l'incident sur le réseau" *CIGRE Session 1988*, paper 34-02.

187. Stalewski, A., Goody, J.L.H., Downes, J.A. "Pole-slipping protection" *2nd Int. Conf. on Developments in Power System Protection*, IEE Conf. Publication No. 185, 1980.

188. Steinbrenner, G., Gelopulos, D.P. "An algorithm for transient and dynamic stability solutions without interface error" *Trans. IEEE PICA Conf.*, pp 113–117, 1973.

189. Sterling, M.J.H. *Power System Control*. Peter Peregrinus, IEE, 1978.

190. Stott, B "Review of load flow calculation methods" *Proc. IEEE*, **62**, pp. 916–929, 1974.

191. Stott, B "Power system dynamic response calculations" *Proc. IEEE*, **67**, pp. 219–241, 1979.

192. Taylor, C.W., Haner, J.M., Hill, L.A., Mittelstad, W.A., Cresap R.L. "A New Out-Of-Step Relay with Rate of Change of Apparent Resistance Augmentation" *IEEE Trans.*, **PAS-102**, No. 3, March 1983.

193. Taylor, C.W., Haner, Laughlin T.D. "Experience with The R-Rdot Out-Of-Step Relay". *IEEE Trans.*, **PWRD-1**, No. 2, April 1986.

194. Taylor, C.W. *Power system voltage stability*. McGraw-Hill, Inc., 1994.

195. Tewerson, R.P. *Sparse Matrices*. Academic Press, New York/London, 1973

196. Tinney, W.F., Walker, J.W. "Direct solutions of sparse network equations by optimally ordered triangular factorization" *Proc. IEEE*, **55**, No. 11, pp. 1801–1809, 1967.

197. Tinney, W.F., Powell, W.L., Peterson, N.M. "Sparsity oriented network reduction" *Proc. PICA Conf.*, pp. 385-390, Minneapolis, May 1973.

198. Tinney, W.F., Bright, J.M. "Adaptive reductions for power equivalents" *IEEE Power Winter Meeting*, paper 86 WM, 1986.

199. Tinney, W.F., Brandwain, V., Chan, S.M. "Sparse vector methods" *IEEE Trans.*, **PAS-104**, pp. 295–301, Feb. 1985.

200. Tiranuchit, A., Thomas, R.J. "A posturing strategy action against voltage instabilities in electric power systems" *IEEE Trans.*, **PWRS-3** No. 1, pp. 87–93, 1988.
201. Undrill, J.M., Turner, A.E. "Construction of power system electromechanical equivalents by modal analysis" *IEEE Trans.*, Vol. **PAS-90**, No. 5, 1971.
202. Ungrad, H., Winkler, W., Wiszniewski, A. *"Protection techniques in electrical energy systems"* Marcel Dekker, Inc., 1995.
203. Vaahedi, E., El-Kady, M.A. Libaque-Esaine, J.A., Carvalho, V.F. "Load models for large-scale stability studies from end-user consumption" *IEEE Trans.*, **PWRS-2**, No. 4, pp. 864–872, November 1987.
204. Van Cutsem, T. "A method to compute reactive power margins with respect to voltage collapse" *IEEE Trans.*, **PWRS-6**, pp. 145–156, 1991.
205. Venikov V.A. *Transient Phenomena in Electrical Power Systems.* Pergamon Press, Oxford, 1964.
206. Venikov V.A. *Transient Processes in Electrical Power Systems.* Moscow USSR, Mir, 1978.
207. Vorley, D.H. *Numerical techniques for analysing the stability of large power systems.* Ph.D. Thesis, University of Manchester (England), 1974.
208. Wache, M. *Verfahren zur Bewertung von Regeleinrichtungen von Synchronengeneratoren.* Ph.D. Thesis, Universität Kaiserslautern, 1994.
209. Wang, H.F., Hao, Y.S., Hogg, B.W., Yang, Y.H. "Stabilization of power systems by governor-turbine control" *Electrical Power & Energy Systems*, **15**, No. 6, pp. 351–361, 1993.
210. Ward, J.B. "Equivalent circuits for power flow studies" *AIEE Trans.*, **PAS**, pp. 373–382, 1949.
211. Watson, W., Coultes, M.E. "Static exciter stabilizing signals on large generators - mechanical problems" *IEEE Trans.*, **PAS-92**, pp. 204–211, 1973.
212. Weedy, B.M. *Underground Transmission of Electric Power.* Wiley, Chichester, 1980.
213. Weedy, B.M. *Electric Power Systems.* 3rd ed. revised, Wiley, Chichester 1987.
214. Weisman, J., Eckart, L.E. *Modern Plant Engineering.* Englewood Cliffs, N.J., Prentice Hall, 1985.
215. Weisman, J., Eckart, L.E. *Modern Plant Engineering.* Englewood Cliffs, N.J., Prentice Hall, 1985.
216. Welfonder E. "Regeldynamisches Zusammenvirken von Kraftwerken und Verbrauchern im Netzverbundbetrieb" *Elektrizitätswirtschaft*, Jg 79, Heft 20, 1980.
217. Westlake, A.J., Bumby, J.R. & Spooner, E. "Damping the power angle oscillations of a permanent magnet synchronous generator with particular reference to wind turbine applications", *IEE-Proc.-Electr. Power Appl.*, **143**, No. 3, May 1996.
218. Willems, J.L. *Stability Theory of Dynamical Systems.* Nelson, London, 1970.
219. Witzke, R.L. Kresser, J.V. and Dillard, J.K. "Influence of ac reactance on voltage regulation of 6-phase rectifiers" *AIEE Transcations*, **72**, pp 244–253, 1953.
220. Wood, A.J., Wollenberg, B.F. *Power Generation Operation and Control.* 2nd ed., Wiley, 1996.
221. Wright, A. and Christopoulos, C. *Electrical Power System Protection.* Chapman & Hall, London, 1993.
222. Yu, Y.N. *Electric Power System Dynamics.* Academic Press, New York, London, 1983.
223. Younkins, T.D., Chow, J.H., Brower, A.S., Kure-Jensen, J., Wagner, J.B. "Fast valving with reheat and straight condensing steam turbines" *IEEE Trans.*, **PWRS-2**, pp. 397–405, 1987.

224. Веников, В.А. *Электромеханические переходные процессы в электрических системах*. Государственное Энергетическое Издательство, Москва — Ленинград, 1958.
225. Веников, В.А. *Переходные электромеханические процессы в электрических системах*, издание третье, Высшая Школа, Москва, 1978.
226. Веников, В.А. *Переходные электромеханические процессы в электрических системах*, издание чствёртое, Высшая Школа, Москва, 1985.
227. Глебов, И.А. *Системы возбуждения генераторов с управляемыми преобразователями* Ленинград, Наука, 1970.
228. Жданов, П.С. *Вопросы устойчивости электрических систем.* Энергия, Москва, 1979.
229. Жданов, П.С. *Устойчивость электрических систем.* Государственное Энергетическое Издательство, Москва — Ленинград, 1948.
230. Жуков, Л.А. *Упрощенное преобразование схем замещание сложных электрических систем. Изв.* Академии Наук СССР, Энергетика и Транспорт, 2, 1964.
231. Казовский, Е.Я., Данилевич, Я.Б., Кашарскийбб Э.Г., Рубисов, Г.В. *Анормалные режимы работы крупных синхронных машин.* Издательство “Наука” — Ленинградское отделение, Ленинград, 1969.
232. Хачатуров А.А. *Несинхронные вклочения и ресинхронизация в энергосистемах.* Москва. Энергия 1969.
233. Ульянов С.А. *Короткие замыкания в электрических системах.* Москва — Ленинград, Госэнергоиздат 1952.

Index